Springer Optimization and Its Applications

Volume 165

Aims and Scope

Optimization has continued to expand in all directions at an astonishing rate. New algorithmic and theoretical techniques are continually developing and the diffusion into other disciplines is proceeding at a rapid pace, with a spot light on machine learning, artificial intelligence, and quantum computing. Our knowledge of all aspects of the field has grown even more profound. At the same time, one of the most striking trends in optimization is the constantly increasing emphasis on the interdisciplinary nature of the field. Optimization has been a basic tool in areas not limited to applied mathematics, engineering, medicine, economics, computer science, operations research, and other sciences.

The series **Springer Optimization and Its Applications (SOIA)** aims to publish state-of-the-art expository works (monographs, contributed volumes, textbooks, handbooks) that focus on theory, methods, and applications of optimization. Topics covered include, but are not limited to, nonlinear optimization, combinatorial optimization, continuous optimization, stochastic optimization, Bayesian optimization, optimal control, discrete optimization, multi-objective optimization, and more. New to the series portfolio include Works at the intersection of optimization and machine learning, artificial intelligence, and quantum computing.

Volumes from this series are indexed by Web of Science, zbMATH, Mathematical Reviews, and SCOPUS.

More information about this series at http://www.springer.com/series/7393

Andrei M. Raigorodskii • Michael Th. Rassias
Editors

Discrete Mathematics and Applications

Springer

Editors
Andrei M. Raigorodskii
Moscow Institute of Physics
and Technology
Dolgoprudny, Russia

Moscow State University
Moscow, Russia

Buryat State University
Ulan-Ude, Russia

Caucasus Mathematical Center
Adyghe State University
Maykop, Russia

Michael Th. Rassias
Institute of Mathematics
University of Zurich
Zurich, Switzerland

Moscow Institute of Physics
and Technology
Dolgoprudny, Russia

Institute for Advanced Study
Program in Interdisciplinary Studies
Princeton, NJ, USA

ISSN 1931-6828 ISSN 1931-6836 (electronic)
Springer Optimization and Its Applications
ISBN 978-3-030-55856-7 ISBN 978-3-030-55857-4 (eBook)
https://doi.org/10.1007/978-3-030-55857-4

Mathematics Subject Classification: 05-XX, 11-XX, 37-XX, 40-XX, 41-XX, 42-XX, 68-XX

This Springer imprint is published by the registered company Springer Nature Switzerland AG
The registered company address is: Gewerbestrasse 11, 6330 Cham, Switzerland

Preface

Discrete Mathematics and Applications is devoted to presenting essential developments in the broad and highly applicable domain of discrete mathematics. This vast domain lies at the intersection and interplay of mathematics and computer science. Classic and modern theories of discrete mathematics have a plethora of applications in a remarkably versatile spectrum of different areas; optimization theory being among the characteristic ones. Combinatorial methods as well as core topics in graph theory offer a wealth of techniques applied in optimization theory, computer science, operations research, etc., with a broad variety of research problems in the latter domains being inspired by beautiful results and theories originating from combinatorics, graph theory, and other facets of discrete mathematics.

The contributions published within this volume survey classical topics in discrete mathematics and also present new results in a variety of problems and applications. Additionally, the book provides an insight into the study of important problems and theories, specifically delving into topics such as automatic sequences, combinatorial identities and inequalities for trigonometric sums, partitions of a set and superelliptic Diophantine equations, special oriented graphs, extremal singular graphs, tree containment, hypergraph Ramsey problems, the game chromatic number of a random hypergraph, randomly colored matchings in random bipartite graphs, disjoint chorded cycles in graphs with high Ore-degree, perfect hash families, the study of prime numbers, exponential variational integrators, the $3x + 1$ dynamical system, a factorization method for solving multipoint problems for second-order difference equations with polynomial coefficients, etc.

The papers have been contributed by eminent experts, all of whom have presented the state of the art in the problems treated. Given the versatility of applications of discrete mathematics, the volume is expected to serve as a valuable resource for graduate students, research mathematicians, physicists, and engineers.

We would like to express our warmest thanks to all the authors of papers in this volume who contributed to this collective effort. Last but not least, we would like to extend our appreciation to the Springer staff for their valuable help throughout the publication process of this work.

Moscow, Russia Andrei M. Raigorodskii
Zurich, Switzerland Michael Th. Rassias

Contents

Automatic Sequences Are Also Non-uniformly Morphic

Jean-Paul Allouche and Jeffrey Shallit

Abstract It is well-known that there exist infinite sequences that are fixed points of non-uniform morphisms, but not k-automatic for any k. In this note we show that every k-automatic sequence is the image of a fixed point of a *non-uniform* morphism.

1 Introduction, Definitions, Notation

Combinatorics on words deals with "alphabets", "words", "languages", and "morphisms of monoids". The first three notions are inspired by the usual meaning of these words in English. Below we recall the precise definitions.

Definition 1 A finite set $\mathcal{A}$ is called an *alphabet*. A *word* over the alphabet $\mathcal{A}$ is a finite (possibly empty) sequence of symbols from $\mathcal{A}$. We let $\mathcal{A}^*$ denote the set of all words on $\mathcal{A}$. A subset of $\mathcal{A}^*$ is called a *language* on $\mathcal{A}$. The *length* of a word w, denoted $|w|$, is the number of symbols that it contains (the length of the empty word ϵ is 0). The *concatenation* of two words $w = a_1 a_2 \cdots a_r$ and $z = b_1 b_2 \cdots b_s$ of lengths r and s, respectively, is the word denoted wz defined by $wz = a_1 a_2 \cdots a_r b_1 b_2 \cdots b_s$ of length $r + s$ obtained by gluing w and z in order. The set $\mathcal{A}^*$ equipped with concatenation is called the *free monoid* generated by $\mathcal{A}$. The concatenation of a word $w = a_1 a_2 \cdots a_r$ and a sequence $(x_n)_{n \geq 0}$ is the sequence $a_1 a_2 \cdots a_r x_0 x_1 \cdots$, denoted $w\ (x_n)_{n \geq 0}$. A word w is called a *prefix* of the word z (or of the infinite sequence $(x_n)_{n \geq 0}$) if there exists a word y with $z = wy$ (respectively, a sequence $(y_n)_{n \geq 0}$ with $(x_n)_{n \geq 0} = w\ (y_n)_{n \geq 0}$).

J.-P. Allouche (✉)
CNRS, IMJ-PRG, Sorbonne, Paris Cedex, France
e-mail: jean-paul.allouche@imj-prg.fr

J. Shallit
School of Computer Science, University of Waterloo, Waterloo, ON, Canada
e-mail: shallit@uwaterloo.ca

A. M. Raigorodskii, M. Th. Rassias (eds.), *Discrete Mathematics and Applications*, Springer Optimization and Its Applications 165,
https://doi.org/10.1007/978-3-030-55857-4_1

Let $(u_\ell)_{\ell\geq 0}$ be a sequence of words in of $\mathcal{A}^*$, and $(a_n)_{n\geq 0}$ be a sequence over the alphabet $\mathcal{A}$. The sequence $(u_\ell)_{\ell\geq 0}$ is said to *converge* to the sequence $(a_n)_{n\geq 0}$ if the length of the largest prefix of u_ℓ that is also a prefix of $(a_n)_{n\geq 0}$ tends to infinity with ℓ.

Remark 2 It is straightforward that $\mathcal{A}^*$ equipped with concatenation is indeed a monoid: concatenation is associative, and the empty word ϵ is the identity element. This monoid is *free*; intuitively, this means that there are no relations between elements, other than the relations arising from the associative property and the fact that the empty word is the identity element. In particular, this monoid is not *commutative* if $\mathcal{A}$ has at least two distinct elements.

Definition 3 Let $\mathcal{A}$ and $\mathcal{B}$ be two alphabets. A *morphism* from $\mathcal{A}^*$ to $\mathcal{B}^*$ is a map φ from $\mathcal{A}^*$ to $\mathcal{B}^*$ such that, for all words u and v, one has $\varphi(uv) = \varphi(u)\varphi(v)$. A morphism of $\mathcal{A}^*$ is a morphism from $\mathcal{A}^*$ to itself.

If there exists a positive integer k such that $\varphi(a)$ has length $k \geq 1$ for all $a \in \mathcal{A}$, the morphism φ is said to be *k-uniform*. If a morphism is k-uniform for some $k \geq 1$, it is called a *uniform morphism*. Otherwise it is *non-uniform*. A 1-uniform morphism is sometimes called a *coding*.

Example 4 The *Thue–Morse morphism* μ sending $0 \to 01$ and $1 \to 10$ is 2-uniform. In contrast, the *Fibonacci morphism* τ sending a to ab and b to a is non-uniform.

Remark 5 A morphism φ from $\mathcal{A}^*$ to $\mathcal{B}^*$ is completely determined by the values of $\varphi(a)$ for $a \in \mathcal{A}$. Namely, if the word u is equal to $a_1a_2\cdots a_n$ with $a_j \in \mathcal{A}$, then $\varphi(u) = \varphi(a_1)\varphi(a_2)\cdots\varphi(a_n)$.

Definition 6 An infinite sequence $(a_n)_{n\geq 0}$ taking values in the alphabet $\mathcal{A}$ is said to be *pure morphic* if there exist a morphism φ of $\mathcal{A}^*$ and a word $x \in \mathcal{A}^*$ such that

- the word $\varphi(a_0)$ begins with a_0; i.e., there exists a word x such that $\varphi(a_0) = a_0x$;
- iterating φ starting from x never gives the empty word, i.e., for each integer ℓ, $\varphi^\ell(x) \neq \epsilon$;
- the sequence of words $(\varphi^\ell(a_0))_{\ell\geq 0}$ converges to the sequence $(a_n)_{n\geq 0}$ when $\ell \to \infty$.

Remark 7 It is immediate that

$$\varphi(a_0) = a_0x$$

$$\varphi^2(a_0) = \varphi(\varphi(a_0)) = \varphi(a_0x) = \varphi(a_0)\varphi(x) = a_0x\varphi(x)$$

$$\varphi^3(a_0) = \varphi(\varphi^2(a_0)) = \varphi(a_0x\varphi(x)) = \varphi(a_0)\varphi(x)\varphi^2(x) = a_0x\varphi(x)\varphi^2(x)$$

and more generally

$$\varphi^\ell(a_0) = a_0x\varphi(x)\varphi^2(x)\cdots\varphi^{\ell-1}(x)$$

for all $\ell \geq 0$.

Definition 8 An infinite sequence $(a_n)_{n\geq 0}$ taking values in $\mathcal{A}$ is said to be *morphic* if there exist an alphabet $\mathcal{B}$ and an infinite sequence $(b_n)_{n\geq 0}$ over the alphabet $\mathcal{B}$ such that

- the sequence $(b_n)_{n\geq 0}$ is pure morphic;
- there exists a coding from $\mathcal{B}^*$ to $\mathcal{A}^*$ sending the sequence $(b_n)_{n\geq 0}$ to the sequence $(a_n)_{n\geq 0}$; i.e., the sequence $(a_n)_{n\geq 0}$ is the pointwise image of $(b_n)_{n\geq 0}$.

If the morphism making $(b_n)_{n\geq 0}$ morphic is k-uniform, then the sequence $(a_n)_{n\geq 0}$ is said to be *k-automatic*. The word "automatic" comes from the fact that the sequence $(a_n)_{n\geq 0}$ can be generated by a finite automaton (see [2] for more details on this topic).

Remark 9 A morphism φ of $\mathcal{A}^*$ can be extended to infinite sequences with values in $\mathcal{A}$ by defining

$$\varphi((a_n)_{n\geq 0}) = \varphi(a_0a_1a_2\cdots) := \varphi(a_0)\varphi(a_1)\varphi(a_2)\cdots .$$

It is easy to see that a pure morphic sequence is a *fixed point* of (the extension to infinite sequences of) some morphism: actually, with the notation above, it is *the* fixed point of φ beginning with a_0. A pure morphic sequence is also called an *iterative fixed point* of some morphism (because of the construction of that fixed point), while a morphic sequence is the pointwise image of an iterative fixed point of some morphism, and a k-automatic sequence is the pointwise image of the iterative fixed point of a k-uniform morphism.

2 The Main Result

Looking at the definitions above, we see that every automatic sequence is also a morphic sequence. We will prove that every automatic sequence can be obtained as a morphic sequence where the involved morphism is *not uniform*.

Definition 10 We say a sequence is *non-uniformly pure morphic* if it is the iterative fixed point of a non-uniform morphism. We say that a sequence is *non-uniformly morphic* if it is the image (under a coding) of a non-uniformly pure morphic sequence.

For example, the sequence $abaababa\cdots$ generated by iterating the morphism τ defined above is non-uniformly pure morphic. This sequence is known as the *(binary) Fibonacci sequence*, since it is also equal to the limit of the sequence of words $(u_n)_{n\geq 0}$ defined by $u_0 := a$, $u_1 := ab$, $u_{n+2} := u_{n+1}u_n$ for each $n \geq 0$.

In order to avoid triviality, we certainly assume (as M. Mendès France once pointed out to us) that the alphabet of the non-uniform morphism involved in the above definition is the same as the minimal alphabet of its fixed point. For example, the fact that the morphism $0 \to 01$, $1 \to 10$, $2 \to 1101$, whose iterative fixed point beginning with 0 is also the iterative fixed point, beginning with 0, of the morphism μ—namely, the Thue–Morse sequence, see, e.g., [1] does not make this sequence non-uniformly morphic.

Although most non-uniformly morphic sequences are not automatic (e.g., the binary Fibonacci sequence is not automatic), some sequences can be simultaneously automatic and non-uniformly morphic. An example is the sequence $\mathcal{Z}$ formed by the lengths of the blocks of 1's between two consecutive zeros in the Thue–Morse sequence.

$$\begin{aligned} &0\,1\,1\,0\,1\,0\,0\,1\,1\,0\,0\,1\,0\,1\,1\,0\,\cdots \\ &0\,(11)\,0\,(1)\,0\,(\,)\,0\,(11)\,0\,(\,)\,0\,(1)\,0\,(11)\,0\,\cdots \\ \mathcal{Z} = &\,2\,1\,0\,2\,0\,1\,2\,\cdots \end{aligned}$$

As is well known [3], this sequence is both the fixed point of the map sending $2 \to 210$, $1 \to 20$, and $0 \to 1$, and also the image, under the coding $0 \to 2$, $1 \to 1$, $2 \to 0$, $3 \to 1$ of the fixed point of the map $0 \to 01$, $1 \to 20$, $2 \to 23$, and $3 \to 02$.

In view of this example, one can ask which non-uniformly morphic sequences are also k-automatic for some integer $k \geq 2$, or which automatic sequences are also non-uniformly morphic. We prove here that *all* automatic sequences are also non-uniformly morphic.

Theorem 11 *Let $(a_n)_{n\geq 0}$ be an automatic sequence taking values in the alphabet $\mathcal{A}$. Then $(a_n)_{n\geq 0}$ is also non-uniformly morphic. Furthermore, if $(a_n)_{n\geq 0}$ is the iterative fixed point of a uniform morphism, then there exist an alphabet $\mathcal{B}$ of cardinality $(3 + \#\mathcal{A})$ and a sequence $(a'_n)_{n\geq 0}$ with values in $\mathcal{B}$, such that $(a'_n)_{n\geq 0}$ is the iterative fixed point of some non-uniform morphism with domain $\mathcal{B}^*$ and $(a_n)_{n\geq 0}$ is the image of $(a'_n)_{n\geq 0}$ under a coding.*

Proof We start with the first assertion. First, we may suppose that the first letter of $(a_n)_{n\geq 0}$ is different from all a_j for $j \geq 1$. If not, take a letter α not in $\mathcal{A}$ and consider the sequence $\alpha a_1 a_2 \cdots$. This sequence is automatic and the morphism $\alpha \to a_0$ and $a \to a$ for all letters a in $\mathcal{A}$ sends it to $(a_n)_{n\geq 0}$.

We may also suppose that the sequence $(a_n)_{n\geq 0}$ is not ultimately periodic (otherwise the result is trivial: if u and v are two words over the alphabet $\mathcal{A}$, the sequence $uvvv\cdots$ is the iterative fixed point of the morphism $\alpha \to u$ and $a \to v^j$ for all $a \in \mathcal{A}$, where j is chosen so that $j|v| \neq |u|$).

Thus we now start with an automatic non-ultimately periodic sequence, still called $(a_n)_{n\geq 0}$, with $a_0 = \alpha \neq a_1$. Since the sequence $(a_n)_{n\geq 0}$ is the pointwise image of the iterative fixed point $(x_n)_{n\geq 0}$ of some uniform morphism, we may suppose, by replacing $(a_n)_{n\geq 0}$ with $(x_n)_{n\geq 0}$, that $(a_n)_{n\geq 0}$ itself is the iterative fixed point beginning with $a_0 = \alpha \neq a_j$ for all $j \geq 1$ of a uniform morphism γ with domain $\mathcal{A}^*$, and still non-ultimately periodic.

We claim that there exists a 2-letter word bc such that $\gamma(bc)$ contains bc as a factor. Namely, since γ is uniform, it has exponential growth (that is, iterating γ on each letter gives words of exponentially growing length). Hence there exists a letter b that is *expanding*; i.e., such that some power of γ maps b to a word that contains at least two occurrences of b (see, e.g., [4]). By replacing γ with this power of γ, we can write $\gamma(b) = ubvbw$ for some words u, v, w. By replacing this new γ with γ^2, we can also suppose that both u and w are non-empty. Let c be the letter following the prefix ub of $ubvbw$. Now there are two cases:

- if $c \neq b$, then $v = cy$ for some word y, and $\gamma(b) = ubcyw$, and $\gamma(bc) = \gamma(b)\gamma(c)$ contains bc as a factor;
- if $c = b$, then $\gamma(b) = ubbz$ for some word z, and $\gamma(bb) = ubbzubbz$ contains bb as a factor.

In both cases, there exist two letters b and c, not necessarily distinct, such that $\gamma(b) = w_1bcw_2$ and $\gamma(bc) = w_1bcw_3$, where w_1, w_2 are non-empty words. Note, in particular, that b can be chosen distinct from a_0 (w_1 is non-empty and $a_0 = \alpha$ is different from all a_j for $j \geq 1$).

Now define a new alphabet $\mathcal{A}' := \mathcal{A} \cup \{b', c'\}$, where b', c' are two new letters not in $\mathcal{A}$. Define the morphism γ' with domain $\mathcal{A}'$ as follows: if the letter y belongs to $\mathcal{A} \setminus \{b\}$, then $\gamma'(y) := \gamma(y)$. If $y = b$, define $\gamma'(b) := w_1b'c'w_2$. Finally, define $\gamma'(b')$ and $\gamma'(c')$ as follows: first recall that $\gamma(bc) = w_1bcw_3$; cut the word w_1bcw_3 into (any) two non-empty words of unequal length, say $w_1bcw_3 := zt$, and define $\gamma'(b') := z$, $\gamma'(c') := t$.

By construction, γ' is not uniform. Its iterative fixed point beginning with a_0 clearly exists, and we denote it by $(a'_n)_{n\geq 0}$. This sequence has the property that each b' in it is followed by a c' and each c' is preceded by a b'. We let D denote the coding that sends each letter of $\mathcal{A}$ to itself, and sends b' to b and c' to c. For every letter x belonging to $\mathcal{A}' \setminus \{b,\ b',\ c'\}$ we have $\gamma(x) = \gamma'(x)$. Hence $D \circ \gamma'(x) = D \circ \gamma(x) = \gamma(x)$. For $x = b$, we have $D \circ \gamma'(b) = D(w_1b'c'w_2) = w_1bcw_2 = \gamma(b)$. Furthermore, we have $D \circ \gamma'(b'c') = D(zt) = zt = w_1bcw_3 = \gamma(bc)$.

Now let P_k be the prefix of the sequence $(a'_n)_{n\geq 0}$ that ends with c' and contains exactly k occurrences of the letter c'. Each occurrence of c' must be preceded by a b', so that P_k can be written $P_k = p_1b'c'p_2b'c' \cdots p_kb'c'$ where the p_i's are words over the alphabet $\mathcal{A}$. We have

$$\begin{aligned}
D \circ \gamma'(P_k) &= D \circ \gamma'(p_1b'c'p_2b'c' \cdots p_kb'c') \\
&= (D \circ \gamma'(p_1))(D \circ \gamma'(b'c'))(D \circ \gamma'(p_2)) \\
&\quad \times (D \circ \gamma'(b'c')) \cdots (D \circ \gamma'(p_k))(D \circ \gamma'(b'c')) \\
&= \gamma(p_1)\gamma(bc)\gamma(p_2)\gamma(bc) \cdots \gamma(p_k)\gamma(bc) \\
&= \gamma(p_1bcp_2bc \cdots p_kbc) \\
&= \gamma \circ D(p_1b'c'p_2b'c' \cdots p_kb'c') \\
&= \gamma \circ D(P_k).
\end{aligned}$$

Letting k go to infinity, we obtain that $D \circ \gamma'((a'_n)_{n\geq 0}) = \gamma \circ D((a'_n)_{n\geq 0})$, but $\gamma'((a'_n)_{n\geq 0}) = (a'_n)_{n\geq 0}$, so that $D((a'_n)_{n\geq 0}) = \gamma \circ D((a'_n)_{n\geq 0})$. Hence $D((a'_n)_{n\geq 0})$ is the iterative fixed point of γ beginning with a_0. Hence it is equal to the sequence $(a_n)_{n\geq 0}$.

The second assertion is a consequence of the fact that we introduced at most only three new letters α, b', c' in the proof above. □

References

1. J.-P. Allouche, J. Shallit, The ubiquitous Prouhet-Thue-Morse sequence, in *Sequences and Their Applications, Proceedings of SETA'98*, ed. by C. Ding, T. Helleseth, H. Niederreiter (Springer, Berlin, 1999), pp. 1–16
2. J.-P. Allouche, J. Shallit, *Automatic Sequences: Theory, Applications, Generalizations* (Cambridge University Press, Cambridge, 2003)
3. J. Berstel, Sur la construction de mots sans carré. Sém. Théor. Nombres Bordeaux Exposé 18, 18-01–18-15 (1978–1979)
4. A. Salomaa, On exponential growth in Lindenmayer systems. Nederl. Akad. Wetensch. Proc. Ser. A **76**, 23–30 (= Indag. Math. **35** (1973))
5. M. Dekking, On the structure of Thue—Morse subwords, with an application to dynamical systems,Theoret. Comput. Sci. **550** (2014), 107–112.
6. F. M. Dekking, M. S. Keane, On the conjugacy class of the Fibonacci dynamical system, Theoret. Comput. Sci. **668** (2017), 59–69

Combinatorial Identities and Inequalities for Trigonometric Sums

Horst Alzer, Omran Kouba, and Man Kam Kwong

Abstract The purpose of this paper is twofold.

1. We present two short and elementary new proofs for the identity

$$(*) \quad \sum_{k=0}^{n} \binom{n}{k}\binom{c}{k+m}(z+1)^k = \sum_{k=0}^{n} \binom{n}{k}\binom{n-k+c}{n+m} z^k,$$

which was recently proved by Chen and Reidys by using combinatorial methods.

2. Inspired by (∗) we obtain identities and inequalities for certain trigonometric sums. Among others, we show that the inequality

$$0 < \sum_{k=0}^{n} \binom{c}{k} \Big/ \binom{n}{k} \cos(kt) \quad (c \in \mathbb{R})$$

holds for all $n \geq 0$ and $t \in [0, \pi]$ if and only if $c \in (c_0, 1)$. Here, $c_0 = -0.97924\ldots$ is the only solution of the cubic equation

$$8x^3 - 16x^2 - 7x + 16 = 0$$

in the interval $[-1, 0]$.

2010 Mathematics Subject Classification 05A19, 26D05, 33B15, 33C45

H. Alzer (✉)
Waldbröl, Germany
e-mail: h.alzer@gmx.de

O. Kouba
Department of Mathematics, Higher Institute for Applied Sciences and Technology, Damascus, Syria
e-mail: omran_kouba@hiast.edu.sy

M. K. Kwong
Department of Applied Mathematics, The Hong Kong Polytechnic University, Hung Hom, Hong Kong
e-mail: mankwong@connect.polyu.hk

A. M. Raigorodskii, M. Th. Rassias (eds.), *Discrete Mathematics and Applications*, Springer Optimization and Its Applications 165,
https://doi.org/10.1007/978-3-030-55857-4_2

1 Introduction and Statement of Results

1.1 The Combinatorial Identity

The work on this paper is inspired by a remarkable article published in 2017 by Chen and Reidys [4]. The authors used combinatorial methods to prove the following identity.

Theorem 1 *For all integers n, m with $n \geq 0$ and $m \geq -1$ and for all complex numbers c, z we have*

$$\sum_{k=0}^{n} \binom{n}{k}\binom{c}{k+m}(z+1)^k = \sum_{k=0}^{n} \binom{n}{k}\binom{n-k+c}{n+m} z^k. \tag{1.1}$$

This identity may be reduced to a particular case of an identity between hypergeometric functions due to Pfaff, namely

$${}_2F_1(-n, \beta; \gamma; x) = \frac{(\gamma-\beta)_m}{(\gamma)_m}\, {}_2F_1(-n, \beta; \beta-\gamma-n+1; 1-x),$$

see Olver et al. [12, 15.8.7]. We need just to take $\beta = m - c$, $\gamma = m + 1$, $x = 1 + z$ and to expand the binomial coefficients. Our aim is to present two elementary new proofs of Theorem 1 and to draw some consequences of (1.1) without recourse to the theory of hypergeometric functions. Then we will use the obtained identities to deduce corresponding results for trigonometric polynomials, and to prove some sharp inequalities.

Chen and Reidys pointed out that (1.1) has some interesting implications. Setting $m = 0$, $c = n$, $z = x - 1$ in (1.1) yields

$$\sum_{k=0}^{n} \binom{n}{k}^2 x^k = \sum_{k=0}^{n} \binom{n}{k}\binom{2n-k}{n}(x-1)^k.$$

This elegant identity is given in Stanley's collection "Bijective proof problems"; see [14].

The Narayana numbers

$$N_{n,k} = \frac{1}{n}\binom{n}{k}\binom{n}{k-1}$$

play a role in several counting problems. For example, $N_{n,k}$ is equal to the number of paths above the x-axis from $(0, 0)$ to $(2n, 0)$ by using steps northeast and southeast, having exactly k peaks; see Petersen [13, chapter 2]. From (1.1) with $m = -1$, $c = n$, $z = x - 1$ we obtain the representation

$$\sum_{k=1}^{n} N_{n,k} x^k = \frac{1}{n} \sum_{k=0}^{n} \binom{n}{k} \binom{2n-k}{n-1} (x-1)^k.$$

If $A(n, k)$ denotes the number of unicellular maps of genus k with n edges, then an application of the Harer–Zagier formula (see Harer and Zagier [7]) and (1.1) with $m = 1, c = x, z = 1$ gives

$$\sum_{k \geq 0} A(n, k) x^{n+1-2k} = (2n-1)!! \sum_{k=0}^{n} \binom{n}{k} \binom{x}{k+1} 2^k$$

$$= (2n-1)!! \sum_{k=0}^{n} \binom{n}{k} \binom{n-k+x}{n+1}.$$

1.2 Vandermonde's Convolution Formula

We denote the sum on the left-hand side of (1.1) by $L_n(m, c, z)$. Then, we have the recurrence formula

$$L_n(m, c, z) = L_{n-1}(m, c, z) + (z+1) L_{n-1}(m+1, c, z)$$

and the differential equation

$$-\frac{z+1}{n} \frac{\partial}{\partial z} L_n(m, c, z) + L_n(m, c, z) = L_{n-1}(m, c, z).$$

The identity (1.1) can be used to deduce certain identities for combinatorial sums involving the product of two or three binomial coefficients. We apply (1.1) and the well-known formulas

$$(z+1)^k = \sum_{j=0}^{k} \binom{k}{j} z^j \quad \text{and} \quad \sum_{k=0}^{n} \sum_{j=0}^{k} a(k, j) = \sum_{k=0}^{n} \sum_{j=k}^{n} a(j, k).$$

Then, we compare the coefficients of z^k. This gives

$$\sum_{j=k}^{n} \binom{n}{j} \binom{j}{k} \binom{c}{j+m} = \binom{n}{k} \binom{n-k+c}{n+m}. \tag{1.2}$$

Using (1.2) and

$$\binom{n}{k+r}\binom{k+r}{k} = \binom{n}{k}\binom{n-k}{r} \tag{1.3}$$

leads after some simplifications to the Vandermonde convolution formula

$$\sum_{j=0}^{q}\binom{n}{j}\binom{c}{m-j} = \binom{n+c}{m} \quad (q = m \text{ or } q = n). \tag{1.4}$$

We note that in Section 2 we apply (1.4) to obtain a short proof of (1.1).

Next, we set $z = x - 1$ in (1.1). A comparison of the coefficients of x^k provides the following companion to (1.2):

$$\sum_{j=k}^{n}(-1)^{j-k}\binom{n}{j}\binom{j}{k}\binom{n-j+c}{n+m} = \binom{n}{k}\binom{c}{k+m}. \tag{1.5}$$

We mention two nice special cases of (1.2) and (1.5). Setting $m = 0$, $c = n$ yields

$$\sum_{j=k}^{n}\binom{n}{j}^2\binom{j}{k} = \binom{n}{k}\binom{2n-k}{n} \quad \text{and} \quad \sum_{j=k}^{n}(-1)^{j-k}\binom{n}{j}\binom{j}{k}\binom{2n-j}{n} = \binom{n}{k}^2.$$

From (1.3) and (1.5) we find that

$$\sum_{j=0}^{n}(-1)^{n-j}\binom{n}{j}\binom{j+c}{m} = \binom{c}{m-n}. \tag{1.6}$$

1.3 Harmonic Numbers

The harmonic numbers H_n ($n \in \mathbb{N}$) are defined by

$$H_n = \sum_{\nu=1}^{n}\frac{1}{\nu} = \psi(n+1) + \gamma,$$

where $\psi = \Gamma'/\Gamma$ is the logarithmic derivative of Euler's gamma function and $\gamma = 0.57721\ldots$ denotes Euler's constant. Since

$$\frac{d}{dc}\binom{N+c}{k} = \binom{N+c}{k}\big(\psi(N+c+1) - \psi(N+c+1-k)\big)$$

and

$$\psi(N+M) - \psi(N) = H_{N+M-1} - H_{N-1} \quad (N, M \in \mathbb{N}),$$

we obtain from (1.4) and (1.6) the following combinatorial identities involving harmonic numbers:

$$\sum_{j=0}^{n} \binom{n}{j} \binom{c}{m-j} [H_c - H_{j+c-m}] = \binom{n+c}{m} [H_{n+c} - H_{n+c-m}],$$

$$\sum_{j=0}^{n} (-1)^{n-j} \binom{n}{j} \binom{j+c}{m} [H_{j+c} - H_{j+c-m}] = \binom{c}{m-n} [H_c - H_{n+c-m}]. \qquad (1.7)$$

Here, c is a natural number with $c + n \geq m$. We display a special case. If we set $n = m$ in (1.7), then

$$\sum_{j=0}^{n} (-1)^{j} \binom{n}{j} \binom{j+c}{n} [H_{j+c} - H_{j+c-n}] = 0.$$

1.4 Trigonometric Sums

From the identities for algebraic polynomials mentioned above we can deduce identities for certain trigonometric polynomials. For example, setting $z = e^{it} - 1$ with $t \in \mathbb{R}$ in (1.1) and by comparing the real and imaginary parts, we obtain for $c \in \mathbb{R}$,

$$\sum_{k=0}^{n} \binom{n}{k} \binom{c}{k+m} \cos(kt) = \sum_{k=0}^{n} \binom{n}{k} \binom{n-k+c}{n+m} \sum_{\nu=0}^{k} (-1)^{k-\nu} \binom{k}{\nu} \cos(\nu t) \qquad (1.8)$$

and

$$\sum_{k=0}^{n} \binom{n}{k} \binom{c}{k+m} \sin(kt) = \sum_{k=0}^{n} \binom{n}{k} \binom{n-k+c}{n+m} \sum_{\nu=0}^{k} (-1)^{k-\nu} \binom{k}{\nu} \sin(\nu t). \qquad (1.9)$$

In the final part of this section, we present some inequalities for trigonometric sums. First, we study the polynomials given in (1.8) and (1.9) with $m = 0$, $c = n$, that is,

$$\sum_{k=0}^{n}\binom{n}{k}^2\cos(kt) \quad \text{and} \quad \sum_{k=0}^{n}\binom{n}{k}^2\sin(kt).$$

Both functions are connected with the classical Legendre polynomials by the identity

$$\sum_{k=0}^{n}\binom{n}{k}^2 x^k = (1-x)^n P_n\Big(\frac{1+x}{1-x}\Big), \tag{1.10}$$

see, for example, Koepf [9, p. 1]. The following theorem offers sharp upper and lower bounds for the cosine polynomial.

Theorem 2 *For all integers $n \ge 0$ and real numbers $t \in [0, \pi]$ we have*

$$a\,(n!)^2 \le \sum_{k=0}^{n}\binom{n}{k}^2\cos(kt) \le b\,(n!)^2 \tag{1.11}$$

with the best possible constant factors $a = -1/2$ and $b = 2$.

Next, we provide sharp bounds for the corresponding sine polynomial.

Theorem 3 *For all integers $n \ge 1$ and real numbers $t \in [0, \pi]$ we have*

$$\alpha\,(n!)^2 \le \sum_{k=1}^{n}\binom{n}{k}^2\sin(kt) \le \beta\,(n!)^2$$

with the best possible constant factors

$$\alpha = -0.06483\ldots \quad \textit{and} \quad \beta = \sqrt{2}\,\sqrt[4]{3}\,\frac{3+\sqrt{3}}{8} = 1.10091\ldots. \tag{1.12}$$

Remark 1 The exact value of α is given by

$$\alpha = \sqrt{1-t^2}\Big(\frac{2}{9}+\frac{1}{2}t+\frac{1}{9}t^2\Big),$$

where t is the only solution of $2x^3 + 6x^2 - 3 = 0$ in $[-1, 0]$. More precisely,

$$\alpha = \sin(t_1)\Big(\frac{2}{9}+\frac{1}{2}\cos(t_1)+\frac{1}{9}\cos^2(t_1)\Big), \tag{1.13}$$

where

$$t_1 = \arccos(x_1) = 2.55304\ldots,$$
$$x_1 = \cos(\tau) - \sqrt{3}\sin(\tau) - 1 = -0.83174\ldots, \quad (1.14)$$
$$\tau = \frac{1}{3}\arctan(\sqrt{15}) = 0.43937\ldots.$$

Remark 2 Using (1.10) and Theorems 2 and 3 gives inequalities for the Legendre polynomials:

$$-\frac{1}{2}(n!)^2 \leq \Re\left((1-e^{it})^n P_n\left(\frac{i\sin(t)}{1-\cos(t)}\right)\right) \leq 2\,(n!)^2$$

and

$$\alpha\,(n!)^2 \leq \Im\left((1-e^{it})^n P_n\left(\frac{i\sin(t)}{1-\cos(t)}\right)\right) \leq \beta\,(n!)^2$$

with the constant factors α and β as given in (1.12). Both double-inequalities are valid for $n \geq 1$ and $t \in (0, \pi]$.

In many papers the authors presented interesting nonnegative trigonometric sums and polynomials and demonstrated that their results have applications in geometric function theory and other branches. We mention the well-known inequality

$$0 < \sum_{k=1}^{n} \frac{\sin(kt)}{k},$$

which holds for $n \geq 1$ and $t \in (0, \pi)$. The validity of this inequality was conjectured by Fejér [5, section 22] in 1910, and the first proofs were published by Jackson [8] and Gronwall [6] in 1911 and 1912, respectively. This result inspired the work of several mathematicians, who provided various proofs and numerous extensions of the Fejér–Jackson–Gronwall inequality. For more information on this subject we refer to the monographs Askey [2], Askey and Gasper [3], Milovanović et al. [11, chapter 4], the survey article Koumandos [10], and the references cited therein.

From (1.8) with $m = 0$ we get the one-parameter class of cosine polynomials

$$F_n(c,t) = \sum_{k=0}^{n} \binom{c}{k}\binom{n}{k}\cos(kt). \quad (1.15)$$

Do there exist real parameters $c \neq 0$ such that $F_n(c,t)$ is positive for all $n \geq 0$ and $t \in [0, \pi]$? The next theorem gives a negative answer to this question.

Theorem 4 *Let $c \in \mathbb{R}$. The inequality $F_n(c,t) > 0$ holds for all integers $n \geq 0$ and real numbers $t \in [0, \pi]$ if and only if $c = 0$.*

We note that in order to prove this result we make use of identity (1.1); see the proof of Lemma 1 given in Section 5.

The following theorem reveals that if we replace in (1.15) the product $\binom{c}{k}\binom{n}{k}$ by the ratio $\binom{c}{k}/\binom{n}{k}$, then there are non-zero parameters c such that the cosine polynomial is positive for $n \geq 0$ and $t \in [0, \pi]$.

Theorem 5 *Let $c \in \mathbb{R}$. The inequality*

$$0 < \sum_{k=0}^{n} \frac{\binom{c}{k}}{\binom{n}{k}} \cos(kt) \tag{1.16}$$

is valid for all integers $n \geq 0$ and real numbers $t \in [0, \pi]$ if and only if $c \in (c_0, 1)$. Here, $c_0 = -0.97924\ldots$ is the only solution of the equation

$$8x^3 - 16x^2 - 7x + 16 = 0$$

in the interval $[-1, 0]$.

Remark 3 A closed-form expression for c_0 is given by

$$c_0 = \frac{2}{3} - \frac{\sqrt{106}}{6}\cos(\rho), \quad \text{where} \quad \rho = \frac{1}{3}\arctan\left(\frac{3}{356}\sqrt{19002}\right) = 0.28667\ldots.$$

In the next section, we present two short and elementary proofs of identity (1.1). In Sections 3 and 4, we prove Theorems 2 and 3, respectively. To prove Theorems 4 and 5 we need some lemmas. They are collected in Section 5. The proofs of Theorems 4 and 5 are given in Sections 6 and 7, respectively.

The numerical and algebraic computations have been carried out using the computer software MAPLE 13.

2 Proofs of Theorem 1

First Proof We use induction on n. Let $L_n(m, c, z)$ and $R_n(m, c, z)$ be the sums on the left-hand side and on the right-hand side of (1.1), respectively. Then,

$$L_0(m, c, z) = R_0(m, c, z) = \binom{c}{m}.$$

Recalling that for $\alpha \in \mathbb{C}$ and $k \in \mathbb{Z}$ with $k \geq 0$ we have

$$\binom{\alpha+1}{k} = \binom{\alpha}{k} + \binom{\alpha}{k-1}, \tag{2.1}$$

we obtain for $m \in \mathbb{Z}$ with $m \geq -1$ and $c, z \in \mathbb{C}$:

$$\begin{aligned} L_{n+1}(m,c,z) &= \sum_{k=0}^{n+1} \binom{n+1}{k} \binom{c}{k+m} (z+1)^k \\ &= \sum_{k=0}^{n} \binom{n}{k} \binom{c}{k+m} (z+1)^k + \sum_{k=1}^{n+1} \binom{n}{k-1} \binom{c}{k+m} (z+1)^k \\ &= L_n(m,c,z) + (z+1) L_n(m+1,c,z). \end{aligned} \tag{2.2}$$

Applying (2.1) again gives

$$\begin{aligned} R_{n+1}(m,c,z) &= \sum_{k=0}^{n+1} \binom{n+1}{k} \binom{n+1-k+c}{n+1+m} z^k \\ &= \sum_{k=0}^{n} \binom{n}{k} \binom{n+1-k+c}{n+1+m} z^k + \sum_{k=1}^{n+1} \binom{n}{k-1} \binom{n+1-k+c}{n+1+m} z^k \\ &= R_n(m+1,c+1,z) + z R_n(m+1,c,z) \end{aligned} \tag{2.3}$$

and

$$\begin{aligned} R_n(m+1,c+1,z) &= \sum_{k=0}^{n} \binom{n}{k} \binom{n-k+c+1}{n+m+1} z^k \\ &= \sum_{k=0}^{n} \binom{n}{k} \binom{n-k+c}{n+m+1} z^k + \sum_{k=0}^{n} \binom{n}{k} \binom{n-k+c}{n+m} z^k \\ &= R_n(m+1,c,z) + R_n(m,c,z). \end{aligned} \tag{2.4}$$

From (2.3) and (2.4) we find

$$R_{n+1}(m,c,z) = R_n(m,c,z) + (z+1) R_n(m+1,c,z). \tag{2.5}$$

Using the induction hypothesis that (1.1) holds we conclude from (2.2) and (2.5) that

$$L_{n+1}(m, c, z) = R_{n+1}(m, c, z).$$

This completes the proof.

Second Proof We apply the Vandermonde convolution formula. Let $0 \leq l \leq n$. We replace in (1.4) (with $q = m$) n by $n - l$ and m by $m + n$. This leads to

$$\sum_{j=0}^{m+n} \binom{n-l}{j} \binom{c}{m+n-j} = \binom{n-l+c}{m+n}.$$

We have

$$\begin{aligned}\sum_{j=0}^{m+n} \binom{n-l}{j} \binom{c}{m+n-j} &= \sum_{j=0}^{m+n} \binom{n-l}{m+n-j} \binom{c}{j} \\ &= \sum_{j=m+l}^{m+n} \binom{n-l}{m+n-j} \binom{c}{j} \\ &= \sum_{k=l}^{n} \binom{n-l}{n-k} \binom{c}{m+k}.\end{aligned}$$

Thus,

$$\sum_{k=l}^{n} \binom{n-l}{n-k} \binom{c}{m+k} = \binom{n-l+c}{m+n}. \tag{2.6}$$

Next, we multiply both sides of (2.6) by $\binom{n}{l}$. Then, we use $\binom{n-l}{n-k}\binom{n}{l} = \binom{n}{k}\binom{k}{l}$ and $\binom{k}{l} = 0$ for $k < l$. This gives

$$\sum_{k=0}^{n} \binom{n}{k} \binom{k}{l} \binom{c}{m+k} = \binom{n}{l} \binom{n-l+c}{m+n}. \tag{2.7}$$

Finally, we multiply both sides of (2.7) by z^l and sum from $l = 0$ to $l = n$. Then,

$$\sum_{k=0}^{n} \binom{n}{k} \binom{c}{m+k} (z+1)^k = \sum_{l=0}^{n} \binom{n}{l} \binom{n-l+c}{m+n} z^l.$$

3 Proof of Theorem 2

Let $t \in [0, \pi]$ and

$$C_n(t) = \frac{1}{(n!)^2} \sum_{k=0}^{n} \binom{n}{k}^2 \cos(kt).$$

Then,

$$C_0(t) = 1 \quad \text{and} \quad 0 \leq C_1(t) = 1 + \cos(t) \leq 2.$$

Let $x \in [-1, 1]$. We define

$$\mu_1(x) = \frac{1}{2}x(x+2) \quad \text{and} \quad \mu_2(x) = \frac{1}{9}(x+1)(x+4)(x-1/2).$$

Then,

$$-\frac{1}{2} = \mu_1(-1) \leq \mu_1(x) \leq \mu_1(1) = \frac{3}{2} \tag{3.1}$$

and

$$-0.23\ldots = \mu_2((\sqrt{7}-3)/2) \leq \mu_2(x) \leq \mu_2(1) = \frac{5}{9}. \tag{3.2}$$

Since

$$C_2(t) = \frac{1}{4} + \cos(t) + \frac{1}{4}\cos(2t) = \mu_1(\cos(t))$$

and

$$C_3(t) = \frac{1}{36} + \frac{1}{4}\cos(t) + \frac{1}{4}\cos(2t) + \frac{1}{36}\cos(3t) = \mu_2(\cos(t)),$$

we conclude from (3.1) and (3.2) that (1.11) with $a = -1/2$ and $b = 2$ holds for $n = 2$ and $n = 3$.

Let $n \geq 4$. Then,

$$|C_n(t)| \leq \frac{1}{(n!)^2} \sum_{k=0}^{n} \binom{n}{k}^2 = \frac{1}{(n!)^2}\binom{2n}{n} < \frac{1}{2}.$$

It follows that (1.11) with $a = -1/2$ and $b = 2$ is valid for $n \geq 4$. Moreover, since $C_1(0) = 2$ and $C_2(\pi) = -1/2$, it follows that the constant factors $a = -1/2$ and $b = 2$ are sharp.

4 Proof of Theorem 3

Let $t \in [0, \pi]$ and

$$S_n(t) = \frac{1}{(n!)^2} \sum_{k=1}^{n} \binom{n}{k}^2 \sin(kt).$$

We have

$$0 \leq S_1(t) = \sin(t) \leq 1 \quad \text{and} \quad S_2(t) = \sin(t) + \frac{1}{4}\sin(2t).$$

From

$$S_2'(t) = \left(\cos(t) - \frac{\sqrt{3}-1}{2}\right)\left(\cos(t) + \frac{\sqrt{3}+1}{2}\right)$$

we conclude that S_2 is increasing $[0, t_0]$ and decreasing on $[t_0, \pi]$, where

$$t_0 = \arccos\left(\frac{\sqrt{3}-1}{2}\right) = 1.196\ldots.$$

Since $S_2(0) = S_2(\pi) = 0$, we obtain

$$0 \leq S_2(t) \leq S_2(t_0) = \sqrt{2}\,\sqrt[4]{3}\,\frac{3+\sqrt{3}}{8} = 1.10091\ldots.$$

Let $x \in [-1, 1]$ and

$$p(x) = \sqrt{1-x^2}\left(\frac{1}{9}x^2 + \frac{1}{2}x + \frac{2}{9}\right).$$

Then,

$$-6\sqrt{1-x^2}\,p'(x) = 2x^3 + 6x^2 - 3 = q(x), \quad \text{say}.$$

Since q is strictly decreasing on $[-1, 0]$ and strictly increasing on $[0, 1]$ with $q(-1) = 1$, $q(0) = -3$, $q(1) = 5$, we conclude that q has precisely one zero on $(-1, 0)$ and one zero on $(0, 1)$. We have $q(x_1) = 0$, where x_1 is given in (1.14). Let $q(x_2) = 0$ with $x_2 \in (0, 1)$. It follows that $q \geq 0$ on $[-1, x_1] \cup [x_2, 1]$ and that $q \leq 0$ on $[x_1, x_2]$. This implies that p is decreasing on $[-1, x_1]$, increasing on $[x_1, x_2]$ and decreasing on $[x_2, 1]$, so that $p(-1) = p(1) = 0$ leads to

$$p(x_1) \leq p(x). \tag{4.1}$$

Moreover,

$$p(x) \leq \frac{1}{9} + \frac{1}{2} + \frac{2}{9} = \frac{5}{6}. \tag{4.2}$$

Since

$$S_3(t) = \frac{1}{4}\sin(t) + \frac{1}{4}\sin(2t) + \frac{1}{36}\sin(3t) = p(\cos(t)),$$

we conclude from (4.1) and (4.2) that

$$\alpha \leq S_3(t) \leq \frac{5}{6},$$

where α is given in (1.12) and (1.13).

We have

$$S_4(t) = \frac{1}{288}\sin(2t)u(t)$$

with

$$u(t) = 18 + 16\cos(t) + \cos(2t).$$

Let $t \in [0, \pi]$. Then,

$$S_4(t) \leq \frac{1}{288}(18 + 16 + 1) = 0.121\ldots < \beta.$$

If $t \in [0, \pi/2]$, then, clearly, $S_4(t) \geq 0$. Let $t \in [\pi/2, \pi]$. Since

$$u'(t) = -4\sin(t)\big(4 + \cos(t)\big) \leq 0,$$

we obtain

$$u(t) \leq u(\pi/2) = 17.$$

If follows that

$$-S_4(t) \leq \frac{1}{288} \cdot 1 \cdot 17 = 0.059\ldots.$$

Thus, $S_4(t) > \alpha$. This settles the case $n = 4$.

Next, let $n \geq 5$. Then,

$$|S_n(t)| \leq \frac{1}{(n!)^2} \sum_{k=1}^{n} \binom{n}{k}^2 = \frac{1}{(n!)^2}\left(\binom{2n}{n} - 1\right) \leq \frac{3}{50} = 0.06.$$

The proof of Theorem 3 is complete.

5 Lemmas

The first seven lemmas are needed to prove Theorem 4. Let $k, n \in \mathbb{Z}$ with $k, n \geq 0$ and $x \in \mathbb{R}$. We define

$$Q_n(x) = \sum_{k=0}^{n} (-1)^k \binom{2n}{2k} T_{2k}(x) \tag{5.1}$$

with

$$T_k(x) = (-1)^k \binom{-x}{k} = \prod_{j=1}^{k} \left(1 - \frac{1-x}{j}\right).$$

Throughout, we maintain these notations.

Lemma 1 *Let $n \geq 0$ be an integer and x be a real number. Then,*

$$Q_n(x) = \sum_{k=0}^{2n} 2^{k/2} \cos(k\pi/4) T_k(x) T_{2n-k}(1-x). \tag{5.2}$$

Proof Let $0 \leq k \leq 2n$. Since

$$\binom{2n}{k}\binom{2n-k-x}{2n} = (-1)^k \binom{-x}{k}\binom{x-1}{2n-k}$$

and

$$\cos(3k\pi/4) = (-1)^k \cos(k\pi/4),$$

we obtain from (1.1) with $2n$ instead of n, $m = 0$, $c = -x$ and $z = -1 + i = \sqrt{2}\exp(3\pi i/4)$,

$$Q_n(x) = \Re\left(\sum_{k=0}^{2n} \binom{2n}{k}\binom{-x}{k} i^k\right)$$

$$= \Re\left(\sum_{k=0}^{2n}(-1)^k\binom{-x}{k}\binom{x-1}{2n-k}2^{k/2}\exp(3k\pi i/4)\right)$$

$$= \sum_{k=0}^{2n}\binom{-x}{k}\binom{x-1}{2n-k}2^{k/2}\cos(k\pi/4)$$

$$= \sum_{k=0}^{2n}2^{k/2}\cos(k\pi/4)T_k(x)T_{2n-k}(1-x).$$

□

Lemma 2 *Let $n \geq 0$ be an integer and x be a real number. Then,*

$$\frac{(-1)^n}{2^{2n}}Q_{2n}(x) = T_{4n}(x) + \sum_{k=1}^{n}\frac{(-1)^k}{2^{2k}}D_{n,k}(x), \tag{5.3}$$

where

$$D_{n,k}(x) = T_{4(n-k)}(x)T_{4k}(1-x) + T_{4(n-k)+1}(x)T_{4k-1}(1-x)$$
$$-2T_{4(n-k)+3}(x)T_{4k-3}(1-x). \tag{5.4}$$

Proof From (5.2) we obtain

$$Q_{2n}(x) = \sum_{k=0}^{4n}2^{k/2}\cos(k\pi/4)T_k(x)T_{4n-k}(1-x).$$

Considering the indexes modulo 4 gives

$$Q_{2n}(x) = \sum_{k=0}^{n}(-1)^k2^{2k}T_{4k}(x)T_{4(n-k)}(1-x)$$
$$+\sum_{k=0}^{n-1}(-1)^k2^{2k}T_{4k+1}(x)T_{4(n-k)-1}(1-x)$$
$$+\sum_{k=0}^{n-1}(-1)^{k+1}2^{2k+1}T_{4k+3}(x)T_{4(n-k)-3}(1-x)$$
$$= (-1)^n2^{2n}T_{4n}(x) + \sum_{k=0}^{n-1}(-1)^k2^{2k}D_{n,n-k}(x).$$

This leads to (5.3). □

Lemma 3 *Let $x \in (0,1)$. The sequence $(m^{1-x}T_m(x))_{m\geq 1}$ is strictly increasing with*

$$\lim_{m\to\infty} m^{1-x}T_m(x) = \frac{1}{\Gamma(x)}. \tag{5.5}$$

Proof Using

$$m^{1-x}T_m(x) = \frac{1}{\Gamma(x)} \cdot m^{1-x}\frac{\Gamma(m+x)}{\Gamma(m+1)} \tag{5.6}$$

and the limit relation

$$\lim_{m\to\infty} m^{b-a}\frac{\Gamma(m+a)}{\Gamma(m+b)} = 1,$$

see Abramowitz and Stegun [1, p. 257], reveals that (5.5) is valid. Since the log function is strictly concave on $(0,\infty)$, we obtain

$$x\log(m+1) + (1-x)\log(m) < \log(m+x). \tag{5.7}$$

From (5.6) and (5.7) we conclude that

$$\begin{aligned}\frac{(m+1)^{1-x}T_{m+1}(x)}{m^{1-x}T_m(x)} &= \frac{(m+1)^{1-x}}{m^{1-x}}\frac{\Gamma(m+1+x)}{\Gamma(m+x)}\frac{\Gamma(m+1)}{\Gamma(m+2)} \\ &= \frac{m+x}{m^{1-x}(m+1)^x} > 1.\end{aligned}$$

Thus, $(m^{1-x}T_m(x))_{m\geq 1}$ is strictly increasing. □

Lemma 4 *Let k, n be integers with $1 \leq k \leq n-1$ and $j \in \{0,1,3\}$. Then, for $x \in (0,1)$,*

$$0 < \frac{(4n)^{1-x}}{2^{2k}}T_{4(n-k)+j}(x)T_{4k-j}(1-x) < \frac{\sin(\pi x)}{2^{(2k+1)x-1}(4k-3)\pi}. \tag{5.8}$$

Proof Since $T_{4(n-k)+j}(x) > 0$ and $T_{4k-j}(1-x) > 0$, we conclude that the left-hand side of (5.8) is valid. From Lemma 3 we obtain for $m \geq 1$,

$$T_m(x) < \frac{1}{m^{1-x}\Gamma(x)}.$$

Using this inequality and the reflection formula

$$\Gamma(x)\Gamma(1-x) = \frac{\pi}{\sin(\pi x)},$$

see [1, p. 256], gives

$$\frac{(4n)^{1-x}}{2^{2k(1-x)}}T_{4(n-k)+j}(x)T_{4k-j}(1-x) < \left(\frac{4n(4k-j)}{2^{2k}\big(4(n-k)+j\big)}\right)^{1-x}\frac{\sin(\pi x)}{(4k-j)\pi}.$$

Since

$$\begin{aligned}\frac{4n(4k-j)}{2^{2k}\big(4(n-k)+j\big)} &\le \frac{4n(4k-j)}{2^{2k}\big(4(n-k)+j\big)} + \frac{4k^2(n-k-1)+j(n+k^2+k)}{2^{2(k-1)}\big(4(n-k)+j\big)}\\ &= \frac{4k(k+1)}{2^{2k}} \le 2,\end{aligned}$$

we get

$$2^{2kx}\frac{(4n)^{1-x}}{2^{2k}}T_{4(n-k)+j}(x)T_{4k-j}(1-x) < \frac{2^{1-x}\sin(\pi x)}{(4k-j)\pi}.$$

This implies the right-hand side of (5.8). □

Lemma 5 *Let $x \in (0,1)$ and*

$$\Theta(x) = \sum_{k=1}^{\infty}\frac{(-1)^k}{2^{2k}}\big(T_{4k}(x) + T_{4k-1}(x) - 2T_{4k-3}(x)\big). \tag{5.9}$$

Then,

$$\Theta(1-x) = 2^{-x/2}\big(\cos(\pi x/4) + \sin(\pi x/4)\big) - 1.$$

Proof We have

$$\begin{aligned}\Theta(x)+1 &= \sum_{k=0}^{\infty}\frac{(-1)^k}{2^{2k}}\binom{-x}{4k} - \sum_{k=1}^{\infty}\frac{(-1)^k}{2^{2k}}\binom{-x}{4k-1} + 2\sum_{k=1}^{\infty}\frac{(-1)^k}{2^{2k}}\binom{-x}{4k-3}\\ &= \sum_{k=0}^{\infty}\frac{(-1)^k}{2^{2k}}\binom{-x}{4k} + \sum_{k=0}^{\infty}\frac{(-1)^k}{2^{2k+2}}\binom{-x}{4k+3} - \sum_{k=0}^{\infty}\frac{(-1)^k}{2^{2k+1}}\binom{-x}{4k+1}\\ &= \Re\left(\sum_{k=0}^{\infty}\binom{-x}{k}\Big(\frac{\exp(3\pi i/4)}{\sqrt{2}}\Big)^k\right)\\ &= \Re\left(\Big(1+\frac{\exp(3\pi i/4)}{\sqrt{2}}\Big)^{-x}\right) = 2^{x/2}\Re\big(\exp(-\pi x i/4)\big)\\ &= 2^{x/2}\cos(\pi x/4).\end{aligned}$$

This leads to

$$\Theta(1-x)+1=2^{(1-x)/2}\cos\big(\pi(1-x)/4\big)=2^{-x/2}\big(\cos(\pi x/4)+\sin(\pi x/4)\big).$$

□

The following limit theorem might be known in the literature. For completeness we offer a short proof.

Lemma 6 *Let $(f_{n,k})_{n\geq k+1\geq 2}$ be a double sequence of complex numbers and let $(h_k)_{k\geq 1}$ be a sequence of positive numbers. Assume that*

(i) *for $k \geq 1$, we have $\lim_{n\to\infty} f_{n,k} = g_k$,*
(ii) *for $k = 1, 2, \ldots, n-1$, we have $|f_{n,k}| \leq h_k$,*
(iii) $\sum_{k=1}^{\infty} h_k < \infty$.

Then, the series $\sum_{k=1}^{\infty} g_k$ is absolutely convergent and

$$\lim_{n\to\infty}\sum_{k=1}^{n-1} f_{n,k} = \sum_{k=1}^{\infty} g_k.$$

Proof From (i) and (ii) we conclude that $|g_k| \leq h_k$ for $k \geq 1$, so that (iii) implies that $\sum_{k=1}^{\infty} g_k$ is absolutely convergent. We define

$$y_n = \sum_{k=1}^{n-1} f_{n,k} - \sum_{k=1}^{n-1} g_k \quad \text{and} \quad z_n = \sum_{k=1}^{n-1} |f_{n,k} - g_k|.$$

Let N be an arbitrary natural number. Then, for $n > N+1$,

$$\begin{aligned}
|y_n| \leq z_n &= \sum_{k=1}^{N} |f_{n,k} - g_k| + \sum_{k=N+1}^{n-1} |f_{n,k} - g_k| \\
&\leq \sum_{k=1}^{N} |f_{n,k} - g_k| + \sum_{k=N+1}^{n-1} |f_{n,k}| + \sum_{k=N+1}^{n-1} |g_k| \\
&\leq \sum_{k=1}^{N} |f_{n,k} - g_k| + 2\sum_{k=N+1}^{\infty} h_k.
\end{aligned}$$

Using (i) gives

$$\overline{\lim_{n\to\infty}}\, z_n \leq 2\sum_{k=N+1}^{\infty} h_k.$$

Since

$$\lim_{N\to\infty} \sum_{k=N+1}^{\infty} h_k = 0,$$

we obtain

$$\overline{\lim_{n\to\infty}} z_n = 0, \quad \underline{\lim}_{n\to\infty} z_n = 0 \quad \text{and} \quad \lim_{n\to\infty} y_n = 0.$$

It follows that

$$\lim_{n\to\infty} \sum_{k=1}^{n-1} f_{n,k} = \lim_{n\to\infty} \left(y_n + \sum_{k=1}^{n-1} g_k \right) = \sum_{k=1}^{\infty} g_k.$$

□

Lemma 7 *Let* $x \in (0, 1)$. *Then, we have the asymptotic formula*

$$Q_{2n}(x) \sim (-1)^n 4^{n-1} n^{x-1} \frac{2^{3x/2}}{\Gamma(x)} \big(\cos(\pi x/4) + \sin(\pi x/4) \big) \quad (n \to \infty). \tag{5.10}$$

Proof Using (5.3) we obtain

$$(-1)^n \frac{(4n)^{1-x}}{2^{2n}} Q_{2n}(x) = (4n)^{1-x} T_{4n}(x) + \frac{(-1)^n}{2^{2n}} (4n)^{1-x} D_{n,n}(x)$$
$$+ \Omega_{n,0}(x) + \Omega_{n,1}(x) - 2\Omega_{n,3}(x), \tag{5.11}$$

where $D_{n,n}(x)$ is defined in (5.4) and

$$\Omega_{n,j}(x) = (4n)^{1-x} \sum_{k=1}^{n-1} \frac{(-1)^k}{2^{2k}} T_{4(n-k)+j}(x) T_{4k-j}(1-x).$$

Since

$$|D_{n,n}(x)| \le T_{4n}(1-x) + T_1(x) T_{4n-1}(1-x) + 2T_3(x) T_{4n-3}(1-x),$$

we conclude from Lemma 3 that $D_{n,n}(x) = O(n^{-x})$. This yields

$$\lim_{n\to\infty} \frac{(-1)^n}{2^{2n}} (4n)^{1-x} D_{n,n}(x) = 0. \tag{5.12}$$

An application of Lemmas 3, 4 and 6 gives that the limit relation

$$\lim_{n\to\infty} \Omega_{n,j}(x) = \frac{1}{\Gamma(x)} \sum_{k=1}^{\infty} \frac{(-1)^k}{2^{2k}} T_{4k-j}(1-x) \tag{5.13}$$

is valid for $j \in \{0, 1, 3\}$.

Using (5.5), (5.12) and (5.13) we conclude from (5.11) that

$$\lim_{n\to\infty} (-1)^n \frac{(4n)^{1-x}}{2^{2n}} Q_{2n}(x) = \frac{1}{\Gamma(x)} + \frac{1}{\Gamma(x)} \Theta(1-x), \tag{5.14}$$

where the function Θ is defined in (5.9). From (5.14) and Lemma 5 we obtain the asymptotic formula (5.10). □

Now, we present three lemmas, which are important for the proof of Theorem 5. The following obvious fact is used repeatedly:

$$a_0 > \sum_{k=1}^{n} |a_k| \quad \Longrightarrow \quad \sum_{k=0}^{n} a_k \cos(kt) > 0 \quad (t \in [0, \pi]). \tag{5.15}$$

Lemma 8 *Let*

$$J(s,t) = \frac{6}{s(s+1)(s+2)} + \frac{2}{(s+1)(s+2)} \cos(t) + \frac{1}{s+2} \cos(2t) + \cos(3t). \tag{5.16}$$

If $t \in [0, \pi]$, then $s \mapsto J(s,t)$ is strictly decreasing on $(0, 1]$.

Proof Let $t \in [0, \pi]$ and $s \in (0, 1]$. We obtain

$$-s^2(s+1)^2(s+2)^2 \frac{\partial}{\partial s} J(s,t) = r_0 + r_1 \cos(t) + r_2 \cos(2t)$$

with

$$r_0 = 12 + 36s + 18s^2, \quad r_1 = (4s+6)s^2, \quad r_2 = s^2(s+1)^2.$$

Since

$$r_0 - |r_1| - |r_2| = 36s + 6(2 - s^3) + s^2(11 - s^2) > 0,$$

we conclude from (5.15) that

$$r_0 + r_1 \cos(t) + r_2 \cos(2t) > 0.$$

Thus, $(\partial/\partial s) J(s,t) < 0$. □

Lemma 9 *Let*

$$\sigma_n(s) = \sum_{k=1}^{n} \eta_{k,n}(s) \quad \textit{with} \quad \eta_{k,n}(s) = \prod_{j=0}^{k-1} \frac{s+j}{j+n+1-k}. \tag{5.17}$$

If $s \in [0,1]$, *then* $n \mapsto \sigma_n(s)$ *is decreasing for* $n \geq 4$.

Proof Let $s \in [0,1]$ and $n \geq 4$. We have

$$\sigma_n(s) - \sigma_{n+1}(s) = \sum_{k=1}^{n} \lambda_{k,n}(s) - \delta_n(s)$$

with

$$\lambda_{k,n}(s) = \frac{k}{n+1} \prod_{j=0}^{k-1} \frac{s+j}{n-j} \quad \text{and} \quad \delta_n(s) = \frac{1}{(n+1)!} \prod_{j=0}^{n} (s+j).$$

Then,

$$\sigma_n(s) - \sigma_{n+1}(s) \geq \lambda_{1,n}(s) + \lambda_{n-1,n}(s) + \lambda_{n,n}(s) - \delta_n(s).$$

Since

$$\begin{aligned}\delta_n(s) - \big(\lambda_{n-1,n}(s) + \lambda_{n,n}(s)\big) &= \frac{s}{(n+1)!}\big(s^2 - (n-1)(1-s)\big) \prod_{j=1}^{n-2}(s+j) \\ &\leq \frac{s}{(n+1)!} \prod_{j=1}^{n-2}(s+j) \leq \frac{s}{n(n+1)} = \lambda_{1,n}(s),\end{aligned}$$

we conclude that $\sigma_n(s) \geq \sigma_{n+1}(s)$. □

Lemma 10 *For* $k, n \in \mathbb{Z}$ *with* $0 \leq k \leq n$ *let*

$$b_{k,n}(s) = \frac{(n-k)!}{\prod_{\nu=k}^{n-1}(s+\nu)} \quad \textit{and} \quad d_{k,n}(s) = b_{k,n}(s) \sum_{\nu=k}^{n-1} \frac{1}{s+\nu}. \tag{5.18}$$

If $s \in (0,1]$ *and* $n \geq 1$, *then*

$$\sum_{k=1}^{n} d_{k,n}(s) < d_{0,n}(s). \tag{5.19}$$

Proof We have

$$b_{k,n}(s) = \frac{1}{\binom{n}{k}} \frac{n!\,\Gamma(s+k)}{k!\,\Gamma(s+n)}.$$

For $s > 0$, we define $\Lambda_0(s) = 0$ and $\Lambda_j(s) = \sum_{\nu=0}^{j-1} 1/(s+\nu)$ for all positive integers j. Then,

$$\frac{d_{k,n}(s)}{d_{0,n}(s)} = \frac{1}{\binom{n}{k}} \frac{\Gamma(s+k)}{k!\,\Gamma(s)} \left(1 - \frac{\Lambda_k(s)}{\Lambda_n(s)}\right).$$

If $s \in (0,1]$ and $0 < k < n$, then we obtain

$$\frac{d_{k,n}(s)}{d_{0,n}(s)} < \frac{1}{\binom{n}{k}} \frac{\Gamma(s+k)}{k!\,\Gamma(s)} \le \frac{1}{\binom{n}{k}} \le \frac{1}{n}.$$

Thus, since $d_{n,n}(s) = 0$, we get

$$\sum_{k=1}^{n} \frac{d_{k,n}(s)}{d_{0,n}(s)} < 1.$$

This yields (5.19). □

6 Proof of Theorem 4

We have $F_n(0,t) = 1$. Next, we assume (for a contradiction) that there exists a parameter $c \neq 0$ such that $F_n(c,t)$ is positive for all $n \ge 0$ and $t \in [0,\pi]$. Since

$$F_1(c,0) = 1 + c > 0 \quad \text{and} \quad F_1(c,\pi) = 1 - c > 0,$$

we find that $c \in (-1,1)$. We consider two cases.

Case 1 $c \in (0,1)$.
We have

$$\begin{aligned} F_n(c,\pi) &= \sum_{k=0}^{n} (-1)^k \binom{n}{k}\binom{c}{k} \\ &= 1 - cn - \sum_{j=1}^{\infty} \binom{n}{2j} \frac{\kappa_j(c)}{2j-c} - \sum_{j=1}^{\infty} \binom{n}{2j+1} \frac{\kappa_j(c)}{2j+1} \end{aligned} \tag{6.1}$$

with

$$\kappa_j(c) = \frac{c}{(2j)!} \prod_{\nu=1}^{j} (c - 2\nu)(c + 1 - 2\nu).$$

Since $\kappa_j(c) > 0$ for $j \in \mathbb{N}$, we conclude from (6.1) that $F_n(c, \pi) < 0$ if $n > 1/c$. A contradiction.

Case 2 $c \in (-1, 0)$.

Let Q_n be the function defined in (5.1) and $x = -c \in (0, 1)$. Then,

$$F_{2n}(c, \pi/2) = Q_n(x).$$

Using (5.10) with $n = 2N + 1$ reveals that $Q_{4N+2}(x)$ is negative for sufficiently large N. A contradiction. Thus, there is no parameter $c \neq 0$ such that $F_n(c, t) > 0$ is valid for all $n \geq 0$ and $t \in [0, \pi]$.

7 Proof of Theorem 5

Let

$$G_n(c, t) = \sum_{k=0}^{n} \frac{\binom{c}{k}}{\binom{n}{k}} \cos(kt).$$

If $G_n(c, t) > 0$ for all $n \geq 0$ and $t \in [0, \pi]$, then

$$G_1(c, 0) = 1 + c > 0 \quad \text{and} \quad G_1(c, \pi) = 1 - c > 0$$

lead to $c \in (-1, 1)$. Next, we set $x = \cos(t) \in [-1, 1]$. Then,

$$0 < G_2(c, t) = 1 - \frac{c(c-1)}{2} + \frac{c}{2}x + c(c-1)x^2 = \phi_c(x), \quad \text{say.}$$

We assume (for a contradiction) that $c \leq c_0$. Let $x_0 = 1/(4(1 - c))$. Then,

$$-1 < c \leq c_0 = -0.97\ldots \quad \text{and} \quad 0 < x_0 < 1.$$

We obtain

$$\phi_c(x_0) = \frac{\varphi(c)}{16(1 - c)}$$

with

$$\varphi(c) = 8c^3 - 16c^2 - 7c + 16. \tag{7.1}$$

Since φ is strictly concave on $[-1, 0]$ with $\varphi(-1) = -1$, $\varphi(c_0) = 0$, $\varphi(0) = 16$, we conclude that $\varphi(c) \le 0$ for $c \in (-1, c_0]$. It follows that $\phi_c(x_0) \le 0$. A contradiction. Thus, $c > c_0$.

Next, we prove that if $c \in (c_0, 1)$, then $G_n(c, t)$ is positive for all $n \ge 0$ and $t \in [0, \pi]$.

We consider two cases.

Case 1 $c \in [0, 1)$.

Let $1 \le k \le n$. Then,

$$\left|\binom{c}{k}\right| = \frac{c}{k!}\prod_{\nu=1}^{k-1}(\nu - c) < \frac{1}{k!}\prod_{\nu=1}^{k-1}\nu = \frac{1}{k} \le \frac{1}{k}\binom{n-1}{k-1} = \frac{1}{n}\binom{n}{k}.$$

Thus,

$$\sum_{k=1}^{n}\frac{|\binom{c}{k}|}{\binom{n}{k}} < 1 = \frac{\binom{c}{0}}{\binom{n}{0}}.$$

From (5.15) we conclude that (1.16) holds for $n \ge 0$ and $t \in [0, \pi]$.

Case 2 $c \in (c_0, 0)$.

We define

$$B_n(s, t) = G_n(-s, \pi - t) = \sum_{k=0}^{n}\eta_{k,n}(s)\cos(kt) \tag{7.2}$$

with $\eta_{k,n}(s)$ as given in (5.17). It remains to show that for $n \ge 0$, $0 < s < s_0 = -c_0$ and $t \in [0, \pi]$ we have

$$B_n(s, t) > 0. \tag{7.3}$$

Since

$$B_0(s, t) = 1 \quad \text{and} \quad B_1(s, t) = 1 + s\cos(t) > 0,$$

we conclude that (7.3) is valid for $n = 0$ and $n = 1$.

Case 2.1 $n = 2$.

We have

$$B_2(s, t) = 1 + \frac{1}{2}s\cos(t) + \frac{1}{2}s(s+1)\cos(2t) = \omega_s(\cos(t))$$

with

$$\omega_s(x) = 1 - \frac{1}{2}s(s+1) + \frac{1}{2}sx + s(s+1)x^2.$$

Then, for $x \in \mathbb{R}$,

$$\omega_s(x) \geq \omega_s\left(-\frac{1}{4(s+1)}\right) = \frac{\varphi(-s)}{16(s+1)}$$

with φ as defined in (7.1). Let $\varphi^*(s) = \varphi(-s)$. Since φ^* is strictly concave on $[0, s_0]$ with $\varphi^*(0) = 16$ and $\varphi^*(s_0) = 0$, we conclude that $\omega_s(x) > 0$ for $s \in (0, s_0)$ and $x \in \mathbb{R}$. This settles (7.3) for $n = 2$.

Case 2.2 $n = 3$.

We have

$$B_3(s,t) = 1 + \frac{1}{3}s\cos(t) + \frac{1}{6}s(s+1)\cos(2t) + \frac{1}{6}s(s+1)(s+2)\cos(3t) = \chi_s(\cos(t))$$

with

$$\chi_s(x) = \left(\frac{2}{3}s^3 + 2s^2 + \frac{4}{3}s\right)x^3 + \left(\frac{1}{3}s^2 + \frac{1}{3}s\right)x^2$$
$$-\left(\frac{1}{2}s^3 + \frac{3}{2}s^2 + \frac{2}{3}s\right)x - \frac{1}{6}s^2 - \frac{1}{6}s + 1.$$

Let $s_1 = 0.9795$. Applying Sturm's theorem (see van der Waerden [15, section 79]) yields that χ_{s_1} has no zero on $[-1, 1]$, so that $\chi_{s_1}(0) > 0$ implies that

$$B_3(s_1, t) > 0 \quad \text{for} \quad t \in [0, \pi]. \tag{7.4}$$

We have

$$\frac{6}{s(s+1)(s+2)} B_3(s,t) = J(s,t), \tag{7.5}$$

where J is defined in (5.16). Using Lemma 8 and (7.4), (7.5) gives for $s \in (0, s_0)$,

$$J(s,t) \geq J(s_0, t) > J(s_1, t) > 0.$$

This implies that (7.3) holds with $n = 3$.

Case 2.3 $4 \leq n \leq 26$.

As in Case 2.2 we set $x = \cos(t)$ and express $B_n(s,t)$ as an algebraic polynomial in x. An application of Sturm's theorem gives that

$$B_n(0.98, t) > 0 \quad \text{for} \quad n = 4, 5, \ldots, 26 \quad \text{and} \quad t \in [0, \pi]. \tag{7.6}$$

Let

$$I_n(s,t) = n!B_n(s,t)\prod_{j=0}^{n-1}\frac{1}{s+j}. \tag{7.7}$$

Then,

$$I_n(s,t) = \sum_{k=0}^{n} b_{k,n}(s)\cos(kt)$$

and

$$-\frac{\partial}{\partial s}I(s,t) = \sum_{k=0}^{n} d_{k,n}(s)\cos(kt), \tag{7.8}$$

where $b_{k,n}(s)$ and $d_{k,n}(s)$ are given in (5.18). Applying (5.15), (5.19) and (7.8) reveals that the function $s \mapsto I_n(s,t)$ is strictly decreasing on $[0,1]$. This leads to

$$I_n(s,t) > I_n(s_0,t) > I_n(0.98,t). \tag{7.9}$$

From (7.6), (7.7) and (7.9) we conclude that $B_n(s,t) > 0$.

Case 2.4 $n \ge 27$.

Since $s \mapsto \eta_{k,n}(s)$ is increasing, we obtain for $s \in (0, s_0)$,

$$\sigma_n(s) = \sum_{k=1}^{n}\eta_{k,n}(s) \le \sum_{k=1}^{n}\eta_{k,n}(s_0) = \sigma_n(s_0).$$

Applying Lemma 9 gives

$$\sigma_n(s_0) \le \sigma_{27}(s_0) = 0.9986\ldots < 1.$$

(We remark that for $n = 4, 5, \ldots, 26$, we have $\sigma_n(s_0) > 1$.) Thus,

$$\sum_{k=1}^{n}\eta_{k,n}(s) < 1 = \eta_{0,n}(s).$$

Using (5.15) and (7.2) yields (7.3). This completes the proof of Theorem 5.

References

1. M. Abramowitz, I.A. Stegun (eds.), *Handbook of Mathematical Functions with Formulas, Graphs and Mathematical Tables* (Dover, New York, 1965)
2. R. Askey, Orthogonal Polynomials and Special Functions. Regional Conference Series in Applied Mathematics, vol. 21 (Society for Industrial and Applied Mathematics, Philadelphia, 1975)
3. R. Askey, G. Gasper, Inequalities for polynomials, in *The Bieberbach Conjecture*, ed. by A. Baernstein II, et al. Mathematical Surveys and Monographs, vol. 21 (American Mathematical Society, Providence, 1986), pp. 7–32
4. R.X.F. Chen, C.M. Reidys, A combinatorial identity concerning plane coloured trees and its applications. J. Integer. Seq. **20** (2017). Article 17.3.7
5. L. Fejér, Einige Sätze, die sich auf das Vorzeichen einer ganzen rationalen Funktion beziehen;.... Monat. Math. Phys. **35**, 305–344 (1928)
6. T.H. Gronwall, Über die Gibbssche Erscheinung und die trigonometrischen Summen $\sin x + \frac{1}{2}\sin 2x + \cdots + \frac{1}{n}\sin nx$. Math. Ann. **72**, 228–243 (1912)
7. J. Harer, D. Zagier, The Euler characteristic of the moduli space of curves. Invent. Math. **85**, 457–485 (1986)
8. D. Jackson, Über eine trigonometrische Summe. Rend. Circ. Mat. Palermo **32**, 257–262 (1911)
9. W. Koepf, *Hypergeometric Summation* (Vieweg, Braunschweig, 1998)
10. S. Koumandos, Inequalities for trigonometric sums, in *Nonlinear Analysis*, ed. by P.M. Pardalos, et al. Springer Optimization and Its Applications, vol. 68 (Springer, Berlin, 2012), pp. 387–416
11. G.V. Milovanović, D.S. Mitrinović, Th.M. Rassias, *Topics in Polynomials: Extremal Problems, Inequalities, Zeros* (World Scientific, Singapore, 1994)
12. F.W.J. Olver, D.W. Lozier, R.F. Boisvert, C.W. Clark (eds.), *NIST Handbook of Mathematical Functions* (Cambridge University Press, New York, 2010)
13. T.K. Petersen, *Eulerian Numbers* (Birkhäuser, Basel, 2015)
14. R.P. Stanley, Bijective proof problems (2009). http://www-math.mit.edu/~rstan/bij.pdf
15. B.L. van der Waerden, *Algebra I* (Springer, Berlin, 1971)

The Number of Partitions of a Set and Superelliptic Diophantine Equations

Dorin Andrica, Ovidiu Bagdasar, and George Cătălin Ţurcaş

Abstract In this chapter we start by presenting some key results concerning the number of ordered k-partitions of multisets with equal sums. For these we give generating functions, recurrences and numerical examples. The coefficients arising from these formulae are then linked to certain elliptic and superelliptic Diophantine equations, which are investigated using some methods from Algebraic Geometry and Number Theory, as well as specialized software tools and algorithms. In this process we are able to solve some recent open problems concerning the number of solutions for certain Diophantine equations and to formulate new conjectures.

2010 AMS Subject Classification Primary 14G05, 05A18; Secondary 11P81, 11Y50

1 Introduction

An integer partition is a way of writing an integer as a sum of natural numbers. The study of partitions was pioneered by Euler (1748) who obtained numerous fundamental results, and then continued by many mathematicians, including Gauss, Cauchy, Jacobi, Weierstrass, MacMahon, Hardy, Ramanujan or Erdös.

D. Andrica (✉)
Department of Mathematics, "Babeş-Bolyai" University, Cluj-Napoca, Romania
e-mail: dandrica@math.ubbcluj.ro

O. Bagdasar
Department of Electronics, Computing and Mathematics, University of Derby, Derby, UK
e-mail: o.bagdasar@derby.ac.uk

G. C. Ţurcaş
Department of Mathematics, "Babeş-Bolyai" University, Cluj-Napoca, Romania

The Institute of Mathematics of the Romanian Academy "Simion Stoilow", Bucharest, Romania
e-mail: george.turcas@math.ubbcluj.ro

A. M. Raigorodskii, M. Th. Rassias (eds.), *Discrete Mathematics and Applications*, Springer Optimization and Its Applications 165,
https://doi.org/10.1007/978-3-030-55857-4_3

Partitions play an important role in many branches of Mathematics, including number theory, combinatorics or group representation theory. They also feature in some key combinatorial optimization problems in Computer Science, as, for example, the Multiprocessor Scheduling Problem (MSP), the Bin Packing Problem (BPP) or the $0-1$ Multiple Knapsack Problem (MKP) [21]. For more details, the interested reader may consult the classical books of Andrews [1, 2], or the ample review paper written by Pak [26].

The number of ways in which a positive integer n can be partitioned as a sum of positive integers (first estimated by Euler) is given by the formula

$$\sum_{n=0}^{\infty} p(n)x^n = \prod_{k=1}^{\infty} \left(\frac{1}{1-x^k} \right). \tag{1}$$

Euler has also considered the partition of number n when the summands belong to certain subsets of $\mathbb{Z}$ (see [32] for more details on generating functions).

The study of partitions of the set $\{1, 2, \ldots, n\}$ in two sets with equal sums presents particular interest and is linked to the *signum equation*. For a given positive integer n, the number of level n solutions, denoted by $S(n)$, represents the number of $+$ and $-$ choices such that $\pm 1 \pm 2 \pm 3 \pm \cdots \pm n = 0$. Indexed as A063865 in the Online Encyclopedia of Integer Sequences (OEIS) [25], this sequence was linked to the Erdös–Surányi problem by Andrica and Ionaşcu [7]. The asymptotic formula for $S(n)$ was conjectured by Andrica and Tomescu [8] in 2002 as

$$\lim_{\substack{n \to \infty \\ n \equiv 0 \text{ or } 3 \pmod 4}} \frac{S(n)}{\frac{2^n}{n\sqrt{n}}} = \sqrt{\frac{6}{\pi}},$$

and was proved to be correct by Sullivan [29] in 2013. Other methods based on the Central Limit Theorem were suggested in [6, 7], while extensions to 2-partitions and 3-partitions with equal sums for multisets were explored in [3] and [11].

In Section 2 we present notations and results concerning ordered k-partitions having equal sums of a multiset M denoted by $S_k(m_1, \ldots, m_n; \alpha_1, \ldots, \alpha_n)$, for which we give the generating function in an integral form. Another important and challenging problem is the study of the number of partitions of multisets having certain properties. Results concerning the number of partitions of multisets, as well as asymptotic formulae for small multiplicity values were obtained in the 1970's by Bender [14] and Bender et al. [15]. In Section 3 we discuss the special case of multisets with equal multiplicity for the set $\{1, 2, \ldots, n\}$, and recurrence formulae involving the coefficients of special polynomials, linked to some special diophantine equations. These results have been formulated in [5].

In Sections 4 and 5, we show that the study of partitions with certain properties gives rise to an infinite family of very interesting Diophantine equations. Given one of these Diophantine equations, it is known (see Siegel's theorem below) that the number of integral solutions is finite. However, finding these solutions is difficult. It will become clear for the reader that, for half of our Diophantine equations, the problem of finding the integral solutions to the equation in question can be related to the problem of finding all the integral points on a hyperelliptic curve.

It is worth mentioning that the equations discussed in Sections 4 and 5 lead to the study of elliptic (see Theorem 6) and hyperelliptic (see Equation (35)) curves. When considered over finite fields, these special curves are of great interest for computer scientists due to their abundance of applications in public-key cryptography.

We start with a brief review of methods coming from Algebraic Geometry and Number Theory needed for our study. In the sections mentioned at the beginning of the previous paragraph, we will present the resolution of three equations in the family. It will be clear that the methods could, in theory, be applied to half of the equations in our infinite family. We will explain the reason why in practice such generalizations would be extremely difficult.

For every bivariate polynomial $f \in \mathbb{Z}[X, Y]$, let $C_f := \{(x, y) \in \overline{\mathbb{Q}}^2 : f(x, y) = 0\}$ be an affine algebraic curve. The points of C_f with coordinates in $\mathbb{Q}$ are called rational and, in general, for any $S \subseteq \overline{\mathbb{Q}}$, we denote by $C_f(S) = C_f \cap S^2$. Curves can be classified by their genus, a non-negative integer associated with their projectivization. The genus is a geometric invariant. A classical result in number theory is the following.

Theorem 1 ([27]) *If $f \in \mathbb{Z}[X, Y]$ defines an irreducible curve C_f of genus $g(C_f) > 0$, then $C_f(\mathbb{Z})$ is finite.*

If additionally $g_f(C_f) \geq 2$, this result is superseded by the notorious Faltings' theorem, which says that $C_f(\mathbb{Q})$ is also finite. Although both Siegels' and Faltings' theorems are milestones in number theory, they are "ineffective" results, meaning that their proof does not even allow one to control the size of the sets known to be finite. Therefore, they cannot be used to explicitly determine $C_f(\mathbb{Z})$ or $C_f(\mathbb{Q})$.

Effectively finding rational points on curves is an incredible difficult task and a very active topic of research. The toolbox for determining $C_f(\mathbb{Z})$ became a lot richer starting with the monumental work of Baker on linear forms in logarithms. As one of the first applications to his theory, Baker proved the following result.

Theorem 2 ([12]) *Consider $f(X, Y) = Y^2 - a_n X^n - a_{n-1} X^{n-1} - \cdots - a_0 \in \mathbb{Z}[X, Y]$, $n \geq 5$, $a_n \neq 0$ such that the polynomial $a_n X^n + \cdots + a_0$ is irreducible. Let $H = \max\{|a_0|, \ldots, |a_n|\}$. Then, any integral point $(x, y) \in C_f(\mathbb{Z})$ satisfies $\max(|x|, |y|) \leq \exp\exp\exp\{(n^{10n} H)^{n^2}\}$.*

Bounds on such solutions have been improved by many authors, but they remain astronomical and often involve inexplicit constants.

For every smooth, projective and absolutely irreducible curve C of genus g defined over $\mathbb{Q}$, the Jacobian J_C is a g-dimensional abelian variety, *functorially* associated with C. Fixing a point $P_0 \in C(\mathbb{Q})$, the curve C can be identified as a subvariety of J_C via the Abel–Jacobi map $\iota : C \hookrightarrow J_C$ with base point P_0. The famous Mordell–Weil theorem gives that, as is the case for elliptic curves, the set of $\mathbb{Q}$ rational points of J_C has the structure of a finitely generated abelian group, i.e.

$$J_C(\mathbb{Q}) \equiv T \oplus \mathbb{Z}^r,$$

where T is a finite abelian group and r is a positive integer, called the *rank*.

A famous theorem due to Chabauty and Coleman [20] is the following.

Theorem 3 *Let C be a smooth, projective and absolutely irreducible curve of genus g over $\mathbb{Q}$, with Jacobian J. Assume that the rank r of the Mordell–Weil group $J_C(\mathbb{Q})$ is strictly less than g. Then, there is an algorithm for determining the set of rational points $C(\mathbb{Q})$. Moreover, if p is a prime of good reduction for C such that $p > 2g$, then*

$$\#C(\mathbb{Q}) \leq \overline{C}(\mathbb{F}_p) + 2g - 2.$$

Here we denoted by $\overline{C}$ the curve obtained by reducing modulo p the coefficients of the equation defining C. An improvement due to Stoll [28] gives the sometimes smaller bound of $\#C(\mathbb{Q}) \leq \overline{C}(\mathbb{F}_p) + 2\,\mathrm{rank}(J_C(\mathbb{Q}))$ if C and p are as above.

Algebraic curves defined by equations of the type $Y^2 = f(X)$, where $f \in \mathbb{Q}[x]$ is a polynomial with distinct roots, are called hyperelliptic. Algorithms for computation in the Jacobian of such curves are described in [18]. These are implemented in the computer algebra package **Magma** [17]. Due to this computational convenience, when we are looking for integral solutions to two Diophantine equations introduced in Section 5, we make the passage to the problem of determining integral points on some hyperelliptic curves. We compute the Jacobian of the latter and apply the algorithm intrinsic in Theorem 3 to find all the rational points on such curves.

2 k-Partitions of Multisets with Equal Sums

The 3-partition problem is a famous strongly NP-complete problems, having the following statement. Let b, m and $a_1, \ldots, a_n$ be positive integers such that $n = 3m$ and $\sum_{s=1}^{n} a_s = mb$. The task is to partition the set $\{a_1, \ldots, a_n\}$ into m subsets, each having exactly three elements, so that the sum is exactly b (see, e.g., [22] and [23]). For instance, the set $\{10, 13, 5, 15, 7, 10\}$ can be partitioned into the two sets $\{10, 13, 7\}$, $\{5, 15, 10\}$, whose sum has the value 30.

For $\alpha_1, \ldots, \alpha_n$ real numbers and $m_1, \ldots, m_n$ positive integers, consider the multiset

$$M = \{\underbrace{\alpha_1, \cdots, \alpha_1}_{m_1 \text{ times}}, \cdots, \underbrace{\alpha_n, \cdots, \alpha_n}_{m_n \text{ times}}\},$$

where the number m_s is called the *multiplicity* of the element α_s, $s = 1, \ldots, n$, while $\sigma(M) = \sum_{s=1}^{n} m_s\alpha_s$ represents the *sum* of the elements of M. For a simple notation we denote $\mathbf{m} = (m_1, \ldots, m_n)$ and $\boldsymbol{\alpha} = (\alpha_1, \ldots, \alpha_n)$.

Definition 1 Let $k \geq 2$ be an integer. Denote by $S_k(\mathbf{m}; \boldsymbol{\alpha})$ the number of ordered k-partitions of M having equal sums, i.e. the number of k-tuples $(C_1, \ldots, C_k)$ of subsets of pairwise disjoint subsets of M such that

(i) $C_1 \cup \cdots \cup C_k = M$;
(ii) $\sigma(C_1) = \cdots = \sigma(C_k) = \frac{1}{k}\sigma(M)$.

Clearly, $S_k(\mathbf{m};\boldsymbol{\alpha}) = k! \cdot N_k(\mathbf{m};\boldsymbol{\alpha})$, where $N_k(\mathbf{m};\boldsymbol{\alpha})$ is the number of non-ordered k-partitions of M. The number $S_k(\mathbf{m};\boldsymbol{\alpha})$ is the constant term of the expansion of a Laurent polynomial $F(X_1,\ldots,X_{k-1}) \in \mathbb{Z}[X_1,\ldots,X_{k-1},X_1^{-1},\ldots,X_{k-1}^{-1}]$, defined by

$$F(X_1,\ldots,X_{k-1}) = \prod_{s=1}^{n}\left(X_1^{\alpha_s}+\cdots+X_{k-1}^{\alpha_s}+\frac{1}{(X_1\cdots X_{k-1})^{\alpha_s}}\right)^{m_s}. \tag{2}$$

Indeed, assume that for $s=1,\ldots,n$, from the term

$$\left(X_1^{\alpha_s}+\cdots+X_{k-1}^{\alpha_s}+\frac{1}{(X_1\cdots X_{k-1})^{\alpha_s}}\right)^{m_s},$$

we have selected c_j^s terms equal to $X_j^{\alpha_s}$, with $j=1,\ldots,k-1$, and c_k^s terms equal to $\frac{1}{(X_1\cdots X_{k-1})^{\alpha_s}}$. Clearly, we must have $c_1^s+\cdots+c_k^s = m_s$.

A selection like this contributes to the free term whenever

$$X_1^{\sum_{s=1}^n c_1^s\alpha_s}\cdots X_{k-1}^{\sum_{s=1}^n c_{k-1}^s\alpha_s}\cdot\frac{1}{(X_1\cdots X_{k-1})^{\sum_{s=1}^n c_k^s\alpha_s}} = 1.$$

This is equivalent to $\sum_{s=1}^n c_1^s\alpha_s = \cdots = \sum_{s=1}^n c_{k-1}^s\alpha_s = \sum_{s=1}^n c_k^s\alpha_s$, therefore the sets

$$C_j = \{\underbrace{\alpha_1,\cdots,\alpha_1}_{c_j^1\ times},\cdots,\underbrace{\alpha_n,\cdots,\alpha_n}_{c_j^n\ times}\},\qquad j=1,2,\ldots,k,$$

represent a partition of M which also satisfies property (ii) in Definition 1.

Ordering (2) after the integer powers of X_j, for $j=1,\ldots,k-1$, one can write

$$F(X_1,\ldots,X_{k-1}) = \sum_{m\in\mathbb{Z}} P_{m,j}(X_1,\ldots,X_{j-1},X_{j+1},\ldots,X_{k-1})X_j^m, \tag{3}$$

where $P_{m,j}(X_1,\ldots,X_{j-1},X_{j+1},\ldots,X_{k-1})$ are Laurent polynomials. Since F is symmetric in its variables, the Laurent polynomials $P_{m,j}$ do not actually depend on j, hence the notation can be simplified to $P_m(X_1,\ldots,X_{j-1},X_{j+1},\ldots,X_{k-1})$. Clearly, the free term of $F(X_1,\ldots,X_{k-1})$ is the free term of $P_0(X_1,\ldots,X_{j-1},X_{j+1},\ldots,X_{k-1})$.

Denoting by $\widetilde{X}_j = (X_1,\ldots,X_{j-1},X_{j+1},\ldots,X_{k-1})$ for simplicity, we have

$$F(X_1,\ldots,X_{k-1}) = \prod_{s=1}^{n}\left(X_1^{\alpha_s}+\cdots+X_{k-1}^{\alpha_s}+\frac{1}{(X_1\cdots X_{k-1})^{\alpha_s}}\right)^{m_s} \tag{4}$$

$$= P_0(\widetilde{X}_j) + \sum_{m\in\mathbb{Z},m\neq 0} P_m(\widetilde{X}_j)X_j^m. \tag{5}$$

Let $X_j = \cos t + i \sin t$ in (4) and integrate with respect to t over the range $[0, 2\pi]$. For $m \neq 0$, the integral of the monomial X_j^m with respect to t over $[0, 2\pi]$, hence the integral representation of the polynomial yields

$$P_0(\widetilde{X}_j) = \frac{1}{2\pi} \int_0^{2\pi} \prod_{s=1}^{n} \left(X_1^{\alpha_s} + \cdots + X_{k-1}^{\alpha_s} + \frac{1}{(X_1 \cdots X_{k-1})^{\alpha_s}} \right)^{m_s} \mathrm{d}t. \tag{6}$$

Setting $X_j = X$, $X_l = 1$ for $l = 1, \ldots, k-1$ and $l \neq j$ in (4), one obtains

$$\prod_{s=1}^{n} \left(X^{\alpha_s} + k - 2 + \frac{1}{X^{\alpha_s}} \right)^{m_s} = P_0(1, \ldots, 1) + \sum_{m \in \mathbb{Z}, m \neq 0} P_m(1, \ldots, 1) X^m. \tag{7}$$

By symmetry in X and X^{-1} we have

$$P_m(1, \ldots, 1) = P_{-m}(1, \ldots, 1), \quad m \in \mathbb{Z}.$$

Also, from (6) we deduce that

$$P_0(1, \ldots, 1) = \frac{1}{2\pi} \int_0^{2\pi} \prod_{s=1}^{n} \left(X^{\alpha_s} + k - 2 + \frac{1}{X^{\alpha_s}} \right)^{m_s} \mathrm{d}t, \tag{8}$$

hence $P_0(1, \ldots, 1)$ depends on k, $\mathbf{m}$ and $\boldsymbol{\alpha}$. Since $X^{\alpha_s} + \frac{1}{X^{\alpha_s}} = 2 \cos \alpha_s t$, we have

$$Q_k(\mathbf{m}; \boldsymbol{\alpha}) := P_0(1, \ldots, 1) = \frac{1}{2\pi} \int_0^{2\pi} \prod_{s=1}^{n} (k - 2 + 2 \cos \alpha_s t)^{m_s} \mathrm{d}t. \tag{9}$$

Furthermore,

$$Q_k(\mathbf{m}; \boldsymbol{\alpha}) = S_k(\mathbf{m}; \boldsymbol{\alpha}) + R_k(\mathbf{m}; \boldsymbol{\alpha}), \tag{10}$$

where $R_k(\mathbf{m}; \boldsymbol{\alpha})$ is the sum of the coefficients of $P_0(\widetilde{X}_j)$, which differ from the free term. Setting $X = 1$ in (7) we get $k^{m_1 + \cdots + m_n} = \sum_{m \in \mathbb{Z}} P_m(1, \ldots, 1)$, hence the sum of all the coefficients in all polynomials P_m is $k^{m_1 + \cdots + m_n}$. By (9), it follows that

$$Q_4(\mathbf{m}; \boldsymbol{\alpha}) = Q_2 \left(2\mathbf{m}; \frac{\boldsymbol{\alpha}}{2} \right).$$

As shown in [11], the integral formula for the number of ordered 2-partitions with equal sum of the multiset M with real elements $\boldsymbol{\alpha}$ and multiplicities $\mathbf{m}$ is

$$c_0(\mathbf{m}; \boldsymbol{\alpha}) = S_2(\mathbf{m}; \boldsymbol{\alpha}) = \frac{2^{m_1 + \cdots + m_n}}{2\pi} \int_0^{2\pi} \prod_{s=1}^{n} (\cos \alpha_s t)^{m_s} \, \mathrm{d}t, \tag{11}$$

which for $m_1 = \cdots = m_n = m$ and $\alpha_s = s$, $s = 1, \ldots, n$, produces the formula

$$c_0^{(m)}(n) = \frac{2^{nm}}{2\pi} \int_0^{2\pi} \prod_{s=1}^{n} (\cos st)^m \, \mathrm{d}t. \tag{12}$$

If $Q_k(n) = Q_k(1, \ldots, 1; 1, \ldots, n)$, then $Q_2(n) = c_0^{(1)}(n)$ and $Q_4(n) = c_0^{(2)}(n)$.

3 k-Partitions with Equal Sums of the Set $[n]$

For $\alpha_s = s$ and $m_s = 1$, for $s = 1, \ldots, n$ we consider the simplified notations $S_k(n)$ for $S_k(\mathbf{m}; \boldsymbol{\alpha})$ and $R_k(n)$ for $R_k(\mathbf{m}; \boldsymbol{\alpha})$. Recurrences can be obtained for the coefficients of $F(X_1, \ldots, X_{k-1})$ defined by (2), indexed by the level n, as in

$$F_n(X_1, \ldots, X_{k-1}) = \prod_{s=1}^{n} \left(X_1^s + \cdots + X_{k-1}^s + \frac{1}{(X_1 \cdots X_{k-1})^s} \right). \tag{13}$$

Writing $F_n(X_1, \ldots, X_{k-1})$ as a Laurent polynomial in $X_1, \ldots, X_{k-1}$, we get

$$F_n(X_1, \ldots, X_{k-1}) = \sum_{j_1, \ldots, j_{k-1} \in \mathbb{Z}} c_{j_1, \ldots, j_{k-1}}(n) X_1^{j_1} X_2^{j_2} \cdots X_{k-1}^{j_{k-1}}. \tag{14}$$

Clearly, we have

$$F_n(X_1, \ldots, X_{k-1}) = \frac{U(X_1, \ldots, X_{k-1})}{V(X_1, \ldots, X_{k-1})},$$

where U and V are polynomials. If $V \neq 0$ at the origin of $\mathbb{R}^{k-1}$, then the coefficients in (14) are given by the Cauchy integral formula

$$c_{j_1, \ldots, j_{k-1}(n)} = \frac{1}{(2\pi i)^{k-1}} \int_T \frac{F_n(X_1, \ldots, X_{k-1})}{x_1 \cdots x_{k-1}} x_1^{-j_1} \cdots x_{k-1}^{-j_{k-1}} \mathrm{d}x_1 \wedge \cdots \wedge \mathrm{d}x_{k-1}, \tag{15}$$

with T a product of small circles around the coordinate axes of $\mathbb{R}^{k-1}$. For effective computations we use the recurrence between the coefficients of levels n and $n-1$.

Theorem 4 *Denote by* $\mathbf{e_1}, \ldots, \mathbf{e_{k-1}}$ *the vectors of the canonical basis in* $\mathbb{Z}^{k-1}$. *The following recurrence relation is valid for* $\mathbf{j} = (j_1, \ldots, j_{k-1}) \in \mathbb{Z}^{k-1}$ *and* $n \geq 1$*:*

$$c_{\mathbf{j}}(n) = c_{\mathbf{j}-n\mathbf{e_1}}(n-1) + \cdots + c_{\mathbf{j}-n\mathbf{e_{k-1}}}(n-1) + c_{\mathbf{j}+n(1,\ldots,1)}(n-1). \tag{16}$$

Proof Indeed, by formula (13) we obtain

$$
\begin{aligned}
&F_n(X_1,\dots,X_{k-1})\\
&= F_{n-1}(X_1,\dots,X_{k-1})\left(X_1^n+\dots+X_{k-1}^n+\frac{1}{(X_1X_2\cdots X_{k-1})^n}\right)\\
&= \left(\sum_{\mathbf{j}\in\mathbb{Z}^{k-1}} c_{\mathbf{j}}(n-1)X_1^{j_1}\cdots X_{k-1}^{j_{k-1}}\right)\left(X_1^n+\dots+X_{k-1}^n+\frac{1}{(X_1\cdots X_{k-1})^n}\right)\\
&= \sum_{\mathbf{j}\in\mathbb{Z}^{k-1}} c_{\mathbf{j}}(n-1)\left(X_1^{j_1+n}\cdots X_{k-1}^{j_{k-1}}+\dots+X_1^{j_1}\cdots X_{k-1}^{j_{k-1}+n}\right.\\
&\qquad\qquad \left.+\,X_1^{j_1-n}\cdots X_{k-1}^{j_{k-1}-n}\right)\\
&= \sum_{\mathbf{j}\in\mathbb{Z}^{k-1}} \left(c_{\mathbf{j}-n\mathbf{e_1}}(n-1)+\dots+c_{\mathbf{j}-n\mathbf{e_{k-1}}}(n-1)+c_{\mathbf{j}+n(1,\dots,1)}(n-1)\right)\\
&\quad\times X_1^{j_1}\cdots X_{k-1}^{j_{k-1}}.
\end{aligned}
$$

□

For each $j\in\{1,\dots,k-1\}$, F_n is written as a Laurent polynomial in X_j as

$$
F_n(X_1,\dots,X_{k-1})=\sum_{m\in\mathbb{Z}} P_{n,m,j}(\widetilde{X}_j)X_j^m, \tag{17}
$$

where $\widetilde{X}_j=(X_1,\dots,X_{j-1},X_{j+1},\dots,X_{k-1})$. Since $F_n(X_1,\dots,X_{k-1})$ is symmetric, the coefficients $P_{n,m,j}$ are independent of j, so the simplified notation $P_{n,m}(\widetilde{X}_j)$ can be used. These polynomials can be obtained recursively as follows.

Theorem 5 *The following recurrence is valid for $m\in\mathbb{Z}$, $j=1,\dots,k-1$ and $n\geq 1$.*

$$
\begin{aligned}
P_{n,m}(\widetilde{X}_j) &= P_{n-1,m-n}(\widetilde{X}_j)+\Big(\sum_{p\neq j}X_p^n\Big)P_{n-1,m}(\widetilde{X}_j)\\
&\quad+\Big(\prod_{p\neq j}X_p\Big)^{-n}P_{n-1,m+n}(\widetilde{X}_j).
\end{aligned}
$$

Also, for $m=0$ we have

$$
P_{n,0}(\widetilde{X}_j)=P_{n-1,-n}(\widetilde{X}_j)+\Big(\sum_{p\neq j}X_p^n\Big)P_{n-1,0}(\widetilde{X}_j)+\Big(\prod_{p\neq j}X_p\Big)^{-n}P_{n-1,n}(\widetilde{X}_j). \tag{18}
$$

Proof The following formula can be established.

$$
\begin{aligned}
F_n(X_1, \dots, X_{k-1}) &= F_{n-1}(X_1, \dots, X_{k-1}) \\
&\times \left(X_1^n + \dots + X_{k-1}^n + \frac{1}{(X_1 \cdots X_{k-1})^n} \right) \\
&= \left(\sum_{m \in \mathbb{Z}} P_{n-1,m}(\widetilde{X}_j) X_j^m \right) \left(X_1^n + \dots + X_{k-1}^n + \frac{1}{(X_1 \cdots X_{k-1})^n} \right) \\
&= \sum_{m \in \mathbb{Z}} \Big[P_{n-1,m-n}(\widetilde{X}_j) + \Big(\sum_{p \neq j} X_p^n \Big) P_{n-1,m}(\widetilde{X}_j) \\
&\quad + \Big(\prod_{p \neq j} X_p \Big)^{-n} P_{n-1,m+n}(\widetilde{X}_j) \Big] X_j^m.
\end{aligned}
$$

Substituting $m = 0$ into this formula, we obtain (18). This ends the proof. □

Setting $X_1 = X$ and $X_p = 1$ for $p = 2, \dots, k-1$ in (13), one obtains

$$
F_n(X, 1, \dots, 1) = \prod_{s=1}^{n} \left(X^s + k - 2 + \frac{1}{X^s} \right) = \sum_{m \in \mathbb{Z}} P_{n,m}(1, \dots, 1) X^m. \tag{19}
$$

Note that the sum of the coefficients of the Laurent polynomial $P_{n,m}(X_1, \dots, X_{k-1})$ is given by $P_{n,m}(1, \dots, 1) = P_{n,-m}(1, \dots, 1)$, $m \in \mathbb{Z}$. Denoting $P_{n,m} := P_{n,m}(1, \dots, 1)$, The terms of the sequence $\{P_{n,0}\}_{n \geq 1}$ can be obtained from the double recurrence:

$$
P_{n,0} = P_{n-1,-n} + P_{n-1,0} + P_{n-1,n} = P_{n-1,0} + 2P_{n-1,n}. \tag{20}
$$

By the integral formula (9), we can directly compute $P_{n,0}$. Since this depends on the value of k, we introduce the notation

$$
Q_k(n) = P_{n,0} = S_k(n) + R_k(n) = \frac{1}{2\pi} \int_0^{2\pi} \prod_{s=1}^{n} (k - 2 + 2\cos st)\, dt. \tag{21}
$$

4 A Family of Diophantine Equations Defined by $Q_k(n)$

An enumerative formula for $Q_k(n)$ is given by the number of ordered partitions of $[n] = \{1, \dots, n\}$ into k disjoint sets $A_1, \dots, A_k$ with the property that $\sigma(A_1) = \sigma(A_k)$, where $\sigma(A)$ denotes the sum of all elements in A. The case $k = 2$ corresponds to the number $S(n)$ of partitions of $[n]$ in two sets with equal sums.

Clearly, $Q_k(n)$ is a monic polynomial of degree n in $k-2$. Moreover, in the paper [5] is proved that

$$Q_k(n) = \sum_{d=0}^{n} N(d,n)(k-2)^{n-d}, \tag{22}$$

where for each $d = 0, \ldots, n$, the coefficient $N(d,n)$ represents the number of ordered partitions of $[n]$ into 3 subsets A, B, C such that $|B| = d$ and $\sigma(A) = \sigma(C)$, where $|B|$ denotes the cardinality of B.

Therefore, $Q_k(n)$ has non-negative integer coefficients, and each coefficient has a combinatorial meaning in terms of partitions of the set $[n]$. A simple direct computation of the integral (21) shows that for $n = 3, 5, 7, 9$ and $k \geq 2$, we have

$$\begin{aligned}
Q_k(3) &= (k-2)^3 + 2;\\
Q_k(5) &= (k-2)^5 + 8(k-2)^2 + 6(k-2);\\
Q_k(7) &= (k-2)^7 + 18(k-2)^4 + 30(k-2)^3 + 18(k-2)^2 + 12(k-2) + 8;\\
Q_k(9) &= (k-2)^9 + 32(k-2)^6 + 82(k-2)^5 + 104(k-2)^4 + 130(k-2)^3\\
&\quad +136(k-2)^2 + 62(k-2).
\end{aligned}$$

The sequence $\{Q_k(3)\}_{k\geq 2}$ is indexed as A084380 in OEIS [25], where it is mentioned that it does not contain any perfect squares, i.e. the elliptic equation $X^3 + 2 = Y^2$ has no solutions in positive integers. Two different proofs for this result were given in [9]. The previous equation is linked to a Catalan-type conjecture related to Pillai's equation $X^U - Y^V = m$, with $X, Y, U, V \geq 2$ integers. The conjecture states that for any given integer m, there are finitely many perfect powers whose difference is m (see [31, Conjecture 1.6]). For $m = 2$, it was computationally checked that the only solution involving perfect powers smaller than 10^{18} is $2 = 3^3 - 5^2$. The number of such solutions is linked to A076427 in OEIS.

We describe here one of the proofs given in [9] for brevity.

Theorem 6 *The only solutions of* $X^3 + 2 = Y^2$ *in the set of integer numbers are* $(-1, 1)$ *and* $(-1, -1)$.

A few remarks are in order before giving the proof of this theorem. Since the genus of (the projectivization of) the curve determined by this equation is 1, we can use Siegel's theorem to deduce that there are finitely many points with integer coordinates. By Theorem 2, we know that if $(x, y) \in \mathbb{Z}^2$ is a point lying on this curve, then

$$\max(|x|, |y|) \leq \exp\exp\exp((3^{30} \cdot 2)^{3^2}).$$

Although theoretically one could now run a for loop through all possible values of x and check for which x^3+2 is a perfect square, the triple exponential bound presented above is astronomical and way out of the current computational limitations. In practice, one could check values of x up to 10^{18}, but could not hope to even get close to the aforementioned triple exponential. We proceed with the proof of our theorem.

We will make use of the following proposition.

Proposition 1 *The only solution* $(a, b) \in \mathbb{Z}^2$ *to the equation*

$$a^3 + 3a^2b + 6ab^2 + 2b^3 = 1 \tag{23}$$

is $(a, b) = (1, 0)$.

Proof Write $f(X) = X^3 + 3X^2 + 6X + 2 \in \mathbb{Q}[X]$. It is an irreducible polynomial and let $\theta \in \overline{\mathbb{Q}}$ be any root of f. Denote by $L = \mathbb{Q}(\theta)$, the number field obtained by adjoining θ to $\mathbb{Q}$ and write $\mathcal{O}_L$ for its ring of integers. L is a degree 3 extension over $\mathbb{Q}$ and has signature $(1, 1)$. We are going to denote by $\sigma_1, \sigma_2, \sigma_3 : L \hookrightarrow \mathbb{C}$ its three different complex embeddings.

It can be checked that ring of integers $\mathcal{O}_L$ is $\mathbb{Z}[\theta, \theta^2]$ and, making use of Dirichelt's unit theorem, one can compute the group of units

$$\mathcal{O}_L^\times = \langle \pm 1 \rangle \cdot \langle -\theta^2 - 3\theta - 1 \rangle \cong (\mathbb{Z}/2\mathbb{Z}) \cdot \mathbb{Z}.$$

The element $\mu := -\theta^2 - 3\theta - 1$ is a fundamental unit, $\mathrm{Norm}_{L/\mathbb{Q}}(\mu) = 1$ and $\mathrm{Norm}_{L/\mathbb{Q}}(-1) = -1$. Equation (23) can be written as

$$\mathrm{Norm}_{L/\mathbb{Q}}(a - b\theta) = \prod_{i=1}^{3} (a - b\sigma_i(\theta)) = 1, \text{ where } a, b \in \mathbb{Z}.$$

The above implies that $a - b\theta$ is a unit of norm 1 in O_L, hence

$$a - b\theta = \mu^n \text{ for some } n \in \mathbb{Z}. \tag{24}$$

We are going to use p-adic analysis to solve this last equation. We first need a local field $\mathbb{Q}_p$ into which there are three distinct embeddings of L, equivalently a prime number p such that the polynomial $X^3 + 3X^2 + 6X + 2$ has three distinct roots in $\mathbb{Q}_p$. We find $p = 79$ to be such a prime and the distinct roots are

$$\left.\begin{aligned} \theta_1 &= 19 - 32 \cdot 79 \pmod{79^2} \\ \theta_2 &= 20 - 7 \cdot 79 \pmod{79^2} \\ \theta_3 &= 37 + 38 \cdot 79 \pmod{79^2} \end{aligned}\right\} \in \mathbb{Q}_{79}.$$

The root θ of f is mapped to r_1, r_2 and r_3, respectively, under the embeddings of L into $\mathbb{Q}_{79}$. Under the same embeddings, the fundamental unit $\mu = -\theta^2 - 3\theta - 1$ maps to

$$\begin{aligned}\mu_1 &= 55 - 37 \cdot 79 \pmod{79^2}\\ \mu_2 &= 13 - 21 \cdot 79 \pmod{79^2} \in \mathbb{Q}_{79}.\\ \mu_3 &= 20 - 22 \cdot 79 \pmod{79^2}\end{aligned}$$

By embedding Equation (24) into $\mathbb{Q}_{79}$, we obtain that $a - b\theta_i = \mu_i^n$ and hence $a = \mu_i^n + b\theta_i$ for $i = 1, 2$ and 3. One obtains the equality

$$(\theta_3 - \theta_2)\mu_1^n + (\theta_1 - \theta_3)\mu_2^n + (\theta_2 - \theta_1)\mu_3^n = 0$$

and since $\mu_1\mu_2\mu_3 = \text{Norm}(\mu) = 1$, we can rewrite this as

$$(\theta_3 - \theta_2) + (\theta_1 - \theta_3)(\mu_2^2\mu_3)^n + (\theta_2 - \theta_1)(\mu_2\mu_3^2)^n = 0. \tag{25}$$

Now $\mu_2^2\mu_3 \equiv 62 \pmod{79}$ and $\mu_2\mu_3^2 \equiv 65 \pmod{79}$. Since the left-hand side of (25) must be equal to zero modulo 79, we can check that n is divisible by 13. Hence $n = 13m$ for some $m \in \mathbb{Z}$.

We have that $(\mu_2^2\mu_3)^{13} \equiv 1 + 8 \cdot 79 \pmod{79^2}$ and $(\mu_2\mu_3^2)^{13} \equiv 1 + 36 \cdot 79 \pmod{79^2}$. We can now use Lemma 5.2 in [19] to expand

$$(\theta_3 - \theta_2) + (\theta_1 - \theta_3)(\mu_2^2\mu_3)^{13\cdot m} + (\theta_2 - \theta_1)(\mu_2\mu_3^2)^{13\cdot m} = \sum_{k=1}^{\infty} a_k m^k,$$

with $\lim_{k\to\infty} \|a_k\|_{79} = 0$ and it can be checked that $\|a_1\|_{79} = 79^{-1}$ and $\|a_k\|_{79} \le 79^{-2}$ for every $k \ge 2$. Using Strassmann's theorem (see Theorem 4.1 in [19]), we obtain that the only value of m for which $\sum_{k=1}^{\infty} a_k m^k$ vanishes is $m = 0$.

This proves that $n = 0$ and replacing in (24) we obtain $(a, b) = (1, 0)$ is the only solution to the equation in the statement, as claimed. □

Remark 1 We have used the computer algebra package **Sage** [30] for basic modular arithmetic computations. Equation (23) is a Thue equation. It was proved that the latter have finitely many solutions and algorithms that find all of them have been implemented in various computer algebra packages. One can consult [16] for a very efficient such algorithm. The known methods for solving general Thue equations are involved, making use of Baker's bounds for linear forms in complex and of complicated reduction methods such as the one in described in loc. cit. In the above proof, we made essential use of the fact that the right-hand side of (23) is 1 and that the ring $\mathcal{O}_L$ has only one fundamental unit to apply p-adic analysis techniques successfully.

4.1 Proof of Theorem 6

We now return to the proof of our theorem. Let $K = \mathbb{Q}(\sqrt{2})$ and denote by $\mathcal{O}_K = \mathbb{Z}[\sqrt{2}]$ its ring of integers. The latter is a Dedekind domain, i.e. it is Noetherian, integrally closed in its field of fractions $\mathrm{Frac}(\mathcal{O}_K) = K$ and all its non-zero prime ideals are maximal. For any element $o \in \mathcal{O}_K$, we are going to denote by $(o) \subseteq O_K$ the principal ideal o generates.

Suppose that $x, y \in \mathbb{Z} \setminus \{0\}$ are such that $y^2 = x^3 + 2$. Therefore, in $\mathcal{O}_K$ we have the factorization $(y - \sqrt{2})(y + \sqrt{2}) = x^3$ and the same holds for the ideals generated by these factors. It is known that ideals of $\mathcal{O}_K$ factor uniquely into prime ideals. Suppose the prime ideal $\mathfrak{p} \subset \mathcal{O}_K$ divides both of the non-zero ideals $(y-\sqrt{2})$ and $(y + \sqrt{2})$. Then, $\mathfrak{p}$ must divide the ideal generated by the difference $y + \sqrt{2} - y + \sqrt{2} = 2\sqrt{2} = \sqrt{2^3}$. As $(\sqrt{2}) \subset \mathcal{O}_K$ is the only prime ideal of $\mathcal{O}_K$ that lies above 2, we must have $\mathfrak{p} = (\sqrt{2})$. Hence, the ideals $(y - \sqrt{2})$ and $(y + \sqrt{2})$ are coprime outside of $(\sqrt{2})$. From the previous factorization, we deduce that for every prime ideal $\mathfrak{p} \neq (\sqrt{2})$, if $\mathfrak{p}$ divides $(y - \sqrt{2})$, then $\mathfrak{p}^3$ divides the same ideal.

To see what happens in the case $\mathfrak{p} = (\sqrt{2})$, let $\mu \in \mathrm{Gal}(K/\mathbb{Q})$ be the non-trivial $\mathbb{Q}$-automorphism of K. Given a rational prime p, $\mathrm{Gal}(K/\mathbb{Q})$ acts naturally on the ideals $\mathfrak{p}$ of $\mathcal{O}_K$ that lie above p. Write $\mathfrak{p}^\mu$ for the ideal obtained from $\mathfrak{p}$ by applying μ to every element in $\mathfrak{p}$. As $\mu(\sqrt{2}) = -\sqrt{2}$, we note that $(\sqrt{2})^\mu = (-\sqrt{2}) = (\sqrt{2})$, i.e. μ stabilizes the ideal above 2. Notice that $(y - \sqrt{2})^\mu = (y + \sqrt{2})$, hence the powers of $(\sqrt{2})$ that divide the ideals $(y - \sqrt{2})$ and $(y + \sqrt{2})$ are equal. Since the product $(y - \sqrt{2})(y + \sqrt{2})$ is a third power, we conclude that the power of $(\sqrt{2})$ dividing $(y - \sqrt{2})$ must be divisible by 3.

It is an easy exercise, using, for example, the Minkowski bound, to prove that the class group of K is trivial. In particular, this means that every ideal of $\mathcal{O}_K$ is principal. Considering the remarks above, we have

$$(y - \sqrt{2}) = (x_0)^3 = (x_0^3), \text{ as ideals, where } x_0 \in \mathcal{O}_K.$$

We deduce that $y - \sqrt{2}$ and x_0^3 are the same up to a unit in the ring $\mathcal{O}_K$, that is, there exists a unit $u \in U(\mathcal{O}_K)$ such that $y - \sqrt{2} = u x_0^3$.

By Dirichlet unit's theorem we know that $U(\mathcal{O}_K)$ is isomorphic to $T \cdot \mathbb{Z}$, where T is the finite group formed by the roots of unity that lie in K. It is an easy exercise to verify that $U(\mathcal{O}_K) = \langle -1 \rangle \langle 1 - \sqrt{2} \rangle$, so $1 - \sqrt{2}$ is the fundamental unit of $\mathcal{O}_K$. Observing that every element $u \in U(U_K)$ can be written as $u = (1 - \sqrt{2})^i (u_0)^3$ where $i \in \{-1, 0, 1\}$ and $u_0 \in U(\mathcal{O}_K) \subseteq \mathcal{O}_K$, we derive that

$$y - \sqrt{2} = (1 - \sqrt{2})^i x_1^3,$$

for some $i \in \{-1, 0, 1\}$ and $x_1 \in \mathcal{O}_K$. The element x_1 is of the form $a + b\sqrt{2}$ for $a, b \in \mathbb{Z}$. For each choice of $i \in \{-1, 0, 1\}$, by equating the coefficients of $\sqrt{2}$ in the left- and right-hand side of the above equation, we obtain an equality of the form

$$f(a, b) = -1, \tag{26}$$

where $f \in \mathbb{Z}[x, y]$ is a homogeneous cubic polynomial. When f is reducible (26) can be easily solved using factorization in $\mathbb{Z}$. If this is not the case and f is irreducible, Equation (26) is a cubic Thue equation. It is known (see, for example, [16]) that the latter have finitely many integral solutions and routines for determining them have been implemented in various computer algebra packages. We will appeal to Equation (23) to find the solutions of the latter type of equations that arise here.

Let us analyse each of the three cases.

Case 1 $i = -1 \Rightarrow y - \sqrt{2} = (1 - \sqrt{2})^{-1}(a + b\sqrt{2})^3$.
Hence,

$$y - \sqrt{2} = -a^3 - 6a^2b - 6ab^2 - 4b^3 + \sqrt{2}\left(-a^3 - 3a^2b - 6ab^2 - 2b^3\right).$$

Using that $1, \sqrt{2}$ are linearly independent over $\mathbb{Q}$, we obtain the following two equations:

$$y = -a^3 - 6a^2b - 6ab^2 - 4b^3 \tag{27}$$

and

$$1 = a^3 + 3a^2b + 6ab^2 + 2b^3. \tag{28}$$

The variable y is an indeterminate and every solution (a, b) to (28) will determine a value for y. From Equation (23), we know that the only solution in integers to the last equation is $a = 1$ and $b = 0$. Substituting, we see that this corresponds to $y = -1$, which implies that $x = -1$.

Case 2 $i = 0 \Rightarrow y - \sqrt{2} = (a + b\sqrt{2})^3$.
Expanding the right-hand side, we see that

$$y - \sqrt{2} = a^3 + 6ab^2 + \sqrt{2}\left(3a^2b + 2b^3\right)$$

and since $1, \sqrt{2}$ are linearly independent over $\mathbb{Q}$ we must have

$$-1 = b(3a^2 + 2b^2).$$

Trying $b = \pm 1$, we see that $3a^2 + 2 = \mp 1$ is not solvable. Hence this case does not give us any solutions.

Case 3 $i = 1 \Rightarrow y - \sqrt{2} = \left(1 - \sqrt{2}\right)(a + b\sqrt{2})^3$.
This gives us

$$y - \sqrt{2} = a^3 - 6a^2b + 6ab^2 - 4b^3 + \sqrt{2}\left(-a^3 + 3a^2b - 6ab^2 + 2b^3\right),$$

which implies that

$$1 = a^3 - 3a^2b + 6ab^2 - 2b^3.$$

By making the substitution $t := -b$ in the last equation we obtain the one discussed in *Case 1*. Therefore, using Equation (23) once again we find $a = 1$, $b = 0$ and hence $y = 1$. Using that $y^2 = x^3 + 2$, we get that $x = -1$. The proof of our theorem is now complete.

In the proof above we made explicit use of the fact that $\mathbb{Q}(\sqrt{2})$ has trivial class group, information that allowed us to pass from factorisations of ideals to nice factorisations of elements in the ring $\mathbb{Z}[\sqrt{2}]$. In general, for $D \in \mathbb{Z}$ the ideal class group of $\mathbb{Q}(\sqrt{D})$ can be arbitrary large so our first strategy will not work for more general Mordell equations. The second proof of our theorem can be adapted to find all the integral solutions of $Y^2 = X^3 + D$ for any fixed $D \in \mathbb{Z}$.

The given problem is one of explicitly determining the integral points on the affine curve given by $Y^2 = X^3 + 2$. This is an elliptic curve and its integral points can be found by exploiting its rich geometric structure. We refer the interested reader to [9] for a proof of Theorem 6 along these lines.

Motivated by the property that the sequence $Q_k(3)$ does not contain any perfect squares, in the papers [4] and [5], the authors suggested the following problem: study if the sequence $\{Q_k(n)\}_{k\geq 2}$ contains any $n-1$ powers for various values of n. These yield, for general n, very difficult Diophantine equations and it seems improbable to develop a general strategy for solving them. We will explain a general method that can sometimes be successfully used to study the aforementioned problem when n is odd, since in this case we can relate the problem to a problem of determining integral (or rational) points on hyperelliptic curves. To demonstrate the versatility of the method, we discuss the problem above for $n = 5$ and 7. We obtain the following Diophantine equations:

$$X^5 + 8X^2 + 6X = Y^4; \tag{29}$$

$$X^7 + 18X^4 + 30X^3 + 18X^2 + 12X + 8 = Y^6. \tag{30}$$

Two of the authors of the present survey have successfully used effective methods for identifying integral points on curves in [10] to prove the following theorems.

Theorem 7 *The only solutions in integers to the equation $Y^4 = X^5 + 8X^2 + 6X$ are given by the pairs $(x, y) = (0, 0)$, $(-1, 1)$ or $(-1, -1)$.*

Since believe that the pairs above are all the rational solutions to (29), we formulated Conjecture 1 which asserts this. Our tools were more efficient when applied to Equation (30) for which we were able to find all the rational solutions.

Theorem 8 *The only rational solutions of the equation $X^7+18X^4+30X^3+18X^2+12X + 8 = Y^2$ are given by $(x, y) = (-1, 1)$ or $(-1, -1)$.*

To keep the material self-contained, we present here the proof of Theorem 7 and we refer the interested reader to the article [10] for the proof of Theorem 8.

We also say a few words about the Diophantine equations one obtains for $n = 8$. Trying to show that $\{Q_k(9)\}_{k\geq 2}$ does not contain any 8-th powers, since we saw that

$$Q_k(9) = (k-2)^9 + 32(k-2)^6 + 82(k-2)^5 + 104(k-2)^4 + 130(k-2)^3 + 136(k-2)^2 + 62(k-2),$$

one is lead to the study of the Diophantine equation

$$X^9 + 32X^6 + 82X^5 + 104X^4 + 130X^3 + 136X^2 + 62X = Y^8. \tag{31}$$

4.2 The Proof of Theorem 7

To settle this, we are going to study points on the affine curve given by

$$Y^4 = X^5 + 8X^2 + 6X. \tag{32}$$

It is an exercise using the Riemann–Hurwitz formula to compute that the genus of this curve is equal to 6, therefore by results mentioned earlier we know that it has finitely many rational points. To find the set of all rational points turns out to be a notorious difficult task from which we choose to detach for now.

We start by proving the following easy result.

Proposition 2 *If x, y are positive integers such that $y^4 = x^5 + 8x^2 + 6x$, then x is divisible by* 6.

Proof Suppose $\gcd(x, 6) = 1$. Then, the numbers x and $x^4 + 8x + 6$ are coprime, and as their product is a fourth power, we can conclude that both numbers must be fourth powers. Hence, we obtain the equation $x^4 + 8x + 6 = z^4$, for some positive integer z.

Since $z > x$ and $(z^2-x^2)(z^2+x^2) = 8x+6$, we get that $8x+6 \geq z^2+x^2 \geq 1+x^2$ implying that $1 \leq x \leq 9$. As x must be a fourth power, the only possibility left is $x = 1$ but then $y^4 = 15$, a contradiction.

Suppose for the sake of contradiction that x is not divisible by 3. The above implies that x is divisible by 2 and so is y. As $4 \leq v_2(y^4) = v_2(6x)$, we have that $v_2(x) \geq 3$. There exist positive integers a, b such that

$$\begin{cases} x = 2^3 a \text{ and} \\ y = 2b \end{cases}.$$

By substituting in the initial equation and dividing by 2^4, we obtain

$$b^4 = a(2^{11}a^4 + 2^5 a + 3).$$

As the terms of the product on the right-hand side are, in this case, coprime we derive that both a and $2^{11}a^4 + 2^5 a + 3$ must be fourth powers of positive integers. But the residue of $2^{11}a^4 + 2^5 a + 3$ when divided by 4 is always 3 and no fourth power has this property. The contradiction implies that 3 divides x.

All that is left is proving that 2 divides x as well. Suppose the contrary, namely x, y are odd and both divisible only by 3. Let c, d be positive integers such that

$$\begin{cases} x = 3c \text{ and} \\ y = 3d \end{cases}.$$

Again, by using the initial equation and dividing by 3^2 we obtain

$$3^2 d^4 = 3^3 c^5 + 8c^2 + 2c.$$

The last assumption implies that c is odd and by looking at the last equation modulo 8, it is easy to deduce that $c = 5 \pmod 8$. Analysing the equation modulo 16, we further find that $c = 5 \pmod{16}$. Now, we get a contradiction by looking modulo 32, since the left-hand side of the equation above is 9 or 25 modulo 32, values that the right-hand side never achieves when $c = 5 \pmod{16}$. This completes the proof of our proposition. □

Remark 2 Naively applied, Baker's theorem tells us that if x, y are integers satisfying the equation $y^4 = x^5 + 8x^2 + 6x$, then $\max(|x|, y^2) \le \exp\exp\exp\{(5^{50} \cdot 8)^{5^2}\}$, which is astronomical and does not help much with our task. Instead, we show that a putative solution to this equation gives rise to a certain point on a projective hyperelliptic curve with special arithmetic properties. Assisted by the computer algebra package **Magma**, we determine the set of all rational points on the hyperelliptic curve and prove that our predicted point does not belong to it.

Finally, we prove our theorem giving a positive answer to the result conjectured by Andrica and Bagdasar [5].

Let $(x, y) \in \mathbb{Z}^2$ be a solution to our Equation (29), i.e. $y^4 = x^5 + 8x^2 + 6x$. It is easy to see that $x = 0$ if and only if $y = 0$. Suppose that $y \neq 0$. If $x < 0$, the positivity of y^4 implies that $-2 < x < 0$ and as x is an integer we obtain $x = -1$. Substituting in the original equation, this gives $y = -1$ and $y = 1$ as the only possibilities for $x < 0$.

We will prove that there are no integral solutions with $x > 0$. Suppose the contrary and let $(x, y) \in \mathbb{Z}_{>0} \times \mathbb{Z}$ be such a solution. Since $(x, -y)$ is also a solution to (29), we can assume without loosing generality that $y > 0$. We proved in Proposition 2 that x and hence y are divisible by 6. Moreover, in the proof of the same proposition, we saw that $v_2(x) \geq 3$. Let a, b be positive integers such that $x = 2^3 \cdot 3 \cdot a$ and $y = 2 \cdot 3 \cdot b$. Substituting into the initial equation and dividing both sides by $2^4 3^2$, we get

$$(3b^2)^2 = 2^{11} 3^3 a^5 + 2^7 a^2 + a.$$

We can now regard $(a, 3b^2)$ as a point on the affine model of the projective hyperelliptic curve

$$C^{\text{proj}} : Y^2 = 2^{11}3^3X^5Z + 2^7X^2Z^4 + Z^6. \tag{33}$$

Notice that the ambient space is not the classical projective plane $\mathbb{P}^2$, but rather the weighted $\mathbb{P}^2_{(1,3,1)}$. The points of $\mathbb{P}^2_{(1,3,1)}$ over $\mathbb{Q}$ are the equivalence classes of triples $[X : Y : Z] \in \mathbb{Q}^3 \setminus \{[0 : 0 : 0]\}$, where two triples $[X_1 : Y_1 : Z_1]$ and $[X_2 : Y_2 : Z_2]$ are equivalent if there exists some $\lambda \in \mathbb{Q} \setminus \{0\}$, such that $[X_2 : Y_2 : Z_2] = [\lambda \cdot X_1 : \lambda^3 \cdot Y_1 : \lambda \cdot Z_1]$. A point with integral coordinates $(a, 3b^2)$ maps to the point with coordinates $[a : 3b^2 : 1] \in \mathbb{P}^2_{(1,3,1)}$ on $C^{\text{proj}}(\mathbb{Q})$.

The latter smooth projective curve has genus 2 and by Falting's theorem we know that $C^{\text{proj}}(\mathbb{Q}) = \{[X : Y : Z] \in \mathbb{P}^2_{(1,3,1)}(\mathbb{Q}) : Y^2 = 2^{11}3^3X^5Z + 2^7X^2Z^4 + Z^6\}$ is a finite set. We pursue the task of explicitly determining $C^{\text{proj}}(\mathbb{Q})$.

The Jacobian J_C is a 2 (equal to the genus) dimensional abelian variety that is "functorially associated" with C^{proj}.

We also know for $P_0 := [1 : 0 : 0] \in C^{\text{proj}}(\mathbb{Q})$, the Abel–Jacobi map associated with P_0 is the embedding $i : C^{\text{proj}}(\mathbb{Q}) \hookrightarrow J_C(\mathbb{Q})$ given by $P \mapsto [P - P_0]$, for every $P \in C^{\text{proj}}(\mathbb{Q})$. Using the computer algebra package **Magma** [17], we compute that

$$J_C(\mathbb{Q}) \cong (\mathbb{Z}/2\mathbb{Z}) \oplus \mathbb{Z}.$$

Moreover, assisted by the same software, we show that the torsion subgroup T is generated by $i([0 : 0 : 1])$ and a generator for $J_C(\mathbb{Q})/T$ is $i([-1 : -5760 : 24])$.

As the rank of the Jacobian is strictly less than the genus of our curve, we can apply the method of the algorithm mentioned in Theorem 3 to determine $C^{\text{proj}}(\mathbb{Q})$. A beautiful presentation of how the method works can be found in the expository article of McCallum and Poonen [24] and an algorithm suitable to our set-up is implemented in **Magma**. For any prime p of good reduction for C^{proj}, the closure of $J_C(\mathbb{Q})$ in $J_C(\mathbb{Q}_p)$ (under the p-adic topology) can be described as the locus where certain power series vanish. It turns out that, under a natural embedding, the image of C^{proj} in J_C meets this closure in a finite set, a set that must contain $C^{\text{proj}}(\mathbb{Q})$.

A description of the **Magma** implementation of the aforementioned algorithm can be found at https://magma.maths.usyd.edu.au/magma/handbook/text/1507.

It takes as input the genus 2 hyperelliptic curve C^{proj} and $i([-1 : -5760 : 24])$, the generator of the torsion-free part of its Jacobian. We remark that the procedure implemented in **Magma** combines the method of Chabauty–Coleman with the Mordell–Weil Sieve at primes $p \in \{5, 11, 29\}$ and, in a few seconds on a personal laptop, it outputs the full set of rational points

$$C^{\text{proj}}(\mathbb{Q}) = \{[0 : 0 : 1], [-1 : 5760 : 24], [1 : 0 : 0], [-1 : -5760 : 24]\}.$$

As we cannot find a point of the form $[a : 3b^2 : 1]$ for a, b positive integers, we get our contradiction and the proof of our theorem is complete.

The reader might wonder why we did not apply Chabauty's method to the hyperelliptic curve defined by $Y^2 = X^5 + 8X^2 + 6X$. It turns out that, although it has genus 2, this curve has a rank 2 Jacobian and does not satisfy the hypothesis required by Theorem 3. We therefore had to work with local methods and to prove Proposition 2 in order to apply Chabauty's method to a different hyperelliptic curve.

As a result of extensive computations performed on a computer, we are confident in formulating the following conjecture.

Conjecture 1 The only rational solutions of the equation $Y^4 = X^5 + 8X^2 + 6X$ are $(x, y) = (0, 0), (-1, 1)$ and $(-1, -1)$.

5 Final Comments on the Family of Diophantine Equations

Let us recall that formula (22) gives

$$Q_k(n) = \sum_{d=0}^{n} N(d, n)(k-2)^{n-d},$$

where for each $d = 0, \dots, n$, the coefficient $N(d, n)$ is the number of ordered partitions of $[n]$ into 3 subsets A, B, C such that $|B| = d$ and $\sigma(A) = \sigma(C)$, where $|B|$ is the cardinality of B. Notice that $Q_k(n)$ is a polynomial of degree n in $k-2$. It was conjectured in [4] and [5] that for every $n \geq 3$, the sequence $\{Q_k(n)\}_{k\geq 2}$ does not contain $n-1$ powers. The strategy presented above, which culminated with the successful resolution of this conjecture in the case $n = 5$ and $n = 7$ was initiated by the authors of [10]. In theory such a strategy could be used for attacking this conjecture for any odd value of n. Indeed, if $n = 2m + 1$ for some positive integer m, then to prove the conjecture one has to show that an equation of the form

$$Y^{2m} = f(X) \tag{34}$$

does not have solutions in positive integers, where $f \in \mathbb{Z}[X]$ is a given polynomial of degree $2m + 1$ in X. Of course, an integral solution (x, y) to (34) gives rise to an integral point (x, y^m) on the affine model of the hyperelliptic curve

$$Y^2 = f(X). \tag{35}$$

Theorem 2 tells us that there are finitely many integral points on this curve and even gives (very weak) upper bounds for the absolute values of these points. Therefore, in theory we could verify whether among these finitely many points there is one of the form (x, y^m) for some positive integers x and y. In practice, the bounds arising in such way are too large to be useful.

However, if we prove that the hyperelliptic curve (35) satisfies the hypothesis of Theorem 3, we know that there exists an effective algorithm for determining the integral points on the hyperelliptic curve. Unfortunately, computer algebra packages such as **Magma** [17] or **Sage** [30] do not include implementations for computing the rank of the Jacobians of hyperelliptic curves with genus $g \geq 4$, equivalently $2m + 1 \geq 9$. Until recently this was also the case for curves of genus $g = 3$, but an implementation for the Jacobian of genus 3 hyperelliptic curves was included in **Sage** by Balakrishnan et al. [13]. The development of such implementations for hyperelliptic curves of large genus is an active topic of research and there is a great interest to include them into computer algebra packages.

As soon as an implementation for the Jacobian of hyperelliptic curves of genus 4 becomes available, one could attack the aforementioned conjecture for $n = 2m + 1 = 9$. The latter leads to the study of the Diophantine Equation (31)

$$X^9 + 32X^6 + 82X^5 + 104X^4 + 130X^3 + 136X^2 + 62X = Y^8.$$

One can first try to find the (finitely many) integral points on the hyperelliptic curve of genus 4

$$X^9 + 32X^6 + 82X^5 + 104X^4 + 130X^3 + 136X^2 + 62X = Y^2.$$

It can be computationally hard to show that the rank of the Jacobian of this hyperelliptic curve is less than or equal to 3, but if successful, one could apply the method of Chabauty, intrinsic in Theorem 3 to find all the integral (even rational) points on this curve. If none of this is of the form (x, y^4) for some positive integers x and y, then the conjecture of Andrica and Bagdasar is true. We should mention that extensive näive search for solutions to (31) in the positive integers did not produce any results.

References

1. G.E. Andrews, *The Theory of Partitions* (Cambridge University Press, Cambridge, 1998), 255 pp.
2. G.E. Andrews, K. Eriksson, *Integer Partitions*, 2nd edn. (Cambridge University Press, Cambridge, 2010), 152 pp.
3. D. Andrica, O. Bagdasar, Some remarks on 3-partitions of multisets. Electron. Notes Discrete Math. **70**, 1–8 (2018). Proceedings of the 2nd IMA TCDM'18
4. D. Andrica, O. Bagdasar, The Cauchy integral formula with applications to polynomials, partitions and sequences, in *Proceedings of the XVth Int. Conf. on Mathematics and Its Applications*, Timisoara, 1–3 Nov 2018 (Editura Politehnică, Timişoara, 2019), pp. 12–25
5. D. Andrica, O. Bagdasar, On k-partitions of multisets with equal sums (submitted)
6. D. Andrica, E.J. Ionascu, Some unexpected connections between Analysis and Combinatorics, in *Mathematics Without Boundaries*, ed. by Th.M. Rassias, P. Pardalos. Topics in Pure Mathematics (Springer, New York, 2014), pp. 1–20
7. D. Andrica, E.J. Ionascu, The signum equation for Erdos-Suranyi sequences. INTEGERS **15A**, 1–9 (2015)
8. D. Andrica, I. Tomescu, On an integer sequence related to a product of trigonometric fuctions, and its combinatorial relevance. J. Integer Seq. **5**, Article 02.2.4 (2002)

9. D. Andrica, G. Ţurcaş, An elliptic Diophantine equation from the study of partitions. Stud. Univ. Babeş-Bolyai Math. **64**(3), 349–356 (2019)
10. D. Andrica, G. Ţurcaş, Hyperelliptic Diophantine equations from the study of partitions. Math. Rep. (to appear)
11. O. Bagdasar, D. Andrica, New results and conjectures on 2-partitions of multisets, in *Proceedings of the 7th ICMSAO'17* (IEEE, Piscataway, 2017), pp. 1–5
12. A. Baker, Bounds for the solutions of the hyperelliptic equations. Proc. Camb. Phil. Soc. **65**, 439–444 (1969)
13. J. Balakrishnan, F. Bianchi, V. Cantoral-Farfán, M. Çiperiani, A. Etropolski, Chabauty-Coleman experiments for genus 3 hyperelliptic curves, in *Research Directions in Number Theory*. Association for Women in Mathematics Series, vol. 19 (Springer, Cham, 2019), pp. 67–90
14. E.A. Bender, Partitions of multisets. Discrete Math. **9**, 301–311 (1974)
15. E.A. Bender, J.S. Devitt, L.S. Richmond, Partitions of multisets II. Discrete Math. **50**, 1–8 (1984)
16. Y. Bilu, G. Hanrot, Solving Thue equations of high degree. J. Number Theory **60**(2), 373–392 (1996)
17. W. Bosma, The Magma algebra system 1. The user language. J. Symb. Comput. **24**, 235–265 (1997)
18. D. Cantor, Computing in the Jacobian of a hyperelliptic curve. Math. Comput. **48**(177), 95–101 (1987)
19. J. Cassels, *Local Fields* (Cambridge University Press, Cambridge, 1986)
20. R. Coleman, Effective Chabauty. Duke Math. J. **52**, 765–770 (1985)
21. M. Dell'Amico, S. Martello, Reduction of the three-partition problem. J. Comb. Optim. **3**(1), 17–30 (1999)
22. M.R. Garey, D.S. Johnson, Complexity results for multiprocessor scheduling under resource constraints. SIAM J. Comput. **4**, 397–411 (1975)
23. M.R. Garey, D.S. Johnson, *Computers and Intractability; A Guide to the Theory of NP-Completeness* (Freeman, San Francisco, 1979)
24. W. McCallum, B. Poonen, *The Method of Chabauty and Coleman*. Explicit Methods in Number Theory, Panor. Synthèses, vol. 36 (Soc. Math. France, Paris, 2012), pp. 99–117
25. OEIS, The online encyclopedia of integer sequences (2018). Published electronically at http://oeis.org
26. I. Pak, Partition bijections, a survey. Ramanujan J. **12**, 5–75 (2006)
27. C.L. Siegel, Über einige Anwendungen diophantischer Approximationen. *Sitzungsberichte der Preussischen Akademie der Wissenschaften* (in German) (1929)
28. M. Stoll, Independence of rational points on twists of a given curve. Compos. Math. **142**(5), 1201–1214 (2006)
29. B.D. Sullivan, On a conjecture of Andrica and Tomescu. J. Integer Seq. **16**, Article 13.3.1 (2013)
30. The Sage Developers, Sagemath, the Sage Mathematics Software system, Version 8.5, 2018
31. M. Waldschmidt, Perfect powers: Pillai's works and their developments. arxiv:0908.4031v1[math.NT], 27 Aug 2009
32. H. Wilf, *Generatingfunctionology* (Academic, New York, 1994)

The Exponent of a Group: Properties, Computations and Applications

Dorin Andrica, Sorin Rădulescu, and George Cătălin Ţurcaş

Abstract This material is a systematic presentation of the exponent of a group with properties, examples, and explicit computations. Some of the results and connexions we present are original or they are presented in a new form. The last section explores general results about the exponent of certain finite groups of matrices. We will explicitly compute the exponent for $\mathrm{SL}_2(\mathbb{Z}/p^n\mathbb{Z})$ and $\mathrm{GL}_2(\mathbb{Z}/p^n\mathbb{Z})$, when $p \in \{2, 3\}$ and n is a positive integer.

2010 AMS Subject Classification 20Axx, 20B05, 20B30, 20D20, 20D45, 20E36

1 Introduction

Even if the concept of exponent of a group has been used for a long time, we are not aware of the existence in the literature of a systematic presentation containing properties, connection with other notions, examples, and explicit computations of the exponent. This reason motivated us to organize such a material. In the present work we discuss various theoretic concepts associated with groups. Along with a comprehensive list of theorems that are new, we incorporated a survey of results in the literature, hoping that our material will prove to be accessible for undergraduate students and, at the same time, an interesting read for seasoned researchers.

D. Andrica (✉)
Department of Mathematics, "Babeş-Bolyai" University, Cluj-Napoca, Romania
e-mail: dandrica@math.ubbcluj.ro

S. Rădulescu
Institute of Mathematical Statistics and Applied Mathematics, Bucharest, Romania

G. C. Ţurcaş
Department of Mathematics, "Babeş-Bolyai" University, Cluj-Napoca, Romania

The Institute of Mathematics of the Romanian Academy "Simion Stoilow", Bucharest, Romania
e-mail: george.turcas@math.ubbcluj.ro

A. M. Raigorodskii, M. Th. Rassias (eds.), *Discrete Mathematics and Applications*,
Springer Optimization and Its Applications 165,
https://doi.org/10.1007/978-3-030-55857-4_4

We start by explaining what we mean by saying that a group has finite exponent and we explore some properties and relations between such groups in Section 2. It will be clear that for a finite group G, the order $|G|$ is a trivial upper bound for the exponent of G. In the same section, we give a nontrivial lower bound for the exponent of G that depends only on the order $|G|$.

The reader will be convinced that computing the exponent of a given group G is, in general, a very difficult problem. The third section is dedicated to proving some theorems that can be helpful in carrying such a computation. We also demonstrated how these theorems can be used to compute the exponent of some concrete groups.

In Section 4 we start by recalling some properties of the automorphism group $\mathrm{Aut}(G)$ for some groups G. We prove some results about the exponent of $\mathrm{Aut}(G)$, when G is abelian and of finite exponent. A subsection is dedicated to the study of $M(G)$, a set of integers associated with a group G, which is closely related to the monoid of endomorphisms of G. We present a series of results which show that properties of the set $M(G)$ imply interesting properties of the group G, such as commutativity. Many interesting connections between the elements of $M(G)$ and the exponent of the group G are also presented.

In Section 5 we study the direct product consisting of the automorphism groups of a finite number of groups. This is always isomorphic to a subgroup of the automorphism group of the direct product of the initial groups and we provide various sufficient criteria for when the latter subgroup is actually the whole group. This section contains many new results concerning the product of the automorphism group for groups with finite exponent that were recently submitted for publication by the same authors.

Section 6 consists only of notations and definitions that are used in further sections. We also listed a few properties that follow from these definition in a straightforward manner.

In Section 7, we show that a correct count for the number of elements of order 2 in the symmetric group S_n gives a formula for the number of endomorphisms of S_n and, more generally, for the number of homomorphisms between S_n and S_m for different values of n and m. A general theorem for counting the number of endomorphisms in groups with a unique proper normal subgroup is also given.

A list of theorems which show that if a finite group G has "many" elements of a given prime order, then G has some special structure is presented in Section 8.

Section 9 consists of a survey of group actions on sets, culminating with the proof of Frobenius theorem (see Theorem 37).

For a given group G, the sequence consisting of the number of elements of a given order is very interesting. In Section 9 we delve into the study of this sequence. We give examples of pairs of non-isomorphic groups which have identical "order-counting" sequences, i.e. they have the same number of elements of any given order. On the other hand, in Section 10 we show that this sequence is an absolute invariant in the class of finite abelian groups. This just means that the "order-counting" sequences characterizes finite abelian groups up to isomorphism.

Finally, in Section 11 we explore general results about the exponent of certain finite groups of matrices. Along with many general theorems, we will present the computation of the exponent for $\mathrm{SL}_2(\mathbb{Z}/p^n\mathbb{Z})$ and $\mathrm{GL}_2(\mathbb{Z}/p^n\mathbb{Z})$, when $p \in \{2, 3\}$

and n is a positive integer. Even for such small values of p, computing the exponent turned out to be a difficult task and we had to use results about p-adic valuations coming from Number Theory to complete it.

2 The Exponent of a Group: General Properties

Recall that, the group G is a **torsion group** if every element $x \in G$ has finite order. A simple example of an infinite such group is $(\mathbb{Z}/2\mathbb{Z}[X], +)$, the additive group of polynomials with coefficients in $\mathbb{Z}/2\mathbb{Z}$, the ring of residue classes modulo 2. Clearly, every nonzero element of this group has order 2.

Let G be a group with the neutral element e and let $n \geq 1$ be a positive integer. We say that G has **exponent** n if n is the smallest positive integer such that $x^n = e$, for every $x \in G$.

We write $\mu(G)$ for the exponent of G. If for a given group G there is no such positive integer n, the convention is to set $\mu(G) = \infty$. The group G is of **finite exponent** if we have $\mu(G) < \infty$. For brevity, in this case we will write that G is a **fe**-group.

Clearly, if G is a **fe**-group, then it is a torsion group. The converse of this implication is not true. In this respect, an example is given by the Prüfer p-group

$$\mathbb{Z}(p^\infty) = \cup_{k=1}^{\infty} \mathbb{Z}(p^k),$$

where p is a prime and $\mathbb{Z}(p^k)$ is the group of the p^k-th roots of unity. The group $\mathbb{Z}(p^\infty)$ is a torsion group with $\mu(\mathbb{Z}(p^\infty)) = \infty$.

The proofs of the following properties constitute easy but instructive exercises. To keep the exposition self-contained, we give brief sketches for their proofs.

1. If G is a group such that $\mu(G)$ is finite, then $\mu(G)$ is the least common multiple of the orders of all the elements of G. Notice that since $g^{\mu(G)} = e$, it follows that $o(g) \mid \mu(G)$ for every $g \in G$. Therefore, $\mu(G)$ is a common multiple of the orders of all elements of G. By definition, it has to be the least such common multiple.
2. If G is a finite group, then $\mu(G) \mid |G|$, where $|G|$ is the order of G. To see this, let us note that from Lagrange's theorem we know that $g^{|G|} = e$, for all $g \in G$. Therefore, $|G|$ is a multiple of $o(g)$, for every $g \in G$. Now $\mu(G) \mid G$ follows from the property 1.
3. The exponent of a finite group is equal to the product of the exponents of its Sylow subgroups. To prove this, one can proceed by induction on the number of prime factors dividing $|G|$.
4. If G is a **fe**-group and H is a subgroup of G, then $\mu(H) \mid \mu(G)$. This statement is just a consequence of the definition.
5. If G_1 is a group, G_2 is a **fe**-group, and $f : G_1 \to G_2$ is an injective homomorphism, then G_1 is a **fe**-group and $\mu(G_1) \mid \mu(G_2)$. To see this, notice that $\mu(G_1)$ is a subgroup of G_2 isomorphic to G_1 and apply the previous property.

6. If G_1 is a **fe**-group, G_2 is a group, and $f : G_1 \to G_2$ is a surjective homomorphism, then G_2 is a **fe**-group and $\mu(G_2) \mid \mu(G_1)$. This is because for every $g_2 \in G_2$, there exists $g_1 \in G_1$ such that $f(g_1) = g_2$. But then, $g_2^{\mu(G_1)} = f(g_1)^{\mu(G_1)} = f(g_1^{\mu(G_1)}) = f(e_1) = e_2$, where e_i is the identity element of G_1, for all $i \in \{1, 2\}$.
7. Let G be a **fe**-group and p be a prime such that $p \mid \mu(G)$. Then there is $x \in G$ with $\sigma(x) = p$. To see this, from the definition of the exponent notice that

$$\mu(G) = \text{lcm}\{o(g) : g \in G\}.$$

As p divides the lcm above, there exists an element $g \in G$ such that $p \mid o(g)$. If $o(g) = p$, there is nothing left to show. Otherwise, the element $g^{o(g)/p}$ has order exactly p.
8. If G is a finite group, then $\text{Aut}(G)$ is finite and $\mu(G/Z(G)) \mid \mu(\text{Aut}(G))$, where $Z(G)$ is the center of G. The finiteness of $\text{Aut}(G)$ follows immediately by observing that the automorphisms can be seen as particular types of permutations of the elements of G. It is known that there are $|G|!$ permutations on the elements of G, therefore $\text{Aut}(G)$ is finite. To prove the second affirmation, consider the injective homomorphism $\phi : G/Z(G) \to \text{Aut}(G)$, where $\phi(g)$ is the conjugation by g automorphism of G. The conclusion follows from property 5.
9. If $\{G_i\}_{i\in I}$ is a finite family of **fe**-groups, then the direct product $\prod\limits_{i\in I} G_i$ is a **fe**-group and $\mu\left(\prod_{i\in I} G_i\right) = \text{lcm}\{\mu(G_i) : i \in I\}$. This can be proved by induction on the size of I.

Theorem 1 *Let G be a finite group. We then have*

1) If $p_1, \ldots, p_r$ are the prime factors of $|G|$, then $\prod_{i=1}^r p_i$ divides $\mu(G)$;
2) $|G|$ divides $\mu(G)^{\lfloor \log_2 |G| \rfloor}$, where $\lfloor\cdot\rfloor : \mathbb{R} \to \mathbb{Z}$ is the floor function.

Proof

1) If p is a prime dividing $|G|$, it follows from Cauchy's theorem that there exists an element $g \in G$, which has order p. It is then obvious that $p \mid \mu(G)$.
2) Let us consider the prime factorizations $|G| = \prod_{i=1}^r p_i^{\alpha_i}$ and $\mu(G) = \prod_{i=1}^r p_i^{\beta_i}$, where $1 \le \beta_i \le \alpha_i$ are integers, for all $i \in \{1, \ldots, r\}$. It is enough to show that the inequality $\alpha_i \le \beta_i \cdot \lfloor \alpha_1 \log_2(p_1) + \cdots + \alpha_r \log_2(p_r) \rfloor$ holds for every i. But it is easy to see that, for each i, $\lfloor \alpha_1 \log_2(p_1) + \cdots + \alpha_r \log_2(p_r) \rfloor \ge \lfloor \alpha_i \log_2(p_i) \rfloor \ge \alpha_i$ and the conclusion follows. □

3 Computing the Exponent

Computing the exponent of some concrete groups is a difficult and challenging problem. Let us begin this discussion with the following list of remarks:

1. Groups of exponent 1 are trivial and groups of exponent 2 are abelian.

2. Groups of exponent 3 are not necessarily abelian, as the discrete Heisenberg group $H(\mathbb{F}_3)$ over the field $\mathbb{F}_3$ shows. This is the multiplicative group of 3×3 upper-triangular matrices, whose elements belong to $\mathbb{F}_3$ and whose diagonal entries are all equal to 1. More precisely, the elements of $H(\mathbb{F}_3)$ are called Heisenberg matrices and are of the form

$$\begin{pmatrix} 1 & a & c \\ 0 & 1 & b \\ 0 & 0 & 1 \end{pmatrix}, \quad a, b, c \in \mathbb{F}_3.$$

Therefore the order of $H(\mathbb{F}_3)$ is 27. By the Cayley–Hamilton theorem, which holds for matrices over arbitrary commutative rings, it follows that the order of every element $A \neq I_3$ of $H(\mathbb{F}_3)$ is 3, that is, $\mu(H(\mathbb{F}_3)) = 3$.

3. In the case of finite cyclic groups, it is easy to see that the exponent of the group is equal to its order. By the fundamental theorem of finite abelian groups we know that any such group G can be written as a direct product of cyclic subgroups. From this we can derive that, if G is finite abelian, then $\mu(G) = |G|$ if and only if G is cyclic.

   Here we present an elementary argument for the property above. Assume that $\mu(G) = |G|$ and we have the prime factorization $|G| = p_1^{\alpha_1} \cdots p_k^{\alpha_k}$. The exponent $\mu(G)$ is the least common multiple of the orders of all elements of G, so for each prime factor p_i there must be an element $x_i \in G$ whose order is divisible by $p_i^{\alpha_i}$, otherwise the exponent would not be divisible by $p_i^{\alpha_i}$. If x_i has order $m_i p_i^{\alpha_i}$, then $y_i = x_i^{m_i}$ has order $p_i^{\alpha_i}$. It is an easy exercise to show that $y = y_1 \cdots y_k$ has order $p_1^{\alpha_1} \cdots p_k^{\alpha_k} = |G|$, hence G is generated by y and it is cyclic.

   In particular, $\mu(\mathbb{Z}/n\mathbb{Z}) = n$, where $\mathbb{Z}/n\mathbb{Z}$ denotes the additive group of integers mod n.

4. Let $U(\mathbb{Z}/n\mathbb{Z})$ be the multiplicative group of integers mod n. The group $U(\mathbb{Z}/n\mathbb{Z})$ consists of residue classes k modulo n, such that $\gcd(k, n) = 1$. The order of the group $U(\mathbb{Z}/n\mathbb{Z})$ is $\varphi(n)$, where φ is the Euler-totient function. A famous theorem of Gauss determines the structure of the group $U(\mathbb{Z}/p^m\mathbb{Z})$, where p is a prime and m is a positive integer. We have $U(\mathbb{Z}/p^m\mathbb{Z}) \simeq \mathbb{Z}/\varphi(p^m)\mathbb{Z} = \mathbb{Z}/p^{m-1}(p-1)\mathbb{Z}$ if $p \geq 3$ and $U(\mathbb{Z}/2^m\mathbb{Z}) \simeq \mathbb{Z}/2\mathbb{Z} \times \mathbb{Z}/2^{m-2}\mathbb{Z}$. Applying property 3 above, we obtain

$$\mu(U(\mathbb{Z}/p^k\mathbb{Z})) = \begin{cases} p^{k-1}(p-1), & \text{if } p \text{ is an odd prime}, \\ 2^{k-2}, & \text{if } p = 2 \text{ and } k \geq 3, \\ 2^{k-1}, & \text{if } p = 2 \text{ and } k \in \{1, 2\}. \end{cases}$$

If $n = \prod_{i=1}^r p_i^{\alpha_i}$ is the prime factorization of the positive integer n, using property 8 of the exponent, it follows $\mu(U(\mathbb{Z}/n\mathbb{Z})) = \operatorname{lcm}\{\mu(U(\mathbb{Z}/p^{\alpha_i}\mathbb{Z})) : i = 1, \ldots, r\}$.

Theorem 2 *Let H be a normal subgroup of G. The following properties are true:*

*1) If G is a **fe**-group, then H and G/H are **fe**-groups and $\mu(G)$ divides the product $\mu(H) \cdot \mu(G/H)$;*

2) *If* H *and* G/H *are* ***fe****-groups, then* G *is a* ***fe****-group and* $\text{lcm}(\mu(H), \mu(G/H))$ *divides* $\mu(G)$. *In particular,* $\gcd(\mu(H), \mu(G/H)) = 1$ *implies the relation* $\mu(G) = \mu(H) \cdot \mu(G/H)$.

Proof For the first statement, let $x \in G$. We obviously have that $x^{\mu(G/H)} \in H$ and therefore $(x^{\mu(G/H)})^{\mu(H)} = e$, where e is the neutral element in G. It follows that $\mu(G)$ divides $\mu(H) \cdot \mu(G/H)$.

For the second one, we saw in the listed properties of the exponent that $\mu(H) \mid \mu(G)$ and $\mu(G/H) \mid \mu(G)$. The conclusion follows. □

Theorem 3 *For every integer* $n \geq 2$, *we have:*

1) *The exponent of the symmetric group* S_n *is equal to*

$$\mu(S_n) = \text{lcm}(1, 2, \ldots, n) = \prod_{i=1}^{r} p_i^{\lfloor \log_{p_i}(n) \rfloor},$$

where $p_1, \ldots, p_r$ *are the prime numbers less than or equal to* n.

2) $\mu(D_{2n}) = \text{lcm}(2, n)$, *where* D_{2n} *is the dihedral group with* $2n$ *elements.*

3)

$$\mu(A_n) = \begin{cases} \text{lcm}(1, 2, \ldots, n-1) \text{ if } n \text{ is even} \\ \text{lcm}(1, 2, \ldots, n-2, n) \text{ if } n \text{ is odd} \end{cases},$$

where $A_n \trianglelefteq S_n$ *is the alternating group on* n *elements.*

Proof

1) Let $\sigma \in S_n$. It is known that σ can be written as a product of $r \geq 1$ disjoint cycles $c_1, \ldots, c_r \in S_n$, each one of them having length less than or equal to n. Since the order of a cycle is equal to its length and disjoint cycles commute, the order of $\sigma \in S_n$ is the least common multiple of the lengths of c_i's, where $i \in \{1, \ldots, r\}$. The claimed result follows.
2) The conclusion follows by noting that $D_{2n} \cong \mathbb{Z}/n\mathbb{Z} \rtimes \mathbb{Z}/2\mathbb{Z}$, where $\mathbb{Z}/2\mathbb{Z}$ is acting on $\mathbb{Z}/n\mathbb{Z}$ by inversion.
3) The result can be verified easily when $n \in \{2, 3, 4\}$. Let us assume that $n \geq 5$. Write $\sigma_k = (1, 2, \ldots, k) \in S_n$, for all $1 \leq k \leq n$. The k-cycle σ_k belongs to A_n if k is odd. Additionally, the product $\delta_k = \sigma_k(n-1, n)$ belongs to A_n if k is even.

For every $\sigma \in S_n$, we write $o(\sigma)$ for its order.

If n is odd, we have that

$$\begin{aligned} &\text{lcm}(o(\sigma_1), o(\sigma_3) \ldots, o(\sigma_n), o(\delta_2), o(\delta_4), \ldots, o(\delta_{n-3})) \\ &= \text{lcm}(1, 3, \ldots, n-2, n, 2, 4, \ldots, n-3), \end{aligned}$$

which divides $\mu(A_n)$.

Similarly, if n is even, we have

$$\begin{aligned} &\text{lcm}(o(\sigma_1), o(\sigma_3) \ldots, o(\sigma_{n-1}), o(\delta_2), o(\delta_4), \ldots, o(\delta_{n-2})) \\ &= \text{lcm}(1, 3, \ldots, n-1, 2, 4, \ldots, n-2) = \text{lcm}(1, 2, \ldots, n-1), \end{aligned}$$

which divides $\mu(A_n)$.

Since in both cases, it is easy to see that the order of every element in A_n divides the lcm expression, the result is proved. □

We mention that the sequence $\mu(S_n), n = 1, 2, \ldots$, is indexed as A003418 in OEIS [29]. Also, the sequence $\mu(D_{2n}), n = 1, 2, \ldots$, is indexed as A109043. It is not difficult to prove the relation

$$\mu(D_{2n}) = \text{lcm}(2, n) = \frac{3 - (-1)^n}{2} n.$$

From the above, it is easy to deduce the following corollary.

Corollary 1 *If $n \geq 2$ is an integer such that $2^k \leq n < 2^{k+1}$, then*

$$\mu(A_n) = \begin{cases} \frac{1}{2}\mu(S_n), & \text{if } n \in \{2^k, 2^k + 1\} \\ \mu(S_n), & \text{if } n \in \{2^k + 2, \ldots, 2^{k+1} - 1\} \end{cases}.$$

Proof If $n = 2^k$, then $\mu(S_n) = \text{lcm}\{j : 1 \leq j \leq 2^k\} = 2 \cdot \text{lcm}\{j : 1 \leq j \leq 2^k - 1\} = 2 \cdot \mu(A_n)$, from Theorem 3.

Similarly, if $n = 2^k + 1$, then we have $\mu(S_n) = \text{lcm}\{j : 1 \leq j \leq 2^k + 1\} = 2 \cdot \text{lcm}\{1, 2, \ldots, 2^k - 1, 2^k + 1\} = 2\mu(A_n)$, where we have used the formulas in the aforementioned theorem.

If $n \in \{2^k + 2, \ldots, 2^{k+1} - 1\}$ is even, then we have $\mu(A_n) = \text{lcm}\{1, 2, \ldots, n - 1\} = \text{lcm}\{1, 2, \ldots, n - 1, n\} = \mu(A_n)$. We used the fact that n is even and that the power of 2 that divides n is less than 2^k and hence does not contribute to the lcm. At the same time, the odd part of n is less than or equal to $n/2$, so it is already contained in $\text{lcm}\{1, 2, \ldots, n - 1\}$. The case in which $n \in \{2^k + 2, \ldots, 2^{k+1} - 1\}$ is odd can be solved similarly. □

Theorem 4 *Let $n \geq 2$ be an integer, G a group, and X a set of generators for G. Then, there is no proper normal subgroup H of G, satisfying the following two properties:*

*a) G/H is **fe** and $\mu(G/H) \mid n$;*
b) $\gcd(n, ord(x)) = 1$, for all $x \in X$.

Proof Let us suppose that there is such a group H. Consider $Y = \{x^n : x \in X\}$. The second condition implies that Y is a set of generators for G as well. Now, using the first condition, we obtain that $Y \subseteq H$. This implies that $G \subseteq H$, which is a contradiction. □

Corollary 2 *Let $n \geq 2$ be an integer, G a group, and X a set of generators for X. Then, there is no normal subgroup H of index n such that $\gcd(n, ord(x)) = 1$ for all $x \in X$.*

Proof This follows from the preceding theorem. H is of finite index n, therefore G/H is finite of order n, in particular a **fe** group, where $\mu(G/H) \mid |G/H| = n$. □

Corollary 3 *The group S_n does not have any odd index normal subgroups.*

Proof Let us choose X to be the set of transpositions in S_n. It is known that all the elements of X have order 2 and that they generate S_n. Now, if H is an odd index normal subgroup of S_n, then S_n/H is finite of odd order, hence $\mu(S_n/H)$ is odd. The conclusion follows from the theorem above. □

Theorem 5 *Let $n \geq 2$ be an integer, G a group, and H_1 a subgroup of index m in G. Suppose X is a set of generators for H_1. Then there is no normal subgroup H of G with the following properties*

1) $\gcd(n, ord(x)) = 1$ *for all $x \in X$;*
*2) G/H is **fe** and $\mu(G/H) \mid n$;*
3) $[G : H] > m$.

Proof Let us suppose, for the sake of contradiction, that there is such a normal subgroup H. We consider $Y = \{x^n : x \in X\}$. The first property implies that Y generates H_1.

The second property implies that $g^n \in H$ for every $g \in G$.

Now, let $x \in X$, $x^n \in H$, so $Y \subseteq H_1$. This implies that $H_1 \subseteq H$, which is a contradiction to the third property. □

Corollary 4 *Let $m \geq 4$ be an integer that is not divisible by* 3. *Then, for any $n \geq 5$, the group S_n does not have normal subgroups of order m.*

Proof It is known that $[S_n : A_n] = 2$ and the set X consisting of 3-cycles in S_n generates A_n for every $n \geq 5$.

Now, suppose H is a normal subgroup of S_n such that S_n/H has order m, as in the hypothesis. We know that $\mu(S_n/H) \mid m$ and $\gcd(m, 3) = 1$. Since $m = [S_n : H]$ was chosen greater than 2, from the previous theorem we deduce that such H does not exist. □

4 The Automorphism Group

Group theory is fundamental in the study of symmetries of objects and configurations. From this point of view, the symmetries of a group G itself are encoded in the automorphism group $\mathrm{Aut}(G)$ of G. Thus, the group of automorphism groups is of great interest in the study of the theory of groups. We refer to the recent monograph [30] for an excellent presentation of the actual stage of this study for finite groups.

4.1 Aut(G) *for Some Concrete Groups*

In this subsection, we describe Aut(G) and some of its properties for some concrete examples of groups G.

4.1.1 Aut($\mathbb{Z}$)

An automorphism of a cyclic group must carry generator to generator. The only generators of $\mathbb{Z}$ are 1 and -1, hence the only automorphisms possible are $f(n) = n$ and $f(n) = -n$. Therefore $|\operatorname{Aut}(\mathbb{Z})| = 2$ and we have $\operatorname{Aut}(\mathbb{Z}) \simeq \mathbb{Z}_2$.

This is a simple example of an infinite group having finite automorphism group.

4.1.2 Aut($\mathbb{Z}/n\mathbb{Z}$)

Consider the positive integer $n \geq 2$ having the prime factorization $n = 2^{n_0} p_1^{n_1} \cdots p_k^{n_k}$, where $n_0 \geq 0, n_i \geq 1$, and p_i are odd primes, $i = 1, \ldots, k$. Note that $U(n)$ is precisely the set of generators of $\mathbb{Z}/n\mathbb{Z}$. Since any automorphism of $\mathbb{Z}/n\mathbb{Z}$ sends 1 to a generator, the evaluation map $\operatorname{Aut}(\mathbb{Z}/n\mathbb{Z}) \to U(n), \phi \mapsto \phi(1)$, is an isomorphism of groups.

In the note [21, Lemma 2.1] it is proved that if $G_1 \times G_2$ is the direct product of finite groups G_1 and G_2, in which the orders of G_1 and G_2 are relatively prime positive integers, then $\operatorname{Aut}(G_1 \times G_2) \simeq \operatorname{Aut}(G_1) \times \operatorname{Aut}(G_2)$. It can be shown, using strong induction that the natural generalization of the above property holds for any finite number of finite groups $G_1, G_2, \ldots, G_m$ with the property $\gcd\big(|G_i|, |G_j|\big) = 1$, for every $1 \leq i < j \leq m$, that is we have

$$\operatorname{Aut}(G_1 \times \cdots \times G_m) \simeq \operatorname{Aut}(G_1) \times \cdots \times \operatorname{Aut}(G_m).$$

Combining these proprieties, we obtain

$$\operatorname{Aut}(\mathbb{Z}/n\mathbb{Z}) \simeq \mathbb{Z}/2\mathbb{Z} \oplus \mathbb{Z}/2^{n_0-2}\mathbb{Z} \times \mathbb{Z}/(p_1-1)p_1^{n_1-1}\mathbb{Z} \times \cdots \times \mathbb{Z}/(p_k-1)p_1^{n_k-1}\mathbb{Z},$$

if $n_0 \geq 2$, and

$$\operatorname{Aut}(\mathbb{Z}/n\mathbb{Z}) \simeq \mathbb{Z}/(p_1-1)p_1^{n_1-1}\mathbb{Z} \times \cdots \times \mathbb{Z}/(p_k-1)p_1^{n_k-1}\mathbb{Z},$$

otherwise. When $n_0 = 2$, the group $\mathbb{Z}/2^{n_0-2}\mathbb{Z}$ is trivial.

As special cases we obtain $\operatorname{Aut}(\mathbb{Z}/p^n\mathbb{Z}) \simeq \mathbb{Z}_{(p-1)p^{n-1}}$, where p is a prime number and n is a positive integer. Also, we have $\operatorname{Aut}(\mathbb{Z}/2^n\mathbb{Z}) \simeq \mathbb{Z}_2 \oplus \mathbb{Z}_{2^{n-2}}$, for every positive integer $n \geq 2$.

4.1.3 Aut(G) for G Finite and Abelian

According to the classification of finite Abelian groups, we have that G is isomorphic to a product of groups of the form

$$H_p = \mathbb{Z}/p^{s_1}\mathbb{Z} \times \cdots \times \mathbb{Z}/p^{s_k}\mathbb{Z},$$

in which p is a prime number and $1 \leq s_1 \leq \ldots s_k$ are positive integers. The group H_p is called elementary abelian group or abelian p-group. Clearly, the order of H_p is $|H_p| = p^{s_1+\cdots+s_k}$.

Now, assume that $G \simeq H_{p_1} \times \cdots \times H_{p_m}$. Because $\gcd\left(|H_{p_i}|, |H_{p_j}|\right) = 1$ for every $1 \leq i < j \leq m$, we obtain

$$\mathrm{Aut}(G) \simeq \mathrm{Aut}(H_{p_1}) \times \cdots \times \mathrm{Aut}(H_{p_m}),$$

hence the problem of determining Aut(G) is reduced to the problem of finding Aut(H_p).

To describe Aut(H_p) is a challenging problem. In the paper [6] it is mention that, when p is an odd prime and $s_1 > s_2 > \cdots > s_k$, Aut(H_p) is isomorphic to a group of $k \times k$ matrices (α_{ij}), where $\alpha_{ij} \in \mathrm{Aut}(\mathbb{Z}/p^{s_i}\mathbb{Z})$ if $i = j$ and $\alpha_{ij} \in \mathrm{Hom}(\mathbb{Z}/p^{s_j}\mathbb{Z}, \mathbb{Z}/p^{s_i}\mathbb{Z})$ if $i \neq j$, $i, j = 1, \ldots, k$.

A special case is given by $s_1 = s_2 = \cdots = s_k = 1$. The elements of this group can be viewed as comprising a vector space of dimension k over the finite field of p elements $\mathbb{F}_p$. The automorphisms of this group are therefore given by the invertible linear transformations, so we have

$$\mathrm{Aut}(H_p) \simeq \mathrm{GL}_k(\mathbb{F}_p),$$

where $\mathrm{GL}_k(\mathbb{F}_p)$ is the general linear group of square matrices of order k with entries in the field $\mathbb{F}_p$. The order of Aut(H_p) is

$$|\,\mathrm{Aut}(H_p)| = (p^k - 1)\cdots(p^k - p^{k-1}).$$

For the general situation the following nice formula for the order of Aut(H_p) is given in the note [21, Theorem 4.1]:

$$|\,\mathrm{Aut}(H_p)| = \prod_{l=1}^{k}(p^{d_l} - p^{l-1}) \prod_{j=1}^{k}(p^{s_j})^{k-d_j} \prod_{i=1}^{k}(p^{s_i-1})^{k-c_i+1},$$

where $d_l = \max\{\alpha : s_\alpha = s_l\}$ and $d_l = \min\{\alpha : s_\alpha = s_l\}$, $l = 1, \ldots, k$.

4.1.4 Aut(S_n) and Aut(A_n)

It is known that if $n \neq 2, 3, 6$, then

$$\mathrm{Aut}(S_n) \simeq S_n \simeq \mathrm{Aut}(A_n).$$

The proofs of this result given in the literature are complicated and involve the use of multiple lemmas. An interesting combinatorial argument is given in the note [33]. In particular, when $n = 6$, every automorphism of S_n is inner and every automorphism of A_n is the restriction of an inner automorphism of S_n. However, Aut(S_6) is not isomorphic to S_6. In fact, $\mathrm{Aut}(S_6) = \mathrm{Aut}(A_6)$ satisfies the relation $[\mathrm{Aut}(S_6) : \mathrm{Inn}(S_6)] = 2$ (see [32] or Chapter 3.2 of [35]) for this fact. Recall that the groups S_n for $n \geq 3$, and A_n for $n \geq 4$ have trivial center.

An inner automorphism of the group G is an automorphism of the form $f_a(x) = a^{-1}xa$ for a fixed $a \in G$. The map f_a is an automorphism for each element $a \in G$. The map $G \to \mathrm{Inn}(G)$, given by $a \mapsto f_a$, is not necessarily injective. If G is commutative, then f_a is always trivial. In general, trivial inner automorphisms of G correspond to the elements of the center $Z(G)$ and we have the following result : the inner automorphisms $\mathrm{Inn}(G)$ form a normal subgroup of $\mathrm{Aut}(G)$ and

$$\mathrm{Inn}(G) \simeq G/Z(G).$$

A group G is **complete** if G is centerless (i.e. it has trivial center) and every automorphism of G is an inner automorphism. Clearly, for every complete group G we have $\mathrm{Aut}(G) \simeq G$.

4.2 *The Automorphism Group and the Exponent*

For a group G, we explore the exponent of $\mathrm{Aut}(G)$ and some of its relations to the exponent of G. A theorem that gives a relation between these two quantities when both are finite is presented. We also provide an example of a group G for which $\mu(G)$ is finite, but $\mu(\mathrm{Aut}(G))$ is infinite.

Theorem 6 *Let G be a group such that $\mu(G) = n$ and p a positive integer. Let $f : G \to G$ be such that $f(x) = x^p$ for all $x \in G$. The following are equivalent:*

1) $\gcd(n, p) = 1$;
2) f is injective;
3) f is surjective;
4) f is bijective.

Proof Suppose 1) holds. We deduce that $x^{p^{\varphi(n)}} = x$, for all $x \in G$. Therefore,

$$\underbrace{(f \circ f \circ \cdots \circ f)}_{\varphi(n) \text{ times}}(x) = x$$

for all $x \in G$, hence f is bijective.

We proved that 1) implies 4). Obviously, 4) implies 2) and 3). To prove the claimed equivalence, we just have to prove that both statements 2) and 3) imply statement 1).

Suppose 3) holds. From the surjectivity of f, it follows that there exists $g : G \to G$ such that $f(g(x)) = x$, for all $x \in G$. We then have that $g(x)^p = x$, for all $x \in G$. If $d = \gcd(p, n)$ and $p = p_1 \cdot d$, $n = n_1 \cdot d$, it follows that $\gcd(p_1, n_1) = 1$. We then have that $e = g^{p_1 n(x)} = x^{n_1}$ for all $x \in G$. This implies that $n = \mu(G) \mid n_1$, hence $d = 1$ and 1) must hold.

If 2) holds, then suppose for the sake of contradiction that $(n, p) = d > 1$. Let q be a prime such that $q \mid d$. From Cauchy's theorem, it follows that there is an element $x \in G$ which has order q. But then $f(x) = x^p = x^{q\frac{p}{q}} = e$. But f is assumed to be injective, which implies that $x = e$, a contradiction. □

Corollary 5 *Suppose G is a finite group and fix p a positive integer. Let $f : G \to G$ such that $f(x) = x^p$. The following are equivalent:*

1) $\gcd(|G|, p) = 1$;
2) f is injective;
3) f is surjective;
4) f is bijective.

Proof We just observe that $\gcd(p, \mu(G)) = 1$ if and only if $\gcd(p, |G|) = 1$. □

Theorem 7 *Let G be an Abelian group with $\mu(G) = n$. Then the following properties hold:*

1. *The function $f : U(\mathbb{Z}/n\mathbb{Z}) \to \mathrm{Aut}(G)$, defined by $f(k)(x) = x^k$, is an injective homomorphism;*
2. *If $\mathrm{Aut}(G)$ is a finite group, then $\varphi(n)$ divides $|\mathrm{Aut}(G)|$;*
3. *If $\mathrm{Aut}(G)$ is a **fe**-group, then $\mu(U(\mathbb{Z}/n\mathbb{Z}))$ divides $\mu(\mathrm{Aut}(G))$.*

Proof

1) It is easy to check that the function f is well-defined and it is a group homomorphism. To prove the property that f is injective, consider $k \in U(\mathbb{Z}/n\mathbb{Z})$ with $f(k)(x) = x$ for every $x \in G$. It follows $x^{k-1} = e$, hence n divides $k - 1$ and we obtain $k = 1$.
2) The property follows from 1).
3) The property follows from 1). □

It is obvious that if the group G is finite, then $\mathrm{Aut}(G)$ is finite. It is a natural problem to ask : if G is a **fe** group, then its automorphism group is also a **fe** group? The following proposition answers this question in the negative.

Proposition 1 *The automorphism group of* $(\mathbb{Z}/2\mathbb{Z}[X], +)$ *has infinite exponent.*

Proof We saw previously that $\mu(\mathbb{Z}/2\mathbb{Z}[X]) = 2$. Suppose, for the sake of contradiction that $\mu\,(\mathrm{Aut}(\mathbb{Z}/2\mathbb{Z}[X])) = n$, for some positive integer n.

Consider the homomorphism $f : \mathbb{Z}/2\mathbb{Z}[X] \to \mathbb{Z}/2\mathbb{Z}[X]$ that maps X to X^2, X^2 to $X^3, \ldots, X^n$ to X^{n+1} and X^{n+1} to X and such that $f(X^k) = X^k$ for every $k \geq n+2$ and for $k = 0$. The definition of f extends linearly to the whole group $Z/2\mathbb{Z}[X]$. It is easy to see that f is an automorphism. Its order is $n + 1$ in $\mathrm{Aut}(Z/2\mathbb{Z}[X])$, a contradiction with the assumption on exponent. □

The group $\mathrm{Aut}(\mathbb{Z}/2\mathbb{Z}[X])$ is not even a torsion group. We can see that the $\mathbb{Z}/2\mathbb{Z}$-linear map which sends $X^i \mapsto X^{i+1}$ for every positive integer i defines an automorphism of $\mathbb{Z}/2\mathbb{Z}$. This automorphism has infinite order.

4.3 *The Power Endomorphisms of a Group*

Following the paper [24], for a group G, we define a set of integers $M(G)$ which is related to the set $\mathrm{End}(G)$ of all endomorphisms of G. We will show that if the set $M(G)$ contains particular numbers, then G must be abelian. The presentation and some of the included results are new.

Definition 1 For a group G and an integer $a \in \mathbb{Z}$, we write $f_a : G \to G$ for the map defined as $f_a(x) = x^a$ for all $x \in G$. Let us define the set

$$M(G) = \{a \in \mathbb{Z} : f_a \in \mathrm{End}(G)\}.$$

Therefore, the integer $a \in M(G)$ if and only if $(xy)^a = x^a y^a$ for all $x, y \in G$. Note that $\{0, 1\} \subseteq M(G)$ for every group G. These values correspond to the trivial and the identity endomorphisms on G.

Theorem 8 *The following properties are equivalent:*

1) G *is an abelian group;*
2) $-1 \in M(G)$*;*
3) $2 \in M(G)$*;*
4) $M(G) = \mathbb{Z}$.

Proof It is easy to see that 1) implies 2). Let us suppose that 2) holds, i.e. that $-1 \in M(G)$. Let $x, y \in G$ be arbitrary. We have that $(x^{-1}y^{-1})^{-1} = xy$, since $-1 \in M(G)$. It follows that $yx = xy$, hence $xyxy = x^2y^2$. We therefore showed that $(xy)^2 = x^2y^2$ for every $x, y \in G$, hence $2 \in M(G)$.

If 2) holds, then for every $x, y \in G$ we have that $(xy)^2 = x^2y^2$, hence $xyxy = x^2y^2$, so $xy = yx$. This proves 1), i.e. G is abelian. We showed so far that 1), 2), and 3) are equivalent.

It is now easy to see that 4) is also equivalent to 1) (and hence 2) and 3)). The fact that 1) implies 4) is trivial. On the other hand $M(G) = \mathbb{Z}$ implies that $2 \in M(G)$, so 4) implies 2), which implies 1), as we saw above. □

Theorem 9 *Let G be a group. We then have that:*

1) *If $a, b \in M(G)$, then $ab \in M(G)$;*
2) *If $a, b \in M(G)$, then $(a-1)(b-1) \in M(G)$;*
3) *If $a \in M(G)$, then $x^a y^{a-1} = y^{a-1} x^a$, for all $x, y \in G$;*
4) *If $a \in M(G)$, then $a(a-1) \in M(G)$;*
5) *If $a, r, s \in \mathbb{Z}$ are integers such that $ar \in M(G)$ and $(a-1)s \in M(G)$, then $ar + (a-1)s \in M(G)$;*
6) *If $a, a+1 \in M(G)$, then $x^a \in Z(G)$, for all $x \in G$;*
7) *If $a, a+1, b \in M(G)$, then $ka + b \in M(G)$, for all $k \in \mathbb{Z}$;*
8) *If $a, a+1, a+2 \in M(G)$, then G is abelian;*
9) *If $\mu(G) < \infty$ and $a \in M(G)$, then $a + k\mu(G) \in M(G)$, for all $k \in \mathbb{Z}$.*

Proof

1) If $a, b \in M(G)$, then $(xy)^{ab} = (x^a y^a)^b = x^{ab} y^{ab}$, for all $x, y \in G$. Therefore $ab \in M(G)$.
2) If $a \in M(G)$, then $(xy)^a = x^a y^a$, for all $x, y \in G$. The latter implies that $(yx)^{a-1} = x^{a-1} y^{a-1}$ for all $x, y \in G$. Analogously $b \in M(G)$, we have $(yx)^{b-1} = x^{b-1} y^{b-1}$, for all $x, y \in G$. We therefore have

$$(xy)^{(a-1)(b-1)} = (y^{a-1} x^{a-1})^{b-1} = x^{(a-1)(b-1)} y^{(a-1)(b-1)},$$

   for all $x, y \in G$. Our claim is proved.
3) If $a \in M(G)$, then $(xy)^a = (xy)(xy)^{a-1} = x^a y^a$, for all $x, y \in G$. Therefore, using a property proved in the second item, we have

$$(xy)(y^{a-1} x^{a-1}) = x(yx)^{a-1} y = x x^{a-1} y^{a-1} y,$$

   for all $x, y \in G$. Hence, $y^a x^{a-1} = x^{a-1} y^a$, for all $x, y \in G$.
4) If $a \in M(G)$, then $(xy)^{a(a-1)} = (x^a y^a)^{a-1} = y^{a(a-1)} x^{a(a-1)}$. Using property 3, we can see that the latter expression is equal to $x^{a(a-1)} y^{a(a-1)}$, proving the claimed result.
5) If $ar \in M(G)$ and $(a-1)s \in M(G)$, then

$$\begin{aligned}(xy)^{ar+(a-1)s} &= x^{ar} y^{ar} x^{(a-1)s} y^{(a-1)s} = x^{ar} x^{(a-1)s} y^{ar} y^{(a-1)s} \\ &= x^{ar+(a-1)s} y^{ar+(a-1)s},\end{aligned}$$

   for all $x, y \in G$, hence $ar + (a-1)s \in M(G)$.
6) If $a, a+1 \in M(G)$, then $(xy)^a = x^a y^a$ and $(xy)^{a+1} = x^{a+1} y^{a+1}$, for all $x, y \in G$. Therefore, $x^{a+1} y^{a+1} = (xy)^{a+1} = xyx^a y^a$. The latter implies that $x^a y = yx^a$, for all $x, y \in G$, i.e. $x^a \in Z(G)$ for all $x \in G$.

7) Given $x^a \in Z(G)$, we obtain

$$(xy)^{ka+b} = x^{ka}y^{ka}(xy)^b = x^{ka}y^{ka}x^by^b = x^{ka+b}y^{ka+b},$$

for all $x, y \in G$. Hence $ka + b \in M(G)$, as claimed.

8) From 6) it follows that for every $x \in G$, we have $x^a, x^{a+1} \in Z(G)$. But $Z(G)$ is a subgroup, hence $x = x^{a+1} \cdot (x^a)^{-1} \in Z(G)$, for all $x \in G$. Hence G is abelian.

9) This is an easy consequence of the fact that $\mu(G)$ and $\mu(G)+1$ belong to $M(G)$. The conclusion follows from part 7). □

Theorem 10 *Let G be a group that is generated by $M_2(G)$, the set of all elements of order 2. If $M(G) \neq \{0, 1\}$, then the following holds:*

1) G is a **fe***-group;*
2) $M(G) = \{k\mu(G), l\mu(G) + 1 : k, l \in \mathbb{Z}\}$.

Proof Let $a \in M(G) \setminus \{0, 1\}$. Without losing generality, we can assume that a is positive, hence $a \geq 2$. For $x \in G$, there exists $x_1, \dots, x_n \in M_2(G)$ such that $x = x_1 \dots x_n$.

If $a \in M(G)$ is even, then $x^a = (x_1 \dots, x_n)^a = x_1^a \dots x_n^a = e$, the neutral element of G. On the other hand, if $a \in M(G)$ is odd, using the same argument one can see that $x^a = x$, hence $x^{a-1} = e$. This implies that $\mu(G)$ is finite and proves 1).

For 2), by the above we notice that for all $b \in M(G)$, either $x^b = e$ or $x^{b-1} = e$, for every $x \in G$. Then for every such $b \in M(G)$, $\mu(G) \mid b$ or $\mu(G) \mid b - 1$. □

Corollary 6 *We have $M(S_n) = \{k\mu(S_n), l\mu(S_n) + 1 : k, l \in \mathbb{Z}\}$ for every $n \geq 2$.*

Definition 2 For every $a \in \mathbb{Z}$, we write

$$Q(a) = \{x \in \mathbb{Z} : x \text{ is congruent to } 0, 1, a \text{ or } (a-1)^2 \text{ modulo } (a^2 - a)\}.$$

Theorem 11 *For every $a \in \mathbb{Z}$ and every group G, if $a \in M(G)$, then $Q(a) \subseteq M(G)$.*

Proof Let us suppose that $a \in M(G)$. From property 4 in the theorem above, we have that $a(a - 1) \in M(G)$. The second property in the theorem above, applied $a = b$, implies that $(a - 1)^2 \in M(G)$. Now, $a \cdot 1 \in M(G)$ and $(a - 1) \cdot (a - 1) \in M(G)$. Using part 5 of the theorem above with $r = 1$ and $s = a - 1$, we obtain that $a^2 - a + 1 = (a - 1)^2 + a \in M(G)$.

Now, using property 7 of Theorem 9, we obtain that $k \cdot (a^2 - a) + b \in M(G)$ for every $k \in \mathbb{Z}$ and for every $b \in \{0, 1, a, (a - 1)^2\}$. Our theorem is now proved. □

Corollary 7 *Let G be a group. The following affirmations are true.*

1) If $\mu(G) = 2$, then $M(G) = \mathbb{Z}$.
2) If $\mu(G) = 3$ and G is not abelian, then $M(G) = \{3k, 3l + 1 : k, l \in \mathbb{Z}\}$.
3) If $\mu(G) = 4$ and G is not abelian, then $M(G) = \{4k, 4l + 1 : k, l \in \mathbb{Z}\}$.
4) If $\mu(G) = 5$ and G is not abelian, then $M(G) = \{5k, 5l + 1 : k, l \in \mathbb{Z}\}$.

Proof The first three affirmations are immediate. To see that 4) holds, notice that $2, 4 \notin M(G)$, as they would both imply the commutativity of G. If $3 \in M(G)$, then $9 \in M(G)$ which means that $x \mapsto x^9 = x^4$ is an endomorphism of G, a contradiction. Therefore, $M(G) = \{5k, 5l + 1 : k, l \in \mathbb{Z}\}$, as claimed. □

Theorem 12 *Let G be a* ***fe****-group and* $a \in M(G)$. *If* $\gcd(\mu(G), a^2 - a) \in \{1, 2\}$, *then G is abelian.*

Proof From the previous theorem we know that $M(G)$ contains all the integers that are congruent with to 1 modulo $a^2 - a$. Suppose first that $\gcd(\mu(G), a^2 - a) = 1$. From the Chinese Remainder theorem, we know that there exists an integer n such that $n = 1 \pmod{a^2 - a}$ and $n = 2 \pmod{\mu(G)}$. We therefore have that

$$n = m \cdot \mu(G) + 2 \in M(G),$$

for some $m \in \mathbb{Z}$. We now know that $f : G \to G$ defined by $f(x) = x^{m\mu(G)+2}$ for all $x \in G$ is an endomorphism of G. But this means that $f(x) = x^2$ for all $x \in G$, hence $2 \in M(G)$ and G is abelian.

Suppose now that $\gcd(\mu(G), a^2 - a) = 2$. We have that $\mu(G)/2$ and $(a^2 - a)/2$ are coprime. By the Chinese Remainder theorem, there exists an integer n such that $n = 1 \pmod{\mu(G)/2}$ and $n = 0 \pmod{(a^2 - a)/2}$. But then $2n = 2 \pmod{\mu(G)}$ and, at the same time, $2n = 0 \pmod{a^2 - a}$. The latter implies that $2n \in M(G)$. Since

$$2n = m \cdot \mu(G) + 2 \in M(G),$$

for some $m \in \mathbb{Z}$, an argument identical to the one in the previous paragraph implies that $2 \in M(G)$, hence G is abelian. □

We remark that this result easily implies the next corollary.

Corollary 8 *If* $\mu(G) = p$, *where* $p \geq 3$ *is a prime, and G is not abelian, then* $M(G) = \{kp, lp + 1 : k, l \in \mathbb{Z}\}$.

Proof To see this, notice that the previous theorem implies that for any $a \in M(G)$, we have $\gcd(p, a^2 - a) = p$. This is equivalent to the statement that $M(G) \subseteq \{kp, kl + 1 : k, l \in \mathbb{Z}\}$. On the other hand, since $p = \mu(G)$, the reverse inclusion is immediate. □

Theorem 13 *For a group G, let* $a \in \mathbb{Z}$ *be such that* $f_a \in \mathrm{Aut}(G)$. *Then* $x^{a-1} \in Z(G)$, *for all* $x \in G$.

Proof If $f_a \in \mathrm{Aut}(G)$, using 7 of Theorem 9, we get

$$x^a y^{a-1} = y^{a-1} x^a,$$

for all $x, y \in G$. As f_a is surjective, the relation above implies that $y^{a-1} \in Z(G)$ for all $y \in G$. □

Theorem 14 *Let G be a group and $a, b \in \mathbb{Z}$. If $f_a, f_b \in \mathrm{Aut}(G)$ and $\gcd(a-1, b-1) = 1$, then G is abelian.*

Proof From the previous theorem, it follows that $x^{a-1}, y^{b-1} \in Z(G)$ for all $x, y \in G$. There exists integers $r, s \in \mathbb{Z}$ such that $1 = r(a-1) + s(b-1)$. We therefore have that $x = (x^r)^{a-1} + (x^s)^{b-1} \in Z(G)$ for all $x \in G$. This clearly implies that the group G is abelian. □

Theorem 15 *Let G be a group, $a, b, k \in \mathbb{Z}$ such that $a, b \in M(G)$. Then $b + ka(a-1) \in M(G)$.*

Proof The result 4 in Theorem 9 implies that $(a-1)^2 \in M(G)$. Now, using the fifth result of the same theorem, we get that $a^2 - a + 1 \in M(G)$, which together with Theorem 9 7), implies the desired result. □

Theorem 16 *Let G be a group such that the positive integers $a_1, a_2, \ldots, a_n \in M(G)$. If $\gcd\left(\binom{a_1}{2}, \binom{a_2}{2}, \ldots, \binom{a_n}{2}\right) = 1$, then G is abelian.*

Proof The condition on the gcd implies that there are $u_1, u_2, \ldots, u_n \in \mathbb{Z}$ such that

$$u_1a_1(a_1 - 1) + u_2a_2(a_2 - 1) + \cdots + u_na_n(a_n - 1) = 2.$$

From Theorem 9 it follows that $a_i(a_i - 1), a_i^2 - a_i + 1 \in M(G)$ for all $i = 1, \ldots, n$. Using the same theorem, we obtain that $u_ia_i(a_i - 1) \in M(G)$ for all $i = 1, \ldots, n$. From statement 5 of the aforementioned theorem, it follows that $2 \in M(G)$, hence G is abelian. □

Remark We have that:

1) $Q(2) = \mathbb{Z}$, $Q(3) = \{x \in \mathbb{Z} : x \equiv 0, 1 \pmod 3\}$ and $Q(4) = \{x \in \mathbb{Z} : x \equiv 0, 1, 4 \text{ or } 9 \pmod{12}\}$.
2) If G is a group and H a subgroup, then $M(G) \subseteq M(H)$. If H is normal in G, then $M(G) \subseteq M(G/H)$.
3) Let $(G_i)_{i \in I}$ be an arbitrary collection of groups. Then

$$M\left(\prod_{i \in I} G_i\right) = \bigcap_{i \in I} M(G_i).$$

Theorem 17 *Let G be a group and $a \geq 3$ an integer such that $a \in M(G)$. If there exists $k \in \mathbb{N}$ such that $\mu(G) = k(a^2 - a) + r$, where $r \in \{1, 2, a+1, a^2 - 2a + 2, a^2 - a - 1\}$, then G is abelian.*

Proof Observe that $\{0, 1, \mu(G), \mu(G) + 1, a, (a-1)^2\} \in M(G)$.

If $r = 1$, then $2 = \mu(G) + 1 - k \cdot (a^2 - a)$, hence $2 \in M(G)$ and G is abelian.

If $r = 2$, then $2 = \mu(G) = k \cdot (a^2 - a)$, hence $2 \in M(G)$ as above.

Similarly, if $r = a + 1$ it follows that $a, a+1, a+2 \in M(G)$, but now $2 = -a + a + 2 \in M(G)$ by part 7 in Theorem 9. Hence G is abelian.

If $r = a^2 - 2a + 2$, it is easy to show that $a^2 - 2a + 1, a^2 - 2a + 2, a^2 - 2a + 3 \in M(G)$. As above, it follows that G is abelian.

If $r = a^2 - a - 1$, then $-1 = \mu(G) - k(a^2 - a) \in M(G)$, hence G is abelian. □

5 The Automorphism Group of a Direct Product of fe-Groups

In the note [21, Lemma 2.1] (see also [12]) the following result is proved : If $G_1 \times G_2$ is the direct product of finite groups G_1 and G_2, in which the orders of G_1 and G_2 are relatively prime positive integers, then

$$\mathrm{Aut}(G_1 \times G_2) \simeq \mathrm{Aut}\,(G_1) \times \mathrm{Aut}\,(G_2). \tag{1}$$

The result above was used by the authors of [7] to describe the automorphism group of a finite Abelian group. In [6] it is mentioned that this result follows from Theorems 3.2 and 3.6 of [7]. The aforementioned results concern the automorphisms group of a product of two finite groups.

It can be shown using strong induction that the natural generalization of (1) holds for any finite number of finite groups $G_1, G_2, \ldots, G_m$ with the property $\gcd\left(|G_i|, |G_j|\right) = 1$, for every $1 \le i < j \le m$.

Let us note that in this context the hypothesis that the groups G_i are finite is essential. It is natural to ask if the isomorphism (1) remains true for some classes of infinite groups. Our main result gives an answer in the case of a special class of torsion groups (see next section for the precise definition).

We note that for any groups $G_1, G_2, \ldots, G_m$, the group $\prod_{i=1}^m \mathrm{Aut}(G_i)$ is isomorphic to a subgroup of $\mathrm{Aut}(\prod_{i=1}^m G_i)$. To explain this, we start by pointing out that elements of $\prod_{i=1}^m \mathrm{Aut}(G_i)$ are just m-tuples $(\phi_1, \ldots, \phi_m)$, where $\phi_i \in \mathrm{Aut}(G_i)$ for every $i = 1, \ldots, m$. On the other hand, elements of $\mathrm{Aut}(\prod_{i=1}^m G_i)$ are of the form $u = (u_1, \ldots, u_m)$, where $u_i : \prod_{i=1}^m G_i \to G_i$ are group homomorphisms for every $i = 1, \ldots, m$.

We now consider the natural map

$$f : \prod_{i=1}^{m} \mathrm{Aut}(G_i) \to \mathrm{Aut}\left(\prod_{i=1}^{m} G_i\right)$$

defined by

$$f(\phi_1, \ldots, \phi_m)(x_1, \ldots, x_m) = (\phi_1(x_1), \ldots, \phi_m(x_m)), \tag{2}$$

where $\phi_i \in \mathrm{Aut}\, G_i$ and $x_i \in G_i$ for all $i = 1, \ldots, m$.

It can be easily verified that f is a well-defined group homomorphism. Moreover, f is injective. Indeed, if $f(\phi_1, \ldots, \phi_m) = 1_{\prod_{i=1}^m G_i}$, the identity automorphism on

$\prod_{i=1}^{m} G_i$, then $(\phi_1(x_1), \ldots, \phi_m(x_m)) = (x_1, \ldots, x_m)$, for all $x_i \in G_i$ and all $i = 1, \ldots, m$. Then $\phi_i = 1_{G_i}$, the identity automorphism on G_i for every $i = 1, \ldots, m$. We just proved that f has trivial kernel, hence it is injective. We therefore have

$$\prod_{i=1}^{m} \mathrm{Aut}(G_i) \simeq \mathrm{Im}\, f \leq \mathrm{Aut}\left(\prod_{i=1}^{m} G_i\right). \tag{3}$$

Now, consider $u = (u_1, \ldots, u_m), \overline{u} = (\overline{u}_1, \ldots, \overline{u}_m) \in \mathrm{Aut}(\prod_{i=1}^{m} G_i)$, with $u \circ \overline{u} = \overline{u} \circ u = 1_{\mathrm{Aut}(\prod_{i=1}^{m} G_i)}$, that is, $\overline{u}$ is the inverse of u. For a fixed index j, define the endomorphisms $\phi_j, \overline{\phi}_j : G_j \to G_j$, where $\phi_j(x_j) = u_j(e_1, \ldots, x_j, \ldots, e_m)$ and $\overline{\phi}_j(x_j) = \overline{u}_j(e_1, \ldots, x_j, \ldots, e_m)$, where e_i is the neutral element of the group G_i, $i = 1, \ldots, m$. The following auxiliary result provides sufficient conditions for ϕ_j and $\overline{\phi}_j$ to be automorphisms of the group G_j with $\phi_j^{-1} = \overline{\phi}_j$.

Lemma 1 *If the relations*

$$u_k(e_1, \ldots, x_j, \ldots, e_m) = \overline{u}_k(e_1, \ldots, x_j, \ldots, e_m) = e_k \tag{4}$$

hold for every $k = 1, \ldots, m$ *with* $k \neq j$, *then* $\phi_j^{-1}, \overline{\phi}_j \in \mathrm{Aut}\, G_j$ *and* $\phi_j^{-1} = \overline{\phi}_j$.

Proof Writing the condition $u \circ \overline{u} = 1_{\mathrm{Aut}(\prod_{i=1}^{m} G_i)}$ in terms of the components of u and $\overline{u}$, and using the hypothesis (4), it follows that for every $x_j \in G_j$, we have

$$\begin{aligned}(\phi_j \circ \overline{\phi}_j)(x_j) &= u_j(e_1, \ldots, \overline{\phi}_j(x_j), \ldots, e_m)\\ &= u_j(e_1, \ldots, \overline{u}_j(e_1, \ldots, x_j, \ldots, e_m), \ldots, e_m) = x_j,\end{aligned}$$

that is, $\phi_j \circ \overline{\phi}_j = 1_{G_j}$. Similarly, we prove that $\overline{\phi}_j \circ \phi_j = 1_{G_j}$, and the conclusion follows. □

In this section we exhibit sufficient conditions on the groups $G_1, G_2, \ldots, G_m$ to derive that the homomorphism f is surjective, i.e. the groups $\prod_{i=1}^{m} \mathrm{Aut}(G_i)$ and $\mathrm{Aut}(\prod_{i=1}^{m} G_i)$ are isomorphic. Our first result generalizes the one for finite groups mentioned at the beginning of this section, but it also applies to some large classes of infinite groups. Other sufficient conditions expressed in terms of the prime divisors of the orders are given in our paper [3].

The main result, based on the paper [3], is the following:

Theorem 18 *Let* $G_1, G_2, \ldots, G_m$ *be* **fe**-*groups. If* $\gcd\big(\mu(G_i), \mu(G_j)\big) = 1$, *for every* $1 \leq i < j \leq n$, *then*

$$\mathrm{Aut}\left(\prod_{i=1}^{m} G_i\right) \simeq \prod_{i=1}^{m} \mathrm{Aut}(G_i).$$

Proof We first remark that the result for general m follows easily via strong induction, if one proves that the same holds for $m = 2$.

We already seen that the group homomorphism f defined by (2) is injective. To prove that f is an isomorphism, it remains to prove surjectivity. Let us first denote by $m_i = \mu(G_i)$, for all $i = 1, 2$. Since m_1, m_2 are coprime, there are integers a_1, a_2 such that $a_1m_1 + a_2m_2 = 1$.

Any element $u \in \mathrm{Aut}(G_1 \times G_2)$ is of the form $u = (u_1, u_2)$ where $u_i : G_1 \times G_2 \to G_i$ are group homomorphisms for all $i = 1, 2$. We have that

$$u_1(x_1, x_2) = u_1(x_1, x_2)^{1-a_1m_1} = u_1(x_1, x_2)^{a_2m_2} = u_1(x_1^{a_2m_2}, x_2^{a_2m_2})$$
$$= u_1(x_1^{a_2m_2}, e_2) = u_1(x_1^{1-a_1m_1}, e_2) = u_1(x_1, e_2),$$

for all $x_i \in G_i, i = 1, 2$.

In a similar way, we can deduce that $u_2(x_1, x_2) = u_2(e_1, x_2)$ for all $x_i \in G_i$, $i = 1, 2$. Let us write $\phi_1 \in \mathrm{Aut}\, G_1$ for the map $\phi_1(x_1) = u_1(x_1, e_2)$, for all $x_1 \in G_1$. Similarly, we write $\phi_2 \in \mathrm{Aut}\, G_2$ for the map defined as $\phi_2(x_2) = u_2(e, x_2)$ for all $x_2 \in G_2$. We observe that $f(\phi_1, \phi_2) = u$, and hence f is surjective. □

Remark If the groups $G_1, \ldots, G_m$ are finite, the hypotheses $\gcd\left(|G_i|, |G_j|\right) = 1$ and $\gcd\left(\mu(G_i), \mu(G_j)\right) = 1$, $1 \le i < j \le m$ are equivalent.

A natural question one can ask is whether there are examples of groups G_1 and G_2, such that not both of them have finite exponent, but for which the relation (1) holds. In the following theorem, we provide a sufficient criterion for the relation (1) to hold and we obtain as a corollary an affirmative answer to the previous question.

Theorem 19 *Let $n \ge 2$ be an integer and G_1, G_2 groups with the following properties:*

1. *If $x \in G_1$ is such that $x^n = e_1$, then $x = e_1$;*
2. *Every normal subgroup H of finite index of G_1 satisfies* $\gcd(n, [G_1 : H]) = 1$;
3. *G_2 is cyclic of order n.*

Then the relation (1) holds and the map f defined by (2) is an isomorphism.

Proof It is sufficient to prove that f is surjective. Let $u = (u_1, u_2), \in \mathrm{Aut}(G_1 \times G_2)$ and let $\overline{u} = (\overline{u}_1, \overline{u}_2)$ be its inverse. If we choose a generator $a \in G_2$, then for every $x_1 \in G_1$ and for every $k = 1, \ldots, n$, the following relations hold

$$u_1(x_1, a^k) = u_1(x_1, e_2) \cdot u_1(e_1, a^k) = u_1(x_1, e_2) \cdot u_1(e_1, a)^k.$$

Taking $k = n$ in the last equality we obtain $u_1(x_1, e_2) = u_1(x_1, e_2) \cdot u_1(e_1, a)^n$, hence $u_1(e_1, a)^n = e_1$. Our hypothesis implies that $u_1(e_1, a) = e_1$, that is, for every $x_2 \in G_2$ we have $u_1(e_1, x_2) = e_1$. In similar way, we can prove the relation $\overline{u}_1(e_1, x_2) = e_1, x_2 \in G_2$.

From Lemma 1, it follows that the map $\phi_1 : G_1 \to G_1$ defined by $\phi_1(x_1) = u_1(x_1, e_2)$, for all $x_1 \in G_1$ is an automorphism of G_1.

We have that $u_2(x_1, x_2) = u_2(x_1, e_2)u_2(e_1, x_2)$ for all $(x_1, x_2) \in G_1 \times G_2$.

Let us define the group homomorphism $\phi : G_1 \to G_2$ by $\phi(x_1) = u_2(x_1, e_2)$ for all $x_1 \in G_1$ and denote by $H = \ker(\phi)$. The subgroup H is normal in G_1 and from the relation $G_1/H \simeq \phi(G_1)$, it follows that it has finite index in G_1 and its index $[G_1 : H]$ divides n. From the second hypothesis, it follows that $[G_1 : H] = 1$ and hence $\phi(x_1) = e_2$ for every $x_1 \in G_1$, that is, we have $u_2(x_1, e_2) = e_2$ for every $x_1 \in G_1$. Similarly, we can prove the relation $\overline{u}_2(x_1, e_2) = e_2$ for every $x_1 \in G_1$. Considering the group endomorphism $\phi_2 : G_2 \to G_2$, defined by $\phi_2(x_2) = u_2(e_1, x_2)$, for all $x_2 \in G_2$, from Lemma 1 we have $\phi_2 \in \mathrm{Aut}(G_2)$.

With the same computation as in the proof of the previous theorem, we obtain that $f(\phi_1, \phi_2) = (u_1, u_2) = u$ and hence f is bijective. The conclusion follows. □

To emphasize the utility of Theorem 19, we present the following two groups that satisfy the first two hypotheses, hence can play the role of G_1. Let us fix a prime number p that is coprime to $n = |G_2|$.

Example 1 It is well-known that the aforementioned Prüfer p-group $\mathbb{Z}(p^\infty)$ has only finite proper subgroups. The only finite index subgroup is $\mathbb{Z}(p^\infty)$ and its index is 1, hence the condition 2 is verified. On the other hand, the condition 1 is obviously satisfied if p does not divide n.

Example 2 The second example is a bit more sophisticated. Let $G_1 = \mathbb{Z}_p = \varprojlim \mathbb{Z}/p^k\mathbb{Z}$ be the projective limit of the finite cyclic groups $\mathbb{Z}_{p^k}$, as k goes to infinity. The reader familiar with p-adics might recognize this group as the unit ball (or the ring of integers) inside the p-adic numbers $\mathbb{Q}_p$. It is an abelian group in which every nontrivial element has infinite order. A result in the theory of profinite groups (see, for example, [10]) tells us that every finite index subgroup N of $\mathbb{Z}_p$ is open, i.e. it contains $p^k\mathbb{Z}_p$ for some large integer $k \geq 0$. In particular, $[\mathbb{Z}_p : N]$ must divide $[\mathbb{Z}_p : p^k\mathbb{Z}_p] = p^k$. We therefore showed that we can choose $G_1 = \mathbb{Z}_p$ in the theorem above.

One can observe a certain similarity between the multiplicative group $\mathbb{Z}(p^\infty)$ and the additive group $\mathbb{Z}_p$. However, they are not isomorphic. The first one is a torsion group, which can be realized as a direct limit $\varinjlim \mathbb{Z}/p^k\mathbb{Z}$, where the homomorphisms $\mathbb{Z}/p^k\mathbb{Z} \to \mathbb{Z}/p^{k+1}\mathbb{Z}$ are induced by multiplication by p. On the other hand, in the additive group $\mathbb{Z}_p$ every element has infinite order.

We note the following known relations between $\mathbb{Z}(p^\infty)$ and $\mathbb{Z}_p$. The first group is isomorphic to the quotient $\mathbb{Q}_p/\mathbb{Z}_p$. The automorphism groups of $\mathbb{Z}(p^\infty)$ and $\mathbb{Z}_p$ are both isomorphic to $U(\mathbb{Z}_p)$, the group of multiplicative units in the ring $\mathbb{Z}_p$. Theorem 19 applied in turns with $G_1 = \mathbb{Z}(p^\infty)$ and $G_1 = \mathbb{Z}_p$ gives

$$\mathrm{Aut}(\mathbb{Z}(p^\infty) \times \mathbb{Z}/n\mathbb{Z}) \simeq U(\mathbb{Z}_p) \times U(\mathbb{Z}/n\mathbb{Z}) \simeq \mathrm{Aut}(\mathbb{Z}_p \times \mathbb{Z}/n\mathbb{Z}).$$

6 Sets and Sequences of Numbers Associated with a Group

In this section we introduce the definition of some sets and sequences associated with a group. We are going to use these notations through the rest of this survey.

For a group G and a positive integer k, define the following quantities:

$$M_k(G) = \{x \in G : o(x) = k\}; \tag{5}$$

$$\text{If } M_k(G) \text{ is finite, we write } m_k(G) = |M_k(G)|; \tag{6}$$

$$T_k(G) = \{x \in G : x^k = e\}; \tag{7}$$

$$\text{If } T_k(G) \text{ is finite, we write } \Theta_k(G) = |T_k(G)|. \tag{8}$$

If G is finite, the set $T_k(G)$ is nonempty and it contains finitely many elements which are the solutions to the "binomial" equation $x^k = e$. For fixed G, the sequences $\{m_k(G)\}_{k\geq 1}$ and $\{\Theta_k(G)\}_{k\geq 1}$ are very interesting. The first one is called the "order-counting" sequence of G and it has been introduced in the paper [16] and extensive studied in [17, 26, 38]. Moreover, in the paper [36] it was proven that a finite simple group and a finite group having equal orders and same sets of element orders are isomorphic. Later we will show that each one of the sequences $\{m_k(G)\}_{k\geq 1}$ and $\{\Theta_k(G)\}_{k\geq 1}$ are absolute invariants for finite abelian groups G. That means they characterize finite abelian groups up to isomorphism. On the other hand, we will see that there are examples of non-isomorphic finite groups for which these sequences are identical.

The proofs of the following propositions are easy consequence of the definition. We encourage the reader to try and prove them.

Proposition 2 *Let G be a finite group. Then*

1) $m_r(G) = 0$ implies that $m_{kr}(G) = 0$, for all $k \in \mathbb{N}^*$;
2) $|G| = \sum\limits_{d \mid |G|} m_d(G)$;
3) $|G| = \sum\limits_{d \mid \mu(G)} m_d(G)$;
4) $|G| = \sum_{k=1}^{\mu(G)} m_k(G)$;
5) $\Theta_k(G) = \sum\limits_{d \mid k} m_d(G)$, for all $k \in \mathbb{N}^*$;
6) $m_k(G) = \sum_{d \mid k} \mu\left(\frac{k}{d}\right) \Theta_d(G)$, for all $k \in \mathbb{N}^*$, where μ is the Möbius function.

A special case of property 5) is given by $k = p^r$, where p is a prime and r a positive integer. We obtain

$$\Theta_{p^r}(G) = 1 + m_p(G) + \cdots + m_{p^r}(G).$$

When $p = 2$ and $r = 1$ we get $\Theta_2(G) = 1 + m_2(G)$.

Proposition 3

1) *If* $(G_i)_{i\in I}$ *are groups and* k *is a positive integer, then* $T_k\left(\prod_{i\in I} G_i\right) = \prod_{i\in I} T_k(G_i)$, *where* I *is arbitrary.*
2) *If* $(G_i)_{i\in I}$ *is a finite family of groups and* k *is a positive integer such that we have* $\Theta_k(G_i) < \infty$ *for all* $i \in I$, *then* $\Theta_k\left(\prod_{i\in I} G_i\right) = \prod_{i\in I} \Theta_k(G_i)$.
3) *If* G *is a* ***fe****-group and* n *a positive integer such that* $\gcd(n, \mu(G)) = 1$, *then* $\Theta_n(G) = 1$.

7 The Set of Elements of Order *k* in a Group G

For a finite group G denote by $M_k(G)$ the set of elements of order k and by $m_k(G) = |M_k(G)|$ the number of elements of order k.

Theorem 20 *Let* G *be a group,* $f \in \mathrm{Aut}(G)$ *and* $A = \{x \in G : f(x) = x^{-1}\}$. *Then, the following holds:*

1) *Suppose additionally that* G *is finite and* p *is the smallest prime that divides* $|G|$. *Then, if* $|A| > \frac{p+1}{2p}|G|$ *implies that* G *is abelian.*
2) *If* G *is an infinite group and* $G \setminus A$ *is finite, then* G *is abelian.*

Proof

1) Observe that if $a, b \in A$, then we have $ab = ba$ if and only if $ab \in A$.

   Denote by $C(x) = \{y \in G \mid yx = xy\}$ the centralizer of the element $x \in G$. We have that $x \in A$ implies $x(A \setminus C(x)) \subseteq G \setminus A$.

   Therefore, $|A \setminus C(x)| = |x(A \setminus C(x))| \leq |G| - |A|$, for all $x \in A$. We hence find that $|C(x)| \geq |A \cap C(x)| \geq 2|A| - |G| > 2 \cdot \frac{p+1}{2p}|G| = \frac{1}{p}|G|$. Since $C(x)$ is a subgroup of G and $|C(x)| > \frac{1}{p}|G|$, for all $x \in A$. Since p is the smallest prime that divides $|G|$, the latter inequality implies $C(x) = G$, for every $x \in A$. In other words, $A \subseteq Z(G)$.

   Since $|A| > \frac{p+1}{2p}|G|$, we have that $|Z(G)| > \frac{1}{2}|G|$, which, using Lagrange's theorem, implies that $Z(G) = G$ hence proving the desired property.
2) Suppose that $a, b \in A$. It is easy to see that $ab \in A$ if and only if $ab = ba$. Therefore,

$$\text{if } x \in A, \text{ then } x(A \setminus C(x)) \subseteq G \setminus A. \tag{9}$$

From (9), it follows that $A \setminus C(x)$ is finite for every $x \in A$. Hence $G \setminus C(x)$ is finite for every $x \in A$. Therefore $C(x) = G$ for every $x \in A$, hence $A \subseteq Z(G)$. The hypothesis implies that $G \setminus A$ is finite, hence $G \setminus Z(G)$ is finite. The center $Z(G)$ is a normal subgroup of G. The fact that $G \setminus Z(G)$ finite implies that the index $|G : Z(G)|$ is equal to 1. Therefore G is abelian. □

We recall (see (7)) that for a positive integer k and a group G, we defined $T_k(G) = \{x \in G : x^k = e\}$ and $M(G) = \{a \in \mathbb{Z} : x \mapsto x^a \in \mathrm{End}(G)\}$ (see Definition 1).

Corollary 9 *Let G be a finite group, n a positive integer such that $n \in M(G)$ and $\gcd(n, \mu(G)) = 1$ and p is the smallest prime dividing $|G|$. Then $|T_{n+1}(G)| > \frac{p+1}{2p}|G|$ implies that G is abelian.*

Proof Apply the previous theorem with $f : G \to G$, $f(x) = x^n$. □

Corollary 10 *If p is a prime and G an infinite group, then the following holds:*

i) If $G \setminus M_2(G)$ is finite, then $G = T_2(G)$.
ii) If G is abelian and $G \setminus M_p(G)$ is finite, then $G = T_p(G)$.

Proof

i) By choosing $f(x) = x$, for all $x \in G$, from part 2) of the previous theorem we obtain that $A = T_2(G)$ and that G is abelian. The latter implies that $T_2(G)$ is a subgroup of G, but since $G \setminus T_2(G)$ is finite, then $G = T_2(G)$.
ii) As above, if G is abelian, then $T_p(G)$ is a subgroup of G. But since $G \setminus T_p(G)$ is finite, it follows that in fact $G = T_p(G)$. □

Corollary 11 *Let G be a finite group. Then the following statements hold:*

i) If $\Theta_2(G) > \frac{3}{4}|G|$, then G is an abelian group. Under the same hypothesis, if G is not trivial, then $|G|$ is divisible by 4.
ii) If $m_2(G) \geq \frac{3}{4}|G|$, then G is an abelian group and $|G|$ is divisible by 4.

Proof

i) By choosing $f(x) = x$, for every $x \in G$ în Theorem 20, we obtain $A = T_2(G)$. Therefore $\Theta_2(G) = |A| > \frac{3}{4}|G|$ implies that G is an abelian group.
   Suppose G is not the trivial group. The hypothesis implies $m_2(G) = \Theta_2(G) - 1 > 0$ and by Cauchy's theorem we have that $|G|$ is even. We know from the previous corollary that G is abelian. If $m_2(G) = 1$, then $\Theta_2(G) = 2 > \frac{3}{4}|G|$, which implies that $|G| \leq 2$. As $|G|$ must be even, we obtain $|G| = 2$. But then $\Theta_2(G) = 1 < \frac{3}{4}|G|$, which is not allowed.
   If $m_2(G) \geq 2$, then let $x, y \in M_2(G)$ be distinct elements. As G is abelian, the set $H = \{e, x, y, xy\}$ is a subgroup of G which has order 4. By Lagrange's theorem, we conclude that $|G|$ is divisible by 4.
ii) Let us just notice that $m_2(G) = \Theta_2(G) - 1$, so $m_2(G) \geq \frac{3}{4}|G|$ implies that $\Theta_2(G) > \frac{3}{4}|G|$. Additionally, as $m_2(G) > 0$, G is not trivial. The conclusion follows from i). □

Theorem 21 *Let G be a group and $n \geq 2$ an integer such that $m_n(G) < \infty$. Then $m_n(G) = \varphi(n) \cdot \tau_n(G)$, where $\tau_n(G)$ is the number of cyclic subgroups with n elements of G, and φ is the Euler-totient function. Consequently, $\varphi(n)$ divides $m_n(G)$.*

Proof If $M_n(G) = \emptyset$, there is nothing to prove. Otherwise, if $a \in M_n(G)$, we write $H(a) = \{a^k : \gcd(k, n) = 1\}$.

Observe that if $H(a) \cap H(b) \neq \emptyset$ for some $a, b \in M_n(G)$, then $H(a) = H(b)$ and $|H(a)| = \varphi(n)$ for every $a \in M_n(G)$. Therefore, we can partition the set $M_n(G)$ into disjoint sets of equal cardinality $\varphi(n)$. This proves the theorem. □

Remark One can also explain why the equality $m_n(G) = \varphi(n) \cdot \tau_n(G)$ holds in the following alternative way. Fix H a cyclic subgroup of order n in G and a generator x of H. All the elements of order n that are contained in H are of the form x^k, where $1 \leq k < n$ and $\gcd(k, n) = 1$. There are $\varphi(n)$ such elements. On the other hand, every element of order n in G is contained in the order n cyclic subgroup of G that it generates.

Theorem 22 *Let G be a group of order $n = 2^k(2l - 1)$, where k, l are positive integers. Then $m_2(G) \leq \frac{(n+1)(2^k-1)-1}{2^k}$.*

Proof Let us denote $m = m_2(G) = |M_2(G)|$. Consider the set $A \subseteq M_2(G)$ of maximal cardinality, having the property that $xy = yx$ for all $x, y \in A$. Observe that $A \cup \{e\}$ is a subgroup of G. Since $x^2 = e$, for all $x \in A \cup \{e\}$. There exists a positive integer t such that $r = |A| = 2^t - 1$. Using Lagrange's theorem, we obtain that $2^t \mid 2^k(2l - 1)$, hence $t \leq k$. If $A = M_2(G)$, then the inequality

$$m_2(G) = |A| \leq 2^k - 1 \leq \frac{(n+1)(2^k - 1) - 1}{2^k}$$

holds for every n and k as in the hypothesis.

Suppose that $A \subsetneq M_2(G)$. From the maximality of A it follows that for every $b \in M_2(G) \setminus A$, there exists an element $a \in A$ such that $ab \neq ba$. By choosing one such element for each $b \in M_2(G) \setminus A$, we can define a function

$$f : M_2(G) \setminus A \to A,$$

with the property that $bf(b) \neq f(b)b$, for all $b \in M_2(G) \setminus A$.

Let $A = \{a_1, a_2, \ldots, a_r\}$ and $q = \max\{f^{-1}(a) \mid a \in A\}$, which implies that $m - r = |M_2(G) \setminus A| = |f^{-1}(a_1) \cup f^{-1}(a_2) \cup \cdots \cup f^{-1}(a_r)| = |f^{-1}(a_1)| + |f^{-1}(a_2)| + \cdots + |f^{-1}(a_r)| \leq q \cdot r$. Hence, $q \geq \lfloor \frac{m-r}{r} \rfloor$.

Let us consider $j \in \{1, 2, \ldots, r\}$ having the property that $q = |f^{-1}(a_j)|$. Observe that $a_j f^{-1}(a_j) \subseteq G \setminus M_2(G)$ and $|a_j f^{-1}(a_j)| = q$. We can therefore write

$$n = |G| \geq |\{e\} \cup M_2(G) \cup a_j f^{-1}(a_j)| = |\{e\}| + |M_2(G)| + |a_j f^{-1}(a_j)|$$

$$= 1 + m + q \geq 1 + m + \lfloor \frac{m - r}{r} \rfloor = m + \lfloor \frac{m}{r} \rfloor > m + \frac{m}{r} - 1.$$

It follows that $nr > mr + m - r$, hence $nr - 1 \geq mr + m - r$. We can write $m \leq \frac{(n+1)r-1}{r+1} \leq \frac{(n+1)(2^k-1)-1}{2^k}$, because $r \leq 2^k - 1$. □

Corollary 12 *If G is a group such that* $n = |G| = 4k + 2$, *for some integer k, then* $m_2(G) \leq \frac{n}{2}$.

Corollary 13 *If G is a group such that* $n = |G| = 8k + 4$, *for some integer k, then* $m_2(G) \leq \frac{3n-2}{4}$.

Theorem 23 *Let G be a group such that* $m_2(G) < \infty$ *and let* $a, b \in M_2(G)$. *Then, there exists a positive integer k such that* $(ab)^k = e$.

Proof Let $A = \{(ab)^p a : p \in \mathbb{N}\}$. Observe that

$$((ab)^p a)^2 = (ab)^p a (ab)^p a = (ab)^p (ba)^p = e, \quad \text{for all } p \in \mathbb{N}.$$

Therefore, $A \subseteq M_2(G)$ is finite. There exist positive integers p, q with $p < q$ such that $(ab)^p a = (ab)^q a$, hence $(ab)^{q-p} = e$. If we write $k = q - p$, then $(ab)^k = e$. □

Theorem 24 *Let G be a finite group,* $a \in Z(G)$ *and p is a prime number such that p divides* $|G|$. *Then, the cardinality of* $A = \{x \in G : x^p = a\}$ *is a multiple of p.*

Proof Let us write $n = |G|$ and we consider the set

$$B = \{(x_1, x_2, \ldots, x_p) \in G^p : x_1 x_2 \ldots x_p = a\}.$$

On the set B, we introduce the equivalence relation

$$(x_1, x_2, \ldots, x_p) \sim (y_1, y_2, \ldots, y_p)$$

if and only if there exists $k \in \mathbb{N}$ such that $y_i = x_{k+i}$, where for all $i \in \mathbb{N}$, where $x_{i+p} = x_i$, for all $i \in \mathbb{N}$. Let us observe that the cardinality of the equivalence class of $\widehat{(x_1, x_2, \ldots, x_p)}$ is equal to p if and only if there exists different $i, j \in \{1, 2, \ldots, p\}$ such that $x_i \neq x_j$. Since $|B| = n^{p-1}$, we have that $k \in \mathbb{N}$ such that $n^{p-1} = |B| = |C| + |B \setminus C| = |A| + k \cdot p$, where $C = \bigcup_{x \in A} \widehat{(x, x, \ldots, x)}$. Since $p \mid n$, the latter implies that p divides $|A|$. □

Theorem 25 *Let G be a finite group and p a prime such that p divides* $|G|$. *Then,* $m_p(G) \equiv p - 1 \pmod{p^2 - p}$.

Proof From Theorem 24 it follows that $m_p(G) \equiv -1 \pmod{p}$. Since $p - 1 = \varphi(p)$, from Theorem 21 we find that $p - 1$ divides $m_p(G)$. It follows that $m_p(G) \equiv p - 1 \pmod{p^2 - p}$, as desired. □

Theorem 26 *Let p, q be prime numbers, such that $p > q$ and $q \nmid p - 1$. If G is a group of order pq, then G is cyclic.*

Proof Suppose that G is not cyclic. Then $m_{pq}(G) = 0$ and we have that $|G| = m_1(G) + m_p(G) + m_q(G)$, and using Theorem 25 we get $pq = 1 + ap(p-1) + p - 1 + bq(q-1) + q - 1$, for some non-negative integers a, b. If $a \geq 1$, then $(p-1)(q-1) \geq p(p-1)$, hence $q - 1 \geq p > q$, which is absurd.

Therefore, $a = 0$ and we have that $(p-1)(q-1) = bq(q-1)$ or $p - 1 = bq$, a contradiction. □

Lemma 2 *Let G be a group, $A \subseteq G$ and H a subgroup of G, generated by the elements of A. If the following conditions are satisfied:*

1. *A is finite and closed under conjugation in G;*
2. *$o(x) < \infty$ for every $x \in A$,*

then H is a finite normal subgroup.

Proof It is easy to see that the hypothesis implies that $[H : C(a)] < \infty$ for all $a \in A$, where $C(a)$ is the centralizer of a in H. If $A = \{a_1, \ldots, a_n\}$, then it follows $\left[H : \bigcap_{i=1}^{n} C(a_i)\right] < \infty$, and notice that $\bigcap_{i=1}^{n} C(a_i) = Z(H)$. This implies that $H/Z(H)$ is finite. Recall the following classical theorem of Schur (see, for instance, [20]): If H is a group such that $H/Z(H)$ is finite, then the commutator subgroup $[H, H]$ is finite. Therefore, $H/[H, H]$ is finite and generated by A. It is also known that the latter quotient is abelian. H is then itself finite, and generated by A, which is closed under conjugation. Therefore H is normal. □

Corollary 14 *Let G be a group and p a prime number. If $1 \leq m_p(G) < \infty$, then $m_p(G) \equiv p - 1 \pmod{p^2 - p}$.*

Proof This follows from Theorem 25 and from the previous Lemma 2. Notice that unlike in the hypothesis of Theorem 25, in the present theorem G does not have to be finite. □

Corollary 15 *Let G be a group such that $1 \leq m_2(G) < \infty$. Then $m_2(G)$ is odd.*

7.1 The Greatest Order in a Torsion Group

Let G be a torsion group. We write $\gamma(G)$ for the greatest order of the elements in G, that is,

$$\gamma(G) = \max\{o(x) : x \in G\}.$$

Theorem 27 *Let G be a torsion group. Then*

1. *If* $\mu(G) < \infty$, *then* $\gamma(G) < \infty$ *and* $\gamma(G) \mid \mu(G)$;
2. *If H is a subgroup of G, then* $\gamma(H) \mid \mu(G)$;
3. *If* G_1 *and* G_2 *are torsion groups and* $f : G_1 \to G_2$ *is a morphism, then*

   i) *If f is injective, then* $\gamma(G_1) \leq \gamma(G_2)$;
   ii) *If f is surjective, then* $\gamma(G_1) \geq \gamma(G_2)$.

4. *Let* G_i *be torsion groups such that* $\gamma(G_i) < \infty$, *for all* $i \in I$. *Then* $\gamma\left(\prod_{i\in I} G_i\right) < \infty$ *and* $\gamma\left(\prod_{i\in I} G_i\right) = \mathrm{lcm}\{\gamma(G_i) : i \in I\}$.
5. *If H is a normal subgroup of G, then* $\gamma(G/H) \leq \gamma(G)$.
6. *If G is an abelian* **fe** *group, then* $\gamma(G) = \mu(G)$.

A special situation is given by the symmetric group S_n. Landau's function $g(n)$ is defined for every natural number n as $g(0) = 1$ and $g(n) = \gamma(S_n)$, that is, for $n \geq 1$ $g(n)$ is the largest possible order of an element of the symmetric group S_n. Equivalently, $g(n)$ is the largest least common multiple of any partition of n, that is,

$$g(n) = \max\left\{\mathrm{lcm}(s_1, \ldots, s_k) : k \geq 1, s_i \geq 1 \text{ and } \sum_{i=1}^{k} s_i = n\right\}.$$

The first terms of the sequence $(g(n))_{n\geq 0}$ are

1, 1, 2, 3, 4, 6, 6, 12, 15, 20, 30, 30, 60, 60, 84, 105, 140, 210, 210, 420, 420, 420, 420,

840, 840, 1260, 1260, 1540, 2310, 2520, 4620, 4620, 5460, 5460, 9240, 9240, 13860,

13860, 16380, 16380, 27720, 30030, 32760, 60060, 60060, 60060, 60060, 120120, . . .

This is the sequence A000793 in the OEIS [29] named after Edmund Landau, who proved in 1902 the asymptotic formula

$$\lim_{n\to\infty} \frac{\ln g(n)}{\sqrt{n \ln n}} = 1,$$

that is, $g(n) = e^{(1+o(1))\sqrt{n \ln n}}$ for $n \to \infty$. A direct elementary proof using the Prime Number Theorem is given in [27].

Let us mention that the statement

$$\ln g(n) < \sqrt{\mathrm{Li}^{-1}(n)}$$

for all sufficiently large n, where Li^{-1} denotes the inverse of the logarithmic integral function $\mathrm{Li}(x) = \int_2^x \frac{1}{\ln t}\,\mathrm{dt}$, is equivalent to the Riemann hypothesis (see [25]).

7.2 $m_2(S_n)$ and $|\operatorname{End}(S_n)|$

In this subsection, we shall compute the number $m_2(S_n)$ for $n \geq 3$ and, as an application we shall obtain a formula for $|\operatorname{End}(S_n)|$, the number of endomorphisms of the symmetric group S_n for such values of n.

Let us begin with the observation that an element of S_n has order two if and only if this element is a product of k disjoint transpositions, where $1 \leq k \leq n/2$. Fix an integer k such that $1 \leq k \leq n/2$. The product of k disjoint transpositions has the form $(i_1 j_1)(i_2 j_2)\cdots(i_k j_k)$ where the numbers $i_1, j_1, i_2, j_2, \ldots, i_k, j_k$ are distinct elements of the set $\{1, \ldots, n\}$. Since $(ij) = (ji)$, the number of ways in which one can choose the first transposition $(i_1 j_1)$ is equal to the number of possible choices of a 2-element subset from a set with n elements. The latter is $\binom{n}{2}$. For each of these choices, the second transposition can be chosen in $(i_2 j_2)$ is $\binom{n-2}{2}$ ways, and so on. This results in $\prod_{s=0}^{k-1}\binom{n-2s}{2}$ ways to choose an ordered product of k disjoint transpositions. However, disjoint permutations commute and thus each reshuffling of such a product gives us the same permutation in S_n. Therefore the number of elements of S_n that are products of k disjoint transpositions is

$$\frac{1}{k!}\prod_{s=0}^{k-1}\binom{n-2s}{2}.$$

Hence the total number of elements of order two in S_n is

$$m_2(S_n) = \sum_{k=1}^{\lfloor n/2 \rfloor}\frac{1}{k!}\prod_{s=0}^{k-1}\binom{n-2s}{2}.$$

The following asymptotic formula for $m_2(S_n)$ was established by the authors of [11]

$$m_2(S_n) \sim 2^{-1/2} n^{n/2} \exp\left(-\frac{n}{2} + n^{1/2} - \frac{1}{4}\right), \text{ as } n \to \infty.$$

In order to compute $|\operatorname{End}(S_n)|$, let us note that for every $f \in \operatorname{End}(S_n)$ we have $S_n/\ker f \simeq \operatorname{Im} f$. Because $\ker f$ is a normal subgroup of S_n and $n \geq 3$, $n \neq 4$ there are not many possible candidates for $\ker f$. In fact, it follows that $\ker f$ must be one of $\{e\}$, A_n, or S_n, where e denotes the identity permutation. This observation defines a partition of $\operatorname{End}(S_n)$ into three subsets I, II, and III. In the first case f is injective, hence $f \in \operatorname{Aut}(S_n)$ and we have $I = \operatorname{Aut}(S_n)$. In the third case f must be the constant endomorphism of S_n, that is, $III = \{e\}$. In the second case it follows that the subgroup $\operatorname{Im} f$ has two elements, hence it is cyclic of order two. Therefore, in each case, $\operatorname{Im} f$ is generated by a permutation of order two. It is easy to see that given $\sigma \in S_n$, an element of order 2, we can define the endomorphism $f_\sigma \in \operatorname{End}(S_n)$, by $f_\sigma(\tau) = e$ if $\tau \in A_n$ and $f_\sigma(\tau) = \sigma$, otherwise. Using this recipe, for every such σ we obtain a different endomorphism f_σ. Similarly, given an endomorphism f such that Im f is a subgroup of order 2, it follows that ker $f = A_n$, the unique normal subgroup of index 2 of S_n. This shows that the choice for Im f determines f. For $n \geq 3, n \neq 4, 6$ we can derive the formula

$$|\operatorname{End}(S_n)| = 1 + n! + m_2(S_n) = 1 + n! + \sum_{k=1}^{\lfloor n/2 \rfloor} \frac{1}{k!} \prod_{s=0}^{k-1} \binom{n-2s}{2}.$$

Using this formula for small values of n we get the sequence

$$|\operatorname{End}(S_3)| = 10, |\operatorname{End}(S_5)| = 146,$$

$$|\operatorname{End}(S_7)| = 5272, |\operatorname{End}(S_8)| = 41{,}084, |\operatorname{End}(S_9)| = 365{,}500,$$

$$|\operatorname{End}(S_{10})| = 3{,}638{,}296.$$

which is not currently indexed in OEIS. For $n = 4$, it is not hard to see that S_4 has an extra normal subgroup isomorphic to the Klein four-group V_4. With the same counting argument, it is easy to compute that $|End(S_4)| = 58$. For $n = 6$ we have that $|\operatorname{Aut}(S_6)| = 2 \cdot 6!$, therefore $|\operatorname{End}(S_6)| = 1 + 2 \cdot 6! + m_2(S_6) = 1516$.

Using the same idea, one can derive a formula for the cardinality of the set $|\operatorname{Hom}(S_n, S_m)|$, where $n, m \geq 3, n \neq 4, m \neq 6$. To be precise, if $f \in \operatorname{Hom}(S_n, S_m)$, then ker f must be a normal subgroup of S_n. Therefore, ker f is equal to one of $\{e\}$, A_n or S_n. So we can partition $\operatorname{Hom}(S_n, S_m)$ into three disjoint subsets I, II, and III, where $I = \{f \in \operatorname{Hom}(S_n, S_m) : f \text{ is injective}\}$, $II = \{f \in \operatorname{Hom}(S_n, S_m) : \operatorname{Im} f \text{ of order 2 in } S_m\}$ and $III = \{e\}$, the set containing only the trivial homomorphism.

It is easy to see that $|II| = m_2(S_m) = \sum_{k=1}^{\lfloor m/2 \rfloor} \frac{1}{k!} \prod_{s=0}^{k-1} \binom{m-2s}{2}$ and $|III| = 1$, as before. On the other hand, if $n > m$ we have $|I| = 0$. If $m \geq n$, then f is in fact an isomorphism onto its image. We can derive the formula

$$|I| = |\operatorname{Aut}(S_n)| \cdot |\{N \text{ subgroup of } S_m : N \simeq S_n\}|.$$

The cardinality of the set $\{N \text{ subgroup of } S_m : N \simeq S_n\}$ is equal to $\binom{m}{n}$. To see this, let N be a subgroup of S_m isomorphic to S_n and $f : S_n \to N$ an isomorphism S_m and hence the subgroup M acts on a set X with m elements, via permutations. Every permutation can be written as a product of disjoint cycles. An isomorphism preserves the cycle-type of a permutation, therefore $N = f(S_n) \subset S_m$ consists of permutations in which only n out of m elements appear in one of their disjoint cycles. In other words, N fixes $m - n$ elements of X and acts as the symmetric group on the other n elements. Therefore, the number of such subgroups N is equal to the number of ways in which we can choose the fixed elements, i.e. $\binom{m}{m-n} = \binom{m}{n}$, as claimed.

Therefore, for $n \geq 3$ $n \neq 4$ and $m \neq 6$ we have

$$|I| = n! \cdot \binom{m}{n} = \frac{m!}{(m-n)!}.$$

Substituting, we get that for $m \geq n \geq 3$, $n \neq 4$ and $m \neq 6$, one has

$$|\operatorname{Hom}(S_n, S_m)| = 1 + \frac{m!}{(m-n)!} + \sum_{k=1}^{\lfloor m/2 \rfloor} \frac{1}{k!} \prod_{s=0}^{k-1} \binom{m-2s}{2}.$$

Let G be a fixed finite group of order n. In [28], the following asymptotic formula for $|\operatorname{Hom}(G, S_m)|$ when $m \to \infty$ is derived. Let $s_G(d)$ be the number of subgroups of index d in G, where $d \mid n$. The author of loc. cit. shows that when $m \to \infty$,

$$|\operatorname{Hom}(G, S_m)| \sim K_G m^{(1-n)m} \exp\left(-\frac{n-1}{n} m + \sum_{d \mid n, d<n} \frac{s_G(d)}{d} m^{d/n}\right)$$

where

$$K_G := \begin{cases} n^{-1/2}, \text{ if n is odd} \\ n^{-1/2} \exp\left(-\frac{(s_G(n/2))^2}{2n}\right), \text{ if n is even.} \end{cases}$$

Moreover in the same article, the following formula in terms of exponential generation functions is given:

$$\sum_{m=0}^{\infty} \frac{|\operatorname{Hom}(G, S_m)|}{m!} z^m = \exp\left(\sum_{d \mid n} \frac{s_G(d)}{d} z^d\right).$$

For completeness, let us remark that for $m \geq 2$, we have

$$|\operatorname{Hom}(S_2, S_m)| = 1 + m_2(S_m) = 1 + \sum_{k=1}^{\lfloor m/2 \rfloor} \frac{1}{k!} \prod_{s=0}^{k-1} \binom{m-2s}{2},$$

since in this case a homomorphism $f : S_2 \to S_m$ is either trivial or injective. To count the injective homomorphisms, we just have to count the number of ways in which we can choose where to send the nontrivial element in S_2. This amounts to counting the number of elements of order 2 in S_m, giving the desired formula.

In fact, we have the following more general result.

Theorem 28 *Let G be a finite group which has a unique proper normal subgroup N. Then,*

$$|\operatorname{End}(G)| = 1 + |\operatorname{Aut}(G)| + \sum_{H \leqslant G, H \simeq G/N} |\operatorname{Aut}(H)|.$$

Proof As before, if $f \in \operatorname{End}(G)$, then $\ker f$ must be a normal subgroup of G. Therefore, $\ker f$ is one of $\{e\}$, N or G. In the first case f must be an automorphism and in the last case f must be the trivial endomorphism. However, if $\ker f = N$, then $\operatorname{Im} f$ is a subgroup of G isomorphic to the quotient G/N. For any such choice of $\operatorname{Im} f$, we can define f as the composition of the natural projection $G \to G/N$, any isomorphism $G/N \to \operatorname{Im} f$ and the natural injection $\operatorname{Im} f \mapsto G$. Hence, there are $|\operatorname{Isom}(G/N, \operatorname{Im} f)| = |\operatorname{Aut}(\operatorname{Im} f)|$ choices for the endomorphism f, for each candidate subgroup $\operatorname{Im} f$. The result follows now. □

Example A_4 has a unique proper normal subgroup of index 3, its Sylow 2-subgroup, which is isomorphic to the Klein four group K_4. We have that $A_4/K_4 \simeq C_3$. The automorphism subgroup of C_3 is cyclic of order 2 and A_4 has 4 distinct Sylow 3 subgroups isomorphic to C_3.

$$|\operatorname{End}(A_4)| = 1 + 4! + 4 \cdot 2 = 33.$$

Remark 1 Given two groups G_1, G_2, let us write $\vartheta(G_1, G_2)$ for the number of subgroups of G_2 which are isomorphic to G_1. The proof of the theorem above can be easily adapted to show that the following more general result holds. Let G_1 and G_2 be two finite subgroups. If $N_1, N_2, \ldots, N_r$ are all the normal subgroups of G_1, then

$$|\operatorname{Hom}(G_1, G_2)| = 1 + \vartheta(G_1, G_2) \cdot |\operatorname{Aut}(G_1)| + \sum_{k=1}^{r} \vartheta\,(G_1/N_k, G_2) \cdot |\operatorname{Aut}(G_1/N_k)|.$$

8 Structure Theorems for Groups with a Prescribed Number of Elements of a Given Order

The philosophy behind the results presented in this section is as follows. The condition of having many elements of a specific prime order imposes very strong restrictions on the structure a group G can have.

Theorem 29 *Let $n \geq 2$ be an integer and G is a group of order $2n$ such that $m_2(G) = n$. Then*

1. *n is odd;*
2. *$H = G \setminus M_2(G)$ is a normal abelian subgroup of G;*
3. *If $a, b \in M_2(G)$ are distinct, then $ab \neq ba$;*
4. *The center $Z(G)$ is trivial and G is generated by $M_2(G)$.*

Proof The statement (1) is exactly Corollary 15.

For (2), let us note that if $x \in H$, then $x^{-1} \in H$. Write $\mathrm{Sym}(G)$ for the group of symmetries of G. Consider the function $f : G \to \mathrm{Sym}(G)$, where $f(x)(y) = xy$, for all $x, y \in G$.

Let us observe that f is an injective group homomorphism and that $x \in M_2(G)$ if and only if $\mathrm{sgn}(f(x)) = -1$ since n is odd. Therefore, $x \in H$ if and only if $\mathrm{sgn}(f(x)) = 1$. It follows that for all $x, y \in H$ we have $\mathrm{sgn}(f(xy)) = \mathrm{sgn}(f(x)f(y)) = \mathrm{sgn}(f(x))\,\mathrm{sgn}(f(y)) = 1$, so $xy \in H$. This proves that H is a subgroup.

H is a normal subgroup since $|G| = 2|H|$.

We will prove that H is abelian. Let $a \in M_2(G)$ and $x \in H$. It follows that $ax \in M_2(G)$ and therefore $axax = e$ or $axa = x^{-1}$. If $x, y \in H$, then we can write

$$(xy)^{-1} = axya = (axa)(aya) = x^{-1}y^{-1} = (yx)^{-1},$$

so $xy = yx$ and hence H is abelian.

To prove (3), let us assume that $a, b \in M_2(G)$ are distinct such that $ab = ba$. It follows that $H_1 = \{e, a, b, ab\}$ is a subgroup of G. From Lagrange's theorem it follows that $4 \mid 2n$, which is a contradiction.

(4) Let $x \in Z(G)$. From (3) it follows that $M_2(G) \cap Z(G) = \emptyset$. This means that $x \in H$. Therefore $axax = e$, for all $a \in M_2(G)$, implying that $x^2 = e$. As H is odd, we obtain that $x = e$. For the second property, let H_2 be the subgroup of G generated by $M_2(G)$. From $T_2(G) \subset H_2$, it follows $n = m_2(G) < \Theta_2(G) \leq |H_2|$, hence the conclusion. □

Corollary 16 *Let G be a group of order $2n$, where n is an odd square-free number. Then $m_2(G) = n$ implies that $G \simeq D_n$.*

Corollary 17 *Let $G = \{x_1, x_2, \ldots, x_n\}$ be a group with n even. If there is a permutation $\sigma \in S_n$ such that $\prod_{i=1}^{n} x_{\sigma(i)} = e$, then 4 divides n.*

Proof Assume, by contradiction, that 4 does not divide n. Then $n = 4k + 2$, for some positive integer k. As in the proof of Theorem 29, write $\mathrm{Sym}(G)$ for the group of symmetries of G and consider the function $f : G \to \mathrm{Sym}(G)$, defined by $f(x)(y) = xy$, for all $x, y \in G$. Let us recall that f is an injective group homomorphism.

Because n is even, there is an element $a \in G, a \neq e$, such that $a^2 = e$. As in the proof of Theorem 29, it follows $\mathrm{sgn}\, f(a) = -1$. Therefore, $n/2$ elements in $\mathrm{Im}\, f$ are of the sign $+1$ and $n/2$ elements in $\mathrm{Im}\, f$ are of the sign -1. From the relation in the hypothesis it follows $f\left(\prod_{i=1}^{n} x_{\sigma(i)}\right) = f(e) = e$, hence $\prod_{i=1}^{n} f(x_{\sigma(i)}) = e$. Considering the sign in both sides, we get $\mathrm{sgn}\left(\prod_{i=1}^{n} f(x_{\sigma(i)})\right) = 1$, that is,

$\prod_{i=1}^{n} \operatorname{sgn}(f(x_{\sigma(i)})) = 1$, hence $(-1)^{n/2} = 1$, a contradiction because $n = 4k + 2$, and the conclusion follows. □

The following result is well-known (see, for instance, [13, 14]). We present here a simple proof.

Theorem 30 *Let G be a group. If* $|Z(G)| = 1$, *then* $|Z(\mathrm{Aut}(G))| = 1$.

Proof If $a \in G$, let us define $f_a : G \to G$ with $f_a(x) = axa^{-1}$, for all $x \in G$.

Let us suppose that $|Z(G)| = 1$ and $f \in Z(\mathrm{Aut}(G))$. Then $f \circ f_a = f_a \circ f$, for all $a \in G$. But then, for all $x, a \in G$, we have that

$$f(axa^{-1}) = af(x)a^{-1}.$$

This means that for all $x, a \in G$,

$$f(a)f(x)f^{-1}(a) = af(x)a^{-1}.$$

Therefore, for all $a, x \in G$, we have

$$a^{-1}f(a)f(x) = f(x)a^{-1}f(a).$$

Since f is surjective, this implies that for all $a \in G$, the element $a^{-1}f(a)$ commutes with every element of G. Hence $a^{-1}f(a) \in Z(G)$, which implies that $f(a) = a$, for all $a \in G$. This proves that f is the trivial automorphism. □

Theorem 31 *Let G be an abelian group such that* $|G| = 2^k \cdot m$, *where* $k, m \in \mathbb{N}\setminus\{0\}$ *and m is odd. Then, the following holds:*

1) If $m_2(G) \geq \frac{1}{2}|G| - 2^{k-1}$, *then* $|G| = 2^k$;
2) If $m_2(G) \geq \frac{1}{2}|G| - 1$, *then* $G \simeq (\mathbb{Z}/2\mathbb{Z})^{k-2s} \times (\mathbb{Z}/4\mathbb{Z})^s$, *where* $s \in \{0, 1\}$;
3) If $m_2(G) \geq \frac{1}{2}|G|$, *then* $G \simeq (\mathbb{Z}/2\mathbb{Z})^k$.

Proof From the structure theorem for finite abelian groups, we know that there are $a_1, a_2, \ldots, a_r \in \mathbb{N}$ such that $k = a_1 + 2a_2 + \cdots + ra_r$ and an abelian group G_1 of order m, such that

$$G \simeq (\mathbb{Z}/2\mathbb{Z})^{a_1} \times (\mathbb{Z}/2^2\mathbb{Z})^{a_2} \times \cdots \times (\mathbb{Z}/2^r\mathbb{Z})^{a_r} \times G_1.$$

We have that $m_2(G) = 2^{a_1+a_2+\cdots+a_r}$ and $|G| = 2^k \cdot m$.

For 1), we see that the hypothesis implies $2^{k-1} - 1 \geq 2^{a_1+\ldots a_r} - 1 \geq 2^{k-1} \cdot m - 2^{k-1}$, which is equivalent to $m \leq 3 - \frac{1}{2^{k-1}} < 3$. But since m is odd, this implies that $m = 1$, or $|G| = 2^k$.

To see that 2) holds, we notice that the hypothesis implies that $m = 1$ and $2^{a_1+\cdots+a_r} - 1 \geq 2^{a_1+2a_2\cdots+ra_r-1} - 1$, therefore $a_1 + a_2 + \cdots + a_r \geq a_1 + 2a_2 + \cdots + ra_r - 1$. It follows that $a_2 \in \{0, 1\}$ and $a_3 = a_4 = \cdots = a_r = 0$ and the conclusion follows.

To prove 3), let us notice that the hypothesis implies that $m = 1$ and $2^{a_1+\cdots+a_r} - 1 \geq 2^{a_1+2a_2+\cdots+ra_r-1}$. Then $a_2 + 2a_3 + \cdots + (r-1)a_r < 1$ and hence $a_2 = a_3 = \cdots = a_r = 0$, which proves our claim. □

Theorem 32 *Let G be a finite abelian group and p a prime number. If* $m_p(G) \geq \frac{1}{2}|G|$*, then there exists* $k \in \mathbb{N} \setminus \{0\}$ *such that* $G \simeq (\mathbb{Z}/p\mathbb{Z})^k$.

Proof If $p = 2$, this follows from the previous theorem. Suppose that $p \geq 3$. As $|G| \geq 1$, we know that $m_p(G) \geq 1$.

From the structure theorem for finite abelian groups, we know that there are $a_1, a_2, \ldots, a_r \in \mathbb{N}$ and a finite abelian group G_1 with $p \nmid |G_1$ such that

$$G \simeq (\mathbb{Z}/p\mathbb{Z})^{a_1} \times (\mathbb{Z}/p^2\mathbb{Z})^{a_2} \times \cdots \times (\mathbb{Z}/p^r\mathbb{Z})^{a_r} \times G_1.$$

We have that $m_p(G) = a_1 + a_2 + \cdots + a_r \geq 1$ and $|G| = p^{a_1+2a_2+\cdots+ra_r} \cdot |G_1|$. Hence,

$$p^{a_1+2a_2+\cdots+ra_r} > p^{a_1+\cdots+a_r} - 1 \geq \frac{1}{2}p^{a_1+2a_2+\cdots+ra_r},$$

which implies that $2 > p^{a_2+2a_3+\cdots+(r-1)a_r}$, from where we find $a_2 = a_3 = \cdots = a_r = 0$, which implies that $G \simeq (\mathbb{Z}/p\mathbb{Z})^{a_1}$. □

Theorem 33 *Let G be a finite group. Then the following statements hold:*

i) If $\Theta_2(G) > \frac{3}{4}|G|$*, then there exists* $k \in \mathbb{N} \setminus \{0\}$ *such that* $G \simeq (\mathbb{Z}/2\mathbb{Z})^k$.
ii) If $m_2(G) \geq \frac{3}{4}|G|$*, then there exists* $k \in \mathbb{N} \setminus \{0\}$ *such that* $G \simeq (\mathbb{Z}/2\mathbb{Z})^k$.

Proof

i) From Corollary 11 it follows that G is abelian. Now, the desired conclusion follows by applying Theorem 32 with $p = 2$.
ii) As in the proof of Corollary 11, let us just notice that $m_2(G) = \Theta_2(G) - 1$, so $m_2(G) \geq \frac{3}{4}|G|$ implies that $\Theta_2(G) > \frac{3}{4}|G|$, and the conclusion follows from i).

□

To see that the previous bound is tight, we present the following illustrative example. For any positive integer n, consider the group $G_n = D_8 \times (\mathbb{Z}/2\mathbb{Z})^n$. Then $m_2(G) = 6 \cdot 2^n - 1$ and $|G_n| = 2^{n+3}$. Hence, $m_2(G_n) < \frac{3}{4}|G_n|$, but $\lim_{n\to\infty} \frac{m_2(G_n)}{|G_n|} = \frac{3}{4}$.

Remark The simplest nonabelian group with $\Theta_2(G) = \frac{3}{4}|G|$ is the dihedral group D_8 of order 8.

9 Frobenius' Theorem

We start by reviewing the theory of group actions on sets.

9.1 Group Actions

We say that group G **acts** on the set X if there is a homomorphism $\varphi : G \to S(X)$ from G into the group $S(X)$ of permutations of X. The pair (X, φ) is sometimes called a G-**set** or a G-**action**. Thus, each $g \in G$ gives a permutation $\varphi(g)$ of X, which sends any $x \in X$ to an element $\varphi(g)x$. If the homomorphism φ is understood or is completely general, we usually omit it from the notation, writing gx instead of $\varphi(g)x$. The notion of a G-set generalizes the notion of a group. For we can regard the action as a map $G \times X \to X$, given by $(g, x) \mapsto gx$ such that $g(g'x) = (gg')x$ for all $g, g' \in G$ and $x \in X$.

If X and Y are two G-sets, a function $f : X \to Y$ is called G-**equivariant** if $f(gx) = gf(x)$ for all $g \in G$ and $x \in X$. We say that X and Y are **equivalent** G-sets if there exists a G-equivariant bijection $f : X \to Y$.

Some standard terminology associated with group actions is as follows.

The **stabilizer** of a point $x \in X$ is the subgroup of G given by

$$G_x = \{g \in G : gx = x\} \leq G.$$

The **orbit** of an element $x \in X$ is the subset of X given by

$$Gx = \{gx : g \in G\} \subseteq X.$$

Orbits are equivalence classes under the equivalence relation $x \equiv y$ if $y = gx$ for some $g \in G$. Hence two orbits are either equal or disjoint and the set of all orbits forms a partition of X. We write X/G for the set of orbits of the G-set X.

The **fixed-point set** of $g \in G$ is the subset of X given by

$$X^g = \{x \in X : gx = x\} \subseteq X.$$

One can check that stabilizers behave well under conjugacy, that is, for all $g \in G$ and $x \in X$ we have

$$G_{gx} = gG_xg^{-1}.$$

In particular, the stabilizers of all elements of the same orbit are conjugate. Similarly, the fixed-point sets behave well under conjugacy, that is, if $g, h \in G$, then

$$hX^g = X^{hgh^{-1}}.$$

Let us recall some auxiliary results involving the group actions on a set.

The kernel of a group action $\varphi : G \to S(X)$ is the normal subgroup $\ker\varphi$ of G consisting of the elements acting trivially on X. We have

$$\ker\varphi = \cap_{x\in X} G_x = \{g \in G : X^g = X\}.$$

A G-action on X is **faithful** if $\ker\varphi$ is trivial. Equivalently, the action is faithful if no nontrivial element of G acts trivially on X. In this case, G is isomorphic to a subgroup of $S(X)$.

Finally, a group action is **free** if $gx = x$ for some $x \in X$ implies $g = 1$. That is, a group action is free if and only if all stabilizers are trivial. Clearly free actions are faithful. An example of a free action is where a subgroup H of a group G acts on G by left multiplication. Here, G is the set and H is the group which is acting. The orbits are the right cosets Hx.

A G-action on X is **transitive** if for all $x, y \in X$ there exists $g \in G$ such that $gx = y$. Equivalently, the action is transitive if and only if X consists of a single G-orbit. For a general group action, each orbit is a transitive G-set. Thus, transitive group actions are the essential ones. An example of a transitive group action is where $X = G/H$, for some subgroup $H \leq G$, and the action is $g(xH) = gxH$. We will see that all transitive G-actions are of this form. More generally, a G-action on X is k-transitive if G is transitive on k-tuples of distinct elements of X. This is a measure of the strength of transitivity.

Theorem 34 (The Main Result of Group Actions) *If a group G acts on a set X, then for each $x \in X$ we have a G-equivariant bijection $f : G/G_x \to Gx$ given by $f(gG_x) = gx$. In particular, any transitive group action is equivalent to an action on cosets.*

Proof The map f is well-defined because for all $h \in G_x$ we have $(gh)x = g(hx) = gx$. The map f is injective because if $gx = g'x$, then $(g^{-1}g')x = x$, so $g^{-1}g' \in G_x$, which means that $gG_x = g'G_x$. The map f is surjective, by the definition of the orbit G_x. Finally, for all $g, g' \in G$ and $x \in X$ we have

$$f(g(g'G_x)) = f(gg'G_x) = (gg')x = g(g'x) = gf(g'G_x)$$

which shows that f is G-equivariant. □

As a corollary, we have one of the most useful formulas in group theory.

Corollary 18 (The Counting Formula) *Let G be a finite group acting on a set X. Then the cardinality of an orbit equals the index of the stabilizer of any point in the orbit. That is, for any $x \in X$ we have*

$$\frac{|G|}{|G_x|} = |Gx|.$$

Note that the right-hand side of this equation depends only on the orbit, while the left side appears to depend on the stabilizer of a particular point in the orbit. However, as we have mentioned, all stabilizers G_x for x in a given orbit are conjugate, hence have the same order.

Let $\mathcal{O}_1, \dots, \mathcal{O}_k$ be the orbits of G in X, and choose $x_i \in \mathcal{O}_i$. Applying the counting formula in Corollary 6 to each orbit, we have the weaker but still useful formula

$$|X| = \sum_{i=1}^{k} |\mathcal{O}_i| = \sum_{i=1}^{k} [G : G_{x_i}],$$

where $[G : G_{x_i}]$ is the number of distinct left cosets of G_{x_i} in G.

9.2 Groups Acting on Themselves by Conjugation: The Class Equation

Now consider the special case in which G acts on itself by conjugation, that is, $(g, x) \to gxg^{-1}$. In this case, the center of G,

$$Z(G) = \{x \in G : xg = gx, \forall g \in G\},$$

is the set of points that are fixed by conjugation. The nontrivial orbits of the action are called the **conjugacy classes** of G. If $x_1, \dots, x_k$ are representatives from each of the nontrivial conjugacy classes of G and $|\mathcal{O}_{x_1}| = n_1, \dots, |\mathcal{O}_{x_k}| = n_k$, then

$$|G| = |Z(G)| + n_1 + \dots + n_k.$$

The stabilizer subgroup of x_i is in this case $C(x_i) = \{g \in G : gx_i = x_i g\}$ and it is called the **centralizer subgroup** of x_i, $i = 1, \dots, k$. Using the Lagrange theorem, it is easy to show that $|\mathcal{O}_{x_i}| = [G : C(x_i)]$, $i = 1, \dots, k$, therefore we obtain the **class equation** of the group G

$$|G| = |Z(G)| + [G : C(x_1)] + \dots + [G : C(x_k)].$$

9.3 The Number of Conjugacy Classes in S_n

Given a permutation $\sigma \in S_n$, we can express it as a product of disjoint cycles $\sigma = \sigma_1\sigma_2 \cdots \sigma_k$. This expression is unique modulo cyclic permutation of elements within cycles and, of course, permuting the order of the cycles. The **cycle-type** of σ is the tuple $(l_1, l_2, \dots, l_k)$, where l_i is the length of σ_i, and $l_1 \le l_2 \le \dots \le l_k$.

A **partition** of n is a tuple $(l_1, l_2, \ldots, l_k)$, where $l_1 \leq l_2 \leq \ldots \leq l_k$ such that $n = l_1 + l_2 + \ldots + l_k$.

The following result is fundamental in the study of the conjugacy classes and it can be found in any basic books (see, for example, [35]).

Theorem 35 *For $\sigma \in S_n$, the conjugacy class of σ consists of all those elements in S_n with the same cycle-type as σ, that is, the cycle-type is an invariant under conjugation of the permutation.*

An important consequence of the above result is that the number of conjugacy classes in S_n is equal to the number of partitions of n.

Theorem 36 *In S_n there are*

$$\frac{n!}{l_1^{k_1} l_2^{k_2} \cdots l_s^{k_s} k_1! k_2! \cdots k_s!}$$

permutations with a cycle-type of k_1 cycles of length l_1, k_2 cycles of length $l_2, \ldots, k_s$ cycles of length l_s. This decomposition includes 1-cycles and the following relation is satisfied $k_1 l_1 + k_2 l_2 + \cdots + k_s l_s = n$.

Proof Put in a fixed order the $k_1 + k_2 + \cdots + k_s$ brackets. There are $n!$ ways of filling the brackets with the numbers $1, \ldots, n$. However, the same permutation can be written as a product of disjoint cycles in many ways. There are l_i ways of cycling the elements of each cycle of length l_i and we have $k_i!$ of permuting the cycles of length l_i. Hence $n!$ is an over count by a factor of $l_1^{k_1} l_2^{k_2} \cdots l_s^{k_s} k_1! k_2! \cdots k_s!$ □

9.4 Burnside's Lemma

The following fundamental result, known as Burnside's lemma, was formulated and proved by Burnside in 1897. Historically it was already discovered in 1845 by Cauchy and in 1887 by Frobenius. Burnside's lemma allows us to count the number of equivalence classes in sets, based on internal symmetry. For a short proof, we refer to [8].

Lemma 3 (Burnside's Lemma) *Let G be a finite group acting on the finite set X. We then have the equality*

$$\sum_{g \in G} |X^g| = |G| \cdot |X/G|.$$

Proof Consider the set $\tilde{X} = \{(g, x) \in G \times X : gx = x\}$. The projections onto G and X give maps $p_G : \tilde{X} \to G$, $p_X : \tilde{X} \to X$, whose fibers are the fixed-point sets and stabilizers, respectively. That is, $p_G^{-1}(g) = X^g$ and $p_X^{-1}(x) = G_x$.

Because G and X are finite, we can compute $|\tilde{X}|$ in two ways, by summing both sets of fibers

$$|\tilde{X}| = \sum_{g\in G} |X^g| = \sum_{x\in X} |G_x|.$$

Since $|G_x|$ is constant for x in an orbit, we can partition the last sum into orbits $\mathcal{O}_1, \ldots, \mathcal{O}_k$, and obtain

$$\sum_{x\in X} |G_x| = \sum_{i=1}^{k} |\mathcal{O}_i| \cdot |G_{x_i}| = k|G|.$$

It follows that the number of orbits is the average size of a fixed-point set:

$$|X/G| = \frac{1}{|G|}\sum_{g\in G} |X^g|,$$

and the desired formula is proved. □

From the Burnside Lemma we obtain that if G is a finite group acting on the finite set X, then we have

$$\sum_{g\in G} |X^g| \equiv 0 \pmod{|G|}.$$

9.5 Frobenius Theorem and Some Applications

Recall the definitions for $M_k(G)$, $m_k(G)$, $T_k(G)$, and $\Theta_k(G)$ given in (5)–(8). In Theorem 24 we have proved that if G is a finite group, $a \in Z(G)$ and p is a prime number such that p divides $|G|$, then the cardinal number of the set $A = \{x \in G : x^p = a\}$ is a multiple of p. Considering $a = e$, the neutral element of G, this result shows that if a prime number p divides $|G|$, then p divides $\Theta_p(G)$. It is a natural question if this property remains true for an arbitrary divisor of $|G|$. The answer is positive and it has been proved by Frobenius in 1903. To present this theorem, we start with two auxiliary results.

Lemma 4 *Let $a, b \in \mathbb{N} \setminus \{0\}$ such that $a \mid b$. We then have*

$$\sum_{c|a} \mu(c) \cdot \binom{\frac{b}{c}}{\frac{a}{c}} \equiv 0 \pmod{b},$$

where μ is the well-known Möbius function.

Proof We give a combinatorial argument. Consider necklaces with b beads, of which a are black and the rest are white. Then $\binom{b/c}{a/c}$ is the number of necklaces which are taken to themselves under a rotation of period c. So the Möbius sum is the number of such necklaces with no nontrivial periodicity. Rotating such a necklace will produce b distinct necklaces, so the number of such aperiodic necklaces is divisible by b. □

Theorem 37 *Let G be a finite group. Then*

$$\sum_{d|n} \binom{\frac{|G|}{d}}{\frac{n}{d}} m_d(G) \equiv 0 \pmod{|G|}.$$

Proof It follows from applying Burnside's Lemma to the set of all subsets of G consisting of n elements. □

Now, we are in position to prove the following important theorem concerning the number of solutions of a binomial equation in a finite group. For various approaches we refer to [5, 9, 15, 22, 23], and [34].

Theorem 38 (Frobenius) *Let $n \in \mathbb{N} \setminus \{0\}$ and G be a finite group. If $n \mid |G|$, then $n \mid \Theta_n(G)$.*

Proof We have that

$$\Theta_n(G) = \sum_{d|n} m_d(G)$$

and, using the Möbius inversion theorem it follows

$$m_d(G) = \sum_{c|d} \mu\left(\frac{d}{c}\right) \Theta_c(G).$$

Therefore

$$\sum_{d|n} \binom{\frac{|G|}{d}}{\frac{n}{d}} m_d(G) = \sum_{d|n} \binom{\frac{|G|}{d}}{\frac{n}{d}} \left(\sum_{c|d} \mu\left(\frac{d}{c}\right) \Theta_c(G)\right) \equiv 0 \pmod{|G|}.$$

Interchanging summation and putting $f = d/c$, we get

$$\sum_{c|n} \Theta_c(G) \left(\sum_{f|n/c} \mu(f) \binom{\frac{|G|}{ef}}{\frac{n}{ef}}\right) \equiv 0 \pmod{|G|}.$$

Considering in Lemma 4 $a = n/c$ and $b = |G|/c$, the inner sum of the previous formula is divisible by $|G|/c$.

Now we use a strong induction argument. Assume that for every divisor $c < n$ of $|G|$, we have c divides $\Theta_c(G)$. So every summand in the formula above, except for possibly the $c = n$ term, is divisible by $c \cdot |G|/c = |G|$. We deduce that the $c = n$ term is also divisible by $|G|$. But this is equivalent to $\Theta_n(G)\binom{|G|/n}{1} \equiv 0 \pmod{|G|}$, and the conclusion follows. □

Remarks

1) This result was greatly generalized in the paper [18]. In the book [19], it is proved the following generalization: If G is a finite group and C is a conjugacy class of cardinality m, then the number of solutions of the equation $x^n = c$ in G, when c ranges over C, is a multiple of $\gcd(mn, |G|)$.
2) The Frobenius conjecture deals with a special case of Theorem 37. It claims that if n is a divisor of the order of the finite group G and if the number $\Theta_n(G)$ of solutions of $x^n = e$ in G is exactly n, then these solutions form a normal subgroup of G. This has been proved as a consequence of the classification of finite simple groups.

The following consequences of Frobenius Theorem are given in the paper [23].

Corollary 19 *Let $|G| = p_1^{\alpha_1} p_2^{\alpha_2} \cdots p_r^{\alpha_r}$, where $p_1 < P_2 < \ldots < p_r$ are primes. If every Sylow p-subgroup of G is cyclic, then a Sylow p_r-subgroup is normal in G (and is thus unique). Moreover, G is solvable. In particular, if $|G|$ is square-free and p is the largest prime divisor of $|G|$, then the Sylow p-subgroup is normal in G and G is solvable.*

Proof We show that $\Theta_d(G) = d$ for every divisor d of $|G|$ that can be written in a particular form, namely $d = p_k^{\beta_k} p_{k+1}^{\alpha_{k+1}} \cdots p_r^{\alpha_r}$, $1 \le k \le r$ and $\beta_k \le \alpha_k$. We proceed by induction on d. For $d = |G|$, the result follows trivially. Assume $d < |G|$ and that the result holds for larger divisors of the given type. Let p be the largest prime divisor of $|G|/d$ and $A = T_{dp}(G) \setminus T_d(G)$. As a Sylow p-subgroup is cyclic, $A \neq \emptyset$. By our assumption, $T_{dp}(G) = dp$ and by the Frobenius theorem, $|T_d(G)| = dt$ for some $1 < t < p$. It follows, $p - 1$ divides $dp - dt = d(p - t)$. As every prime divisor of d is greater than or equal to p, $\gcd(p - 1, d) = 1$ and so $p - l \mid p - t$, implying that $t = 1$. Thus $|T_d(G)| = d$ and, in particular, $T_{p_r^{\alpha_r}}(G) = p_r^{\alpha_r}$ implying that a Sylow p_r-subgroup N is normal. Now by induction on the size of the group, N and G/N are solvable and thus G is solvable. □

Corollary 20 *Let n be a positive integer. Then a group of order n is cyclic if and only if* $\gcd(n, \varphi(n)) = 1$.

Proof One can easily check that $\gcd(n, \varphi(n)) = 1$ if and only if n is square-free and $p \nmid q - 1$, where p and q are prime divisors of n.

"$\Rightarrow$" For each n with $\gcd(n, \varphi(n)) \neq 1$, we will construct a noncyclic group of order n. If $p^2 \mid n$, for some prime p, then $\mathbb{Z}/p\mathbb{Z} \times \mathbb{Z}/q\mathbb{Z} \times \mathbb{Z}/\frac{n}{pq}\mathbb{Z}$ is a noncyclic group of order n. Now, suppose n is square-free and $p < q$ are two prime divisors of n such that $p \mid q - 1$. As $\mathbb{Z}/q\mathbb{Z} \setminus \{0\}$ is group under multiplication modulo q and $p \nmid q-1$, there exists a subgroup say H, of order p. Define the operation on $\mathbb{Z}/q\mathbb{Z} \times H$ by $(x,h)(y,k) = (x+hy, hk)$. Then $G = \mathbb{Z}/q\mathbb{Z} \times H$ is a group with identity $(0,1)$ in which $(x,h)^{-1} = (-h^{-1}x, h^{-1})$. Notice that if $h \neq 1$, then $(1,h)(1,1) \neq (1,1)(1,h)$ showing that $G = \mathbb{Z}/q\mathbb{Z} \times H$ is nonabelian. Thus $G \times \mathbb{Z}/pq\mathbb{Z}$ is a nonabelian group of order n.

"$\Leftarrow$" We show that $\Theta_d(G) = d$ for every divisor d of $|G|$. We proceed by induction on d. For $d = |G|$, the result follows trivially. Assume $d < |G|$ and that the result holds for all divisors greater than d. Let p be any prime divisor of $|G|/d$ and $A = T_{dp}(G) \setminus T_d(G)$. Clearly $A \neq \emptyset$. By our assumption, $|T_{dp}(G)| = dp$ and by the theorem, $|T_d(G)| = dt$ for some $1 < t < p$. Arguing just as in Application 1, we see that $t = 1$ and so $|T_d(G)| = d$. In particular, $|T_p(G)| = p$ for every prime divisor of $|G|$, which implies that every Sylow p-subgroup is normal. Thus G, being direct product of its cyclic Sylow p-subgroups of coprime order, is cyclic. □

Corollary 21 *For any prime p and any natural number $n > p$, we have*

$$\sum_{k=1}^{t} \frac{n!}{p^k (n-kp)!k!} \equiv -1 \pmod{p},$$

where t is the largest positive integer such that $tp < n$.

Proof As $T_p(S_n)$ contains only those elements that are products of p-cycles and 1-cycles (fixed points), then by Theorem 35, we have

$$\Theta_n(G) = 1 + \sum_{k=1}^{t} \frac{n!}{p^k (n-kp)!k!},$$

where the summand counts those permutations that are the product of k p-cycles and $n-kp$ fixed points, and the initial 1 counts the identity permutation. Thus, the result follows from the Frobenius theorem. □

Note that by putting $n = p$ in Corollary 22, we get Wilson's Theorem, that is, $(p-1)! \equiv -1 \pmod{p}$, for any prime p.

10 The Order-Counting Sequence

In what follows, G will be a finite group. We will consider the "order-counting" sequence of non-negative integers $(m_n(G))_{n \geq 1}$. Remark that for every such G, $m_1(G) = 1$ and $m_n(G) = 0$ for all but finitely many n. It is not hard to show that if G is abelian, then the sequence $(m_n(G))_{n \geq 1}$ is a fundamental invariant for G. This means that $(m_n(G))_{n \geq 1}$ characterizes the group up to isomorphism.

In what follows, we present some examples of non-isomorphic finite groups which have the same order-counting sequence. The smallest order of a pair consisting of such groups is 16.

Example 1 Let $G_1 = C_4 \times C_4$, the direct product of cyclic group of order 4 and $G_2 = C_4 \rtimes C_4$, the nontrivial semidirect product. A presentation for $G_2 = \langle x, y \mid x^4 = y^4 = 4, xyx^{-1} = y^3 \rangle$. It is easy to see that G_1 and G_2 have the same number of elements of a given order, but they are not isomorphic as G_1 is abelian and G_2 is not.

Example 2 One can construct infinitely many such examples of pairs of non-isomorphic groups. To emphasize this, let $p \geq 3$ be a prime number. Consider $G_1 = C_p \times C_p \times C_p$, where C_p is the cyclic group of order p. Now, let G_2 be the 3×3 Heisenberg matrix group $H(\mathbb{F}_p)$ over $\mathbb{F}_p$. We have that

$$G_2 = \left\{ \begin{pmatrix} 1 & a & b \\ 0 & 1 & c \\ 0 & 0 & 1 \end{pmatrix} : a, b, c \in \mathbb{F}_p \right\}.$$

Both groups have exponent p, hence we have $m_1(G_1) = m_1(G_2) = 1$, $m_p(G_1) = m_p(G_2) = p^3 - 1$, and $m_n(G_1) = m_n(G_2) = 0$ for every other value of n. On the other hand, G_1 is abelian an G_2 is not.

The following table shows the nonzero elements in the sequence $(m_n(G))_{n \geq 1}$, when G is one the symmetric groups S_n, for $2 \leq n \leq 8$.

$G \backslash m_i(G)$	1	2	3	4	5	6	7	8	10	12	15
S_2	1	1	–	–	–	–	–	–	–	–	–
S_3	1	3	2	–	–	–	–	–	–	–	–
S_4	1	9	8	6	–	–	–	–	–	–	–
S_5	1	25	20	30	24	20	–	–	–	–	–
S_6	1	75	80	180	144	240	–	–	–	–	–
S_7	1	231	350	840	504	1470	720	–	504	420	–
S_8	1	763	1232	5460	1344	10640	5760	5040	4032	3360	2688

This table was produced by Jim Belk in an answer on the forum **Mathstackexchange** [4] and represents entry A057731 at OEIS [29].

10.1 The m_k and Θ_k Invariances for Finite Abelian Groups

In this subsection we show that the sequences $\{m_k(G)\}_{k \geq 1}$ and $\{\Theta_k(G)\}_{k \geq 1}$ are fundamental invariants for finite abelian groups G. It is easy to see that if two groups G_1 and G_2 are such that $\{m_k(G_1)\}_{k \geq 1}$ and $\{m_k(G_2)\}_{k \geq 1}$ are identical, then so are $\{\Theta_k(G_1)\}_{k \geq 1}$ and $\{\Theta_k(G_2)\}_{k \geq 1}$.

Theorem 39 *Let G_1, G_2 be finite abelian groups. Then*

1) $\Theta_k(G_1) = \Theta_k(G_2)$, for every $k \in \mathbb{N}$, implies $G_1 \simeq G_2$.
2) $m_k(G_1) = m_k(G_2)$ for every $k \in \mathbb{N}$, implies that $G_1 \simeq G_2$.

Proof Since every abelian group is a direct product of its Sylow subgroups, it follows that it is enough to consider the case in which G_1 and G_2 are p-groups, for some prime number p.

It follows that $G_1 \simeq (\mathbb{Z}/p\mathbb{Z})^{a_1} \times (\mathbb{Z}/p^2\mathbb{Z})^{a_2} \times \cdots \times (\mathbb{Z}/p^r\mathbb{Z})^{a_r}$, for some non-negative integers $a_1, a_2, \dots, a_r$ and $G_2 \simeq (\mathbb{Z}/p\mathbb{Z})^{b_1} \times (\mathbb{Z}/p^2\mathbb{Z})^{b_2} \times \cdots \times (\mathbb{Z}/p^r\mathbb{Z})^{b_r}$, for some non-negative integers $b_1, b_2 \dots, b_r$.

We have that $\Theta_{p^k}(\mathbb{Z}/p^s\mathbb{Z}) = p^{\min(k,s)}$, for every positive integers k, s. At the same time, our hypothesis implies that $\Theta_{p^k}(G_1) = \Theta_{p^k}(G_2)$, for every non-negative integer k. It follows that

$$\sum_{i=1}^{r} a_i \min(i,k) = \sum_{i=1}^{r} b_i \min(i,k),$$

for every $k \in \mathbb{N} \setminus \{0\}$. Hence $a_i = b_i$ for every $i \in \{1, 2, \dots, r\}$. The second affirmation follows easily from 1). □

11 The Exponent of the Group $\mathrm{GL}_n(R)$

We start by giving a few general results concerning the exponent of groups of matrices defined over a ring R. The proofs of the next theorem follow immediately from the definition of the exponent.

Theorem 40 *The following holds:*

1) If R is a ring with $\mu(U(R)), \mu(U(M_n(R))) < \infty$, then $\mu(U(R)) \mid \mu(U(M_n(R)))$.
2) Let R_1, R_2 be rings and let $f : R_1 \to R_2$ be a surjective homomorphism. If $\mu(U(R_1)), \mu(U(R_2)) < \infty$ and $\mu(U(M_n(R_1))), \mu(U(M_n(R_2))) < \infty$, then

   i) $\mu(U(R_2)) \mid \mu(U(R_1))$;
   ii) $\mu(U(M_n(R_2))) \mid \mu(U(M_n(R_1)))$.

Corollary 22 *If $m_1 \mid m_2$, then*

$$\mu(U(M_n(\mathbb{Z}/m_1\mathbb{Z}))) \mid \mu(U(M_n(\mathbb{Z}/m_2\mathbb{Z}))).$$

Theorem 41 *If $\{R_i\}_{i \in I}$ is a family of rings, then*

$$M_n\left(\prod_{i \in I} R_i\right) \simeq \prod_{i \in I} M_n(R_i)$$

as rings.

In what follows we denote by $\mathrm{GL}_n(R)$ the group of units of the ring $M_n(R)$. The following result motivates the computation of the exponent of $\mathrm{GL}_n(R)$, where R is a finite ring.

Theorem 42 *For every integer $n, k \geq 2$, the following relation holds:*

$$\mu(\mathrm{GL}_n(\mathbb{F}_{p^k})) = p^{\lceil \log_p(n) \rceil} \prod_{i=1}^{n} \Phi_i(p^k),$$

where Φ_i is the i-th cyclotomic polynomial.

A proof of this result is given by Marc Van-Leeuwen in his beautifully written answer on **Mathstackexchange** [37]. In this situation, one could find such a formula, because one knows the order of a matrix from its characteristic and minimal polynomials, via the theory of Jordan Canonical Forms. This theory is not available if one replaces the field $\mathbb{F}_{p^k}$ with a finite ring.

If $n = 2$ in the theorem above, we have the following closed formula

$$\mu(\mathrm{GL}_2(\mathbb{F}_{p^k})) = p(p^{2k} - 1).$$

11.1 The Group $\mathrm{GL}_n(\mathbb{Z}/m\mathbb{Z})$

In this subsection we derive some results concerning the exponent of $\mathrm{GL}_n(\mathbb{Z}/m\mathbb{Z})$. In this respect, we have the following results.

The order of the group $\mathrm{GL}_n(\mathbb{Z}/m\mathbb{Z})$ can be expressed in terms of the Jordan totient functions J_k, as follows

$$|\mathrm{GL}_n(\mathbb{Z}/m\mathbb{Z})| = m^{\frac{n(n-1)}{2}} \prod_{k=1}^{n} J_k(m).$$

Also, we have

$$|\mathrm{SL}_n(\mathbb{Z}/m\mathbb{Z})| = m^{\frac{n(n-1)}{2}} \prod_{k=2}^{n} J_k(m)$$

and

$$\mathrm{Sp}_{2n}(\mathbb{Z}/m\mathbb{Z}) = m^{n^2} \prod_{k=1}^{n} J_{2k}(m),$$

where $\mathrm{SL}_n(\mathbb{Z}/m\mathbb{Z})$ and $\mathrm{Sp}_{2n}(\mathbb{Z}/m\mathbb{Z})$ are the special linear group of degree n and the symplectic group of degree $2n$ with elements in the ring $\mathbb{Z}/m\mathbb{Z}$.

If $m = p_1^{\alpha_1} \dots p_r^{\alpha_r}$ is the prime factorization of m, then it is well-known that (see, for instance, [2] or [31])

$$J_k(m) = m^k \prod_{j=1}^{r} \left(1 - \frac{1}{p_i^k}\right), \ k = 1, 2, \dots \tag{10}$$

Clearly, we have $J_1 = \varphi$, the classical Euler's totient function.

In particular, if $m = p^j$ is a prime power, then

$$\left|\mathrm{GL}_n(\mathbb{Z}/p^j\mathbb{Z})\right| = p^{\frac{jn(n-1)}{2}+3j-3} \cdot (p-1) \cdot \left(p^2 - 1\right) \tag{11}$$

and

$$\left|\mathrm{SL}_n(\mathbb{Z}p^j/\mathbb{Z})\right| = p^{\frac{jn(n-1)}{2}+2j-2} \cdot \left(p^2 - 1\right). \tag{12}$$

Theorem 43 *If $m = p_1^{\alpha_1} \cdot p_r^{\alpha_r}$ is the prime factorization of the positive integer m, then*

1) $M_n(\mathbb{Z}/m\mathbb{Z}) \simeq \prod_{i=1}^r M_n(\mathbb{Z}/p_i^{\alpha_i}\mathbb{Z})$,
2) $\mu(GL_n(\mathbb{Z}/m\mathbb{Z})) = \mathrm{lcm}\{\mu(M_n(\mathbb{Z}/p_i^{\alpha_i}\mathbb{Z})) : 1 \le i \le r\}$.

Let us denote by $UT_n(\mathbb{Z}/m\mathbb{Z})$ the subgroup of $\mathrm{GL}_n(\mathbb{Z}/m\mathbb{Z})$ consisting of the upper-triangular matrices.

Theorem 44 *For $A \in UT_n(\mathbb{Z}/m\mathbb{Z})$, we have*

$$A^{\alpha_m \beta_{m,n}} = I_n,$$

where $\alpha_m = \mu(U(\mathbb{Z}/m\mathbb{Z}))$ and $\beta_{m,n}$ is the smallest positive integer $k \ge n$ with $m | \binom{k}{j}$, for every $j \in \{1, 2, \dots, n-1\}$.

Proof We have $A^{\alpha_m} = I_n + B$ with $B^n = O_n$. Therefore,

$$A^{k\alpha_m} = I_n + \binom{k}{1} B + \binom{k}{2} B^2 + \dots + \binom{k}{n-1} B^{n-1},$$

hence $A^{\alpha_m \beta_{m,n}} = I_n$. □

Corollary 23 $\mu(UT_n(\mathbb{Z}/m\mathbb{Z})) \mid \alpha_m \beta_{m,n}$.

11.2 The Exponent of $\mathrm{SL}_2(\mathbb{Z}/2^n\mathbb{Z})$ and $\mathrm{GL}_2(\mathbb{Z}/2^n\mathbb{Z})$

Applying formulas (11) and (12) we get

$$\left|\mathrm{GL}_2(\mathbb{Z}/2^n\mathbb{Z})\right| = 3 \cdot 2^{4n-3}$$

and

$$\left|\mathrm{SL}_2(\mathbb{Z}/2^n\mathbb{Z})\right| = 3 \cdot 2^{3n-2}.$$

The following result can be found, for instance, in the book [1].

Theorem 45 (Kummer) *Let n and i be positive integers with $i \le n$, and let p be a prime. Then the largest non-negative exponent t for which p^t divides $\binom{n}{i}$ is equal to the number of carries in the base p addition of $n-i$ and i.*

Using the above, we prove that the following formulae hold.

Theorem 46 *For every $n \ge 2$, we have*

$$\mu(\mathrm{SL}_2(\mathbb{Z}/2^n\mathbb{Z})) = \mu(\mathrm{GL}_2(\mathbb{Z}/2^n\mathbb{Z})) = 3 \cdot 2^n.$$

Proof Let $A \in \mathrm{GL}_2(\mathbb{Z}/2^n\mathbb{Z})$. By the Cayley–Hamilton theorem, which holds for matrices over arbitrary commutative rings, we know that

$$A^2 = \mathrm{Tr}(A) \cdot A - \det(A) \cdot I_2,$$

where $\mathrm{Tr}(A) \in \mathbb{Z}/2^n\mathbb{Z}$ and $\det(A) \in U(\mathbb{Z}/2^n\mathbb{Z}) \cong C_2 \times C_{2^{n-2}}$.

Suppose that $\mathrm{Tr}(A)$ is not a unit in $\mathbb{Z}/2^n\mathbb{Z}$. Using Kummer's theorem [1, Theorem 10.2.2] for computing the 2-adic valuation of binomial coefficients, from

$$\begin{aligned} A^{2^n} &= (\mathrm{Tr}(A) \cdot A - \det(A) \cdot I_2)^{2^{n-1}} \\ &= \sum_{k=0}^{2^{n-1}} (-1)^k \binom{2^{n-1}}{k} \det(A)^k \, \mathrm{Tr}(A)^{2^{n-1}-k} \cdot A^{2^{n-1}-k}, \end{aligned}$$

we see that all but the last term in the expansion are equal to zero due to the characteristic of the ring.

We saw above that $\mu(U(\mathbb{Z}/2^n\mathbb{Z})) = 2^{n-2}$, hence $A^{2^n} = (-\det(A) \cdot I_2)^{2^{n-1}} = I_2$.

If $\mathrm{Tr}(A) \in U(\mathbb{Z}/2^n\mathbb{Z})$, then we observe that

$$A^3 = Tr(A) \cdot A^2 - \det(A) \cdot A = \left(\mathrm{Tr}(A)^2 - \det(A)\right) \cdot A - \mathrm{Tr}(A) \cdot \det(A) \cdot I_2.$$

Since $\mathrm{Tr}(A) \cdot \det(A) \in U(\mathbb{Z}/2^n\mathbb{Z})$ and $\mathrm{Tr}(A)^2 - \det(A)$ is not a unit in $\mathbb{Z}/2^n\mathbb{Z}$, performing a computation analogous to the one above, one easily derives that $A^{3\cdot 2^{n-1}} = I_2$.

The matrix $\begin{pmatrix} 1 & 1 \\ 0 & 1 \end{pmatrix}$ has order 2^n in $\mathrm{GL}_2(\mathbb{Z}/2^n\mathbb{Z})$. On the other hand, since 3 divides $|\mathrm{GL}_2(\mathbb{Z}/2^n\mathbb{Z})|$, we know by Cauchy's theorem that there is $B \in \mathrm{GL}_2(\mathbb{Z}/2^n\mathbb{Z})$ of order 3. Together with the above, this implies that $\mu(\mathrm{GL}_2(\mathbb{Z}/2^n\mathbb{Z})) = 3 \cdot 2^n$.

We know that $\mu(\mathrm{SL}_2(\mathbb{Z}/2^n\mathbb{Z})) \mid \mu(\mathrm{GL}_2(\mathbb{Z}/2^n\mathbb{Z})) = 3 \cdot 2^n$. But the order of $\mathrm{SL}_2(\mathbb{Z}/2^n\mathbb{Z})$ is divisible by 3, hence by Cauchy's theorem there exists $B \in \mathrm{SL}_2(\mathbb{Z}/2^n\mathbb{Z})$ of order 3. Additionally, the matrix $\begin{pmatrix} 1 & 1 \\ 0 & 1 \end{pmatrix}$ has order 2^n and belongs to $\mathrm{SL}_2(\mathbb{Z}/2^n\mathbb{Z})$. The theorem is hence proved. □

11.3 The Groups $\mathrm{SL}_2(\mathbb{Z}/3^n\mathbb{Z})$ *and* $\mathrm{GL}_2(\mathbb{Z}/3^n\mathbb{Z})$

Using formulas (11) and (12) we obtain

$$\left|\mathrm{GL}_2(\mathbb{Z}/3^n\mathbb{Z})\right| = 16 \cdot 3^{4n-3}$$

and

$$\left|\mathrm{SL}_2(\mathbb{Z}/3^n\mathbb{Z})\right| = 8 \cdot 3^{3n-2}.$$

We first prove the following theorem.

Theorem 47 *For every $n \geq 2$, we have*

$$\mu(\mathrm{SL}_2(\mathbb{Z}/3^n\mathbb{Z})) = 2^2 \cdot 3^n$$

and

$$\mu(\mathrm{GL}_2(\mathbb{Z}/3^n\mathbb{Z})) = 2^3 \cdot 3^n.$$

Proof Let $A \in \mathrm{SL}_2(\mathbb{Z}/3^n\mathbb{Z})$. Suppose that $\mathrm{Tr}(A)$ is not a unit in $\mathbb{Z}/3^n\mathbb{Z}$. Since

$$A^2 = \mathrm{Tr}(A) \cdot A - I_2,$$

we have

$$A^{2\cdot 3^{n-1}} = (\mathrm{Tr}(A) \cdot A - I_2)^{3^{n-1}}$$
$$= \sum_{k=0}^{3^{n-1}} (-1)^k \binom{3^{n-1}}{k} \det(A)^k \,\mathrm{Tr}(A)^{3^{n-1}-k} \cdot A^{3^{n-1}-k}.$$

Using again Kummer's theorem for computing 3-adic valuations of binomial coefficients, we see that all but the last term in the expansion above are zero in $\mathbb{Z}/3^n\mathbb{Z}$. One has $A^{2\cdot 3^{n-1}} = -I_2$, hence $A^{4\cdot 3^{n-1}} = I_2$.

If $\mathrm{Tr}(A) \in U(\mathbb{Z}/3^n\mathbb{Z})$, then

$$A^3 = \left(\mathrm{Tr}^2(A) - 1\right) \cdot A - \mathrm{Tr}(A) \cdot I_2.$$

We observe that $\mathrm{Tr}^2(A) - 1$ is not a unit in $\mathbb{Z}/3^n\mathbb{Z}$, therefore by a computation analogous to the one above we see that $A^{3^n} = (-\mathrm{Tr}(A))^{3^{n-1}}$. It is known that $U(\mathbb{Z}/3^n\mathbb{Z}) \cong C_{2\cdot 3^{n-1}}$, hence $A^{2\cdot 3^n} = I_2$.

The matrices $\begin{pmatrix} 1 & 1 \\ 0 & 1 \end{pmatrix}$ and $\begin{pmatrix} 0 & 1 \\ -1 & 0 \end{pmatrix}$ have orders 3^n and 4, respectively. Together with the above, this implies that $\mu(\mathrm{SL}_2(\mathbb{Z}/3^n\mathbb{Z})) = 2^2 \cdot 3^n$, as claimed.

Let us now consider $A \in \mathrm{GL}_2(\mathbb{Z}/3^n\mathbb{Z})$. If $\mathrm{Tr}(A)$ is not a unit, i.e. $\mathrm{Tr}(A) \in 3 \cdot \mathbb{Z}/3^n\mathbb{Z}$, then by a computation identical to the above, we can prove that $A^{2\cdot 3^{n-1}} = I_2$.

Suppose now that $\mathrm{Tr}(A) \in U(\mathbb{Z}/3^n\mathbb{Z})$. We have

$$A^3 = (\mathrm{Tr}^2(A) - \det(A)) \cdot A - \det(A) \cdot I_2.$$

One can see that if $\det(A) \in 1 + 3 \cdot \mathbb{Z}/3^n\mathbb{Z}$, then as previously, by raising everything to the power 3^{n-1}, we get that $A^{3^n} = (-\det(A))^{3^{n-1}} \cdot I_2$, so $A^{2\cdot 3^n} = I_2$.

On the other hand, if $\det(A) \in 2 + 3 \cdot \mathbb{Z}/3^n\mathbb{Z}$, then by considering successive powers of A, we observe that

$$A^{24} = x \cdot A + y \cdot I_2,$$

where $x \in 3 \cdot \mathbb{Z}/3^n\mathbb{Z}$ and $y \in U(\mathbb{Z}/3^n\mathbb{Z})$. As before, we conclude that

$$A^{8\cdot 3^n} = A^{24\cdot 3^{n-1}} = y^{3^{n-1}} \cdot I_2.$$

As $U(\mathbb{Z}/3^n\mathbb{Z}) \cong C_{2\cdot 3^{n-1}}$, it follows that $A^{16\cdot 3^n} = I_2$.

Using the first listed property of the exponent, by what we proved above we know that

$$2^2 \cdot 3^n = \mu(\mathrm{SL}_2(\mathbb{Z}/3^n\mathbb{Z})) \mid \mu(\mathrm{GL}_2(\mathbb{Z}/3^n\mathbb{Z})) | 16 \cdot 3^n.$$

It can be easily verified that the order of $\begin{pmatrix} 1 & 1 \\ 1 & 0 \end{pmatrix}$ in $\mathrm{GL}_2(\mathbb{Z}/3^2\mathbb{Z})$ is 24, therefore for every $n \geq 2$ the order of $\begin{pmatrix} 1 & 1 \\ 1 & 0 \end{pmatrix}$ in $\mathrm{GL}_2(\mathbb{Z}/3^n\mathbb{Z})$ must be a multiple of 24.

From the above, it follows that

$$2^3 \cdot 3^n \mid \mu(\mathrm{GL}_2(\mathbb{Z}/3^n\mathbb{Z})) \mid 2^4 \cdot 3^n. \tag{13}$$

We have the short exact sequence

$$1 \longrightarrow \mathrm{SL}_2(\mathbb{Z}/3^n\mathbb{Z}) \longrightarrow \mathrm{GL}_2(\mathbb{Z}/3^n\mathbb{Z}) \longrightarrow \mathrm{GL}_2(\mathbb{Z}/3^n\mathbb{Z})/\,\mathrm{SL}_2(\mathbb{Z}/3^n\mathbb{Z}) \longrightarrow 1.$$

It is easy to see that $\mathrm{GL}_2(\mathbb{Z}/3^n\mathbb{Z})/\,\mathrm{SL}_2(\mathbb{Z}/3^n\mathbb{Z}) \cong U(\mathbb{Z}/3^n\mathbb{Z}) \cong C_{2\cdot 3^{n-1}}$. Using part 3) of Theorem 2, we get that

$$\mu(G) \mid \mu(C_{2\cdot 3^{n-1}}) \cdot \mu(\mathrm{SL}_2(\mathbb{Z}/3^n\mathbb{Z})) = 2^3 \cdot 3^{2n-1}. \tag{14}$$

It follows from (13) and (14) that $\mu(\mathrm{GL}_2(\mathbb{Z}/3^n\mathbb{Z})) = 2^3 \cdot 3^n$, as desired. □

References

1. T. Andreescu, D. Andrica, *Number Theory. Structures, Examples, and Problems* (Birkhauser Verlag, Boston, 2009)
2. D. Andrica, M. Piticari, On some extensions of Jordan's arithmetic functions, in *Proceedings of the International Conference on Theory and Applications of Mathematics and Informatics – ICTAMI 2003*, Alba Iulia, vol. 7. Acta Univ. Apulensis Math. Inform. (2004), pp. 13–22
3. D. Andrica, S. Rădulescu, G. C. Ţurcaş, On the automorphism group of direct products of groups with finite exponent, Am. Math. Mon., to appear
4. J. Belk, *Finding the number of elements of particular order in the symmetric group*, URL (version: 2015-06-26): https://math.stackexchange.com/q/1340368
5. E.A. Bender, J.R. Goldman, On the applications of Möbius inversion in combinatorial analysis. Am. Math. Mon. **82**(8), 789–803 (1975)
6. J.N.S. Bidwell, M.J. Curran, Automorphisms of finite Abelian groups. Math. Proc. R. Ir. Acad. **110A**(1), 57–71 (2010)
7. J.N.S. Bidwell, M.J. Curran, D.J. McCaughan, Automorphisms of direct products of finite groups. Arch. Math. **86**, 481–489 (2006)
8. K.P. Bogart, An obvious proof of Burnside's lemma. Am. Math. Mon. **98**(10), 927–928 (2018)
9. R. Brauer, On a theorem of Frobenius. Am. Math. Mon. **76**(1), 12–15 (1969)
10. J.W.S. Cassels, A. Fröhlich, *Algebraic Number Theory* (London Mathematical Society, London, 2010)
11. S. Chowla, I. N. Herstein and W. K. Moore, On recursions connected with symmetric groups I, Can. J. Math. **3**, 328–334 (1951)
12. M.J. Curran, Automorphisms of products of finite groups, in *Groups St Andrews 2009 in Bath*, vol. 1, ed. by C.M. Campbell. London Mathematical Society Lecture Notes Series 387 (Cambridge University Press, Cambridge, 2011), pp. 213–223
13. M. Deaconescu, G.L. Walls, On the group of automorphisms of a group. Am. Math. Mon. **118**(5), 452–455 (2011)

14. M. Deaconescu, G.L. Walls, Remarks on commutators in finite groups. J. Reine Angew. Math. **2017**(732), 247–253 (2017)
15. H. Finkelstein, Solving equations in groups. Period. Math. Hungar. **9**, 187–204 (1978)
16. R. Freud, P.P. Pálfy, On the possible number of elements of given order in a finite group. Isr. J. Math. **93**, 345–358 (1996)
17. D. Gupta, V.D. Mazurov, On groups with small orders of elements. Bull. Aust. Math. Soc. **60**, 197–205 (1999)
18. P. Hall, On a theorem of Frobenius. Proc. Lond. Math. Soc. **7**(3), 1–42 (1956)
19. M. Hall, *The Theory of Groups* (Macmillan, New York, 1959)
20. P. Hall, *The Edmonton Notes on Nilpotent Groups*. Queen Mary College Mathematics Notes (Queen Mary College Department of Mathematics, London, 1969)
21. C.J. Hillar, D.L. Rhea, Automorphisms of finite abelian groups. Am. Math. Mon. **114**(10), 917–923 (2007)
22. I.M. Isaacs, G.R. Robinson, On a theorem of Frobenius: solutions of $x^n = 1$ in finite groups. Am. Math. Mon. **99**(4), 352–354 (1992)
23. D. Khurana, A. Khurana, A theorem of Frobenius and its applications. Math. Mag. **78**(3), 220–225 (2005)
24. D. MacHale, Power mappings and group morphisms. Proc. R. Ir. Acad. Sect. A: Math. Phys. Sci. **74**, 91–93 (1974)
25. J.-P. Massias, J.-L. Nicolas, G. Robin, Evaluation asymptotique de l'ordre maximum d'un él ement du groupe symétrique. Acta Arith. **50**, 221–242 (1988)
26. D.V. Mazurov, Characterizations of finite groups by sets of orders of their elements. Algebra Logika **36**(1), 23–32 (1997)
27. W. Miller, The maximum order of an element of a finite symmetric group. Am. Math. Mon. **94**, 497–506 (1987)
28. T. Müller, Finite group actions and asymptotic expansion of $e^{P(z)}$. Combinatorica, **17**(4), 523–554 (1997)
29. OEIS, The online encyclopedia of integer sequences (2018). Published electronically at http://oeis.org
30. I.B.S. Passi, M. Singh, M.K. Yadav, *Automorphisms of Finite Groups* (Springer, Berlin, 2018)
31. S. Rădulescu, M. Rădulescu, Positive-semidefinite matrices and the Jordan totient function. Ann. Univ. Craiova Math. Comput. Sci. Ser. **43**(1), 82–87 (2016)
32. J.J. Rotman, *An Introduction to the Theory of Groups*, 4th edn. (Springer, New York, 1995)
33. I.E. Segal, The automorphisms of the symmetric group. Bull. Am. Math. Soc. **46**(6), 565 (1940)
34. D.E. Speyer, A counting proof of a theorem of Frobenius. Am. Math. Mon. **124**(4), 357–359 (2017)
35. M. Suzuki, *Group Theory I* (Springer, Berlin, 1982)
36. A.V. Vasil'ev, M.A. Grechkoseeva, V.D. Mazurov, Characterization of the finite simple groups by spectrum and order. Algebra Logika **48**(6), 685–728 (2009)
37. M. van Leeuwen, *Exponent of* $GL(n, q)$, URL (version: 2017-04-13): https://math.stackexchange.com/q/294524
38. A.V. Zavarnitsin, V.D. Mazurov, Element orders in coverings of symmetric and alternating groups. Algebra Logika **38**(3), 159–170 (1999)

Hankel Tournaments and Special Oriented Graphs

Richard A. Brualdi and Lei Cao

Abstract A Hankel tournament T of order n (an $n \times n$ Hankel tournament matrix $T = [t_{ij}]$) is a tournament such that $i \to j$ an edge implies $(n+1-j) \to (n+1-i)$ is also an edge ($t_{ij} = t_{n+1-j,n+1-i}$) for all i and j. Hankel tournament matrices are $(0, 1)$-matrices which are combinatorially antisymmetric about the main diagonal and symmetric about the Hankel diagonal (the antidiagonal). Locally transitive tournaments are tournaments such that the in-neighborhood and the out-neighborhood of each vertex are transitive. Tournaments form a special class of oriented graphs. The score vectors of Hankel tournaments and of locally transitive tournaments have been characterized where each score vector of a locally transitive tournament is also a score vector of a Hankel tournament. In this paper we continue investigations into Hankel tournaments and locally transitive tournaments. We investigate Hankel cycles in Hankel tournaments and show in particular that a strongly connected Hankel tournament contains a Hankel Hamilton cycle and, in fact, is Hankel "even-pancyclic" or Hankel "odd-pancyclic." We show that a Hankel score vector can be achieved by a Hankel "half-transitive" tournament, extending the corresponding result for score vectors of tournaments. We also consider some results on oriented graphs and the question of attainability of prescribed degrees by oriented graphs. Finally, we extend some results on 2-tournaments to Hankel 2-tournaments. In some instances we rely on the reader to extend arguments already in the literature. We illustrate our investigations with many examples.

Mathematics Subject Classifications 05C07, 05C20, 05C38, 05C50, 15B05

R. A. Brualdi (✉)
Department of Mathematics, University of Wisconsin, Madison, WI, USA
e-mail: brualdi@math.wisc.edu

L. Cao
Department of Mathematics, Halmos College of Arts and Natural Sciences, Nova Southeastern University, Fort Lauderdale, FL, USA
e-mail: lcao@nova.edu

A. M. Raigorodskii, M. Th. Rassias (eds.), *Discrete Mathematics and Applications*, Springer Optimization and Its Applications 165,
https://doi.org/10.1007/978-3-030-55857-4_5

1 Introduction

A *tournament* of order n is a directed graph obtained by giving a direction to each of the $\binom{n}{2}$ edges of the complete graph K_n of order n. With a labeling of the vertices using the integers $1, 2 \ldots, n$, associated with a tournament there is a *tournament matrix*, that is, an $n \times n$ (0, 1)-matrix $T = [t_{ij}]$ where $t_{ij} = 1$ if and only if there is a directed edge from vertex i to vertex j $(1 \leq i, j \leq n)$. Thus $t_{ii} = 0$ for all i, and for all $i \neq j, t_{ij}+t_{ji} = 1$. The tournament matrix T satisfies $T+T^t = J_n - I_n$ where J_n is the $n \times n$ matrix of all 1's and the superscript t denotes transpose. In general, we do not distinguish between a tournament as a directed graph with vertices labeled as $1, 2, \ldots, n$ (a tournament of order n) and its associated tournament matrix (an $n \times n$ tournament). We move freely back and forth between these two interpretations.

Let $A = [a_{ij}]$ be an $n \times n$ matrix. Recall that the *main diagonal* of A is the set of positions $\{(i, i) : 1 \leq i \leq n\}$. The *Hankel diagonal*[1] of A is the set of positions $\{(i, n+1-i) : 1 \leq i \leq n\}$ running from the lower left of A to the upper right. The *Hankel transpose* of A, denoted as $A^h = [a^h_{ij}]$, is the matrix obtained by reflecting over the Hankel diagonal. Thus $a^h_{ij} = a_{n+1-j,n+1-i}$ for all i and j. In analogy with a symmetric matrix, the matrix A is called *Hankel symmetric* provided $A^h = A$, that is, $a_{ij} = a_{n+1-j,n+1-i}$ for all i and j. Note that if A is a Hankel symmetric matrix, then $a_{ji} = a_{n+1-i,n+1-j}$ and hence the transpose A^t is also Hankel symmetric. Similarly, the Hankel transpose of a symmetric matrix is symmetric. Since the inverse of a permutation matrix is its transpose, it follows that the inverse of a Hankel symmetric permutation matrix is also Hankel symmetric. Just as the product of two symmetric matrices need not be symmetric, the product of two Hankel symmetric matrices need not be Hankel symmetric. For instance,

$$\left[\begin{array}{c|c|c|c} & 1 & & \\ \hline & & 1 & \\ \hline & & & 1 \\ \hline 1 & & & \end{array}\right] \cdot \left[\begin{array}{c|c|c|c} & & & 1 \\ \hline & & 1 & \\ \hline & 1 & & \\ \hline 1 & & & \end{array}\right] = \left[\begin{array}{c|c|c|c} & & 1 & \\ \hline & 1 & & \\ \hline 1 & & & \\ \hline & & & 1 \end{array}\right],$$

where, here and elsewhere, unspecified entries are to be interpreted as 0's.

In general, the following holds: Let $A = [a_{ij}]$ and $B = [b_{ij}]$ be $n \times n$ Hankel symmetric matrices. Then the (i, j)-entry of AB is given by

$$\begin{aligned} (AB)_{ij} &= \sum_k a_{ik} b_{kj} \\ &= \sum_k a_{n+1-k,n+1-i} b_{n+1-j,n+1-k} \\ &= \sum_k b_{n+1-j,n+1-k} a_{n+1-k,n+1-i} \end{aligned}$$

[1] Often called the *antidiagonal*.

$$= \sum_p b_{n+1-j,p} a_{p,n+1-i}$$

$$= (BA)_{n+1-j,n+1-i}.$$

Thus $(AB)^h = BA$ if A and B are Hankel symmetric.

A tournament of order n is called a *Hankel tournament* provided its vertices can be labeled $1, 2, \ldots, n$ in such a way that for all $i, j \in \{1, 2, \ldots, n\}$, $i \to j$ is an edge if and only if $(n+1-j) \to (n+1-i)$ is also an edge. With such a labeling, the adjacency matrix of a Hankel tournament is a Hankel matrix. Thus an $n \times n$ tournament matrix $T = [t_{ij}]$ is a *Hankel tournament matrix* [9] provided it is Hankel symmetric. A Hankel tournament matrix is combinatorially antisymmetric about the main diagonal ($a_{ij} = 0$ if and only if $a_{ji} = 1$ for $i \neq j$) and symmetric about the Hankel diagonal ($a_{ij} = a_{n+1-j,n+1-i}$ for all i and j).

Example 1.1 The following matrices are Hankel tournament matrices:

$$\begin{bmatrix} 0 & 1 & 0 \\ 0 & 0 & 1 \\ 1 & 0 & 0 \end{bmatrix} \text{ and } \begin{bmatrix} 0 & 1 & 1 & 0 \\ 0 & 0 & 0 & 1 \\ 0 & 1 & 0 & 1 \\ 1 & 0 & 0 & 0 \end{bmatrix}.$$

□

The *score vector* of a tournament of order n (with vertices labeled $1, 2, \ldots, n$) is the outdegree sequence of its vertices, and this is the same as the row sum vector $R = (r_1, r_2, \ldots, r_n)$ of its associated $n \times n$ tournament matrix. For an arbitrary tournament, there is no loss of generality in assuming that $r_1 \leq r_2 \leq \cdots \leq r_n$, since this can always be achieved by a relabeling of the vertices, equivalently by a simultaneous permutation of the rows and columns of the associated tournament matrix. The column sum vector $S = (s_1, s_2, \ldots, s_n)$ then satisfies $s_1 \geq s_2 \geq \cdots \geq s_n$.

The following results are known.

1. *Landau's Theorem*: Let $R = (r_1, r_2, \ldots, r_n)$ be a nondecreasing sequence of nonnegative integers. Then R *is the score vector of a tournament if and only if Landau's condition (1) holds*:

$$\sum_{i=1}^{k} r_i \geq \binom{k}{2} \quad (1 \leq k \leq n), \text{ with equality for } k = n. \tag{1}$$

2. A simultaneous permutation of the rows and columns of a Hankel tournament matrix may result in a tournament matrix that is not Hankel. For example, taking the rows and columns of the 4×4 Hankel tournament matrix in Example 1.1 in the order $2, 4, 1, 3$ results in a non-Hankel tournament. Nonetheless, the following important property holds [9]:

   - Given a Hankel tournament matrix, there is a simultaneous permutation of its rows and columns (that is, a relabeling of the vertices of the corresponding tournament of order n) that results also in a Hankel tournament matrix whose score vector is nondecreasing. Thus in investigating Hankel tournament matrices, one can generally assume that the row sum vector is nondecreasing.

In [9] Theorem 1.2 below is proved.

Theorem 1.2 *Let $R = (r_1, r_2, \ldots, r_n)$ be a nondecreasing sequence of nonnegative integers. Then R is the score vector of a Hankel tournament if and only if*

(a) *R is the score vector of a tournament, that is, R satisfies Landau's condition* (1), *and*

(b) $r_i + r_{n+1-i} = n - 1 \quad (1 \leq i \leq n)$. □

That (b) holds for the score vector of a Hankel tournament is a direct consequence of the definition of a tournament and the property of being Hankel. That the score vector R of a tournament which satisfies (b) implies the existence of a Hankel tournament with that score vector is not obvious. In addition, an algorithm is given in [9] for construction of a Hankel tournament with score vector R when R satisfies (a) and (b). This algorithm is adapted in Section 6 for Hankel 2-tournaments.

Example 1.3 An $n \times n$ *regular tournament matrix* is a tournament in which all the scores are equal. In this case, n must be odd with all scores equal to $(n-1)/2$. For each odd integer n, there are regular tournaments T_n^* which are also Hankel. This is illustrated by T_7^* for $n = 7$ below:

$$T_7^* = \left[\begin{array}{c|c|c|c|c|c|c} & 1 & 1 & 1 & & & \\ \hline & & 1 & 1 & 1 & & \\ \hline & & & 1 & 1 & 1 & \\ \hline & & & & 1 & 1 & 1 \\ \hline 1 & & & & & 1 & 1 \\ \hline 1 & 1 & & & & & 1 \\ \hline 1 & 1 & 1 & & & & \end{array}\right].$$

In general, if P_n denotes the $n \times n$ permutation matrix corresponding to the permutation cycle $(2, 3, \ldots, n, 1)$ (thus 1's in positions $(1, 2), (2, 3), \ldots, (n-1, n), (n, 1)$), then for n odd, $T_n^* = P_n + P_n^2 + \cdots + P_n^{\lfloor n/2 \rfloor}$ is a Hankel regular tournament matrix. □

It follows easily from the Perron–Frobenius theory of nonnegative matrices that the regular tournament matrices have the largest spectral radius, namely $(n-1)/2$, among all $n \times n$ tournament matrices. If n is even, then a *near-regular tournament matrix* is one for which the scores are as equal as possible, and so $n/2$ scores are equal to $n/2$ and $n/2$ scores are equal to $(n-2)/2$. For each even integer n, there are near-regular tournaments T_n^* which are Hankel tournaments. This is illustrated by T_8^* for $n = 8$ below and, in general, for n even equals $P_n + P_n^2 + \cdots + P_n^{(n-2)/2} + Q_n$ where Q_n is obtained from $P_n^{n/2}$ by replacing its first $n/2$ 1's with 0's:

$$T_8^* = \left[\begin{array}{c|c|c|c||c|c|c|c} & 1 & 1 & 1 & & & & \\ \hline & & 1 & 1 & 1 & & & \\ \hline & & & 1 & 1 & 1 & & \\ \hline & & & & 1 & 1 & 1 & \\ \hline\hline 1 & & & & & 1 & 1 & 1 \\ \hline 1 & 1 & & & & & 1 & 1 \\ \hline 1 & 1 & 1 & & & & & 1 \\ \hline 1 & 1 & 1 & 1 & & & & \end{array}\right].$$

It had been conjectured [10] that for n even, T_n^* has the largest spectral radius of all $n \times n$ tournament matrices and this was finally proved by Drury [14], who also proved that every tournament matrix with largest spectral radius is permutation similar to the Hankel tournament matrix T_n^*. Thus the maximal spectral radius of $n \times n$ tournament matrices is always achieved by a Hankel tournament matrix. This discussion gives some evidence that, in addition to their structural properties investigated in this paper, Hankel tournaments are of some interest. More evidence is given in the next section. Finally, we remark that what we call "Hankel symmetric" is called "antidiagonally symmetric" in [4] and other places. We prefer to use the term "Hankel symmetric."

We now briefly summarize the contents of this paper. In the next section we discuss some results on locally transitive tournaments and their score vectors, and explore their connection with Hankel tournaments. In Section 3 we investigate cycles in Hankel tournaments that have the Hankel property (Hankel cycles) and show that strongly connected Hankel tournaments always have a Hankel Hamilton cycle, indeed either have Hankel cycles of all even lengths (even-pancyclic) or have Hankel cycles of all odd lengths (odd-pancyclic). In Section 4 we show that given a score vector of a Hankel tournament, there always exists a special half-transitive Hankel tournament with that score vector. In Section 5, we consider, more generally, oriented graphs and their score vectors and establish some existence theorems. We note that the concept of the score vectors of tournaments and oriented graphs is investigated in [20] for another class of digraphs. Finally, in Section 6 we extend some results on 2-tournaments (tournaments that are allowed to result in ties) to Hankel 2-tournaments. Many illustrative examples are given throughout.

2 Locally Transitive Tournaments

A tournament is *locally transitive* (see [5, 6, 13]) provided that the in-neighborhood and the out-neighborhood of every vertex are transitive (so they induce transitive tournaments of smaller orders). As noted in [13], a tournament T is locally transitive if and only if neither the in-neighborhood nor the out-neighborhood of a vertex contains a 3-cycle, equivalently T does not contain a subtournament of order 4 with either of the score sequence $(1, 1, 1, 3)$ and $(0, 2, 2, 2)$. Unlike the Hankel property, the property of being locally transitive does not depend on the labeling of the vertices. Every subtournament of a locally transitive tournament is also locally transitive. As noted in [13], for each odd integer n, up to isomorphism there is a unique locally transitive, regular tournament T_n^* of order n and such a tournament is called a *carousel tournament*. This tournament is the regular Hankel tournament T_n^* (n odd) described in Example 1.3 in the previous section.

For a vertex x of a tournament T, let $T_{\to x}$ and $T_{x\to}$ be the tournaments induced on the in-neighborhood and out-neighborhood of x, respectively. Thus T is locally transitive if and only if, for all vertices x, $T_{\to x}$ and $T_{x\to}$ are transitive tournaments.

Let $k \geq 1$, and let Q_k denote the $k \times k$ *standard transitive tournament matrix* with decreasing score vector $(k-1, \ldots, 1, 0)$; thus Q_k has all 1's above the main diagonal and all 0's elsewhere. Let $S_{k,n-1-k}$ denote a $k \times (n-1-k)$ $(0, 1)$-matrix with a nondecreasing row sum vector and with 1's left-justified. Thus $S_{k,n-1-k}$ has a *staircase pattern*. We call such a matrix $S_{k,n-1-k}$ a *Ferrers matrix*, since such matrices arise as conjugates of partitions of integers.[2] For each such k, let

$$T(S_{k,n-1-k}) = \left[\begin{array}{ccc||c||ccc} & & & 1 & & & \\ & Q_k & & \vdots & & S_{k,n-1-k} & \\ & & & 1 & & & \\ \hline\hline 0 & \cdots & 0 & 0 & 1 & \cdots & 1 \\ \hline\hline & & & 0 & & & \\ & J_{n-1-k,k} - S^t_{k,n-1-k} & & \vdots & & Q_{n-k-1} & \\ & & & 0 & & & \end{array}\right], \tag{2}$$

where $J_{n-1-k,k}$ is the $(n-1-k) \times k$ matrix of all 1's. Then $T(S_{k,n-1-k})$ is an $n \times n$ tournament matrix. The $n \times n$ standard tournament matrix can be partitioned to have the form (2) for each choice of $k = 1, 2, \ldots, n-1$ with $S_{k,n-1-k}$ equal to $J_{k,n-1-k}$.

The definition of $T(S_{k,n-1-k})$ in (2) immediately implies that the in-neighborhood and the out-neighborhood of vertex $(k+1)$ are both transitive. In the next lemma we show that this holds for every vertex of $T(S_{k,n-1-k})$, that is, that $T(S_{k,n-1-k})$ is a locally transitive tournament.

Lemma 2.1 *The tournament $T(S_{k,n-1-k})$ is locally transitive.*

Proof Let p be any vertex with $1 \leq p \leq k$. We simultaneously permute the rows and columns of $T(S_{k,n-1-k})$ by moving to the top the rows that contain 1's in the last positions of column p and moving the corresponding columns to become the initial columns. The resulting matrix is a $T(S_{p,n-1-p})$ for some $S_{p,n-1-p}$ and can be repartitioned as

[2] In partition theory, the Ferrers matrices usually have nondecreasing row sum vectors.

$$T(S_p) = \left[\begin{array}{ccc||c||ccc} & Q_p & & \begin{array}{c}1\\ \vdots\\ 1\end{array} & & S_{p,n-1-p} & \\ \hline\hline 0 & \cdots & 0 & 0 & 1 & \cdots & 1 \\ \hline\hline & J_{n-1-p,p} - S^t_{n-1-p,p} & & \begin{array}{c}0\\ \vdots\\ 0\end{array} & & Q_{n-1-p} & \end{array}\right].$$

Hence the out-neighborhood and the in-neighborhood of p are transitive. The same conclusion holds if $k < p \leq n$. Thus $T(S_{k,n-1-k})$ is locally transitive. □

Example 2.2 With $n = 8$, an example of a locally transitive tournament $T(S_{4,3})$ is

$$\left[\begin{array}{c|c|c|c||c||c|c|c} 0 & 1 & 1 & 1 & 1 & 1 & & \\ \hline & 0 & 1 & 1 & 1 & 1 & & \\ \hline & & 0 & 1 & 1 & 1 & 1 & \\ \hline & & & 0 & 1 & 1 & 1 & 1 \\ \hline\hline 0 & 0 & 0 & 0 & 0 & 1 & 1 & 1 \\ \hline\hline & & & & 0 & 0 & 1 & 1 \\ \hline 1 & 1 & & & 0 & & 0 & 1 \\ \hline 1 & 1 & 1 & & 0 & & & 0 \end{array}\right].$$

If in the proof of Lemma 2.1 we take $p = 3$, and permute row 8 to become row 1 and column 8 to become column 1, we get

$$\left[\begin{array}{c|c|c||c||c|c|c|c} 0 & 1 & 1 & 1 & & & & \\ \hline & 0 & 1 & 1 & 1 & 1 & 1 & \\ \hline & & 0 & 1 & 1 & 1 & 1 & \\ \hline\hline 0 & 0 & 0 & 0 & 1 & 1 & 1 & 1 \\ \hline\hline 1 & & & 0 & 0 & 1 & 1 & 1 \\ \hline 1 & & & 0 & & 0 & 1 & 1 \\ \hline 1 & & & 0 & & & 0 & 1 \\ \hline 1 & 1 & 1 & 0 & & & & 0 \end{array}\right].$$

□

If n is odd, and $k = (n-1)/2$, and $S_{\frac{n-1}{2},\frac{n-1}{2}}$ is the lower triangular matrix with all 1's below the main diagonal (so, in particular, its 1's are left-justified), then the standard transitive tournament $T_{\frac{n-1}{2},\frac{n-1}{2}}$ is the carousel tournament of order n.

Theorem 2.3 *A tournament T is locally transitive if and only if its vertices can be labeled so that for some k and some $k \times (n-1-k)$ Ferrers matrix $S_{k,n-1-k}$, T has the form $T(S_{k,n-1-k})$ given in (2) where Q_k and Q_{n-1-k} are $k \times k$ and $(n-1-k) \times (n-1-k)$, respectively, triangular matrices with all 1's above the main diagonal.*

Proof That every tournament $T(S_{k,n-1-k})$ is locally transitive follows from Lemma 2.1. Now suppose that T is a locally transitive tournament. Choosing any vertex u of T, we see that, after suitable relabeling of the vertices, T has the form

$$T = \left[\begin{array}{c||c||c} Q_k & \begin{matrix}1\\ \vdots\\ 1\end{matrix} & X_{k,n-1-k} \\ \hline\hline 0 \quad \cdots \quad 0 & 0 & 1 \quad \cdots \quad 1 \\ \hline\hline J_{n-1-k,k} - X^t_{k,n-1-k} & \begin{matrix}0\\ \vdots\\ 0\end{matrix} & Q_{n-1-k} \end{array}\right], \tag{3}$$

where vertex u corresponds to row $k+1$ and $X_{k,n-1-k}$ is a $k \times (n-1-k)$ matrix. Suppose that a 1 lies above a 0 in $X_{k,n-1-k}$, say, that in column $q > k+1$ of T, there is a 1 in row $p < k$ and a 0 in row $p+1$ (so $p+1 < k$ and both the 1 and 0 are in $X_{k,n-1-k}$). The 0 implies that the $(q, p+1)$-entry of T equals 1. Thus in the out-neighborhood of p we have the vertices $p+1, k+1$, and q. But $p+1 \to k+1 \to q \to p+1$ is a 3-cycle in the out-neighborhood of p, a contradiction since T is locally transitive. Thus there does not exist a 1 above a 0 in $X_{k,n-1-k}$ and similarly there does not exist a 0 to the left of a 1 in $X_{k,n-1-k}$. This implies that the 1's in $X_{k,l}$ are left-justified and that the row sums of $X_{k,n-1-k}$ are nonincreasing. Hence $X_{k,n-1-k}$ has the form of an $S_{k,n-1-k}$.[3] □

The following theorem is from [5].

Theorem 2.4 *Let $R = (r_1, r_2, \ldots, r_n)$ be a nondecreasing sequence of nonnegative integers. Then R is the score vector of a locally transitive tournament if and only if*

(a) $r_i \le r_{i+1} \le r_i + 1 \quad (1 \le i \le n-1)$, *and*
(b) $r_i + r_{n+1-i} = n-1 \quad (1 \le i \le n)$. □

Remark 2.5 Before giving a proof we note that condition (b) implies Landau's condition for the existence of a tournament with nondecreasing score vector $R = (r_1, r_2, \ldots, r_n)$, namely that $\sum_{i=1}^{k} r_i \ge \binom{k}{2}$ for all k with equality when $k = n$. This

[3]This theorem in a different form (and to us not so clearly stated) is quoted in [5] as being proved in [6]. Since it is not so easily accessible, and since we have been unable to obtain a copy of [6], we have given a proof.

is a simple verification. Also observe that conditions (a) and (b) in Theorem 2.4 imply that if n is odd, then $r_{(n+1)/2} = (n-1)/2$, and if n is even, then $r_{n/2} = n/2-1$ and $r_{(n+2)/2} = n/2$. Thus the score vector R of a locally transitive tournament of order n is determined by

$$(r_1, r_2, \ldots, r_{(n-1)/2}) \quad (n \text{ odd}), \text{ and } (r_1, r_2, \ldots, r_{(n-2)/2}) \quad (n \text{ even}).$$

If n is odd, then we have two choices for $r_{(n-1)/2}$, namely $(n-1)/2$ or $(n-1)/2-1$, two choices for $r_{(n-3)/2}$, and so on giving a total of $2^{(n-1)/2}$ possible score vectors for locally transitive tournaments of order n. Similarly, if n is even, there are $2^{(n-2)/2}$ possible score vectors for locally transitive tournaments of order n. □

Proof We now prove Theorem 2.4. We first show that (a) and (b) are satisfied by the score vector R of a locally transitive tournament T. Properties (a) and (b) are proved by induction on n. If $n \leq 3$, these are trivially true. Now assume that $n \geq 4$. Let x be a vertex of the minimal outdegree r_1. Then, since $T_{\to x}$ is transitive, there is a vertex y of $T_{\to x}$ of outdegree at least $n - r_1 - 1$. If y had outdegree $> n - r_1 - 1$, then y has indegree less than r_1, a contradiction to the minimality of r_1. Thus y has outdegree $n - r_1 - 1$ and hence $r_1 + r_n = n - 1$. Since $T_{\to x}$ has a vertex z of outdegree 0, z has outdegree at most $r_1 + 1$ in T, implying that $r_2 \leq r_1 + 1$, and hence that $r_2 - 1, \ldots, r_{n-1} - 1$ is nondecreasing. The locally transitive tournament T' of order $n - 2$ obtained from T by removing vertices x and y has score vector $R' = (r_2 - 1, \ldots, r_{n-1} - 1)$. Applying induction using R', we get $r_i + r_{n+1-i} = n-1$ for all i with $2 \leq i \leq n-2$ and that $r_{i+1} \leq r_i + 1$ for $2 \leq i \leq n-1$. Thus if R is the score vector of a locally transitive tournament, then R satisfies (a) and (b).

We now assume that $R = (r_1, r_2, \ldots, r_n)$ satisfies (a) and (b), and show that R is the score vector of a locally transitive tournament. We do this by constructing a tournament T with score vector R which has the structure as given in Theorem 2.3 and so is locally transitive. It follows from (a) and (b) that $r_i \geq i-1$ and $r_{n-i+1} \leq n-i$ for $1 \leq i \leq n/2$.

Let $R' = (0, 1, 2, \ldots, n-1)$, then Q_n is a transitive tournament with score vector R'. Let

$$D = (d_1, \ldots, d_n) = R - R' = (r_1, r_2 - 1, \ldots, r_i - (i-1), \ldots, r_{n-1} - (n-2), \\ r_n - (n-1))$$

for $i = 1, 2, \ldots, n$. According to (b),

- $d_i = -d_{n+1-i}$ for $i = 1, 2, \ldots, n$,
- $d_i \geq 0$ for $i \leq \lfloor \frac{n}{2} \rfloor$;
- $d_i \leq 0$ for $i \geq \lceil \frac{n+1}{2} \rceil$;
- $d_i = 0$ for $i = \frac{n+1}{2}$ for odd n ; for $i = \frac{n+2}{2}$ and $\frac{n}{2}$ for even n;
- $d_1 \geq d_2 \geq d_3 \geq \cdots \geq d_k > 0$.

Thus $D = (d_1, d_2, \ldots, d_k, 0, \ldots, 0, -d_k, \ldots, -d_2, -d_1)$. Let $\tilde{D}$ be the $k \times d_1$ (0, 1)-matrix with d_i right-justified 1's in each row i. Then removing $\tilde{D}$ from the right top corner of Q_n and putting $\tilde{D}^t$ in the left bottom corner, we obtain a tournament with score vector R of the form given in Theorem 2.3 which is therefore locally transitive. □

We now give several examples of the construction used in the proof of Theorem 2.4.

Example 2.6 Let $R = (1, 2, 2, 3, 4, 5, 6, 6, 7)$, then $D = R - R' = (1, 1, 0, 0, 0, 0, 0, -1, -1)$ and let

$$\tilde{D} = \begin{bmatrix} 1 \\ 1 \end{bmatrix}.$$

To obtain a locally transitive tournament which has score vector R, we remove $\tilde{D}$ from the right top corner of Q_9 and put $\tilde{D}^t$ at the left bottom corner:

$$\begin{bmatrix} 0&1&1&1&1&1&1&1&1 \\ &0&1&1&1&1&1&1&1 \\ &&0&1&1&1&1&1&1 \\ &&&0&1&1&1&1&1 \\ &&&&0&1&1&1&1 \\ &&&&&0&1&1&1 \\ &&&&&&0&1&1 \\ &&&&&&&0&1 \\ &&&&&&&&0 \end{bmatrix} \longrightarrow T = \begin{bmatrix} 0&1&1&1&1&1&1&1& \\ &0&1&1&1&1&1&1& \\ &&0&1&1&1&1&1&1 \\ &&&0&1&1&1&1&1 \\ &&&&0&1&1&1&1 \\ &&&&&0&1&1&1 \\ &&&&&&0&1&1 \\ &&&&&&&0&1 \\ 1&1&&&&&&&0 \end{bmatrix}.$$

Now let $R = (2, 2, 3, 3, 4, 5, 5, 6, 6)$, then $D = R - Q = (2, 1, 1, 0, 0, 0, -1, -1, -2)$ and let

$$\tilde{D} = \begin{bmatrix} 1&1 \\ &1 \\ &1 \end{bmatrix}.$$

To obtain a locally transitive tournament T with score vector R, we remove $\tilde{D}$ from the right top corner of Q_9 and put $\tilde{D}^t$ at the left bottom corner.

$$\begin{bmatrix} 0&1&1&1&1&1&1&1&1 \\ &0&1&1&1&1&1&1&1 \\ &&0&1&1&1&1&1&1 \\ &&&0&1&1&1&1&1 \\ &&&&0&1&1&1&1 \\ &&&&&0&1&1&1 \\ &&&&&&0&1&1 \\ &&&&&&&0&1 \\ &&&&&&&&0 \end{bmatrix} \longrightarrow T = \begin{bmatrix} 0&1&1&1&1&1&1&& \\ &0&1&1&1&1&1&1& \\ &&0&1&1&1&1&1& \\ &&&0&1&1&1&1&1 \\ &&&&0&1&1&1&1 \\ &&&&&0&1&1&1 \\ &&&&&&0&1&1 \\ 1&&&&&&&0&1 \\ 1&1&1&&&&&&0 \end{bmatrix}.$$

□

Example 2.7 Let $n = 9$ and consider the score vector $R = (4, 4, 4, 4, 4, 4, 4, 4, 4)$ of a regular tournament. We can use the construction defined in Theorem 2.4 to get

$$\begin{bmatrix} 0 & 1 & 1 & 1 & 1 & & & & \\ & 0 & 1 & 1 & 1 & 1 & & & \\ & & 0 & 1 & 1 & 1 & 1 & & \\ & & & 0 & 1 & 1 & 1 & 1 & \\ 0 & 0 & 0 & 0 & 0 & 1 & 1 & 1 & 1 \\ 1 & & & & 0 & 0 & 1 & 1 & 1 \\ 1 & 1 & & & 0 & & 0 & 1 & 1 \\ 1 & 1 & 1 & & 0 & & & 0 & 1 \\ 1 & 1 & 1 & 1 & 0 & & & & 0 \end{bmatrix},$$

the classical carousel tournament of order 9.

Now let $R = (2, 3, 3, 4, 4, 4, 5, 5, 6)$. We can use the construction defined in Theorem 2.4 to get

$$\begin{bmatrix} 0 & 1 & 1 & 1 & 1 & 1 & 1 & & \\ & 0 & 1 & 1 & 1 & 1 & 1 & & \\ & & 0 & 1 & 1 & 1 & 1 & 1 & \\ & & & 0 & 1 & 1 & 1 & 1 & \\ & & & & 0 & 1 & 1 & 1 & 1 \\ & & & & & 0 & 1 & 1 & 1 \\ & & & & & & 0 & 1 & 1 \\ 1 & 1 & & & & & & 0 & 1 \\ 1 & 1 & 1 & 1 & & & & & 0 \end{bmatrix}.$$

Here the vertex of outdegree 2 is joined to vertices of degrees 3 and 4.

Consider again $R = (2, 3, 3, 4, 4, 4, 5, 5, 6)$. Suppose we join vertex of outdegree 2 to the two vertices of outdegree 3. Then this gives

$$\begin{bmatrix} 0 & 1 & 1 & 1 & 1 & 1 & 1 & & \\ & 0 & 1 & 1 & 1 & 1 & 1 & & \\ & & 0 & 1 & 1 & 1 & 1 & 1 & \\ & & & 0 & 1 & 1 & 1 & 1 & 1 \\ & & & & 0 & 1 & 1 & 1 & 1 \\ & & & & & 0 & 1 & 1 & 1 \\ & & & & & & 0 & 1 & 1 \\ 1 & 1 & & & & & & 0 & 1 \\ 1 & 1 & 1 & & & & & & 0 \end{bmatrix}.$$

But then the outdegrees are: $6, 5, 5, 5, 4, 3, 2, 3, 3$, so we do not get R: we have three 5's instead of two, one 4 instead of three, and three 3's instead of two. □

From Theorems 1.2 and 2.3, it follows that if there exists a locally transitive tournament with a given score vector, then there exists a Hankel tournament with that score vector.

Example 2.8 Let $n = 5$ and let $R = (1, 1, 2, 3, 3)$. By Theorems 1.2 and 2.3 there exists a Hankel tournament with score vector R and also a locally transitive tournament with score vector R. The following is the adjacency matrix of a tournament with score vector R that is both Hankel and locally transitive:

$$\left[\begin{array}{c|c|c|c|c} 0&1&0&0&0\\ \hline 0&0&0&1&0\\ \hline 1&1&0&0&0\\ \hline 1&0&1&0&1\\ \hline 1&1&1&0&0 \end{array}\right].$$

Now let $n = 9$. The following example from [5] is an example of a regular tournament which is not locally transitive but, as it turns out, is Hankel:

$$\left[\begin{array}{c|c|c|c|c|c|c|c|c} &1&&1&1&1&&&\\ \hline &&1&1&1&1&&&\\ \hline 1&&&1&1&1&&&\\ \hline &&&&1&&1&1&1\\ \hline &&&&&1&1&1&1\\ \hline &&&1&&&1&1&1\\ \hline 1&1&1&&&&&1&\\ \hline 1&1&1&&&&&&1\\ \hline 1&1&1&&&&1&& \end{array}\right].$$

It is not locally transitive since vertex 4 contains a 3-cycle in its in-neighborhood (and vertex 6 contains a 3-cycle in its out-neighborhood). □

Example 2.9 The following is an example of a locally transitive tournament for which there does not exist a labeling of its vertices to give a Hankel tournament (although there must exist a Hankel tournament with the same score vector) (Figure 1):

$$D = \left[\begin{array}{c|c|c|c|c|c} 0&1&0&0&0&0\\ \hline 0&0&1&1&0&0\\ \hline 1&0&0&1&1&1\\ \hline 1&0&0&0&1&1\\ \hline 1&1&0&0&0&1\\ \hline 1&1&0&0&0&0 \end{array}\right]$$

□

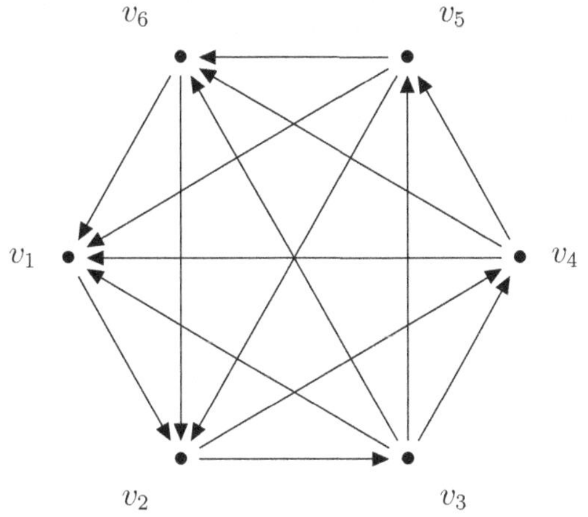

Fig. 1 A locally transitive tournament that cannot be relabeled to a Hankel tournament

In Examples 2.8 and 2.9, we have seen examples of score vectors satisfying (a) and (b) of Theorem 2.4 for which there is a tournament which is (1) both locally transitive and Hankel and (2) locally transitive but not Hankel. It is easy to construct score vectors with a corresponding tournament which is (3) Hankel but not locally transitive (just violate (a) of Theorem 2.4). We now show that there is a score vector R satisfying (a) and (b) of Theorem 2.4 (thus for which there exists a tournament with score vector R which is Hankel and a tournament with score vector R which is locally transitive), for which there does not exist a locally transitive, Hankel tournament with score vector R. In fact, the score vector used in Example 2.9, made monotone nondecreasing, works.

Example 2.10 Let $R = (1, 2, 2, 3, 3, 4)$ be a score vector which satisfies (a) and (b) of Theorem 2.4. By Theorem 2.3, there are only two possible locally transitive tournaments with this score vector R, namely, (1) and (2) below:

(1)

$$\begin{bmatrix} 0&1&1&1&1&1\\ 0&0&1&1&1&1\\ 0&0&0&1&1&1\\ 0&0&0&0&1&1\\ 0&0&0&0&0&1\\ 0&0&0&0&0&0 \end{bmatrix} \Longrightarrow T = \begin{bmatrix} 0&1&1&1&1&0\\ 0&0&1&1&1&0\\ 0&0&0&1&1&1\\ 0&0&0&0&1&1\\ 0&0&0&0&0&1\\ 1&1&0&0&0&0 \end{bmatrix}$$

with corresponding digraph (Figure 2) with outdegrees and indegrees of vertices given by

$$v_1, (4, 1);\; v_2, (3, 2);\; v_3, (3, 2);\; v_4, (2, 3);\; v_5, (1, 4);\; v_6, (2, 3).$$

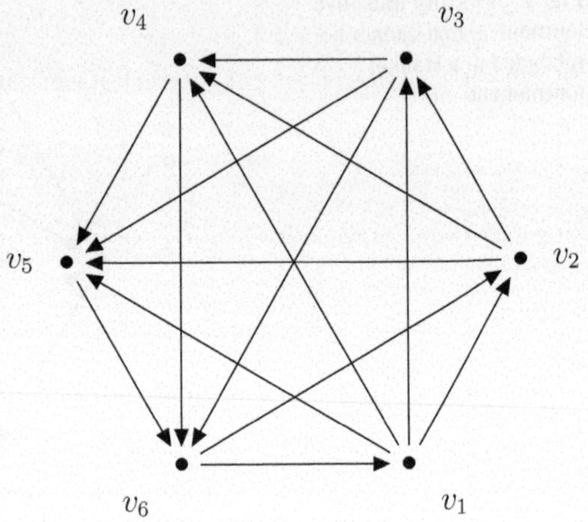

Fig. 2 A locally transitive tournament

(2)

$$\begin{bmatrix} 0&1&1&1&1&1\\ 0&0&1&1&1&1\\ 0&0&0&1&1&1\\ 0&0&0&0&1&1\\ 0&0&0&0&0&1\\ 0&0&0&0&0&0 \end{bmatrix} \Longrightarrow T' = \begin{bmatrix} 0&1&1&1&0&0\\ 0&0&1&1&1&1\\ 0&0&0&1&1&1\\ 0&0&0&0&1&1\\ 1&0&0&0&0&1\\ 1&0&0&0&0&0 \end{bmatrix}$$

The tournament T' cannot be simultaneously permuted to the tournament T. The reason is that in T, the two rows with sum equal to 3 have 1's in two common columns while in T they have 1's in only one common column. A necessary condition for a tournament T' to be Hankel is that the vertices can be paired up $\{x, y\}$ so that the outdegree of x equals the indegree of y. Given such a pairing, let x be u_i and y being u_{n+1-i} and forming the incidence matrix B of T', then T' is a Hankel tournament if and only if B is a Hankel matrix, that is, the Hankel property does not depend on the choice of i and $n+1-i$ for each vertex pair. (Changing the choice is a simultaneous Hankel permutation.)

In this case the possible pairings are:

$$\{v_2, v_6\}, \{v_1, v_4\}, \{v_3, v_5\} \text{ and } \{v_2, v_6\}, \{v_1, v_5\}, \{v_3, v_4\}.$$

Thus there are two possible adjacency matrices neither of which is Hankel:

$$\begin{bmatrix} 0&1&1&0&0&1\\ 0&0&1&1&1&1\\ 0&0&0&1&1&1\\ 1&0&0&0&1&0\\ 1&0&0&0&0&0\\ 0&0&0&1&1&0 \end{bmatrix} \text{ and } \begin{bmatrix} 0&1&1&1&0&0\\ 0&0&1&1&1&1\\ 0&0&0&1&1&1\\ 1&0&0&0&1&0\\ 0&0&0&0&1&1\\ 1&0&0&0&0&0 \end{bmatrix}.$$

□

From Theorem 2.3, it follows that a locally transitive tournament of order n results from the transitive tournament of order n by choosing a partition of an integer m with $0 \leq m \leq \binom{n}{2}$ into k parts $m_1 \geq m_2 \geq \cdots \geq m_k$ where $k \leq n-1$ and $m_j \leq n-j-1$ for $j = 1, 2, \ldots, k$. Let the set of such partitions be Π_n.

Example 2.11 Let $n = 6$ and consider the partition $\pi : 6 = 2+2+2$ with conjugate partition $\pi' : 6 = 3+3$. These give the locally transitive tournaments

$$T = \begin{bmatrix} 0&1&1&1&0&0\\ 0&0&1&1&0&0\\ 0&0&0&1&0&0\\ 0&0&0&0&1&1\\ 1&1&1&0&0&1\\ 1&1&1&0&0&0 \end{bmatrix} \text{ and } T' = \begin{bmatrix} 0&1&1&0&0&0\\ 0&0&1&0&0&0\\ 0&0&0&1&1&1\\ 1&1&0&0&1&1\\ 1&1&0&0&0&1\\ 1&1&0&0&0&0 \end{bmatrix}.$$

The score vector of T is $(3, 2, 1, 2, 4, 3)$ and of T' is $(2, 1, 3, 4, 3, 2)$ which arranged in nondecreasing order is $(1, 2, 2, 3, 3, 4)$ in both cases. Thus different partitions may give the same score vector, so there is not a bijection between score vectors of locally transitive tournaments of order n and partitions Π_n. These two tournaments are not isomorphic. In T the vertex of outdegree 1 has its outedge to a vertex of indegree 3, while in T' the vertex of outdegree 1 has its outedge to a vertex of indegree 2. □

3 Hankel Cycles

As a directed graph (digraph), a tournament T of order n is strongly connected if and only if there does not exist a *directed cut*, that is, disjoint nonempty sets $K_1, K_2 \subseteq \{1, 2, \ldots, n\}$ partitioning $\{1, 2, \ldots, n\}$ such that all the edges between the vertex sets K_1 and K_2 are directed from K_2 to K_1 and none is directed from K_1 to K_2. As a tournament matrix, this is equivalent to the $|K_1| \times |K_2|$ submatrix (with rows indexed by K_1 and columns indexed by K_2) $T[K_1, K_2] = O$ and the $|K_2| \times |K_1|$ submatrix $T[K_2, K_1] = J$, an all 1's matrix. Suppose that T is a Hankel tournament. For any subset K of $\{1, 2, \ldots, n\}$, let $K^c = \{n+1-i : i \in K\}$. Then

$T[K_1, K_2] = O$ implies that $T[K_2^c, K_1^c] = O$. Thus if $[K_1, K_2]$ is a directed cut of T so is $[K_2^c, K_1^c]$.

A strongly connected tournament always has a Hamilton cycle, in particular, a strongly connected Hankel tournament has a Hamilton cycle. The question naturally arises as to whether or not a strongly connected Hankel tournament always has a *Hankel Hamilton cycle*, that is, a Hamilton cycle γ such that whenever $i \to j$ is an edge of γ so is $(n+1-j) \to (n+1-i)$ an edge. Thus the adjacency matrix of a Hankel Hamilton cycle is, in particular, a permutation matrix that is Hankel symmetric. More generally, we define a *Hankel cycle* to be a cycle γ of any length with the property that if (i, j) is an edge so is $(n+1-j, n+1-i)$. The adjacency matrix C of a Hankel cycle, in particular a Hankel Hamilton cycle, is invariant under the Hankel transpose: $C^h = C$.

Example 3.1 An example of a Hankel Hamilton cycle and its Hankel symmetric adjacency matrix A is $1 \to 3 \to 5 \to 2 \to 4 \to 6 \to 1$, and

$$A = \left[\begin{array}{c|c|c|c|c|c} & & 1 & & & \\ \hline & & & 1 & & \\ \hline & & & & 1 & \\ \hline & & & & & 1 \\ \hline & 1 & & & & \\ \hline 1 & & & & & \end{array}\right].$$

□

Let T be a Hankel tournament of odd order $n = 2k-1$. Then vertex k is called the *center vertex* of T, and row k and column k are the *center row* and *center column*, respectively, of T regarded as a tournament matrix. Note that $n+1-k = k$ and thus $t_{k,n+1-k} = t_{kk} = 0$. Deleting row k and column k from T produces a $2(k-1) \times 2(k-1)$ Hankel tournament (matrix) T_{*k}, called the *central subtournament* of T. Every (Hankel) cycle γ of T that does not use the center vertex k gives a (Hankel) cycle of T_{*k} with the same length. Thus in investigating properties of Hankel cycles in Hankel tournaments of order n, it suffices to consider Hankel cycles of any length when n is even (see Lemma 3.2 below) and Hankel cycles that contain the central vertex when n is odd.

Consider a Hankel tournament of order n and a cycle γ of some length. Then replacing each edge (i, j) of γ with the edge $(n+1-j, n+1-i)$ produces another cycle γ^h of the same length as γ. We call γ^h the *Hankel image* of γ. A natural question is whether or not a Hankel cycle of the same or larger length can be constructed out of γ and γ^h? Unfortunately, this is not the case. For example, let $n = 10$ and consider the cycle

$$\gamma : 1 \to 3 \to 4 \to 8 \to 6 \to 1$$

of length 6 where

$$\gamma^h : 8 \to 10 \to 5 \to 3 \to 7 \to 8.$$

Then putting γ and γ^h together will not produce any Hankel cycles.

Lemma 3.2 *Let n be even and let T be a Hankel tournament of order n. Then T does not contain a Hankel cycle of any odd length.*

Proof Suppose that γ is a Hankel cycle of T of odd length $2p+1$. Choose any path $\pi = i_1 \to i_2 \to \cdots \to i_{p+1}$ within γ having $p+1$ vertices and length p. Then $\pi_h = (n+1-i_{p+1}) \to \cdots \to (n+1-i_2) \to (n+1-i_1)$ is also a path within γ having $p+1$ vertices and length p. Thus π and π^h must have a common vertex. Let $n+1-i_j = i_k$. Since $j = k$ implies that n is odd, we must have $j \neq k$. This implies that there are two edges of γ going into one of its vertices, a contradiction. □

A strongly connected tournament of order n not only contains a Hamilton cycle but contains a cycle of each length $k = 3, 4, \ldots, n$, that is, is *pancyclic*. In light of Lemma 3.2, a Hankel tournament T of order n is called *Hankel even-pancyclic* provided T contains Hankel cycles of all even lengths between 4 and n and *Hankel odd-pancyclic* provided T contains Hankel cycles of all odd lengths between 3 and n.

Now assume that $A = [a_{ij}]$ is an arbitrary $n \times n$ Hankel symmetric (0, 1)-matrix. Then A is the adjacency matrix of a digraph D called here a *Hankel digraph*. Thus if $i \to j$ is an edge of the Hankel digraph D so is $(n+1-j) \to (n+1-i)$. If $\pi : i_i \to i_2 \to \cdots \to i_k$ is a path of D, then as with cycles, $\pi^h : n+1-i_k \to \cdots \to n+1-i_2 \to n+1-i_1$ is also a path called the *Hankel image* of π. An edge of D of the form $i \to n+1-i$, corresponding to the entry $a_{i,n+1-i} = 1$ on the Hankel diagonal of A, will be called an *hd-edge* of γ. A Hankel tournament contains exactly $\lfloor n/2 \rfloor$ hd-edges since, by being a tournament, a Hankel tournament matrix must contain exactly $\lfloor n/2 \rfloor$ 1's on its Hankel diagonal.

Lemma 3.3 *Let T be a Hankel tournament. Then a Hankel cycle of odd length contains at most one hd-edge, and a Hankel cycle of T of even length contains at most two hd-edges.*

Proof First suppose that γ is a Hankel cycle containing two hd-edges, $a \to n+1-a$ followed later by $b \to n+1-b$. Then, since γ is a Hankel-cycle, the path of γ from $n+1-a$ to b has the same length as the path of γ from $n+1-b$ to a, implying that γ has even length. Thus a Hankel cycle of odd length contains at most one hd-edge.

Now suppose that γ is a Hankel cycle containing three or more hd-edges. Let $a \to n+1-a, b \to n+1-b$, and $c \to n+1-c$ be three distinct hd-edges of γ which occur successively. Then γ contains a path of the form

$$a \to n+1-a \to i_1 \to \cdots \to i_s \to b \to n+1-b \to j_1 \to \cdots \to j_t$$
$$\to c \to n+1-c.$$

Since γ is a Hankel cycle, γ also contains

$$c \to n+1-c \to n+1-j_t \to \cdots \to n+1-j_1 \to b$$

and

$$b \to n+1-b \to n+1-i_s \to \cdots \to n+1-i_1 \to a \to n+1-a.$$

Thus within γ there are two cycles, contradicting that γ is itself a cycle. □

We can say more.

Lemma 3.4 *Let T be a Hankel tournament of order n, and let γ be a Hankel cycle of T of length k. If k is even, then γ contains exactly two hd-edges while if k is odd, then γ contains exactly one hd-edge.*

Proof By Lemma 3.3, γ contains at most two hd-edges if k is even and at most one if k is odd. If $i \to j$ is an edge of γ, then so is $n+1-j \to n+1-i$. Hence if γ does not contain an hd-edge, then k must be even and, if k is odd, γ contains exactly one hd-edge.

Now suppose that $k = 2m$ is even. First observe that because γ has even length, γ cannot contain exactly one hd-edge. Thus we need only rule out that γ does not contain any hd-edges. So suppose that γ does not contain an hd-edge. Let

$$\gamma_1 : i_1 \to i_2 \to \cdots \to i_m$$

be any path within γ of length $m-1$. Then its Hankel image

$$\gamma_1^h : (n+1-i_m) \to \cdots \to (n+1-i_2) \to (n+1-i_1)$$

is also a path within γ of length $m-1$. If γ_1 and γ_1^h do not have any vertices in common, then $i_m \to (n+1-i_m)$ and $(n+1-i_1) \to i_1$ must both be edges of γ and thus are hd-edges of γ. Thus γ contains exactly two hd-edges. So we assume that γ_1 and γ_1^h have a vertex in common, say vertex $(n+1-i_q)$ equals vertex i_p. There are two possibilities.

- $q = p$, that is, $n+1-i_p = i_p$ or equivalently $i_p = \frac{n+1}{2}$. In this case, the union of γ and γ' looks as shown in Figure 3 in which vertex i_p has indegree and outdegree both equal to 2, contradicting γ is a cycle.
- $p \neq q$. If $i_p = n+1-i_q$, then $i_q = n+1-i_p$. Without loss of generality, suppose $p < q$, and there is no other common vertex between p and q. Then the union of γ and γ' looks as shown in Figure 4 and again we contradict that γ is a cycle. □

If n is odd, we can say even more.

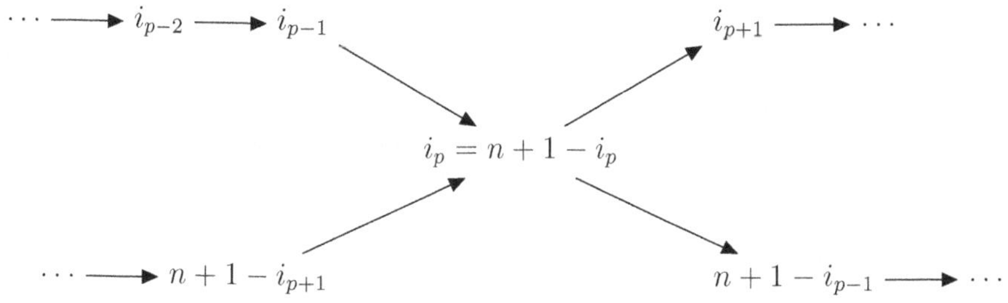

Fig. 3 Union of γ and γ' giving a contradiction in case $q = p$

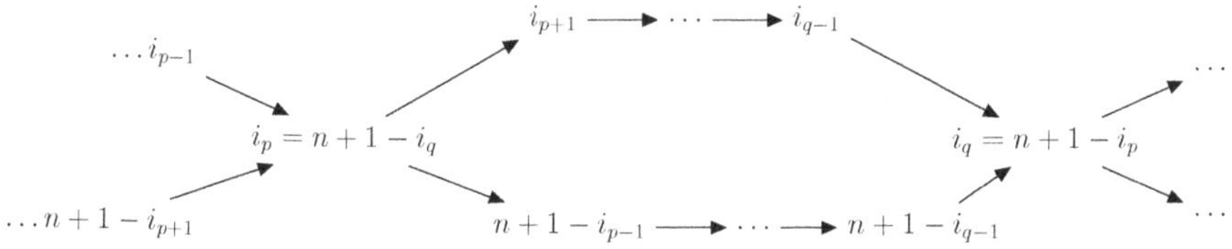

Fig. 4 Union of γ and γ' giving a contradiction in case $q \neq p$

Corollary 3.5 *Let $n = 2k+1$ and T be a Hankel tournament of order n. Then every cycle of T of odd length has exactly one hd-edge and contains the center vertex k.*

Proof Let γ be a cycle of odd length of T. Then by Lemma 3.4, γ contains an hd-edge $i \to n+1-i$ where $i \neq n+1-i$. Following the path of γ from $n+1-i$ and its Hankel image in the reverse order from i, these two paths meet at a vertex k such that $n+1-k = k$. Thus k is the central vertex of T. □

Example 3.6 Let $n = 5$ and let T be the 5×5 Hankel tournament

$$\begin{bmatrix} 0 & 0 & 1 & 0 & 1 \\ 1 & 0 & 0 & 0 & 0 \\ 0 & 1 & 0 & 0 & 1 \\ 1 & 1 & 1 & 0 & 0 \\ 0 & 1 & 0 & 1 & 0 \end{bmatrix}.$$

Then the central vertex of T is vertex 3 and the two hd-edges of T are $1 \to 5$ and $4 \to 2$. Since there does not exist an edge from 5 to 3 or from 2 to 3, using Lemma 3.4 we see that T does not have a Hankel cycle of length 3. But T does have a Hankel cycle of length 5, namely the Hankel Hamilton cycle

$$1 \to 5 \to 4 \to 3 \to 2 \to 1$$

corresponding to the shading in the matrix. □

We show that a strongly connected Hankel tournament contains a Hankel Hamilton cycle but first we show there is a Hankel cycle of some length.

Lemma 3.7 *Let T be a strongly connected Hankel tournament of order n. If n is even, T has a Hankel cycle γ of some even length (necessarily containing two hd-edges), and if n is odd, T has a Hankel cycle of some odd length (necessarily containing one hd-edge and the center vertex).*

Proof If $n = 3$, T itself is a Hankel cycle. Now assume that $n \geq 4$. Since we have a tournament, for each vertex a we have $a \to n + 1 - a$ or $n + 1 - a \to a$ unless $a = n + 1 - a$ which implies that n is odd and a is the central vertex.

Let $i \to n+1-i$ and $n+1-p \to p$ be any two hd-edges of T. There is a path

$$\pi : p \to g \to f \to \cdots \to c \to b \to a \to i$$

from p to i with Hankel image π^h from $n+1-i$ to p. If π and π^h do not intersect, we get the Hankel cycle of even length

$$(\pi) \to n + 1 - i \to (\pi^h) \to n + 1 - p \to p$$

containing the two hd-edges $i \to n+1-i$ and $n+1-p \to p$. If π and π^h intersect, either we get a Hankel cycle of even length with two hd-edges, or n is odd and we get a Hankel cycle of odd length containing one hd-edge and the center vertex. □

Theorem 3.8 *Let $T = [t_{ij}]$ be a strongly connected Hankel tournament of order n. Then T has a Hankel Hamilton cycle.*

Proof By Lemma 3.7, we know that T contains a Hankel cycle

$$\gamma : i_1 \to i_2 \to \cdots \to i_m \to i_1$$

of some length m where we assume that m is as large as possible. If $m = n$, we are done so we assume that $m < n$.

First suppose that there is a vertex k not on γ such that for some i_p there is an edge from i_p to k and from k to i_{p+1}. Since T is a Hankel tournament, there are edges in T from $n + 1 - k$ to $n + 1 - i_p$ and from $n + 1 - i_{p+1}$ to $n + 1 - k$. Thus in T,

$$i_p \to k \to i_{p+1} \text{ and } n + 1 - i_{p+1} \to n + 1 - k \to n + 1 - i_p,$$

where the edge $n + 1 - i_{p+1} \to n + 1 - i_p$ is also an edge of γ. This gives the Hankel cycle γ':

$$\begin{aligned} i_1 \to i_2 \to \cdots \to i_p \to k \to i_{p+1} \to \cdots \to n + 1 - i_{p+1} \\ \to n+1-k \to n+1-i_p \to \cdots \to i_1 \end{aligned}$$

contradicting the choice of m.

Now suppose that no such vertex k exists. Thus for each vertex $k \neq i_1, i_2, \ldots, i_m$, either all edges of T between k and vertices of γ go from k to γ with such k defining a set X, or all such edges go from γ to k with such k defining a set Y. Since T is strongly connected, there must be an edge $r \to s$ from a vertex $r \in Y$ to a vertex $s \in X$, and so an edge from $n+1-s$ to $n+1-r$. We thus have edges $s \to i_1$ and $i_m \to r$, and so also edges $n+1-i_1 \to n+1-s$ and $n+1-r \to n+1-i_m$. This gives the Hankel cycle

$$r \to s \to i_1 \to i_2 \to \cdots \to n+1-i_1 \to n+1-s \to n+1-r$$
$$\to n+1-i_m \to \cdots \to i_m \to r$$

of larger length, again a contradiction. We conclude that $m = n$ and a Hamilton Hankel cycle exists. □

Corollary 3.9 *Let $n \geq 4$ be even and let T be an $n \times n$ strongly connected Hankel tournament of order n. Then T is Hankel even-pancyclic, that is, T contains Hankel cycles of all even lengths between 4 and n.*

Proof By Theorem 3.8, T has a Hamilton cycle γ and by Lemma 3.7, γ has exactly two hd-edges. Thus γ is of the form

$$i_1 \to n+1-i_1 \to n+1-i_2 \to \ldots \to n+1-i_k \to i_k \to i_{k-1} \to \ldots \to i_2 \to i_1,$$

as pictured in Figure 5.

Since T is a tournament, for each $p = 2, 3, \ldots, k-1$, either $i_p \to n+1-i_p$ or $n+1-i_p \to i_p$. Since $i_1 \to n+1-i_1$ and $n+1-i_k \to i_k$, there exists q with $1 \leq q \leq k-1$ such that $i_q \to n+1-i_q$ and $n+1-i_{q+1} \to i_{q+1}$. This gives the Hankel cycle

$$i_q \to n+1-i_q \to n+1-i_{q+1} \to i_{q+1} \to i_q$$

of length 4. Using the same technique as in the proof of Theorem 3.8 we can get Hankel cycles of all even lengths from 4 to n. □

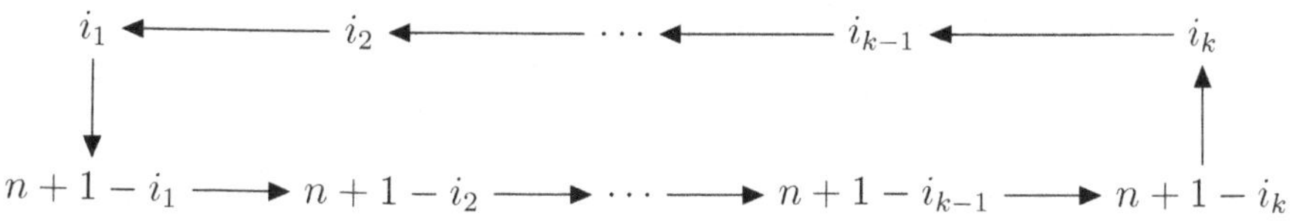

Fig. 5 A Hamilton cycle with two hd-edges

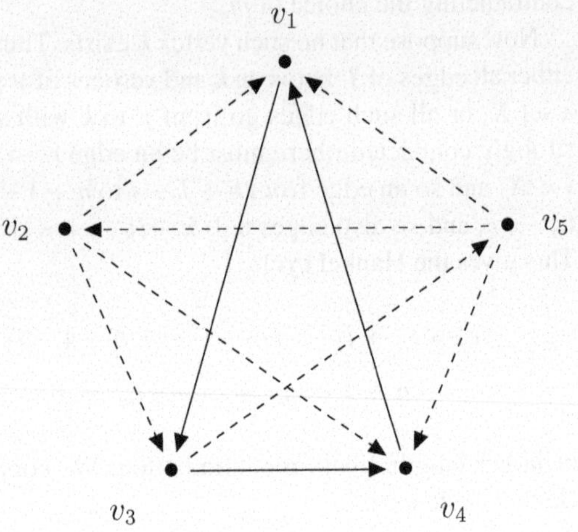

Fig. 6 Hankel tournament with a Hankel 3-cycle but no Hankel 4-cycle

Corollary 3.10 *Let $n \geq 3$ be odd and let T be a strongly connected Hankel tournament of order n. Then T is either Hankel odd-pancyclic or Hankel even-pancyclic, that is, either T contains Hankel cycles of all odd lengths between 3 and n, or T contains Hankel cycles of all even lengths between 3 and n. There are Hankel tournaments with n odd that are Hankel even-pancyclic but not Hankel odd-pancyclic, and there are Hankel tournaments with n odd that are Hankel odd-pancyclic but not Hankel even-pancyclic.*

Proof The proof of the first assertion is similar to the proof used in Corollary 3.9 with the only difference being that one of the hd-edges is replaced by the central vertex and the conclusion now being that either T contains a Hankel cycle of length 4 or a Hankel cycle of length 3. As in the proof of Theorem 3.8 either we can get Hankel cycles of all odd lengths from 3 to n or we can get Hankel cycles of all even lengths from 4 to n.

Let $n = 5$. Consider the $n \times n$ Hankel tournaments given in Figures 6 and 7. The first has a Hankel 3-cycle (undashed edges) but no Hankel 4-cycle. The second has a Hankel 4-cycle (undashed edges) but no Hankel 3-cycle. □

4 A Special Hankel Tournament

Let $R = (r_1, r_2, \ldots, r_n)$ be a nondecreasing sequence of nonnegative integers. We now let $\mathcal{T}(R)$ be the set of $n \times n$ tournament matrices with row sum vector R and so column sum vector $S = (n-1, n-1, \ldots, n-1) - R$. A theorem due to Ao and Hansen [1] and independently to Guiduli et al. [18] (see [12] for a short proof which is also given in [7] (pages 230–234)) guarantees the existence of a special half-transitive tournament (matrix) in $\mathcal{T}(R)$ whenever $\mathcal{T}(R) \neq \emptyset$. Here a *half-transitive*

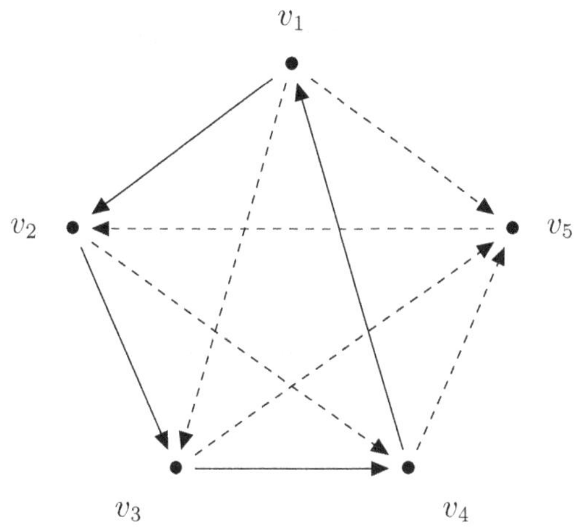

Fig. 7 Hankel tournament with a Hankel 4-cycle but no Hankel 3-cycle

tournament matrix is a tournament matrix such that the two subtournaments on the odd-indexed vertices and the even-indexed vertices are transitive.

Let $\mathcal{T}_H(R)$ be the set of Hankel tournament matrices in $\mathcal{T}(R)$. By Theorem 1.2, $\mathcal{T}_H(R) \neq \emptyset$ if and only if $\mathcal{T}(R) \neq \emptyset$ and $r_i + r_{n+1-i} = n-1$ for all $1 \leq i \leq n$. We show that $\mathcal{T}_H(R) \neq \emptyset$ implies the existence of a half-transitive Hankel tournament matrix in $\mathcal{T}_H(R)$ by adapting the proof in the non-Hankel case. This proof requires the use of the following lemma (Lemma 6.3.1 in [11]).

Lemma 4.1 *Let $R = (r_1, r_2, \ldots, r_n)$ be a nonnegative integral vector such that there exists an $n \times n$ $(0, 1)$-matrix with row and column sum vector equal to R. Then there exists such a matrix which is also symmetric.*

First we illustrate this with an example.

Example 4.2 Let $n = 8$ and $R = (1, 2, 3, 3, 4, 4, 5, 6)$, and let

$$T = \begin{bmatrix} 0&0&0&0&0&1&0&0 \\ 1&0&1&0&0&0&0&0 \\ 1&0&0&0&0&1&0&1 \\ 1&1&1&0&\mathbf{0}&0&0&0 \\ 1&1&1&1&0&0&0&0 \\ 0&1&0&1&1&0&1&0 \\ 1&1&1&1&1&0&0&0 \\ 1&1&0&1&1&1&1&0 \end{bmatrix} \in \mathcal{T}_H(R).$$

Then the two 4×4 principal submatrices $T[\{1, 3, 5, 7\}]$ (rows and columns indexed by $\{1, 3, 5, 7\}$ and cells in light gray above) and $T[\{2, 4, 6, 8\}]$ (cells in dark gray above) equal the transitive tournament matrix

$$\begin{bmatrix} 0 & 0 & 0 & 0 \\ 1 & 0 & 0 & 0 \\ 1 & 1 & 0 & 0 \\ 1 & 1 & 1 & 0 \end{bmatrix}.$$

For the complementary submatrices we have

$$T[\{1,3,5,7\}|\{2,4,6,8\}] = \begin{bmatrix} 0 & 0 & 1 & 0 \\ 0 & 0 & 1 & 1 \\ 1 & 1 & 0 & 0 \\ 0 & 1 & 0 & 0 \end{bmatrix}, \text{ and}$$

$$T[\{2,4,6,8\}|\{1,3,5,7\}] = \begin{bmatrix} 1 & 1 & 0 & 1 \\ 1 & 1 & 0 & 0 \\ 0 & 0 & 1 & 1 \\ 1 & 0 & 1 & 1 \end{bmatrix},$$

where $T[\{1,3,5,7\}|\{2,4,6,8\}] + T[\{2,4,6,8\}|\{1,3,5,7\}] = J_4$, the 4×4 matrix of all 1's. □

Let L_m denote the $m \times m$ transitive tournament matrix with all 1's below the main diagonal and all 0's on and above the main diagonal.

Theorem 4.3 *Let $R = (r_1, r_2, \ldots, r_n)$ be a nonnegative, nondecreasing integer vector such that $\mathcal{T}_H(R) \neq \emptyset$. Then there exists a Hankel tournament matrix T with score vector R such that*

$$T[\{1,3,\ldots,\lceil n/2\rceil] = L_{\lceil \frac{n}{2} \rceil} \text{ and } T[\{2,4,\ldots,\lfloor n/2\rfloor] = L_{\lfloor \frac{n}{2} \rfloor}.$$

Proof The proof follows from the proof given in [7] for the non-Hankel case but there is a slight twist since we have to show the existence of a matrix which is symmetric about the Hankel diagonal. Since $\mathcal{T}_H(R) \neq \emptyset$, for $k = 1, 2, \ldots, n$, we have

$$r_k + r_{n+1-k} = n-1 \quad (1 \leq k \leq n), \text{ and } \sum_{i=1}^{k} r_i \geq \binom{k}{2} \text{ with equality when } k = n.$$

After inserting $L_{\lceil \frac{n}{2} \rceil}$ as $T[\{1,3,\ldots,\lceil n/2\rceil]$ and $L_{\lfloor \frac{n}{2} \rfloor}$ as $T[\{2,4,\ldots,\lfloor n/2\rfloor]$, we consider two cases depending on the parity of n.

Case 1 ($n = 2m+1$): In this case we require an $(m+1) \times m$ (0, 1)-matrix B (corresponding to rows $1, 3, \ldots, 2m+1$ and columns $2, 4, \ldots, 2m$) with row sum vector $(r_{2i} - i + 1 : 1 \leq i \leq m)$ and column sum vector $(m + i - r_{2i-1} : 1 \leq i \leq m+1)$. Inserting that matrix in T and then reflecting it around the Hankel diagonal of T to complete T gives a half-transitive Hankel tournament matrix with score vector R.

Case 2 ($n = 2m$): In this case we require an $m \times m$ *Hankel symmetric* (0, 1)-matrix (corresponding to rows $1, 3, \ldots, 2m-1$ and columns $2, 4, \ldots, 2m$) with row sum vector $\hat{R} = (r_{2i-1} - i + 1 : 1 \leq i \leq m)$. Equivalently, after reordering the columns from last to first, we require an $m \times m$ symmetric (0, 1)-matrix B with row sum vector $\hat{R}$, that is, a loopy graph with degree sequence $\hat{R}$ where loops contribute 1 to a vertex degree. Then the entries corresponding to rows $2, 4, \ldots, 2m$ and columns $1, 3, \ldots, 2m-1$ are determined by the skew-symmetry property of a tournament. By Lemma 6.3.1 in [11] it is enough to find a (0, 1)-matrix B with row and column sum vector $\hat{R}$, since such a matrix can be modified to obtain a symmetric (0, 1)-matrix with row sum vector $\hat{R}$.

Thus for both Cases 1 and 2, the argument given in [12] (see also the proof of Theorem 5.4.2 in [7]) applies and completes the proof. □

5 Oriented Graphs

Tournaments are oriented graphs obtained by orienting the edges of a complete graph. But one may orient the edges of any graph to obtain an oriented graph. In this section we explore some properties and existence questions concerning oriented graphs which are related to our discussions in the previous sections.

Let G be a graph with vertex set $V = \{1, 2, \ldots, n\}$ and let the degree sequence of G be $D = (d_1, d_2, \ldots, d_n)$. Let $\vec{G}$ be obtained from G by orienting each of its edges. Let $R = (r_1, r_2, \ldots, r_n)$ be the *outdegree sequence* of $\vec{G}$ and let $S = (s_1, s_2, \ldots, s_n)$ be its *indegree sequence*. We have $R + S = D$ so that any two of D, R, S determine the other.

The book [3] on pages 446–447 discusses indegree sequences and outdegree sequences of oriented graphs (see also [15] and [16]). In particular, given a graph G, a certain network is defined to determine whether or not G has an orientation with a prescribed indegree sequence (then the outdegree sequence is determined by the degree sequence of G and the prescribed indegree sequence). It is remarked that from this one can derive necessary and sufficient conditions for the existence of such an orientation, but that the conditions are not especially illuminating.

The following necessary and sufficient conditions for a graph to have an orientation with a prescribed outdegree sequence (and so a prescribed indegree sequence) follows from more general results in [17].

Theorem 5.1 *Let $G = (V, E)$ be a graph with degree sequence $D = (d_1, d_2, \ldots, d_n)$. Let $R = (r_1, r_2, \ldots, r_n)$ be a sequence of nonnegative integers with $R \leq D$ (entrywise) and $\sum_{i=1}^{n} r_i = (\sum_{i=1}^{n} d_i)/2$. Then the following are equivalent:*

(a) *There is an orientation of G so that the resulting oriented graph has outdegree sequence R (and so indegree sequence D − R).*

(b) *Where E_X denotes the set of edges of G both of whose vertices are in X,*

$$|E_X| \leq \sum_{i\in X} r_i \quad (X \subseteq V).$$

(c) *Where $\widetilde{E}_X$ denotes the set of edges of G with at least one of its vertices in X,*

$$|\widetilde{E}_X| \geq \sum_{i\in X} r_i \quad (X \subseteq V).$$

In contrast to the above situation where a graph G is specified and then oriented, in [19] a theorem is proved concerning the existence of an oriented graph $\vec{G}$ with prescribed outdegree sequence $R = (r_1, r_2, \ldots, r_n)$ and indegree sequence $S = (s_1, s_2, \ldots, s_n)$ *but with G not specified.* This theorem does not directly give arithmetical conditions on R and S to guarantee the existence of G, since the conditions involve another existential criterion. Nonetheless this theorem, we think, is of sufficient interest. One way to view it is: Start with the complete graph K_n and put an arrow on some edges and remove other edges to get a prescribed indegree sequence and prescribed outdegree sequence. In this theorem one seeks an $n \times n$ asymmetric (0,1)-matrix with zeros on the main diagonal with prescribed row sum and column sum vectors. In [8] it is shown that an $R = (r_1, r_2, \ldots, r_n)$ is the row sum vector of a symmetric and Hankel symmetric (0, 1)-matrix if and only if R is palindromic and the well-known Gale–Ryser conditions hold.[4]

The proof in [19] is not easy to follow and not easily accessible. We give a proof of this theorem using similar ideas to those in [19] but presented differently.

We first prove the following lemma of some interest in itself.

Lemma 5.2 *Let $X = [x_{ij}]$ be an $n \times n$ symmetric matrix with entries taken from $\{0, r, b\}$ satisfying the following properties:*

(i) *X has only 0's on its main diagonal.*

(ii) *For each row (and therefore for each column) the number of entries equal to r is the same as the number of entries equal to b.*

Then the entries of X equal to r or b can be partitioned to give an $n \times n$ asymmetric matrix Y such that $X = Y + Y^t$. (Here $r+0$ and $s+0$ in $X = Y + Y^t$ are interpreted as r and s, respectively.)

Proof The matrix X determines a *colored graph* $G(X)$ with vertex set $V = \{1, 2, \ldots, n\}$ and edges colored red or blue: $\{i, j\}_r$ is a *red edge* provided $x_{ij} = r$ and $\{i, j\}_b$ is a *blue edge* provided $x_{ij} = b$ $(1 \leq i \neq j \leq n)$. It follows from (ii)

[4] The Gale–Ryser conditions tell when a (0, 1)-matrix exists with prescribed row sum and column sum vectors. See e.g. [7, 11].

by a standard argument that the edges of $G(X)$ can be partitioned into closed walks whose edges are alternatingly red and blue. Let W be such an *alternating walk*. Then W corresponds to a symmetric matrix X_W obtained from X by replacing with 0 all those entries which do not correspond to edges of W. (Thus X_W determines the colored graph whose edges are the edges of the walk W.) We follow the walk in its given direction. If the edge $\{i, j\}$ is an edge of W traversed from i to j, then we choose the entry (red or blue) at position (i, j) and discard the entry at position (j, i). This gives an asymmetric matrix X_W^* with the same row and column sums as X_W. To make this clear we give an example.

Consider the walk $1-3-5-8-6-7-2-4-1$ alternating between r and b, where

$$X_W = \left[\begin{array}{c|c|c|c|c|c|c|c} & & r & b & & & & \\ \hline & & & r & & & b & \\ \hline r & & & & b & & & \\ \hline b & r & & & & & & \\ \hline & & b & & & & & r \\ \hline & & & & & & r & b \\ \hline & b & & & & r & & \\ \hline & & & & r & b & & \end{array}\right] \longrightarrow X_W^* = \left[\begin{array}{c|c|c|c|c|c|c|c} & & r & & & & & \\ \hline & & & r & & & & \\ \hline & & & & b & & & \\ \hline b & & & & & & & \\ \hline & & & & & & & r \\ \hline & & & & & & r & \\ \hline & b & & & & & & \\ \hline & & & & & b & & \end{array}\right].$$

We do this for each such walk W and arrive at the asymmetric matrix Y. $\square$

Now let $R = (r_1, r_2, \ldots, r_n)$ and $S = (s_1, s_2, \ldots, s_n)$ be vectors of nonnegative integers, and define $D = (d_1, d_2, \ldots, d_n)$ by $d_i = r_i + s_i$ for $i = 1, 2, \ldots, n$. Let

1. $\mathcal{A}_0(R, S)$ be the set of $n \times n$ (0, 1)-matrices with zeros on the main diagonal having row sum vector R and column sum vector S (equivalently, digraphs without loops with outdegree sequence R and indegree sequence S).
2. Let $\mathcal{A}_0^*(R, S)$ be the subset of $\mathcal{A}_0(R, S)$ of those matrices that are asymmetric (equivalently, oriented graphs with outdegree sequence R and indegree sequence S).
3. Let $\mathcal{A}_0(D)$ be the set of $n \times n$ symmetric matrices with row sum vector D with zeros on the main diagonal (equivalently, graphs without loops, having degree sequence D).

The following theorem is given in [19] in the language of graphs which we translate using the above definitions.

Theorem 5.3 *The set $\mathcal{A}_0^*(R, S)$ is nonempty if and only if there is a matrix $A \in \mathcal{A}_0(R, S)$ and a matrix $B \in \mathcal{A}_0(D)$ with $A \leq B$ (entrywise).*

Proof If there exists $A \in \mathcal{A}_0^*(R, S)$, then since $\mathcal{A}_0^*(R, S) \subseteq \mathcal{A}_0(R, S)$, we have $A \in \mathcal{A}_0(R, S)$ and $B = A + A^t \in \mathcal{A}_0(D)$ where $A \leq B$.

Conversely, suppose there exists $A \in \mathcal{A}_0(R, S)$ and a matrix $B \in \mathcal{A}_0(D)$ with $A \leq B$. The matrix $C = A + A^t$ is an $n \times n$ symmetric (0, 1, 2)-matrix with row sum vector D and all 0's on its main diagonal. We construct an $n \times n$ symmetric $(0, r, b)$-matrix X as follows (it is convenient to think of the positions occupied by 0's in X as empty positions since they play no role in what follows):

(0) Put a 0 in all pairs of symmetric positions which equal 1 in C or are 0 in B. (Thus a 0 in X either corresponds to a pair of symmetric positions which are 0 in both A and B, or a pair of symmetric positions exactly one of which is 1 in A and both of which are 1 in B.)

(r) Put an r in all positions of X which equal 2 in C. (Thus an r corresponds to a pair of symmetric positions both of which are 1 in A and 1 in B.)

(b) Put a b in all pairs of symmetric positions of X which equal 1 in B and equal 0 in A.

In each row and column of the symmetric matrix X, the number of r's equals the number of b's. By Lemma 5.2 the entries of X equal to r or b can be partitioned to give an asymmetric matrix Y such that $X = Y + Y^t$. Let Y' be the $(0, 1, -1)$-matrix obtained from Y by replacing the r's with -1's and the s's with 1's. Then $A + Y$ is an asymmetric $(0, 1)$-matrix with $A + Y \in \mathcal{A}_0^*(R, S)$ and $A + Y \le B$. □

Two basic results in the theory of digraphs are: (1) A digraph $\Gamma = (V, E)$ is strongly connected if and only if for every subset U of V with $\emptyset \ne U \ne V$, there is an edge from U to its complement $\overline{U}$ (that is, Γ does not have a directed cut), and (2) A graph has a strongly connected orientation if and only if it is connected and does not have a bridge (an edge whose removal leaves a graph that is not connected). For completeness we mention the following theorem from [17].

Theorem 5.4 *Let $G = (V, E)$ be a bridgeless graph with degree sequence $D = (d_1, d_2, \ldots, d_n)$. Let $R = (r_1, r_2, \ldots, r_n)$ be a sequence of nonnegative integers with $R \le D$ (entrywise) and $\sum_{i=1}^n r_i = (\sum_{i=1}^n d_i)/2$. Then G has a strongly connected orientation with outdegree sequence R (and so indegree sequence $D - R$) if and only if*

$$|E_X| + \#G(\overline{X}) \le \sum_{i \in X} r_i \quad (X \subseteq V),$$

where $\#G(\overline{X})$ equals the number of connected components in the subgraph $G(\overline{X})$ with vertex set equal to the complement $\overline{X}$ of X. □

Theorem 5.4 can be phrased in terms of irreducible matrices. One begins with an irreducible symmetric $n \times n$ $(0, 1)$-matrix A (that is, PAP^t is not a nontrivial direct sum for any permutation matrix P) such that replacing a pair of symmetric 1's with 0's does never result in a reducible matrix, and then one wants an asymmetric, irreducible $(0, 1)$-matrix $B \le A$ such that B has a specified row sum vector.

Remark 5.5 In Theorem 5.3 we can replace $\mathcal{A}_0^*(R, S)$ with the set $\mathcal{A}_0^h(R, S)$ of asymmetric Hankel $(0, 1)$-matrices with 0's on the Hankel diagonal and with row sum vector R and column sum vector S. We only need to take a matrix $A \in$

$\mathcal{A}_0^h(R, S)$, reorder the columns from last to first to get a matrix $A' \in \mathcal{A}_0^*(R, S')$ where S' is S reordered from last component to first component. So this is not more general or really different. But suppose we consider $n \times n$ (0, 1)-matrices $A = [a_{ij}]$, with all 0's on both the main and Hankel diagonal, which are asymmetric both with respect to the main diagonal and Hankel diagonal (so no two 1's are opposites relative to the main or Hankel diagonals). For example,

$$\begin{bmatrix} 0 & 0 & 0 & 1 & 0 \\ 1 & 0 & 1 & 0 & 0 \\ 0 & 0 & 0 & 0 & 1 \\ 0 & 0 & 1 & 0 & 1 \\ 0 & 1 & 0 & 0 & 0 \end{bmatrix}. \tag{4}$$

The row sum vector and column sum vector are $R = (1, 2, 1, 2, 1)$ and $S = (1, 1, 2, 1, 2)$. Thus we have a graph G whose adjacency matrix (5) has 0's on both the main and Hankel diagonals, and is both symmetric and Hankel symmetric.

$$\begin{bmatrix} 0 & 1 & 0 & 1 & 0 \\ 1 & 0 & 1 & 0 & 1 \\ 0 & 1 & 0 & 1 & 1 \\ 1 & 0 & 1 & 0 & 1 \\ 0 & 1 & 0 & 1 & 0 \end{bmatrix}, \tag{5}$$

and an orientation of G (with respect to the main diagonal) and Hankel orientation (with respect to the Hankel diagonal). (Here Hankel orientation means: if $i \to j$ is an edge, then $n - j + 1 \to n - i + 1$ is not an edge. Just as an oriented graph is a "partial tournament," an oriented and Hankel oriented digraph is a partial Hankel tournament (with the restriction of 0's on the Hankel diagonal). □

We conclude this section with some questions and remarks.

Question 5.6 *Recall that Theorem 1.2 from [9] characterizes score vectors of Hankel tournaments. So one might consider the following. Let $\mathcal{A}_0^b(R, S)$ be the set of all* (0, 1)*-matrices with all 0's on the main diagonal which are both asymmetric and Hankel asymmetric with row sum vector R and column sum vector S. As before let $D = R + S$. Let $\mathcal{A}_0^\#(D)$ be the set of symmetric and Hankel symmetric $n \times n$* (0, 1)*-matrices with row sum vector D and* 0*'s on the main diagonal. Is the following true? $\mathcal{A}_0^b(R, S)$ is nonempty if and only if there is a matrix $A \in \mathcal{A}_0(R, S)$ and a matrix $B \in \mathcal{A}_0^\#(D)$ with $A \leq B$.* □

Remark 5.7 Let A be an $n \times n$ (0, 1)-matrix with all 0's on both the main diagonal and Hankel diagonal. Consider the following three possibilities:

I. A is both symmetric and Hankel symmetric.
II. A is both asymmetric and Hankel asymmetric.

III. A is both asymmetric and Hankel symmetric (or symmetric and Hankel asymmetric).

By using the Hankel diagonal of A in place of the main diagonal (equivalently reordering the columns from n to 1 to obtain A^*), one obtains a graph in case of a Hankel symmetric matrix and an oriented graph in case of a Hankel asymmetric matrix. In Case I., A is the adjacency matrix of a graph G and a graph H defined using A^*. In Case II., A is the adjacency matrix of an oriented graph G' and an oriented graph H' defined using A^*. In Case III. A is the adjacency matrix of an oriented graph and a graph defined using A^*. For example, regarding I., consider the symmetric and Hankel symmetric matrix

$$A = \begin{bmatrix} 0&1&0&1&0 \\ 1&0&1&0&1 \\ 0&1&0&1&0 \\ 1&0&1&0&1 \\ 0&1&0&1&0 \end{bmatrix},$$

where the Hankel graph is the graph of the matrix A^* obtained from A by reading the columns in the reverse order 5, 4, 3, 2, 1

$$A^* = \begin{bmatrix} 0&1&0&1&0 \\ 1&0&1&0&1 \\ 0&1&0&1&0 \\ 1&0&1&0&1 \\ 0&1&0&1&0 \end{bmatrix}.$$

Notice that in this case $A = A^*$ (since the columns of A form a palindromic sequence of vectors, and because A is symmetric, so do the rows); in this case, the two graphs are the same as labeled graphs. In fact, as we know symmetric and Hankel symmetric imply centrosymmetric. Since the columns of A (and also rows) are invariant under a 180° rotation, A^* must equal A. In general, column i and column $n+1-i$ would be reverses of one another. □

Remark 5.8 If $A = [a_{ij}]$ is symmetric and has palindromic rows, then it has palindromic columns and is Hankel symmetric as well:

$$a_{ij} = a_{i,n+1-j} = a_{n+1-j,i} = a_{n+1-j,n+1-i}.$$

The graph of A is the edge-disjoint union of 4-cycles, with a path of length 2 if n is odd. As an example, using symmetric and palindromic rows:

$$\begin{bmatrix} 0 & a & & & a & 0 \\ a' & 0 & b & b & 0 & a' \\ & b' & 0 & 0 & b' & \\ & b' & 0 & 0 & b' & \\ a' & 0 & b & b & 0 & a' \\ 0 & a & & & a & 0 \end{bmatrix} \qquad (a = a' = b = b' = 1).$$

The degree sequence of such a graph must itself be palindromic; if n is even the degrees are also even. □

Question 5.9 *Some questions related to the previous discussion:*

(a) *Characterize the degree sequences $D = (d_1, d_2, \ldots, d_n)$ of graphs whose adjacency matrices have palindromic rows (so symmetric, Hankel symmetric, and palindromic columns). So D must be the degree sequence of a graph and must be palindromic. If n is even, then the d_i's must be even as well.*

More generally, one can ask: Let $R = (r_1, r_2, \ldots, r_m)$ and $S = (s_1, s_2, \ldots, s_n)$ be palindromic, nonnegative integral vectors. When does there exist a palindromic matrix in $\mathcal{A}(R, S)$, that is, a matrix in $\mathcal{A}(R, S)$ which has palindromic rows and palindromic columns? Thus $\mathcal{A}(R, S)$ must be nonempty, and R and S must be palindromic vectors.

(b) *When does there exists an $n \times n$ symmetric $(0, 1)$-matrix A with all zeros on the main diagonal and the Hankel diagonal, having row sum vector R? The well-known Erdős–Gallai theorem takes care of the case where only zeros on the main diagonal are required. So we require a graph with a prescribed degree sequence R so that there are no "hd-edges."*

(c) *Referring to (a) suppose we also insist that A is Hankel symmetric.*

(d) *When does there exist an $n \times n$ symmetric and Hankel symmetric $(0, 1)$-matrix A with all zeros on the main diagonal and on the Hankel diagonal having palindromic row sum vector R?*

□

6 Hankel 2-Tournaments

Let n be a positive integer. A 2-*tournament* (see [2, 7]) of order n is a round-robin tournament in which games may now end in ties. As with tournaments we use "2-tournament" and "2-tournament matrix" interchangeably. If we record 0 for a loss, 1 for a tie, and 2 for a win, a 2-tournament is an $n \times n$ $(0, 1, 2)$-matrix $T = [t_{ij}]$ with 0's on the main diagonal, where the 1's occur in symmetric positions across the main diagonal, and the 0's and 2's occur opposite to each other in symmetric positions. In particular, once one knows the positions of the 2's (the matrix $T^{(2)}$ below), then a 2-tournament T is uniquely determined:

$$T = T^{(1)} + 2T^{(2)}; \tag{6}$$

where $T^{(1)}$ is an $n \times n$ symmetric (0, 1)-matrix with all 0's on its main diagonal, and $T^{(2)}$ is an asymmetric (0, 1)-matrix with all 0's on its main diagonal and with the sum of symmetrically opposite elements equal to 0 or 1. Thus T_1 is the adjacency matrix of a graph G, and $T^{(2)}$ is the adjacency matrix of a digraph Γ on vertices $\{1, 2, \ldots, n\}$ such that if $i \neq j$ and $i \to j$ is an edge of Γ, then $j \to i$ is not an edge of Γ, that is, $T^{(2)}$ is the adjacency matrix of an oriented graph. If $\{i, j\}$ is an edge of G, then neither $i \to j$ nor $j \to i$ is an edge of Γ. Thus, *a 2-tournament can be viewed as the result of orienting some of the edges of a complete graph* K_n.

A 2-tournament matrix T has a *score vector* $R = (r_1, r_2, \ldots, r_n)$ where r_i records the total number of points received by player i. Thus r_i is the sum of the elements in row i of T, where the sum of the elements in column i of T equals $2(n-1) - r_i$. The 2-tournament T (its adjacency matrix) is a Hankel matrix if and only if the symmetric matrix $T^{(1)}$ and the asymmetric matrix $T^{(2)}$ are both Hankel matrices. We denote by $\mathcal{T}_2^h(R)$ the set of all Hankel 2-tournaments with score vector R.

Example 6.1 Let $n = 7$, and let T, T_1, and T_2 be given by

$$\begin{bmatrix} 0&1&2&1&1&0&1\\ 1&0&1&2&1&1&0\\ 0&1&0&1&2&1&1\\ 1&0&1&0&1&2&1\\ 1&1&0&1&0&1&2\\ 2&1&1&0&1&0&1\\ 1&2&1&1&0&1&0 \end{bmatrix} = \begin{bmatrix} 0&1&0&1&1&0&1\\ 1&0&1&0&1&1&0\\ 0&1&0&1&0&1&1\\ 1&0&1&0&1&0&1\\ 1&1&0&1&0&1&0\\ 0&1&1&0&1&0&1\\ 1&0&1&1&0&1&0 \end{bmatrix} + 2 \begin{bmatrix} 0&0&1&0&0&0&0\\ 0&0&0&1&0&0&0\\ 0&0&0&0&1&0&0\\ 0&0&0&0&0&1&0\\ 0&0&0&0&0&0&1\\ 1&0&0&0&0&0&0\\ 0&1&0&0&0&0&0 \end{bmatrix}.$$

Then T is a Hankel 2-tournament. The score vector of T is (6, 6, 6, 6, 6, 6, 6) and thus T is a *regular 2-tournament*. This example can be generalized to all odd $n \geq 3$ as follows. Let P_n denote the $n \times n$ permutation matrix corresponding to the full cycle permutation $(2, 3, \ldots, n, 1)$. Thus $P_n^n = I_n$, and hence P^k and P^{n-k} are inverses of one another. Consider the set of pairs $U = \{\{1, n-1\}, \{2, n-2\}, \ldots, \{(n-1)/2, (n+1)/2\}\}$ and let $V_1 \subseteq U$ with $V_2 = U \setminus V_1$. Let V_2' be obtained from V_2 by choosing one integer in each of its pairs. Then

$$\sum_{\{i,n-i\} \in V_1} (P_n^i + P_n^{n-i}) + 2 \sum_{j \in V_2'} P^j$$

is a Hankel regular 2-tournament. □

Recall that the score vectors R of Hankel tournaments are characterized by Landau's conditions and the conditions that $r_i + r_{n+1-i} = n - 1$ for $i = 1, 2, \ldots, n$. Assuming that R is nondecreasing, then in [2] (see also Theorem 5.9.1 of [7]) the score vectors of 2-tournaments are characterized by the Landau-like inequalities

$$\sum_{i=1}^{k} r_i \geq k(k-1) \quad (k = 1, 2, \ldots, n) \text{ with equality for } k = n.$$

For Hankel 2-tournaments, we must have in addition that

$$r_i + r_{n+1-i} = 2(n-1) \quad (i = 1, 2, \ldots, n).$$

Note that if T_2 is a (Hankel) asymmetric (0, 1)-matrix with all 0's on its main diagonal, then necessarily the matrix $T_1 = (J_n - I_n) - (T_2 + T_2^t)$ is a (Hankel) symmetric (0, 1)-matrix with all 0's on its main diagonal.

In this section, we modify the proof given in [9] for the existence of a Hankel tournament with a prescribed score vector in order to obtain the corresponding existence theorem for Hankel 2-tournaments. We also adapt the algorithm in [9] for Hankel tournaments to apply to Hankel 2-tournaments and discuss other similar considerations. Except for the proof of existence given by Theorem 6.3, and that of Lemma 6.4, we shall omit details.

As in Lemma 10 of [9] for tournaments, we have the following lemma for 2-tournaments.

Lemma 6.2 *Let R be a vector of nonnegative integers and let R' be its nondecreasing rearrangement. If there exists a Hankel 2-tournament with score vector R, then by a simultaneous permutation of rows and columns we obtain a Hankel 2-tournament with score vector R'.*

Proof The proof is as in [9] requiring only to replace "tournament" with "2-tournament." □

From Lemma 6.2 it follows that in investigating Hankel 2-tournaments, it suffices to assume the score vector is nondecreasing.

Theorem 6.3 *Let $R = (r_1, r_2, \ldots, r_n)$ be a nondecreasing, nonnegative integral vector. Then there exists a Hankel 2-tournament matrix A whose row sum vector is R if and only if*

$$\sum_{i=1}^{k} r_i \geq k(k-1) \text{ for } k = 1, 2, \ldots, n \text{ with inequality } k = n, \tag{7}$$

and

$$r_i + r_{n+1-i} = 2(n-1) \text{ for } i = 1, 2, \ldots, n. \tag{8}$$

Proof It is clear that both (7) and (8) hold for a Hankel 2-tournament with score vector R. Now suppose there exists a vector R satisfying (7) and (8) and there does not exist a 2-tournament such that the row sum vector is R. Choose such an R with n smallest and, for that n, with r_1 smallest.

Case 1. Suppose for each $1 \le k \le n-1$, we have strict inequality in (7). In this case, we have $r_1 > 0$ and we consider $R' = (r'_1, r'_2, \ldots, r'_n) = (r_1 - 1, r_2, \ldots, r_n + 1)$ where R' therefore also satisfies (7) and (8). By the minimality condition on r_1, there exists a Hankel 2-tournament matrix $T' = [t'_{ij}]$ with score vector R'. We have $r'_n - r'_1 = (r_n + 1) - (r_1 - 1) \ge 2$. If $t_{n1} = 2$, then replacing $t_{n1} = 2$ with 1 and $t_{1n} = 0$ with 1 gives a 2-tournament with row sum vector R.

Now assume that $t_{n1} \ne 2$ so that there exists an integer p with $2 \le p \le n-1$, such that one of the three possibilities, (i), (ii), and (iii) below, occurs. We focus on the principal submatrix $T'[1, p, q, n]$ of T' determined by rows and columns $1, p, q, n$ where $q = n + 1 - p$.

(i) $t_{1p} = 0$ and $t_{np} = 1$. Since T is a Hankel 2-tournament, we have $t_{n+1-p,n} = 0$, $t_{n+1-p,1} = 1$, $t_{p1} = 2$, and $t_{pn} = 1$. Thus with $q = n + 1 - p$, we have

$$T'[1,p,q,n] = \left[\begin{array}{c|c|c|c} 0 & 0 & 1 & 2-x \\ \hline 2 & 0 & a & 1 \\ \hline 1 & 2-a & 0 & 0 \\ \hline x & 1 & 2 & 0 \end{array}\right] \quad \text{where } x \ne 2,$$

and the a on the Hankel diagonal in position (p, q) is 0, 1, or 2. Replacing as shown below according to the value of a, we obtain the desired 2-tournament:

$[a = 0:]$

$$T'[1,p,q,n] = \left[\begin{array}{c|c|c|c} 0 & 0 & 1 & 2-x \\ \hline 2 & 0 & 0 & 1 \\ \hline 1 & 2 & 0 & 0 \\ \hline x & 1 & 2 & 0 \end{array}\right] \rightarrow \left[\begin{array}{c|c|c|c} 0 & 1 & 1 & 2-x \\ \hline 1 & 0 & 1 & 1 \\ \hline 1 & 1 & 0 & 1 \\ \hline x & 1 & 1 & 0 \end{array}\right].$$

$[a = 1:]$

$$T'[1,p,q,n] = \left[\begin{array}{c|c|c|c} 0 & 0 & 1 & 2-x \\ \hline 2 & 0 & 1 & 1 \\ \hline 1 & 1 & 0 & 0 \\ \hline x & 1 & 2 & 0 \end{array}\right] \rightarrow \left[\begin{array}{c|c|c|c} 0 & 1 & 1 & 2-x \\ \hline 1 & 0 & 2 & 1 \\ \hline 1 & 0 & 0 & 1 \\ \hline x & 1 & 1 & 0 \end{array}\right].$$

$[a = 2:]$

$$T'[1,p,q,n] = \left[\begin{array}{c|c|c|c} 0 & 0 & 1 & 2-x \\ \hline 2 & 0 & 2 & 1 \\ \hline 1 & 0 & 0 & 0 \\ \hline x & 1 & 2 & 0 \end{array}\right] \rightarrow \left[\begin{array}{c|c|c|c} 0 & 0 & 2 & 2-x \\ \hline 2 & 0 & 1 & 2 \\ \hline 0 & 1 & 0 & 0 \\ \hline x & 0 & 2 & 0 \end{array}\right].$$

(ii) $t'_{1p} = 0$ and $t'_{np} = 2$. Since T' is a Hankel 2-tournament, we have $t'_{n+1-p,n} = 0$, $t'_{n+1-p,1} = 2$, $t'_{p1} = 2$, and $t'_{pn} = 0$. Thus with $q = n-p+1$, we have

$$T'[1,p,q,n] = \left[\begin{array}{c|c|c|c} 0 & 0 & 0 & 2-x \\ \hline 2 & 0 & a & 0 \\ \hline 2 & 2-a & 0 & 0 \\ \hline x & 2 & 2 & 0 \end{array}\right] \quad \text{where } x \neq 2,$$

and the a on the Hankel diagonal is 0 or 1 or 2. Replacing as shown below according to the value of a, we obtain the desired 2-tournament:
[$a = 0$:]

$$T'[1,p,q,n] = \left[\begin{array}{c|c|c|c} 0 & 0 & 0 & 2-x \\ \hline 2 & 0 & 0 & 0 \\ \hline 2 & 2 & 0 & 0 \\ \hline x & 2 & 2 & 0 \end{array}\right] \to T'[1,p,q,n] = \left[\begin{array}{c|c|c|c} 0 & 1 & 0 & 2-x \\ \hline 1 & 0 & 1 & 0 \\ \hline 2 & 1 & 0 & 1 \\ \hline x & 2 & 1 & 0 \end{array}\right].$$

[$a = 1$:]

$$T'[1,p,q,n] = \left[\begin{array}{c|c|c|c} 0 & 0 & 0 & 2-x \\ \hline 2 & 0 & 1 & 0 \\ \hline 2 & 1 & 0 & 0 \\ \hline x & 2 & 2 & 0 \end{array}\right] \to T'[1,p,q,n] = \left[\begin{array}{c|c|c|c} 0 & 1 & 0 & 2-x \\ \hline 1 & 0 & 2 & 0 \\ \hline 2 & 0 & 0 & 1 \\ \hline x & 2 & 1 & 0 \end{array}\right].$$

[$a = 2$:]

$$T'[1,p,q,n] = \left[\begin{array}{c|c|c|c} 0 & 0 & 0 & 2-x \\ \hline 2 & 0 & 2 & 0 \\ \hline 2 & 0 & 0 & 0 \\ \hline x & 2 & 2 & 0 \end{array}\right] \to T'[1,p,q,n] = \left[\begin{array}{c|c|c|c} 0 & 0 & 1 & 2-x \\ \hline 2 & 0 & 1 & 1 \\ \hline 1 & 1 & 0 & 0 \\ \hline x & 1 & 2 & 0 \end{array}\right].$$

(iii) $t'_{1p} = 1$ and $t'_{np} = 2$. Since T' is a Hankel 2-tournament, we have $t'_{n+1-p,n} = 1$, $t'_{n+1-p,1} = 2$, $t'_{p1} = 1$, and $t'_{pn} = 0$. Thus with $q = n+1-p$, we have

$$T'[1,p,q,n] = \left[\begin{array}{c|c|c|c} 0 & 1 & 0 & 2-x \\ \hline 1 & 0 & a & 0 \\ \hline 2 & 2-a & 0 & 1 \\ \hline x & 2 & 1 & 0 \end{array}\right] \quad \text{where } x \neq 2,$$

and the a on the Hankel diagonal in position (p, q) is 0 or 1 or 2. Replacing as shown below according to the value of a, we obtain the desired 2-tournament:

$[a = 0{:}]$

$$T'[1, p, q, n] = \left[\begin{array}{c|c|c|c} 0 & 1 & 0 & 2-x \\ \hline 1 & 0 & 0 & 0 \\ \hline 2 & 1 & 0 & 1 \\ \hline x & 2 & 1 & 0 \end{array}\right] \rightarrow \left[\begin{array}{c|c|c|c} 0 & 2 & 0 & 2-x \\ \hline 0 & 0 & 1 & 0 \\ \hline 2 & 1 & 0 & 2 \\ \hline x & 2 & 0 & 0 \end{array}\right].$$

$[a = 1{:}]$

$$T'[1, p, q, n] = \left[\begin{array}{c|c|c|c} 0 & 1 & 0 & 2-x \\ \hline 1 & 0 & 1 & 0 \\ \hline 2 & 1 & 0 & 1 \\ \hline x & 2 & 1 & 0 \end{array}\right] \rightarrow \left[\begin{array}{c|c|c|c} 0 & 2 & 0 & 2-x \\ \hline 0 & 0 & 2 & 0 \\ \hline 2 & 0 & 0 & 2 \\ \hline x & 2 & 0 & 0 \end{array}\right].$$

$[a = 2{:}]$

$$T'[1, p, q, n] = \left[\begin{array}{c|c|c|c} 0 & 1 & 0 & 2-x \\ \hline 1 & 0 & 2 & 0 \\ \hline 2 & 0 & 0 & 1 \\ \hline x & 2 & 1 & 0 \end{array}\right] \rightarrow \left[\begin{array}{c|c|c|c} 0 & 1 & 1 & 2-x \\ \hline 1 & 0 & 1 & 1 \\ \hline 1 & 1 & 0 & 1 \\ \hline x & 1 & 1 & 0 \end{array}\right].$$

This finishes Case 1.

Case 2. Suppose that there exists a k with $1 \le k \le n$ such that $\sum_{i=1}^{k} r_i = k(k-1)$. We have

$$\begin{aligned}\sum_{i=1}^{n-k} r_i &= n(n-1) - \sum_{i=n-k+1}^{n} r_i \\ &= n(n-1) - \sum_{i=1}^{k}(2(n-1) - r_i) \\ &= n(n-1) - 2k(n-1) + k(k-1) = (n-k)(n-k-1).\end{aligned}$$

Hence we may assume that $k \le \frac{n}{2}$, and $\sum_{i=1}^{n-k} r_i = (n-k)(n-k-1)$. Since

$$r_{k+1} = \sum_{i=1}^{k+1} r_i - \sum_{i=1}^{k} r_i \ge (k+1)k - k(k-1) = 2k,$$

the monotonicity assumption implies that $r_i \geq 2k$ for $k+1 \leq i \leq n$. Similarly

$$r_k = \sum_{i=1}^{k} r_i - \sum_{i=1}^{k-1} r_i \leq k(k-1) - (k-1)(k-2) = 2(k-1).$$

It follows from Avery's theorem that there exists a 2-tournament T' with score vector $R_1 = (r_1, r_2, \ldots, r_k)$. Since $r_i + r_{n+1-i} = 2(n-1)$ for all i and $r_k \leq 2(k-1)$, we have $r_{n-k+1} \geq 2(n-1) - 2(k-1) = 2(n-k)$. The monotonicity assumption on R now implies that $R_2 = (r_{n-k+1} - 2(n-k), r_{n-k+2} - 2(n-k), \ldots, r_n - 2(n-k))$ is a vector of nonnegative integers. Since for $1 \leq i \leq k$, we have

$$r_i + (r_{n+1-i} - 2(n-k)) = r_i + r_{n+1-i} - 2(n-k) = 2(n-1) - 2(n-k) = 2(k-1),$$

then $s_i = r_{n+1-i-2(n-k)}$ is the ith column sum of T'. Hence the row sum vector of T'^h equals R_2. Thus

$$T = \left[\begin{array}{c|c|c} T' & O_{k,n-2k} & O_{k,k} \\ \hline 2J_{n-2k,k} & \hat{T} & O_{n-2k,k} \\ \hline 2J_{k,k} & 2J_{k,n-2k} & T'^h \end{array}\right]$$

is a Hankel 2-tournament with score vector R, provided we can choose $\hat{T}$ as an $(n-2k) \times (n-2k)$ Hankel 2-tournament with score vector $(r_{k+1} - 2k, r_{k+2} - 2k, \ldots, r_{n-k} - 2k)$, where $0 \leq r_{k+1} - k \leq r_{k+2} - k \leq \ldots \leq r_{n-k} - k$. In addition, for $1 \leq l \leq n-2k$,

$$\sum_{j=1}^{l} (r_{k+j} - 2k) = \left(\sum_{i=1}^{k+l} r_i - \sum_{i=1}^{k} r_i\right) - 2lk \geq (k+l)(k+l-1) - k(k-1)$$
$$-2lk = l(l-1).$$

Since $r_{k+j} + r_{n+1-(k+j)} = 2(n-1)$, we have

$$(r_{k+j} - 2k) + (r_{n+1-(k+j)} - 2k) = 2(n-1) - 4k = 2(n-2k-1).$$

The minimality assumption on n now implies that the Hankel 2-tournament $\hat{T}$ exists, and hence we have a Hankel 2-tournament with score vector R. This contradiction completes this case, and the theorem now holds. □

Let $T = T^{(1)} + 2T^{(2)}$ be an $n \times n$ Hankel 2-tournament. We define the *2-digraph* $\Gamma_2(T)$ of T to be the digraph with vertex set $\{1, 2, \ldots, n\}$ in which there is an edge from i to j provided the (i, j)-entry of T equals 2 (thus the (i, j)-entry of $T^{(2)}$ equals 1). Note that $\Gamma_2(T)$ is an oriented graph. We next show that if there exists a

Hankel 2-tournament with score vector R, then there exists such a 2-tournament for which the 2-digraph is transitive.

Lemma 6.4 *Let $R = (r_1, r_2, \ldots, r_n)$ be such that $\mathcal{T}_2^h(R) \neq \emptyset$. Then there exists a 2-tournament in $\mathcal{T}_2^h(R)$ whose 2-digraph is transitive.*

Proof We need to show that there is a 2-tournament T' in $\mathcal{T}_2^h(R)$ such that $\Gamma_2(T')$ does not have a 3-cycle nor a path $i \to j \to k$ for which $i \not\to k$.

Case 1. Let $T \in \mathcal{T}_2^h(R)$. If (i) does not hold, then T has a principal submatrix equal to

$$T[i, j, k] = \left[\begin{array}{c|c|c} 0 & 2 & 0 \\ \hline 0 & 0 & 2 \\ \hline 2 & 0 & 0 \end{array}\right].$$

There are three essentially different possibilities: $\{i, j, k\} \cap \{n+1-i, n+1-j, n+1-k\} = \emptyset$, $k = n+1-i$, and $k = n+1-k$. In the first instance we also have

$$T[n+1-k, n+1-j, n+1-i] = \left[\begin{array}{c|c|c} 0 & 2 & 0 \\ \hline 0 & 0 & 2 \\ \hline 2 & 0 & 0 \end{array}\right],$$

where $T[i, j, k]$ and $T[n+1-k, n+1-j, n+1-i]$ do not have a common position. Replacing each of them with

$$\left[\begin{array}{c|c|c} 0 & 1 & 1 \\ \hline 1 & 0 & 1 \\ \hline 1 & 1 & 0 \end{array}\right]$$

we obtain a 2-tournament in $\mathcal{T}_2^h(R)$ with three fewer 2's. In the second instance, we have a 4×4 principal submatrix of the form

$$T[i = n+1-k, j, k = n+1-i, n+1-j] = \left[\begin{array}{c|c|c|c} 0 & 2 & 0 & * \\ \hline 0 & 0 & 2 & 0 \\ \hline 2 & 0 & 0 & 2 \\ \hline * & 2 & 0 & 0 \end{array}\right]$$

with two 3-cycles. Replacing this submatrix with

$$\begin{bmatrix} 0 & 1 & 1 & * \\ 1 & 0 & 0 & 1 \\ 1 & 2 & 0 & 1 \\ * & 1 & 1 & 0 \end{bmatrix}$$

we obtain a 2-tournament in $\mathcal{T}_2^h(R)$ with four fewer 2's. In the third instance, we have a 5×5 principal submatrix of the form

$$T[i, j, k = n+1-k, n+1-j, n+1-i] = \begin{bmatrix} 0 & 2 & 0 & * & * \\ 0 & 0 & 2 & * & * \\ 2 & 0 & 0 & 2 & 0 \\ * & * & 0 & 0 & 2 \\ * & * & 2 & 0 & 0 \end{bmatrix}.$$

Replacing this submatrix with

$$\begin{bmatrix} 0 & 1 & 1 & * & * \\ 1 & 0 & 1 & * & * \\ 1 & 1 & 0 & 1 & 1 \\ * & * & 1 & 0 & 1 \\ * & * & 1 & 1 & 0 \end{bmatrix}$$

we obtain a 2-tournament in $\mathcal{T}_2^h(R)$ with six fewer 2's.

Case 2. Let $T \in \mathcal{T}_2^h(R)$. Suppose that T has a path $i \to j \to k$ for which $i \not\to k$. Then T has a principal submatrix equal to

$$T[i, j, k] = \begin{bmatrix} 0 & 2 & 1 \\ 0 & 0 & 2 \\ 1 & 0 & 0 \end{bmatrix}.$$

Replacing this submatrix with

$$\begin{bmatrix} 0 & 1 & 2 \\ 1 & 0 & 1 \\ 0 & 1 & 0 \end{bmatrix},$$

we obtain a 2-tournament in $\mathcal{T}_2^h(R)$ with one fewer 2.

It follows that the 2-digraph of some 2-tournament in $\mathcal{T}_2^h(R)$ is transitive. □

We call a nondecreasing, nonnegative integral vector $R = (r_1, r_2, \ldots, r_n)$ a *Hankel 2-tournament score vector* provided it satisfies (7) and (8).

Let k be a nonnegative integer. A sequence $(r_1, r_2, \ldots, r_n)$ of integers is *k-nearly nondecreasing* provided that $r_j \geq r_i - k$ $(1 \leq i < j \leq k)$. As in [9] for the Landau inequalities, we have the following whose proof follows the corresponding proof in [9].

Lemma 6.5 *Let* $R = (r_1, r_2, \ldots, r_n)$ *be a 2-nearly nondecreasing sequence of nonnegative integers satisfying* (7). *Then the nondecreasing rearrangement of R also satisfies* (7).

Proof Let i and j be such that $i < j$ and $r_i - 2 \leq r_j \leq r_i - 1$. Then it follows easily that there exists p with $i \leq p$ such that $r_{p+1} = r_p - l$ where $l = 1$ or 2. Let R' be obtained from R by switching r_p and r_{p+1}. Suppose that $(\sum_{i=1}^{p-1} r_i) + r_{p+1} < p(p-1)$. Since $\sum_{i=1}^{p} r_i \geq p(p-1)$, we have that $\sum_{i=1}^{p} r_i = p(p-1) + h$ where $0 \leq h \leq l-1$. Since R satisfies (7), we compute that

$$r_{p+1} = \left(\sum_{i=1}^{p+1} r_i\right) - \left(\sum_{i=1}^{p} r_i\right) \geq (p+1)p - p(p-1) + h = 2p - h,$$

and

$$r_p = \left(\sum_{i=1}^{p} r_i\right) - \left(\sum_{i=1}^{p-1} r_i\right) \leq p(p-1) + h - (p-1)(p-2) = 2(p-1) + h.$$

Thus

$$p - h + l \leq r_{p+1} + l = r_p \leq p - 1 + h \text{ and so } l \leq 2h - 1.$$

Since $h \leq l-1$, this gives the contradiction $l \geq 3$. Thus switching r_p and r_{p+1} gives a sequence also satisfying (7). Continuing like this we see that the nondecreasing rearrangement of R also satisfies (7). □

Lemma 6.5 and Theorem 6.3 give the following.

Corollary 6.6 *Let* $R = (r_1, r_2, \ldots, r_n)$ *be a 2-nearly nondecreasing, nonnegative integral vector. Then there exists a Hankel 2-tournament matrix A whose row sum vector is R if and only if (7) and (8) hold.*

In [9] an algorithm is given for obtaining a Hankel tournament with a given Hankel score vector. In [2] (see also [7]) an algorithm is given for obtaining a 2-tournament with a given score vector. We now adapt these algorithms to obtain an algorithm for obtaining a Hankel 2-tournament matrix with a given row sum vector. Because of similarities with these two algorithms, we shall be brief and the interested reader should consult [9].

Algorithm 2-HT for a 2-Tournament in $\mathcal{T}_2(R)$ When R Is Nondecreasing
Let $R = (r_1, r_2, \ldots, r_n)$ be a nondecreasing vector of nonnegative integers satisfying $r_i + r_{n+1-i} = 2(n-1)$ $(1 \leq i \leq n)$ and the inequalities $\sum_{i=1}^{k} r_i \geq k(k-1)$ $(1 \leq k \leq n)$ with equality for $k = n$. Let $S = (s_1, s_2, \ldots, s_n)$ where $s_i = 2(n-1) - r_i$ $(1 \leq i \leq n)$. If n is even, then it follows that $r_n \geq n-1$ and hence $s_n \leq n-1$; if n is odd, then it follows that $r_{(n+1)/2} = (n-1)$, $r_n \geq n-1$ and $s_n \leq n-1$. We construct an $n \times n$ 2-tournament matrix T as follows:

(n even)
1. Let the last column of T be $(n-1-s_n)$ 0's, followed by s_n 1's, followed by one 0 (on the main diagonal). For a Hankel 2-tournament, the first row, first column, and last column are also now determined, and we put those in T.
2. We now modify the middle $n-2$ components of R and of S as determined by the entries inserted in rows 1 and n and columns 1 and n, resulting in vectors R' and S' of $n-2$ components which have been (simultaneously) permuted in the same way so that R' is nondecreasing and R' and S' satisfy the Hankel property.
3. We repeat inductively, using R' and S' in place of R and S and then reverse the simultaneous permutations, resulting in a 2-tournament T' which we take as the middle $(n-2) \times (n-2)$ submatrix of T with row sum vector R.

(n odd)
1. We have $r_{(n+1)/2} = s_{(n+1)/2} = (n-1)$. In row $(n+1)/2$ and in column $(n+1)/2$ we put all 1's except in the position on the main diagonal where we put a 0.
2. Deleting row $(n+1)/2$ and column $(n+1)/2$ we are left with a $(n-1) \times (n-1)$ matrix with $(n-1)$ even and, after a simultaneous permutation to achieve a nondecreasing Hankel row sum vector, we apply the even part of the algorithm to the resulting row sum vector.

□

Remark 6.7 In the first application of the algorithm, 2's are inserted only in row n and column 1. Since rows and columns 1 and n are not permuted in the next application (although their entries in positions $2, 3, \ldots, n-1$ may be), it follows inductively that the all 2's of the final 2-tournament matrix are below the main diagonal. □

We give several instances of the algorithm in the next examples.

Example 6.8 Let $n = 6$ and let $R = (1, 3, 3, 7, 7, 9)$. Applying the algorithm we get the Hankel 2-tournament with row sum vector R and ten 2's:

$$\begin{bmatrix} 0&1&0&0&0&0 \\ 1&0&1&0&1&0 \\ 2&1&0&0&0&0 \\ 2&2&2&0&1&0 \\ 2&1&2&1&0&1 \\ 2&2&2&2&1&0 \end{bmatrix}.$$

If we apply Avery's algorithm [2, 7] for the construction of a 2-tournament with score vector R, we get

$$\begin{bmatrix} 0&1&0&0&0&0\\ 1&0&1&1&0&0\\ 2&1&0&0&0&0\\ 2&1&2&0&1&1\\ 2&2&2&1&0&0\\ 2&2&2&1&2&0 \end{bmatrix},$$

a non-Hankel tournament with row sum vector R and also ten 2's. Thus Avery's algorithm may not give a Hankel tournament with score vector R even though such a Hankel 2-tournament exists. □

Example 6.9 First let $n = 6$ and $R = (3, 4, 4, 6, 6, 7)$. Applying the algorithm we get

$$\left[\begin{array}{c||cccc||c} 0&1&1&1&0&0\\ \hline\hline 1&&&&&0\\ 1&&&&&1\\ 1&&&&&1\\ 2&&&&&1\\ \hline\hline 2&2&1&1&1&0 \end{array}\right] \rightarrow \left[\begin{array}{c||cccc||c} 0&1&1&1&0&0\\ \hline\hline 1&0&1&1&1&0\\ 1&1&&&1&1\\ 1&1&&&1&1\\ 2&1&1&1&0&1\\ \hline\hline 2&2&1&1&1&0 \end{array}\right] \rightarrow \left[\begin{array}{c||cccc||c} 0&1&1&1&0&0\\ \hline\hline 1&0&1&1&1&0\\ 1&1&0&0&1&1\\ 1&1&2&0&1&1\\ 2&1&1&1&0&1\\ \hline\hline 2&2&1&1&1&0 \end{array}\right].$$

Now let $n = 7$ and $R = (3, 4, 4, 6, 8, 8, 9)$. Then applying the algorithm we get

$$\left[\begin{array}{ccc||c||ccc} 0&1&1&1&0&0&0\\ 1&0&1&1&0&1&0\\ 1&1&0&1&1&0&0\\ \hline\hline 1&1&1&0&1&1&1\\ \hline\hline 2&2&1&1&0&1&1\\ 2&1&2&1&1&0&1\\ 2&2&2&1&1&1&0 \end{array}\right].$$

□

Theorem 6.10 *Given a nondecreasing, nonnegative integral vector* $R = (r_1, r_2, \ldots, r_n)$ *satisfying the criteria* (7) *and* (8), *the algorithm 2-HT outputs a Hankel 2-tournament matrix with row sum vector* R.

Proof The proof is very similar to the proof of Theorem 14 in [9] that shows the validity of the corresponding algorithm for a Hankel tournament with a prescribed row sum vector. It relies on the fact that, in the case of even n, after one application of the algorithm applied to a nondecreasing vector satisfying (7) and (8), the row sum vector used in the next step is 2-nearly nondecreasing because, as is easily seen, the components of the row sum vector do not change by more than 2, and this is enough by Corollary 6.6. The reason that we have 2-nearly nondecreasing (and not 3-nearly nondecreasing) as we proceed in the inductive algorithm is that if some component of R decreases by 3, then the component preceding it decreases by at least 1, and hence the difference is at most 2. □

Lemma 6.11 *Let $R = (r_1, r_2, \ldots, r_n)$ be a nondecreasing, nonnegative integral vector satisfying the criteria (7) and (8) and let $T = T^{(1)} + 2T^{(2)}$ be the Hankel 2-tournament with row sum vector R produced by algorithm 2-HT. Then the oriented graph $\Gamma(T^{(2)})$ is transitive.*

Proof By Remark 6.7, all the 1's of the (0, 1)-matrix $T^{(2)}$ are below its main diagonal. We prove the lemma by induction on n where initial case $n = 2$ is obvious. Referring to the algorithm, we have that inductively the $(n-2) \times (n-2)$ 2-tournament $T' = T'^{(1)} + 2T'^{(2)}$ has the property that the oriented graph $\Gamma(T'^{(2)})$ is transitive.

First assume that n is even. Thus any intransitive triple of $\Gamma(T^{(2)})$ must include at least one of vertices 1 and n. In fact, any intransitive triple including one of vertices $\{1, n\}$ gives an intransitive triple using the other. So suppose we have an intransitive triple u, w, n where $u \to w$ and $w \to n$ where necessarily $1 \leq u < w < n$. Since T is a Hankel tournament, then T has the form as illustrated below with $n = 8$ and $u = 6$ and $v = 3$:

$$\left[\begin{array}{cccccccc} 0 & & & & & & & \\ & 0 & & & & & & \\ & & 0 & & & & & \\ & & & 0 & & & & \\ & & & & 0 & & & \\ & & 2 & & & 0 & & \\ & & & & & & 0 & \\ & & 1 & & & 2 & & 0 \end{array}\right] \to \left[\begin{array}{cccccccc} 0 & & 0 & & & 1 & & \\ & 0 & & & & & & \\ 2 & & 0 & & & & & 1 \\ & & & 0 & & & & \\ & & & & 0 & & & \\ 1 & & 2 & & & 0 & & 0 \\ & & & & & & 0 & \\ & & 1 & & & 2 & & 0 \end{array}\right].$$

This contradicts the algorithm whereby the 1's in the last column have only the diagonal 0 below them. Thus $\Gamma(T^{(2)})$ is transitive.

Now assume that n is odd. Since the middle row and column, that is, row and column $(n+1)/2$, of T contains all 1's except for the 0 on the main diagonal, it is easy to say that vertex $(n+1)/2$ does not belong to any intransitive triple. Hence, in view of our algorithm, the transitivity of $\Gamma(T^{(2)})$ follows from the even case. □

Theorem 6.12 *Let $R = (r_1, r_2, \ldots, r_n)$ be a nondecreasing, nonnegative integral Hankel score vector. Then the algorithm* 2-HT *constructs a Hankel 2-tournament with the smallest number of 2's in Hankel 2-tournaments in $\mathcal{T}_2^h(R)$.*

Proof Both our algorithm 2-HT and Avery's algorithm construct a 2-tournament with score vector R having only 0's and 1's above the main diagonal. Avery's algorithm constructs a 2-tournament with the smallest number of 2's in $\mathcal{T}_2(R)$. The proof given in [2] (see also [7]) can be adapted to apply to our algorithm for the case of Hankel 2-tournaments. Note that the number of 1's above the main diagonal determines the number of 2's below the main diagonal, and hence the total number of 2's. □

Acknowledgments We are grateful to a referee for a thorough reading of our manuscript.

References

1. S. Ao, D. Hanson, Score vectors and tournaments with cyclic chromatic number 1 or 2. Ars Comb. **49**, 185–191 (1998)
2. P. Avery, Score sequences of oriented graphs. J. Graph Theory **15**, 252–257 (1991)
3. J. Bang-Jensen, G. Gutin, *Digraphs: Theory, Algorithms and Applications* (Springer, Berlin, 2001)
4. R.E. Behrend, I. Fischer, M. Konvalinka, Diagonally and antidiagonally symmetric alternating sign matrices of odd order. Adv. Math. **315**, 324–365 (2017)
5. A. Belkillani, Scores des tournois localement transitifs. Ars Comb. **117**, 75–83 (2014)
6. A. Belkillani, R. Ouannas, Construction des tournois localement transitifs. Symétries et périodicité. Commun. Math. Anal. **3**(2), 69–82 (2007). This paper is referenced in [5] but we have been unable to get a copy
7. R.A. Brualdi, *Combinatorial Matrix Classes* (Cambridge University Press, Cambridge, 2006)
8. R.A. Brualdi, S.-M. Ma, Centrosymmetric, and symmetric and Hankel-symmetric matrices, in *Mathematics Across Contemporary Sciences*. Springer Proc. Math. Stat., vol. 190 (Springer, Cham, 2017), pp. 17–31
9. R.A. Brualdi, E. Fritscher, Loopy, Hankel, and combinatorially skew-Hankel tournaments. Discrete Appl. Math. **194**, 37–59 (2015)
10. R.A. Brualdi, Q. Li, Problem 31. Discrete Math. **43**, 329–330 (1983)
11. R.A. Brualdi, H.J. Ryser, *Combinatorial Matrix Theory* (Cambridge University Press, Cambridge, 1991)
12. R.A. Brualdi, J. Shen, Landau'sinequalities for tournament scores and a short proof of a theorem on transitive sub-tournaments. J. Graph Theory **82**, 244–254 (2000)
13. L.N. Coregliano, R.F. Parente, C.M. Sato, On the maximum density of fixed strongly connected subtournaments. Electr. J. Comb. (2019). https://doi.org/10.37236/6557
14. S.W. Drury, Solution of the conjecture of Brualdi and Li. Linear Algebra Appl. **436**, 3392–3399 (2012)
15. R.C. Entringer, L.K. Tolman, Characterizations of graphs having orientations satisfying local degree restrictions. Czechoslov. Math. J. **28**, 108–119 (1978)
16. A. Frank, On the orientation of graphs. J. Comb. Theory B **28**(3), 251–261 (1980)
17. A. Frank, A. Gárfás, How to orient the edges of a graph? in *Combinatorics* (*Proc. Fifth. Hungarian Colloq,. Keszthely 1976*. Colloq. Math. Soc. János Bolyai, vol. 18 (North-Holland, Amsterdam, 1978), pp. 353–364
18. B. Giuduli, A. Gyárfás, S. Thomassé, P. Weidl, 2-Partitionable transitive tournaments. J. Comb. Theory B **72**, 181–196 (1998)
19. J.S. Li, K. Yang, Degree sequence of oriented graphs. J. Math Study **35**(2), 140–146 (2002)
20. S. Pirzada, U. Samee, Mark sequences in digraphs. Semin. Lotharingen Comb. **55**(B55c), 13 pp. (2006)

The Game Chromatic Number of a Random Hypergraph

Debsoumya Chakraborti, Alan Frieze, and Mihir Hasabnis

Abstract We consider the following game, played on a k-uniform hypergraph H. There are q colors available and two players take it in turns to color vertices. A partial coloring is proper if no edge is mono-chromatic. One player, A, wishes to color all the vertices and the other player, B, wishes to prevent this. The *game chromatic number* $\chi_g(H)$ is the minimum number of colors for which A has a winning strategy. We consider this in the context of a random k-uniform hypergraph and prove upper and lower bounds that hold w.h.p.

1 Introduction

Let $G = (V, E)$ be a graph and let q be a positive integer. Consider the following game in which two players A(lice) and B(ob) take turns in coloring the vertices of G with q colors. Each move consists of choosing an uncolored vertex of the graph and assigning to it a color from $\{1, \ldots, k\}$ so that the resulting coloring is *proper*, i.e., adjacent vertices get different colors. A wins if all the vertices of G are eventually colored. B wins if at some point in the game the current partial coloring cannot be extended to a complete coloring of G, i.e., there is an uncolored vertex such that each of the q colors appears at least once in its neighborhood. We assume that A goes first (our results will not be sensitive to this choice). The *game chromatic number* $\chi_g(G)$ is the least integer q for which A has a winning strategy.

This parameter is well defined, since it is easy to see that A always wins if the number of colors is larger than the maximum degree of G. Clearly, $\chi_g(G)$ is at least as large as the ordinary chromatic number $\chi(G)$, but it can be considerably more. The game was first considered by Brams about 25 years ago in the context of coloring planar graphs and was described in Martin Gardner's column [7] in

D. Chakraborti · A. Frieze (✉) · M. Hasabnis
Department of Mathematical Sciences, Carnegie Mellon University, Pittsburgh, PA, USA
e-mail: alan@random.math.cmu.edu

A. M. Raigorodskii, M. Th. Rassias (eds.), *Discrete Mathematics and Applications*, Springer Optimization and Its Applications 165,
https://doi.org/10.1007/978-3-030-55857-4_6

Scientific American in 1981. The game remained unnoticed by the graph-theoretic community until Bodlaender [3] re-invented it. For a survey see Bartnicki et al. [2].

The papers by Bohman et al. [4], Frieze et al. [6], and Keusch and Steger [9] discuss the game chromatic number of random graphs. In this paper we discuss the game chromatic number of random hypergraphs. Given a hypergraph $H = (V, E)$ we can consider basically the same game. Here A, B color the vertices of H consecutively and a coloring is proper if there is no $e \in E$ such that all vertices in e have the same color. This problem has hardly been studied, even in a deterministic setting, but we feel it is of interest to extend the results of [4, 6] and [9] to this setting.

In this paper we will restrict our attention to the random k-uniform hypergraph $H_{n,p;k}$ where each of the $\binom{n}{k}$ edges appears independently with probability $p = \frac{d}{n^{k-1}}$, where d is a large constant. Now Krivelevich and Sudakov [10] have shown that

$$\chi(H_{n,p;k}) \approx \left(\frac{d}{k(k-2)!\log d}\right)^{1/(k-1)} \tag{1}$$

Here we use the notation $A_n \approx B_n$ for sequences $A_n, B_n, n \geq 1$ to mean that $A_n = (1 + o(1))B_n$ as $n \to \infty$.

Our first theorem shows that w.h.p. the game chromatic number χ_g is significantly larger than the chromatic number.

Theorem 1 *There exists a constant* $\varepsilon > 0$ *such that w.h.p.,*

$$\chi_g(H_{n,p;k}) \geq (1+\varepsilon)\left(\frac{d}{k(k-2)!\log d}\right)^{1/(k-1)}, \qquad \text{if } d \text{ is sufficiently large.}$$

We also prove an upper bound in the case $k = 3$ that is somewhat far from that implied by (1).

Theorem 2 *Let* $\delta > 0$ *be arbitrary. Then w.h.p.,*

$$\chi_g(H_{n,p;3}) \leq d^{2/3+\delta}, \qquad \text{if } d \text{ is sufficiently large.}$$

It is natural to state the following:

Conjecture W.h.p. $\chi_g(H_{n,p;k}) = O\left(\frac{d}{k!\log d}\right)^{1/(k-1)}$.

We often refer to the following Chernoff-type bounds for the tails of binomial distributions (see, e.g., [1] or [8]). Let $X = \sum_{i=1}^{n} X_i$ be a sum of independent indicator random variables such that $\mathbf{Pr}(X_i = 1) = p_i$ and let $p = (p_1 + \cdots + p_n)/n$. Then

$$\mathbf{Pr}(X \leq (1-\varepsilon)np) \leq e^{-\varepsilon^2 np/2}, \tag{2}$$

$$\mathbf{Pr}(X \geq (1+\varepsilon)np) \leq e^{-\varepsilon^2 np/3}, \qquad \varepsilon \leq 1, \tag{3}$$

$$\mathbf{Pr}(X \geq \mu np) \leq (e/\mu)^{\mu np}. \tag{4}$$

2 Lower Bound

Let

$$D = \left(\frac{d}{k! \log d}\right)^{1/(k-1)}$$

and suppose that there are $q = \alpha D$ colors available.

Bob's strategy is to choose the same color as Alice, say i, but assign it randomly to one of the set of available vertices for color i.

Notation Let $C_i = C_i(t)$ be the set of vertices that have been colored i after t rounds. Let $S_i = S_i(t)$ be the set of vertices that were colored by B. Let $C = C(t) = \cup_{i=0}^{q} C_i$ denote the partial coloring of the vertex set.

Lemma 1 *Suppose we run this process for $t = \theta n, \theta < 1/2$ many rounds and that $|C_i(t)| = 2\beta n/D$. We show that if d is sufficiently large and*

$$2(2\beta + \gamma)^k - (2\beta)^k > \frac{2(\beta + \gamma)}{k - 1}, \tag{5}$$

then with probability $1 - o(1/n)$, there exists no set T such that (1) $C_i \cap T = \emptyset$, (2) $C_i \cup T$ is independent and $|T| = \gamma n/D$.

The reader can easily check that (5) is satisfied for $k \geq 3$ and

$$\beta = \frac{1 - 2\varepsilon}{2(k-1)^{1/(k-1)}}, \quad \gamma = \frac{\varepsilon}{(k-1)^{1/(k-1)}}$$

when $\varepsilon > 0$ is sufficiently small. The proof of the lemma is deferred to Section 2.1.

If the event $\{\exists i : C_i, T\}$ does not occur, then because no color class has size greater than $(2\beta + \gamma)n/D$ the number ℓ of colors i for which $|S_i| \geq \beta n/D$ by this time satisfies

$$\frac{\ell(2\beta + \gamma)}{D} + \frac{2(q - \ell)\beta}{D} \geq 2\theta \text{ or } \frac{\ell\gamma + 2q\beta}{D} \geq 2\theta.$$

We choose $\alpha = (1 + \varepsilon)(k - 1)^{1/(k-1)}$ and $\theta = \alpha(2\beta + \gamma)/2 < 1/2$. Since $q \geq \ell$, this implies that

$$\frac{q}{D} \geq \frac{2\theta}{2\beta + \gamma} = \alpha.$$

This completes the proof of Theorem 1, after replacing $(1+\varepsilon)$ by $(1+\varepsilon)^{1/(k-1)}$ for esthetic purposes.

2.1 Proof of Lemma 1

For expressions X, Y we sometimes use the notation $X \leq_O Y$ in place of $X = O(Y)$ when the bracketing is "ugly."

Now, (explanations for (6)–(9) below), if d is sufficiently large, then

$$\mathbf{Pr}(\exists i, C_i, T)$$

$$\leq_O q \binom{n}{\frac{\beta n}{D}} \binom{n}{\frac{\gamma n}{D}} \sum_{|S|=\beta n/D} P(S_i = S)(1-p)^{\frac{(2\beta+\gamma)^k n^k}{k! D^k}} \tag{6}$$

$$\leq q \binom{n}{\frac{\beta n}{D}} \binom{n}{\frac{\gamma n}{D}} \sum_{|S|=\beta n/D} \left(\frac{\beta n}{D}\right)! (1-p)^{\frac{(2\beta+\gamma)^k n^k}{k! D^k}}$$

$$\times \prod_{j=1}^{\beta n/D} \frac{7}{(1-2\theta)n(1-p)^{\left(\frac{\beta n}{D}-j+1\right)\frac{2(2j-1)^{k-2}}{(k-2)!}}} \tag{7}$$

$$\leq_O q \binom{n}{\frac{\beta n}{D}}^2 \binom{n}{\frac{\gamma n}{D}} \left(\frac{\beta n}{D}\right)! \frac{7^{\frac{\beta n}{D}} (1-p)^{\frac{(2\beta+\gamma)^k n^k}{k! D^k}}}{((1-2\theta)n)^{\beta n/D} (1-p)^{\sum_{j=1}^{\beta n/D} \left(\frac{\beta n}{D}-j+1\right)\frac{2(2j-1)^{k-2}}{(k-2)!}}}$$

$$\leq_O q \binom{n}{\frac{\beta n}{D}}^2 \binom{n}{\frac{\gamma n}{D}} \left(\frac{\beta n}{D}\right)! \frac{7^{\frac{\beta n}{D}} (1-p)^{\frac{(2\beta+\gamma)^k n^k}{k! D^k}}}{((1-2\theta)n)^{\beta n/D} (1-p)^{\frac{(2\beta n)^k}{2k! D^k}}} \tag{8}$$

$$\leq_O q \binom{n}{\frac{\beta n}{D}}^2 \binom{n}{\frac{\gamma n}{D}} \left(\frac{\beta n}{D}\right)! (1-p)^{\left(\frac{2n^k((2\beta+\gamma)^k)-(2\beta)^k}{2k! D^k}\right)} \frac{7^{\frac{\beta n}{D}}}{((1-2\theta)n)^{\frac{\beta n}{D}}}$$

$$\leq_O q n^{1/2} \left(\frac{(eD)^{2\beta+\gamma}}{\beta^{2\beta}\gamma^{\gamma}} \left(\frac{7\beta}{De(1-2\theta)}\right)^{\beta} \times \exp\left\{-\frac{d}{D^{k-1}}\left(\frac{2(2\beta+\gamma)^k-(2\beta)^k}{2k!}\right)\right\}\right)^{\frac{n}{D}}$$

$$\leq \left(cD^{\beta+\gamma} \exp\left\{-\left(\frac{2(2\beta+\gamma)^k-(2\beta)^k}{2} \log d\right)\right\}\right)^{\frac{n}{D}}, \tag{9}$$

(where $c = c(\theta, \beta, \gamma) = O(1)$)

$$= o(1/n),$$

assuming (5).

Justifying (8)

$$\sum_{j=1}^{\beta n/D}\left(\frac{\beta n}{D}-j+1\right)\frac{2(2j-1)^{k-2}}{(k-2)!}$$
$$\approx \int_{x=1}^{\beta n/D}\left(\frac{\beta n}{D}-x+1\right)\frac{2(2x-1)^{k-2}}{(k-2)!}dx$$
$$=\frac{1}{(k-2)!}\int_{x=1}^{\beta n/D}\left(\left(\frac{2\beta n}{D}+1\right)(2x-1)^{k-2}-(2x-1)^{k-1}\right)dx$$
$$=\left(\frac{2\beta n}{D}+1\right)\left(\frac{(2\beta n/D-1)^{k-1}-1}{2(k-1)!}\right)-\frac{(2\beta n/D-1)^{k}-1}{2k(k-2)!}$$
$$\le\frac{(2\beta n/D)^{k}}{2k!}.$$

Justifying (9) We used the asymptotic formula for summation of k-th power of first n natural numbers, i.e., $\sum_{i=1}^{n} i^k \approx \frac{n^{k+1}}{k+1}$.

Justifying (6) There are q choices for color i. Then we take the union bound over all $\binom{n}{\beta n/D}\binom{n}{\gamma n/D}$ possible choices of $C_i \setminus S_i$ and T. In some sense we are allowing Alice to simultaneously choose all possible sets of size $\beta n/D$ for $C_i \setminus S_i$. The union bound shows that w.h.p. all choices fail. We do not sum over orderings of $C_i \setminus S_i$. We instead compute an upper bound on $\mathbf{Pr}(S_i = S)$ that holds regardless of the order in which Alice plays. We consider the situation after θn rounds. That is, we think of the following random process: pick a k-uniform hypergraph $H \sim H(n, p; k)$, let Alice play the coloring game on H with q colors against a player who randomly chooses an available vertex to be colored by the same color as Alice. Stop after θn moves. At this point Alice played with color i and there are $\beta n/D$ vertices that were colored i by Alice and the same number that were colored i by Bob. We bound the probability that at this point there are $\gamma n/D$ vertices that form an independent set with the color class C_i. We take a union bound over all the possible sets for Alice's vertices and for the vertices in T. The probability of Bob choosing a certain set is computed next.

Justifying (7) Consider a sequence of random variables $X_1 = N = (1 - 2\theta)n$, $X_j = Bin(X_{j-1}, p_j)$, where $p_j = (1-p)^{\frac{2(2j-1)^{k-2}}{(k-2)!}}$, $2 \le j \le t$. X_j is a lower bound for the number of vertices that Bob can color i and p_j is a lower bound on the probability that a vertex v that was i-available at step $j-1$ is also i-available after step j. The probability that a vertex v was i-available at time $j-1$ and is still i-available now is at least $(1-p)^{\binom{2j-2}{k-3}}(1-p)^{2\binom{2j-2}{k-2}} \ge p_j$. Our estimate for p_j arises as follows: there are at most $2(j-1)$ vertices $x_1, \ldots, x_{2(j-1)}$ of color i and each of the vertices $z = y, y'$ colored i in round j yields possible edges $\{v, z, x_i, \ldots\}$ that could remove v from C_i. There are also the edges $\left\{v, y, y', x_i, \ldots\right\}$ to account for.

We need to estimate $\mathbf{E}(Y_t)$, where $Y_t = 1/(X_1X_2\dots X_t)$. $1/X_j$ is an upper bound for the probability that Bob chooses a particular vertex at step j and then $Y_{\beta n/D}$ is an upper bound on the probability that Bob's sequence of choices is $x_1, x_2, \dots .x_{\beta n/D}$, where $S = \{x_1, x_2, \dots x_{\beta n/D}\}$. The following lemma is proven in [6]:

Lemma 2 *If $B = Bin(v, \rho)$, then $\mathbf{E}(\prod_{i=1}^{q} \frac{1}{B+i-1}) \leq \frac{7}{\rho^q} \prod_{i=1}^{q} \frac{1}{v+i}$.*

Using Lemma 2 we see that

$$\begin{aligned}
\mathbf{Pr}(S_i = S) &\leq \mathbf{E}\left(\frac{1}{X_1X_2\dots X_t}\right) \\
&\leq \mathbf{E}\left(\frac{7}{X_1\cdots X_{t-1}(X_{t-1}+1)p_t}\right) \\
&\leq \mathbf{E}\left(\frac{7^2}{X_1\cdots X_{t-2}(X_{t-2}+1)(X_{t-2}+2)p_{t-1}^2 p_t}\right) \\
&\vdots \\
&\leq \prod_{j=1}^{t} \frac{7}{(N+j-1)(1-p)^{(t-j+1)\frac{2(2j-1)^{k-2}}{(k-2)!}}} \\
&\leq \prod_{j=1}^{t} \frac{7}{N(1-p)^{(t-j+1)\frac{2(2j-1)^{k-2}}{(k-2)!}}}.
\end{aligned}$$

This completes the justification of (7).

3 Upper Bound

3.1 Simple Density Properties

For $S \subseteq [n]$ and $k = 2, 3$ we let $e_3(S) = |\{f \in E : f \subseteq S\}$ and

$$e_2(S) = \{\{x, y\} \subseteq S : \exists y \notin S \text{ such that } \{x, y, z\} \in E\}.$$

Lemma 3 *If $\theta > 1$ and*

$$\left(\frac{\sigma ed}{2\theta}\right)^{\theta} \leq \frac{\sigma}{2e},$$

then w.h.p. there does not exist $S \subset [n]$, $|S| \leq \sigma n$ such that $e_2(S) \geq \theta|S|$.

Proof

$$\begin{aligned}
\mathbf{Pr}(\exists S, |S| \leq \sigma n, e_2(S) \geq \theta|S|) &\leq \sum_{s=2\theta}^{\sigma n} \binom{n}{s} \binom{\binom{s}{2}}{\theta s} \left(1 - \left(1 - \frac{d}{n^2}\right)^{n-2}\right)^{\theta s} \\
&\leq \sum_{s=2\theta}^{\sigma n} \left(\frac{ne}{s}\right)^s \left(\frac{es}{2\theta}\right)^{\theta s} \left(1 - \left(1 - \frac{d}{n}\right)\right)^{\theta s} \\
&\leq \sum_{s=2\theta}^{\sigma n} \left(\left(\frac{ne}{s}\right) \left(\frac{es}{2\theta} \frac{d}{n}\right)^{\theta}\right)^s \\
&= \sum_{s=2\theta}^{\sigma n} \left(e \left(\frac{s}{n}\right)^{\theta-1} \left(\frac{ed}{2\theta}\right)^{\theta}\right)^s \\
&= O\left(\frac{d^\theta}{n^{\theta-1}}\right) = o(1)
\end{aligned}$$

□

Lemma 4 *If $\theta > 1/2$ and*

$$\left(\frac{\sigma^2 ed}{6\theta}\right)^{\theta} \leq \frac{\sigma}{2e},$$

then w.h.p. there does not exist $S \subseteq [n]$, $|S| \leq \sigma n$ such that $e_3(S) \geq \theta|S|$.

Proof

$$\begin{aligned}
\mathbf{Pr}(\exists S, |S| \leq \sigma n, e_3(S) \geq \theta|S|) &\leq \sum_{s=\sqrt{\theta}}^{\sigma n} \binom{n}{s} \binom{\binom{s}{3}}{\theta s} \left(\frac{d}{n^2}\right)^{\theta s} \\
&\leq \sum_{s=\sqrt{\theta}}^{\sigma n} \left(\frac{ne}{s} \cdot \left(\frac{es^2 d}{6\theta n^2}\right)^{\theta}\right)^s \\
&= \sum_{s=\sqrt{\theta}}^{\sigma n} \left(e \left(\frac{s}{n}\right)^{2\theta-1} \cdot \left(\frac{ed}{6\theta}\right)^{\theta}\right)^s \\
&= O\left(\frac{d^\theta}{n^{2\theta-1}}\right) \\
&= o(1)
\end{aligned}$$

□

For $S \subseteq [n]$ and $k = 1, 2$ and vertex v, we let $d_{S,k}(v)$ denote the number of edges $\{v, x, y\}$ such that $|\{x, y\} \cap S| = k$.

Lemma 5 *Let σ and θ be as in Lemma 3. If $(\Delta - 2\theta)\tau > 1$ and*

$$\left(\frac{\sigma ed}{(\Delta - 2\theta)\,\tau}\right)^{(\Delta-2\theta)\tau} \leq \frac{\sigma}{4e},$$

then w.h.p. there does not exist $S \supseteq T$ such that $|S| = s \leq \sigma n, |T| \geq \tau s$, and $d_{S,1}(v) \geq \Delta, \forall v \in T$.

Proof In the light of Lemma 3, the assumptions imply that w.h.p. $|e_1(T : S \setminus T)| \geq (\Delta - 2\theta)\tau s$. In which case,

$$\mathbf{Pr}(\exists S \supseteq T,\ |S| \leq \sigma n,\ |T| \geq \tau s : |e_1(T : S \setminus T)| \geq (\Delta - 2\theta)\tau s)$$

$$\leq \sum_{s=2\theta}^{\sigma n} \sum_{t=\tau s}^{s} \binom{n}{s}\binom{s}{t}\binom{st}{(\Delta - 2\theta)\tau s}\left(1 - \left(1 - \frac{d}{n^2}\right)^{n-2}\right)^{(\Delta-2\theta)\tau s} \tag{10}$$

$$\leq \sum_{s=2\theta}^{\sigma n} \sum_{t=\tau s}^{s} \left(\frac{ne}{s}\right)^s \cdot 2^s \cdot \left(\frac{eds}{(\Delta - 2\theta)\tau n}\right)^{(\Delta-2\theta)\tau s}$$

$$= \sum_{s=2\theta}^{\sigma n} \sum_{t=\tau s}^{s} \left(\frac{2ne}{s} \cdot \left(\frac{eds}{(\Delta - 2\theta)\tau n}\right)^{(\Delta-2\theta)\tau}\right)^s$$

$$= \sum_{s=2\theta}^{\sigma n} \sum_{t=\tau s}^{s} \left(2e\left(\frac{s}{n}\right)^{(\Delta-2\theta)\tau-1} \cdot \left(\frac{ed}{(\Delta - 2\theta)\tau}\right)^{(\Delta-2\theta)\tau}\right)^s \tag{11}$$

$$= O\left(\frac{d^{(\Delta-2\theta)\tau}}{n^{(\Delta-2\theta)\tau-1}}\right) = o(1).$$

□

Lemma 6 *Let σ and θ be as in Lemma 4 and $(\Delta - 3\theta)\tau > 1$ and*

$$\left(\frac{\sigma^2 ed}{2(\Delta - 3\theta)\tau}\right)^{(\Delta-3\theta)\tau} \leq \frac{\sigma}{4e}$$

then w.h.p. there does not exist $S \supseteq T$ such that $|S| = s \leq \sigma n, |T| \geq \tau s$, and $d_{S,2}(v) \geq \Delta, \forall v \in T$

Proof In the light of Lemma 4, the assumptions imply that w.h.p. $|e_2(T : S \setminus T)| \geq (\Delta - 3\theta)\tau s$. In which case,

$$\mathbf{Pr}(\exists S \supseteq T,\ |S| \leq \sigma n,\ |T| \geq \tau s : |e_1(T : S \setminus T)| \geq (\Delta - 3\theta)\tau s)$$

$$\leq \sum_{s=2\theta}^{\sigma n} \sum_{t=\tau s}^{s} \binom{n}{s}\binom{s}{t}\binom{s^2 t/2}{(\Delta - 3\theta)\tau s}\left(\frac{d}{n^2}\right)^{(\Delta-3\theta)\tau s}$$

$$
\begin{aligned}
&\leq \sum_{s=2\theta}^{\sigma n} \sum_{t=\tau s}^{s} \left(\frac{ne}{s}\right)^s \cdot 2^s \cdot \left(\frac{eds^2}{2(\Delta-3\theta)\tau n^2}\right)^{(\Delta-3\theta)\tau s} \\
&= \sum_{s=2\theta}^{\sigma n} \sum_{t=\tau s}^{s} \left(\frac{2ne}{s} \cdot \left(\frac{eds^2}{2(\Delta-3\theta)\tau n^2}\right)^{(\Delta-3\theta)\tau}\right)^s \\
&= \sum_{s=2\theta}^{\sigma n} \sum_{t=\tau s}^{s} \left(2e\left(\frac{s}{n}\right)^{2(\Delta-3\theta)\tau-1} \cdot \left(\frac{ed}{2(\Delta-3\theta)\tau}\right)^{(\Delta-3\theta)\tau}\right)^s \\
&= O\left(\frac{d^{(\Delta-3\theta)\tau}}{n^{(\Delta-3\theta)\tau-1}}\right) = o(1).
\end{aligned}
$$

□

Now let

$$
q = d^{2/3+\delta} \text{ and } \beta = \frac{q}{3} \text{ and } \gamma = \frac{14\log d}{q},
$$

for some small absolute constant $\delta > 0$.

We will now argue that w.h.p. A can win the game if q colors are available.

A's initial strategy will be the same as that described in [4]. Let $\mathscr{C} = (C_1, C_2, \ldots, C_q)$ be a collection of pairwise disjoint subsets of $[n]$, i.e., a (partial) coloring. Let $\bigcup \mathscr{C}$ denote $\bigcup_{i=1}^{q} C_i$. For a vertex v let

$$
A(v, \mathscr{C}) = \{i \in [q] : \ v \text{ is not in an edge } \{v, x, y\} \text{ such that } x, y \in C_i\},
$$

and set

$$
a(v, \mathscr{C}) = |A(v, \mathscr{C})|.
$$

Note that $A(v, \mathscr{C})$ is the set of colors that are available at vertex v when the partial coloring is given by the sets in $\mathscr{C}$ and $v \notin \bigcup \mathscr{C}$. A's initial strategy can now be easily defined. Given the current color classes $\mathscr{C}$, A chooses an uncolored vertex v with the smallest value of $a(v, \mathscr{C})$ and colors it by any available color.

As the game evolves, we let u denote the number of uncolored vertices in the graph. So, we think of u as running "backward" from n to 0.

We show next that w.h.p. every q-coloring (proper or improper) of the full vertex set has the property that there are at most γn vertices with less than $\beta/2$ available colors. For this we need the following lemma:

Lemma 7 *$p = d/n^2$ and let $x_0 = (2/(-\log(1-p)))^{1/2} \approx (2/d)^{1/2} n$ and*

$$
f(C) = \sum_{i=1}^{q} (1-p)^{c_i^2/2} \text{ where } \sum_{i=1}^{q} c_i = n.
$$

Then we have that for n sufficiently large,

$$f(C) \geq \begin{cases} q(1-p)^{n^2/2q^2} & qx_0 \leq n \\ q(1-p)^{x_0^2/2} & qx_0 > n \end{cases}.$$

Proof We have that the function $\phi(x) = (1-p)^{x^2/2}$ is convex in the interval $[x_0, \infty]$. It follows from convexity that if $I = \{i : c_i \leq x_0\}$, then

$$f(C) \geq |I|(1-p)^{x_0^2/2} + (q-|I|)(1-p)^{(n-|I|x_0)^2/2(q-|I|)^2}. \tag{12}$$

Suppose now that $x_0 \leq n/q$. Then $\frac{n-|I|x_0}{q-|I|} \geq \frac{n}{q}$. And then convexity and (12) implies

$$f(C) \geq q\left(\frac{|I|}{q}\phi(x_0) + \frac{q-|I|}{q}\phi\left(\frac{n-|I|x_0}{q-|I|}\right)\right) \geq q\phi\left(\frac{n}{q}\right). \tag{13}$$

If $x_0 > n/q$, then $\frac{n-|I|x_0}{q-|I|} < \frac{n}{q}$ and then (12) implies that

$$f(C) \geq q(1-p)^{x_0^2/2}.$$

□

Let

$$B(\mathscr{C}) = \{v : a(v, \mathscr{C}) < \beta/2\}.$$

Lemma 8 *W.h.p., for all collections* $\mathscr{C}$,

$$|B(\mathscr{C})| \leq \gamma n.$$

Proof We first note that if $|S| = \gamma n$, then w.h.p. $e_2(S) \leq 4\gamma^2 dn$. This follows from Lemma 3 with $\sigma = \gamma$ and $\theta = 4\gamma d$. It follows that for any $\varepsilon > 0$ that there is a set $S_1' \subseteq S$ of size at least $(1-\varepsilon)\gamma n$ such that if $v \in S_1'$, then $d_{S,1}(v) \leq 8\varepsilon^{-1}\gamma d$. Furthermore, Lemma 4 with $\sigma = \gamma$ and $\theta = 3$ implies that w.h.p. $e_3(S) \leq 3\gamma n$. Therefore there is a set $S_1'' \subseteq S$ of size at least $(1-\varepsilon)\gamma n$ such that if $v \in S_1''$, then $d_{S,2}(v) \leq \varepsilon^{-1}$. Let $S_1 = S_1' \cup S_1''$.

Fix $\mathscr{C}$ and suppose that $|B(\mathscr{C})| \geq \gamma n$. Choose $S \subseteq B(\mathscr{C})$ and let S_1 be as defined above. For $v \in S_1$ let

$$b(v, \mathscr{C}) = |\{i \in [q] : \ v \text{ is not in an edge } \{v, x, y\} \text{ such that } \ x, y \in C_i \setminus S\}|.$$

Thus $a(v, \mathscr{C}) \geq b(v, \mathscr{C}) - 8\varepsilon^{-1}\gamma d - \varepsilon^{-1}$. $b(v, \mathscr{C})$ is the sum of independent indicator variables X_i, where $X_i = 1$ if v is not in a hyperedge (v, x, y) such that $x, y \in C_i \setminus S$ in $G_{n,p}$. Then $\mathbf{Pr}(X_i = 1) \geq (1-p)^{\binom{|C_i|}{2}}$ and since $(1-p)^t$ is a convex function of t and using the Lemma 7 we get

$$\mathbf{E}(b(v,\mathscr{C})) \geq \sum_{i=1}^{q}(1-p)^{\binom{|C_i|}{2}} \geq \beta.$$

It follows from the Chernoff bound (2) that

$$\mathbf{Pr}(b(v,\mathscr{C}) \leq 0.51\beta) \leq e^{-\beta/9}.$$

Now, when $\mathscr{C}$ is fixed, the events $\{b(v,\mathscr{C}) \leq 0.51\beta\}$, $v \in S_1$ are independent. Thus, because $a(v,\mathscr{C}) \leq \beta/2$ implies that $b(v,\mathscr{C}) \leq 0.51\beta$ we have

$$\begin{aligned}
&\mathbf{Pr}(\exists \mathscr{C} : |B(\mathscr{C})| \geq \gamma n) \\
&\leq q^n \binom{n}{(1-\varepsilon)\gamma n} e^{-(1-\varepsilon)\gamma\beta n/9} \\
&\leq q^n \left(\frac{e}{(1-\varepsilon)\gamma}\exp\left\{-\frac{\beta}{9}\right\}\right)^{(1-\varepsilon)\gamma n} \qquad (14)\\
&= \exp\left\{n\left(\log q + \frac{14(1-\varepsilon)\log d}{q}\left(\log\left(\frac{e}{1-\varepsilon}\right) - \frac{q}{27}\right)\right)\right\} \\
&= o(1),
\end{aligned}$$

for large d and small enough ε. □

Let u_0 to be the last time for which A colors a vertex with at least $\beta/2$ available colors, i.e.,

$$u_0 = \min\left\{u : a(v,\mathscr{C}_u) \geq \beta/2, \text{ for all } v \notin \bigcup \mathscr{C}_u\right\},$$

where $\mathscr{C}_u$ denotes the collection of color classes when u vertices remain uncolored.

If u_0 does not exist, then A will win. It follows from Lemma 8 that w.h.p. $u_0 \leq 2\gamma n$ and that at time u_0, every vertex still has at least $\beta/2$ available colors. Indeed, consider the final coloring $\mathscr{C}^*$ in the game that would be achieved if A follows her current strategy, even if she has to improperly color an edge. Let $U = \{v \notin \mathscr{C}_{u_0} : a(v,\mathscr{C}^*) < \beta/2\}$. Now we can assume that $|U| \leq \gamma n$. Because the number of colors available to a vertex decreases as vertices get colored, from u_0 onward, every vertex colored by A is in U. Therefore $u_0 \leq 2\gamma n$. Next let $G_U = (U, F)$ be the graph with vertex set U and edges F where $\{x, y\} \in F$ if there exists z such that $\{x, y, z\} \in E$.

Now let u_1 be the first time that there are at most $2\gamma n$ uncolored vertices and $a(v,\mathscr{C}_u) \geq \beta/2$, for all $v \notin \bigcup \mathscr{C}_u$. By the above, w.h.p. $u_1 \leq u_0$, so in particular w.h.p. u_1 exists. A can determine u_1 but not u_0, as u_0 depends on the future.

A will follow a more sophisticated strategy from u_1 onward. A will however play the remainder of the game on the graph G_U. By this we mean that she will ensure that if $\{x, y\}$ is a G_U-edge and x has color i at some stage, then she will not color y with color i even though this is strictly admissible.

This weakens A and explains why our upper bound does not match our lower bound. On the other hand, if she can properly color G_U, then she will have succeeded in properly coloring $H = H_{n,p;3}$. B, of course, does not play by these rules. We will show next that we can find a sequence $U = U_0 \supseteq U_1 \supseteq \cdots \supseteq U_\ell$ with the following properties: The G_U-edges of $U_i : (U_{i-1} \setminus U_i)$ between U_i and $U_{i-1} \setminus U_i$ will be divided into two classes, *heavy* and *light*. Vertex w is a heavy (resp. light) G_U-neighbor of vertex v if the edge (v, w) is G_U-heavy (resp. G_U-light).

(P1) Each vertex of $U_i \setminus U_{i+1}$ has at most one light G_U-neighbor in U_{i+1}, for $0 \leq i < \ell$.
(P2) All $U_i : (U_{i-1} \setminus U_i)$ G_U-edges are light for $i \geq 3$.
(P3) Each vertex of U_i has at most $3\beta/50$ G_U-heavy neighbors in $U_{i-1} \setminus U_i$ for $i = 1, 2$.
(P4) Each vertex of $U_i \setminus U_{i+1}$ has at most $\beta/3$ G_U-neighbors in U_i, for $0 \leq i < \ell$.
(P5) U_ℓ contains at most one G_U-cycle.

From this, we can deduce that the G_U-edges of U_0 can be divided up into the G_U-heavy edges E_H, G_U-light edges F_L, the G_U-edges inside U_ℓ, and the rest of the G_U-edges. Assume first that U_ℓ does not contain a G_U-cycle. $\Phi = (U, F_L)$ is a forest and the strategy in [5] can be applied. When attempting to color a vertex v of Φ, there are never more than three Φ-neighbors of v that have been colored. Since there are at most $\beta/3 + 2 \cdot 3\beta/50$ non-Φ neighbors, A will succeed since she has an initial list of size $\beta/2$.

If U_ℓ contains a G_U-cycle C, then A can begin by coloring a vertex of C. This puts A one move behind in the tree coloring strategy, in which case we can bound the number of Φ-neighbors by four.

It only remains to prove that the construction P1–P5 exists w.h.p. Remember that d is sufficiently large here.

We can assume without loss of generality that $|U_0| = 2\gamma n$. This will not decrease the sizes of the sets $a(v, U_0)$.

3.2 *The Verification of P1–P4: Constructing U_1*

The general strategy will be as follows : We will consider two separate types of edge listed below. To tackle each type, we will formulate corresponding lemmas that will be presented subsequently.

Type 1: The edges $\{x, y\}$ in G_U such that $\{x, y, z\} \notin E$ for all $z \in U$.
Type 2: The remaining edges where for $\{x, y\}$ in G_U, there is $z \in U$ such that $\{x, y, z\} \in E$.

Note that $d_U(v) \leq d_{U,1}(v) + 2d_{U,2}(v)$. Recall their definition just before Lemma 5.

Let $L = 100$. Applying Lemmas 5 and 6 separately with

$$\sigma = 2\gamma \text{ and } \theta = \frac{ed^{\frac{1}{3}-\delta}\log^2 d}{14} \text{ and } \Delta = 3\theta + \frac{\beta}{L} \text{ and } \tau = \frac{\theta}{\beta},$$

we see that w.h.p.

$$S_1 = \left\{v \in U_0 : d_{U_0,1}(v) \geq 3\theta + \beta/L\right\} \text{ satisfies } |S_1| \leq 2\tau\gamma n = \frac{6e\log^3 d}{d^{1+3\delta}}n.$$

$$S_2 = \left\{v \in U_0 : d_{U_0,2}(v) \geq 3\theta + \beta/L\right\} \text{ satisfies } |S_2| \leq 2\tau\gamma n = \frac{6e\log^3 d}{d^{1+3\delta}}n.$$

$$U'_{1,a} = \left\{v \in U_0 : d_{U_0}(v) \geq 3\Delta\right\} \text{ satisfies } |U'_{1,a}| \leq 4\tau\gamma n = \frac{12e\log^3 d}{d^{1+3\delta}}n.$$

We then let $U_{1,a} \supseteq U'_{1,a}$ be the subset of U_0 consisting of the vertices with the $4\tau\gamma n$ largest values of d_{U_0}.

Let $A_0 = U_0 \setminus U_{1,a}$ and $B_0 = \left\{v \in U_{1,a} : |d_{A_0}(v)| \geq \frac{3\beta}{L}\right\}$. Iteratively we define

$$A_i := \left\{v \in A_{i-1} : |d_{B_{i-1}}(v)| \geq 2\right\} \tag{15}$$

$$B_i := \left\{v \in B_{i-1} : |d_{A_i}(v)| \geq \frac{3\beta}{L}\right\} \tag{16}$$

Lemma 9 *W.h.p., $\nexists$ disjoint sets $S, T \subseteq V(G)$, $G = G_U$ such that*

$$t = |T| \leq t_0 = \frac{100\log d}{d^{\frac{2}{3}+\delta}}n \text{ and } |S| \geq s_0 = \frac{8L|T|}{\beta} \text{ and } d_{T,1}(v) \geq \frac{\beta}{L} \text{ for all } v \in S.$$

Proof We observe that if S, T exist, then one of the following two cases must occur:

C1: $f(v) = |\{u \in T : \exists w \in V(G) \setminus (S \cup T), \{u, w, v\} \in E(G)\}| \geq \frac{\beta}{2L}$ for at least $\frac{s_0}{2}$ vertices $v \in S$.

C2: There are at least $t = \frac{\beta s_0}{8L}$ hyperedges $\{u, v, w\}$ such that $u, v \in S$ and $w \in T$.

$$\mathbb{P}(\exists S, T \text{ satisfying C1}) \leq \sum_{t=\frac{\beta}{L}}^{t_0} \binom{n}{t}\binom{n}{\frac{8Lt}{\beta}}\left(\binom{t}{\frac{\beta}{2L}}\left(\frac{d}{n}\right)^{\frac{\beta}{2L}}\right)^{\frac{8Lt}{\beta}}$$

$$\leq \sum_{t=\frac{\beta}{L}}^{t_0} \left(\frac{ne}{t}\right)^t \left(\frac{ne\beta}{8Lt}\right)^{\frac{8Lt}{\beta}} \left(\frac{2Lted}{\beta n}\right)^{4t}$$

$$= \sum_{t=\frac{\beta}{L}}^{t_0} \left(\left(\frac{t}{n}\right)^{3-\frac{8L}{\beta}} \frac{16e^{5+8l/\beta}\beta^{8L/\beta-4}d^4}{L^{8l/\beta-4}} \right)^t$$

$$\leq n \left(\left(\frac{t_0}{n}\right)^{3-\frac{8L}{\beta}} \frac{16e^{5+8l/\beta}\beta^{8L/\beta-4}d^4}{L^{8l/\beta-4}} \right)^{\beta/L}$$

$$= o(1).$$

$$\mathbb{P}(\exists S, T \text{ satisfying C2})$$

$$\leq \sum_{t=\frac{\beta}{L}}^{t_0} \binom{n}{t}\binom{n}{\frac{8Lt}{\beta}}\binom{\binom{\frac{8Lt}{\beta}}{2}t}{t}\left(\frac{d}{n^2}\right)^t$$

$$\leq \sum_{t=\frac{\beta}{L}}^{t_0} \left(\frac{ne}{t}\right)^t \left(\frac{ne\beta}{8Lt}\right)^{\frac{8Lt}{\beta}} \left(\frac{32L^2t^2ed}{\beta n^2}\right)^t$$

$$\leq \sum_{t=\frac{\beta}{L}}^{t_0} \left(\left(\frac{t}{n}\right)^{1-8L/\beta} \frac{32e^{1+8L/\beta}\beta^{8L/\beta-1}d}{L^{8L/\beta-2}} \right)^t$$

$$\leq \sum_{t=\frac{\beta}{L}}^{\log^2 n} \left(\left(\frac{\log^2 n}{n}\right)^{1-8L/\beta} d^{1/3} \right)^t + \sum_{t=\log^2 n}^{t_0} \left(\left(\frac{t_0}{n}\right)^{1-8L/\beta} d^{1/3} \right)^{\log^2 n}$$

$$= o(1). \qquad \square$$

Thus if $B' = \left\{ v \in U_{1,a} : d_{A_0,1}(v) \geq \frac{\beta}{L} \right\}$, then w.h.p.

$$|B'| \leq \frac{8L}{\beta}|U_0| = \frac{16L\gamma}{\beta} n = \frac{2016L \ \log d}{d^{\frac{4}{3}+2\delta}} n.$$

Lemma 10 *W.h.p. $\nexists$ disjoint S, T s.t*

$$t = |T| \leq t_0 = \frac{30n \ \log d}{d^{\frac{2}{3}+\delta}} \text{and } |S| \geq \frac{2L|T|}{\beta} \text{ and } d_{T,2}(v) \geq \frac{\beta}{L} \text{ for all } v \in S. \tag{17}$$

Proof

$$
\begin{aligned}
&\mathbb{P}(\exists S, T \text{ satisfying (17)})\\
&\leq \sum_{t=\frac{\beta}{L}}^{t_0} \binom{n}{t}\binom{n}{\frac{2Lt}{\beta}}\left(\binom{\binom{t}{2}}{\frac{\beta}{L}}\left(\frac{d}{n^2}\right)^{\frac{\beta}{L}}\right)^{\frac{2Lt}{\beta}}\\
&\leq \sum_{t=\frac{\beta}{L}}^{t_0} \left(\frac{ne}{t}\right)^t \left(\frac{ne\beta}{2Lt}\right)^{\frac{2Lt}{\beta}} \left(\left(\frac{t^2dLe}{2\beta n^2}\right)^{\beta/L}\right)^{\frac{2Lt}{\beta}}\\
&= \sum_{t=\frac{\beta}{L}}^{t_0} \left(\frac{ne}{t}\right)^t \left(\frac{ne\beta}{2Lt}\right)^{\frac{2Lt}{\beta}} \left(\frac{t^2dLe}{2\beta n^2}\right)^{2t}\\
&= \sum_{t=\frac{\beta}{L}}^{t_0} \left(\left(\frac{t}{n}\right)^{3-2L/\beta} \frac{e^{3+2L/\beta}\beta^{2L/\beta-2}d^2}{4L^{2L/\beta-2}}\right)^t\\
&\leq \sum_{t=\frac{\beta}{L}}^{\log^2 n} \left(\left(\frac{t}{n}\right)^{3-2L/\beta} d^{2/3}\right)^{\beta/L} + \sum_{t=\log^2 n}^{t_0} \left(\left(\frac{t_0}{n}\right)^{3-2L/\beta} d^{2/3}\right)^{\log^2 n}\\
&= o(1)
\end{aligned}
$$

□

Thus if $B'' = \left\{v \in U_{1,a} : d_{A_0,2}(v) \geq \frac{\beta}{L}\right\}$, then w.h.p.

$$|B''| \leq \frac{2}{\beta}L|U_0| = \frac{4L\gamma}{\beta}n = \frac{504L\log d}{d^{\frac{4}{3}+2\delta}}n.$$

Clearly, $B_0 \subseteq B' \cup B''$. Hence,

$$|B_0| \leq |B' \cup B''| \leq \frac{10L}{\beta}|U_0| \leq \frac{3000L\log d}{d^{\frac{4}{3}+2\delta}}n.$$

Let $D_2(S) = \left\{v : d_{S,1}(v) \geq 2\right\}$ for $S \subseteq V(G)$.

Lemma 11 *W.h.p.,* $|D_2(S)| < 3K|S|, \quad \forall\ |S| \leq s_0 = \frac{3000L\log d}{d^{\frac{4}{3}+2\delta}}n$, *where* $K = d^{\frac{2}{3}-2\delta}\log^2 d$.

Proof Suppose that there exist S and T with $|S| \leq s_0$ and $|T| = 3 \cdot K|S|$ such that $d_{s,1}(v) \geq 2$ for all $v \in T$. Then for $v \in T$, one of the following can occur.

D1: There are $x, y \in S$ and $a, b \in T$ such that $\{v, x, a\}, \{v, y, b\} \in E(G)$.

D2: There are $x, y \in S$ and $a \in T$, $b \in V(G) \setminus (S \cup T)$ such that $\{v, x, a\}, \{v, y, b\} \in E(G)$.

D3: There are $x, y \in S$ and $a, b \in V(G) \setminus (S \cup T)$ such that $\{v, x, a\}, \{v, y, b\} \in E(G)$.

Now we construct $T' \subseteq T$ with $|T'| \geq K|S|$ such that if $v \in T$, then there exist $x, y \in S$ and $a, b \in V(G) \setminus (S \cup T')$ such that D1 holds. First, for every vertex of type D1, put v in T' and remove a, b from further consideration. Second, for every vertex of type D2, put v in T' and remove a from further consideration. Finally, for every vertex of type D3, put v in T'. We observe that for every $v \in T'$ we have thrown away at most 2 vertices of T and hence $|T'| \geq K|S|$. We will now estimate the probability of the existence of S, T'.

$$\begin{aligned}
\mathbb{P}(\exists\, |S| \leq s_0, |D_2(S)| \geq 3K|S|) &\leq \sum_{s=2}^{s_0} \binom{n}{s}\binom{n}{Ks}\left(\binom{s}{2} n^2 \left(\frac{d}{n^2}\right)^2\right)^{Ks} \\
&\leq \sum_{s=2}^{s_0} \left(\frac{ne}{s}\right)^s \left(\frac{ne}{Ks}\right)^{Ks} \left(\frac{s^2 d^2}{2n^2}\right)^{Ks} \\
&= \sum_{s=2}^{s_0} \left(\left(\frac{s}{n}\right)^{K-1} \frac{e^{K+1} d^{2K}}{K^K 2^K}\right)^s \\
&= o(1)
\end{aligned}$$

□

Thus if $A' = \{v \in A_0 : v \in D_2(B_0)\}$, then w.h.p.

$$|A'| \leq \frac{9000 \log^3 d}{d^{\frac{2}{3}+4\delta}} n.$$

Let $D_2'(S) = \left\{v : d_{S,2}(v) \geq 1\right\}$ for $S \subseteq V(G)$.

Lemma 12 *W.h.p.* $|D_2'(S)| \leq K|S|, \ \forall\ |S| \leq \frac{3000L \log d}{d^{\frac{4}{3}+2\delta}} n$, *where* $K = d^{\frac{2}{3}-2\delta} \log^2 d$.

Proof

$$\begin{aligned}
\mathbb{P}(\exists\, S \leq s_0, |D_2(S) \geq K|S|) &\leq \sum_{s=2}^{s_0} \binom{n}{s}\binom{n}{Ks}\left(\binom{s}{2}\frac{d}{n^2}\right)^{Ks} \\
&\leq \sum_{s=2}^{s_0} \left(\frac{ne}{s}\right)^s \left(\frac{ne}{Ks}\right)^{Ks} \left(\frac{s^2 d}{2n^2}\right)^{Ks}
\end{aligned}$$

$$= \sum_{s=2}^{s_0} \left(\left(\frac{s}{n}\right)^{K-1} \frac{e^{K+1} d^K}{2^K K^K} \right)^s$$

$$= o(1). \qquad \square$$

So w.h.p. $|A_1| = |A' \cup A''| \leq \frac{12000 \log^3 d}{d^{\frac{2}{3}+4\delta}} n$. From (15), (16), Lemmas 9 and 10 we see that

$$|B_1| \leq \frac{10L}{\beta} |A_1|.$$

$$|A_2| \leq 4K|B_1|.$$

$$|B_i| \leq \frac{10L}{\beta} |A_i| \leq \frac{40KL}{\beta} |B_{i-1}| \leq \left(\frac{40KL}{\beta}\right)^i |B_0|.$$

Using Lemmas 11 and 12,

$$|A_{i+1}| \leq 4K|B_i| \leq 4K \left(\frac{40KL}{\beta}\right)^i |B_0| \leq 4K \left(\frac{40KL}{\beta}\right)^i \cdot \frac{3000L \log d}{d^{\frac{4}{3}+2\delta}} n,$$

where $\frac{KL}{\beta} = 150 d^{-3\delta} \log^2 d$.

Let $\zeta = \left\lceil \frac{2}{\delta} \right\rceil$ and let $Y = N(B_\zeta) \cap A_\zeta$. Then,

$$|Y| \leq |A_\zeta| \leq 4K \left(\frac{40KL}{\beta}\right)^{\zeta-1} \cdot \frac{3000 \log d}{d^{4/3+2\delta}} n$$

$$= 12000(120L)^{\zeta-1} \cdot \frac{(\log d)^{1+2\zeta}}{d^{2/3+\delta+\delta\zeta}} n$$

$$\leq \frac{n}{d^{\zeta\delta}} \leq \tau\gamma n.$$

Let $U_1 = U_{1,a} \cup Y$. Then

$$|U_1| \leq 5\tau\gamma n = \frac{15e \log^3 d}{d^{1+3\delta}}.$$

We now define the light and heavy edges in the following fashion:

Q1: The edges between B_i and $A_i \setminus A_{i+1}$ are light
Q2: The edges between $B_i \setminus B_{i+1}$ and A_{i+1} are heavy
Q3: The edges between $U_1 \setminus U_{1,a}$ and $U_0 \setminus U_1$ are heavy

We now check that **P1–P4** hold. First consider the light edges. For every vertex $v \in U_0 \setminus U_1$ there is at most one light neighbor in U_1. Because if $v \in A_i$ and

$v \notin A_{i+1}$ and there are 2 light neighbors x, w of v in U_1, by Q1, $x, w \in B_i$ and that would contradict the fact that $v \notin A_{i+1}$. This implies that **P1** holds.

We will argue next that for all $v \in U_1$, $d_{U_0 \setminus U_1}(v) \leq 3\Delta \leq \frac{3\beta}{50}$. For $v \notin U_{1,a}$ this is true from the definition of $U_{1,a}$. Similarly, for $v \notin B_0$. Now consider $v \in B_i \setminus B_{i+1}, i \geq 0$. It only has light neighbors in $A_i \setminus A_{i+1}$ and if v has more than $\frac{3\beta}{L}$ heavy neighbors in A_{i+1}, then v should be in B_{i+1}, which is a contradiction. Because it is also in $B_j, j \leq i-1$ it only has light neighbors in $(A_{i-1} \setminus A_i) \cup (A_{i-2} \setminus A_{i-3}) \cup \cdots = A_0 \setminus A_i$. Clearly **P3, P4** hold.

3.3 The Verification of P1–P4: Constructing U_2

Applying Lemmas 5 and 6 separately with

$$\sigma = \frac{15e \log^3 d}{d^{1+3\delta}} \text{ and } \theta = \frac{L}{\delta} \text{ and } \Delta = 3\theta + \frac{\beta}{L} \text{ and } \tau = \frac{\theta}{\beta},$$

we see that w.h.p.

$$S_1 = \left\{v \in U_1 : d_{U_1,1}(v) \geq 3\theta + \beta/L\right\} \text{ satisfies } |S_1| \leq \tau\sigma n = \frac{45eL \log^3 d}{\delta d^{5/3+4\delta}} n.$$

$$S_2 = \left\{v \in U_1 : d_{U_1,2}(v) \geq 3\theta + \beta/L\right\} \text{ satisfies } |S_2| \leq \tau\sigma n = \frac{45eL \log^3 d}{\delta d^{5/3+4\delta}} n.$$

$$U'_{2,a} = \left\{v \in U_1 : d_{U_1}(v) \geq 3\Delta\right\} \text{ satisfies } |U'_{2,a}| \leq 2\tau\sigma n = \frac{90eL \log^3 d}{\delta d^{5/3+4\delta}} n.$$

We then let $U_{2,a} \supseteq U'_{2,a}$ be the subset of U_1 consisting of the vertices with the $2\tau\sigma n$ largest values of d_{U_1}. As in Section 3.2, define $A_0 = U_1 \setminus U_{2,a}$ and let $B_0 = \left\{v \in U_{2,a} : |d_{A_0}(v)| \geq \frac{3\beta}{L}\right\}$. Iteratively we define

$$A_i := \left\{v \in A_{i-1} : \ |d_{B_{i-1}}(v)| \geq 2\right\}$$

$$B_i := \left\{v \in B_{i-1} : \ |d_{A_i}(v)| \geq \frac{3\beta}{L}\right\}$$

Let $B' = \left\{v \in U_{2,a} : d_{A_0,1}(v) \geq \frac{\beta}{L}\right\}$. Using Lemma 9, we see that w.h.p., $|B'| \leq \frac{8L|U_1|}{\beta} \leq \frac{360eL \ \log^3 d}{d^{\frac{5}{3}+4\delta}} n$.

Let $B'' = \left\{v \in U_{2,a} : d_{A_1,2}(v) \geq \frac{\beta}{L}\right\}$. Using Lemma 10, we see that w.h.p. $|B''| \leq \frac{90eL\log^3 d}{d^{\frac{5}{3}+4\delta}}n$.

Clearly, $B_0 \subseteq B' \cup B''$. Therefore, w.h.p.,

$$|B_0| \leq |B' \cup B''| \leq \frac{450eL\log^3 d}{d^{\frac{5}{3}+4\delta}}n.$$

Arguing as in Section 3.2 we see that w.h.p.

$$B_i| \leq \frac{10L}{\beta}|A_i| \leq \frac{40KL}{\beta}|B_{i-1}| \leq \left(\frac{40KL}{\beta}\right)^i |B_0|.$$

$$|A_{i+1}| \leq 4K|B_i| \leq 4K\left(\frac{40KL}{\beta}\right)^i |B_0|$$

Remember that $\zeta = \left\lceil \frac{2}{\delta} \right\rceil$ and let $Y = N(B_\zeta) \cap A_\zeta$. Then,

$$|Y| \leq |A_\zeta| \leq 4K\left(\frac{40KL}{\beta}\right)^{\zeta-1} \cdot \frac{3000\log d}{d^{4/3+2\delta}}n \leq \frac{n}{d^2}.$$

Let $U_1 = U_{1,a} \cup Y$. Then w.h.p.

$$|U_1| \leq 5\tau\gamma n = \frac{15e\log^3 d}{d^{1+3\delta}}.$$

We now define the light and heavy edges in the following fashion:

Q1: The edges between B_i and $A_i \setminus A_{i+1}$ are light
Q2: The edges between $B_i \setminus B_{i+1}$ and A_{i+1} are heavy
Q3: The edges between $U_1 \setminus U_{1,a}$ and $U_0 \setminus U_1$ are heavy

We now check that **P1–P4** hold. First consider the light edges. For every vertex $v \in U_0 \setminus U_1$ there is at most one light neighbor in U_1. Because if $v \in A_i$ and $v \notin A_{i+1}$ and there are 2 light neighbors x, w of v in U_1, by Q1, $x, w \in B_i$ and that would contradict the fact that $v \notin A_{i+1}$. This implies that **P1** holds.

We will argue next that for all $v \in U_1$, $d_{U_0 \setminus U_1}(v) \leq 3\Delta \leq \frac{3\beta}{50}$. For $v \notin U_{1,a}$ this is true from the definition of $U_{1,a}$. Similarly, for $v \notin B_0$. Now consider $v \in B_i \setminus B_{i+1}$, $i \geq 0$. It only has light neighbors in $A_i \setminus A_{i+1}$ and if v has more than $\frac{3\beta}{L}$ heavy neighbors in A_{i+1}, then v should be in B_{i+1}, which is a contradiction. Because it is also in B_j, $j \leq i-1$ it only has light neighbors in $(A_{i-1} \setminus A_i) \cup (A_{i-2} \setminus A_{i-3}) \cup \cdots = A_0 \setminus A_i$. Clearly **P3, P4** hold.

3.4 The Verification of P1–P4: Constructing U_2

Applying Lemmas 5 and 6 separately with

$$\sigma = \frac{15e\log^3 d}{d^{1+3\delta}} \text{ and } \theta = \frac{L}{\delta} \text{ and } \Delta = 3\theta + \frac{\beta}{L} \text{ and } \tau = \frac{\theta}{\beta},$$

we see that w.h.p.

$$S_1 = \{v \in U_1 : d_{U_1,1}(v) \geq 3\theta + \beta/L\} \text{ satisfies } |S_1| \leq \tau\sigma n = \frac{45eL\log^3 d}{\delta d^{5/3+4\delta}}n.$$

$$S_2 = \{v \in U_1 : d_{U_1,2}(v) \geq 3\theta + \beta/L\} \text{ satisfies } |S_2| \leq \tau\sigma n = \frac{45eL\log^3 d}{\delta d^{5/3+4\delta}}n.$$

$$U'_{2,a} = \{v \in U_1 : d_{U_1}(v) \geq 3\Delta\} \text{ satisfies } |U'_{2,a}| \leq 2\tau\sigma n = \frac{90eL\log^3 d}{\delta d^{5/3+4\delta}}n.$$

We then let $U_{2,a} \supseteq U'_{2,a}$ be the subset of U_1 consisting of the vertices with the $2\tau\sigma n$ largest values of d_{U_1}. As in Section 3.2, define $A_0 = U_1 \setminus U_{2,a}$ and let $B_0 = \left\{v \in U_{2,a} : |d_{A_0}(v)| \geq \frac{3\beta}{L}\right\}$. Iteratively we define

$$A_i := \{v \in A_{i-1} : |d_{B_{i-1}}(v)| \geq 2\} \tag{18}$$

$$B_i := \left\{v \in B_{i-1} : |d_{A_i}(v)| \geq \frac{3\beta}{L}\right\} \tag{19}$$

Let $B' = \left\{v \in U_{2,a} : d_{A_0,1}(v) \geq \frac{\beta}{L}\right\}$. Using Lemma 9, we see that w.h.p., $|B'| \leq \frac{8L|U_1|}{\beta} \leq \frac{360eL\log^3 d}{d^{\frac{5}{3}+4\delta}}n$.

Let $B'' = \left\{v \in U_{2,a} : d_{A_1,2}(v) \geq \frac{\beta}{L}\right\}$. Using Lemma 10, we see that w.h.p. $|B''| \leq \frac{90eL\log^3 d}{d^{\frac{5}{3}+4\delta}}n$.

Clearly, $B_0 \subseteq B' \cup B''$. Therefore, w.h.p.,

$$|B_0| \leq |B' \cup B''| \leq \frac{450eL\log^3 d}{d^{\frac{5}{3}+4\delta}}n.$$

Arguing as in Section 3.2 we see that w.h.p.

$$B_i| \leq \frac{10L}{\beta}|A_i| \leq \frac{40KL}{\beta}|B_{i-1}| \leq \left(\frac{40KL}{\beta}\right)^i |B_0|.$$

$$|A_{i+1}| \leq 4K|B_i| \leq 4K\left(\frac{40KL}{\beta}\right)^i |B_0|$$

With $\zeta = \left\lceil \frac{2}{\delta} \right\rceil$ and $K = d^{\frac{2}{3}-2\delta} \log^2 d$ as before and $Y = N(B_\zeta) \cap A_\zeta$ we get

$$|Y| \leq |A_\zeta| \leq 4K\left(\frac{40KL}{\beta}\right)^{\zeta-1} \cdot \frac{450eL\ \log^3 d}{d^{5/3+4\delta}} n \leq \frac{n}{d^2}.$$

Letting $\gamma_2 = \frac{500Le\log^3 d}{\delta d^{5/3+4\delta}}$ and $U_2 = U_{2,a} \cup Y$ we see that w.h.p. $|U_2| \leq \gamma_2 n$.
We can define heavy and light edges as in U_1 and **P1–P4** follows.

3.5 *The Verification of P1–P4: Constructing U_3*

Applying Lemmas 5 and 6 separately with

$$\sigma = \frac{500eL\log^3 d}{\delta d^{5/3+4\delta}} \text{ and } \theta = \frac{5}{2} \text{ and } \Delta = 3\theta + \frac{\beta}{L} \text{ and } \tau = \frac{L\theta}{\beta},$$

we see that w.h.p.

$$S_1 = \left\{v \in U_2 : d_{U_2,1}(v) \geq 3\theta + \beta/L\right\} \text{ satisfies } |S_1| \leq \tau\sigma n = \frac{1250Le\log^3 d}{\delta d^{7/3+5\delta}} n.$$

$$S_2 = \left\{v \in U_2 : d_{U_2,2}(v) \geq 3\theta + \beta/L\right\} \text{ satisfies } |S_2| \leq \tau\sigma n = \frac{1250Le\log^3 d}{\delta d^{7/3+5\delta}} n.$$

$$U_3' = \left\{v \in U_2 : d_{U_2}(v) \geq 3\Delta\right\} \text{ satisfies } |U_3'| \leq 2\tau\sigma n = \frac{2500Le\log^3 d}{\delta d^{7/3+5\delta}} n.$$

We now construct $U_3 \supseteq U_3'$ by repeatedly adding vertices $y_1, y_2, ..y_s$ of $U_2 \setminus U_3'$ such that y_j is the lowest numbered vertex not in $Y_j = U_3' \cup \left\{y_1, y_2 \ldots . y_{j-1}\right\}$ that has at least two neighbors in Y_j in G. W.h.p., this process terminates with $j = t \leq 39|U_3'|$. We can apply Lemma 4 to see that w.h.p. this does not happen. Indeed, let $S_0 = U_3'$. We add vertices to S_{j-1} to create the set S_j iteratively. In this procedure we encounter two cases.

- $\exists x, w \in Y_j$ such that $(y_j, x, w) \in E(G)$. Then $S_j = S_{j-1} \cup \{y_j\}$.
- $\exists x, w \in Y_j$ and $a, b \notin Y_j$ such that $(x, a, y_j), (w, b, y_j) \in E(G)$. Then $S_j = S_{j-1} \cup \left\{y_j, a, b\right\}$

Note that we are adding at least 2 hyperedges for every 3 vertices added to S_t. If $s \geq 39|U_3'|$, then $e_3(S_s) \geq \frac{13}{20}|S_s|$. Apply Lemma 4 with $\theta = \frac{13}{20}$ and $\sigma = 120 \cdot \frac{2500Le\log^3 d}{\delta d^{7/3+5\delta}}$ to conclude that $t \leq 39|U_3'|$.

Putting $U_3 = U_3' \cup S_t$ we see that each vertex in $U_2 \setminus U_3$ has at most one G-neighbor in U_3. We can, therefore, make the $U_3 : (U_2 \setminus U_3)$ edges light and satisfy P1, P2, and P4.

3.6 The Verification of P1–P5: Construction of U_i, $i \geq 4$

We repeat the argument of Section 3.5 to construct the rest of the sequence $U_0 \supseteq U_1 \supseteq U_2 \supseteq \ldots \supseteq U_l$. One can check that $|U_i| \leq \frac{200L}{\beta}|U_{i-1}|$. We choose l so that $|U_l| \leq \log n$. We can then easily prove that w.h.p. S contains at most $|S|$ edges of G whenever $|S| \leq \log n$, implying P5.

4 Final Remarks

We have shown lower bounds for the game chromatic number of random k-uniform hypergraphs and upper bounds for random 3-uniform hypergraphs. The lower bound is satisfactory in that it is within a constant factor of the chromatic number. The upper bound is most likely not tight, but it is still non-trivial in that it is much smaller than d.

We conjecture that the upper bound can be reduced to within a constant factor of the lower bound. It would also be of interest to consider upper bounds for k-uniform hypergraphs, $k \geq 4$.

Acknowledgment The author "Alan Frieze's" research was supported in part by NSF grant DMS1362785.

References

1. N. Alon, J.H. Spencer, *The Probabilistic Method*, 3rd edn. (Wiley, New York, 2008)
2. T. Bartnicki, J.A. Grytczuk, H.A. Kierstead, X. Zhu, The map coloring game. Am. Math. Mon. **114**, 793–803 (2007)
3. H.L. Bodlaender, On the complexity of some coloring games. Int. J. Found. Comput. Sci. **2**, 133–147 (1991)
4. T. Bohman, A.M. Frieze, B. Sudakov, The game chromatic number of random graphs. Random Struct. Algorithms **32**, 223–235 (2008)
5. U. Faigle, U. Kern, H. Kierstead, W.T. Trotter, On the game chromatic number of some classes of graphs. Ars Comb. **35**, 143–150 (1993)

6. A.M. Frieze, S. Haber, M. Lavrov, On the game chromatic number of sparse random graphs. SIAM J. Discret. Math. **27**, 768–790 (2013)
7. M. Gardner, Mathematical games. Sci. Am. **244**(4), 18–26 (1981)
8. S. Janson, T. Łuczak, A. Ruciński, *Random Graphs* (Wiley, New York, 2000)
9. R. Keusch, A. Steger, The game chromatic number of dense random graphs. Electron. J. Comb. **21** (2014). https://doi.org/10.37236/4391
10. M. Krivelevich, B. Sudakov, The chromatic number of random hypergraphs. Random Struct. Algorithms **12**, 381–403 (1998)

Perfect Hash Families: The Generalization to Higher Indices

Ryan E. Dougherty and Charles J. Colbourn

Abstract Perfect hash families are often represented as combinatorial arrays encoding partitions of k items into v classes, so that every t or fewer of the items are completely separated by at least a specified number of chosen partitions. This specified number is the index of the hash family. The case when each t-set must be separated at least once has been extensively researched; they arise in diverse applications, both directly and as fundamental ingredients in a column replacement strategy for a variety of combinatorial arrays. In this paper, construction techniques and algorithmic methods for constructing perfect hash families are surveyed, in order to explore extensions to the situation when each t-set must be separated by more than one partition.

1 Introduction

Suppose that there are k items, and each is assigned one of v values. Our objective is to ensure that each set of t items receives t different values; when this occurs, the t items are *separated*. Evidently if $v \geq k$, each item can be assigned a value that is different from all others assigned, so that every set of t items is separated. However, when $v < k$, some two items necessarily receive the same value; then any t-set containing these two cannot be separated. When this occurs, suppose that N assignments of values to items are chosen, rather than one. Then one can ask: How small can N be so that every t-set of items is separated in at least λ of the

R. E. Dougherty
Department of Electrical Engineering and Computer Science, United States Military Academy, West Point, NY, USA
e-mail: ryan.dougherty@westpoint.edu

C. J. Colbourn (✉)
Computing, Informatics, and Decision Systems Engineering, Arizona State University, Tempe, AZ, USA
e-mail: harles.colbourn@asu.edu

A. M. Raigorodskii, M. Th. Rassias (eds.), *Discrete Mathematics and Applications*, Springer Optimization and Its Applications 165,
https://doi.org/10.1007/978-3-030-55857-4_7

assignments? This easily stated combinatorial question is challenging, and many open problems remain despite substantial research effort. It is an important question as well, with applications described next.

Mehlhorn [47] originally examined this question to provide an efficient way to store and retrieve frequently used information; in that context, the assignment of values to the items is treated as a *hash function* [30], and hence the question is phrased as one about families of hash functions. Applications to derandomization [5], circuit complexity [49] , and cryptography [17, 36, 59] arose. Subsequently, Stinson et al. [60] established applications of such families (with $\lambda = 1$) to construct numerous other combinatorial objects, such as separating systems, key distribution patterns, cover-free families, and secure frameproof codes. A general strategy, column replacement, has extended their range of applications into testing and measurement [22] and compressive sensing [27], among others.

In many of these applications, error correction through redundancy in the separation is needed; a few examples are given in [1, 42, 55]. Despite this, there has been little examination of such hash families with $\lambda > 1$, with the notable exception of [3, 4]. In this paper, we therefore survey a number of main construction methods for such hash families, with an eye to extending them to treat cases with $\lambda > 1$ when possible. Our emphasis is on fixed values of $\lambda \geq 1$; we only treat cases when λ increases as a function of N in the concluding remarks. We focus on combinatorial aspects, discussing in particular constructive approaches to produce explicit examples for use in applications.

In order to develop and extend these ideas formally, we extend the presentation in [22], employing the very general language of t-restrictions [7]. We denote the set $\{1, \ldots, n\}$ by $[n]$. Let $N, k, v_1, \cdots, v_N, x_1, \cdots, x_k, t$, and λ be positive integers. An *abstract simplicial complex* (ASC), $\mathcal{A}$, is a family of non-empty finite subsets of a vertex set Γ that is closed under taking non-empty subsets; the *dimension* of an ASC, $\dim(\mathcal{A})$, is the maximum of $|X| - 1$ overall $X \in \mathcal{A}$. Let $\mathcal{H}$ be an abstract simplicial complex on k vertices such that the maximum cardinality of any set in $\mathcal{H}$ is t; label the vertices of $\mathcal{H}$ as $c_1, \cdots, c_k$. Let Σ_i be a v_i-ary alphabet not containing $\star$ for all $1 \leq i \leq N$, and let Δ_j be an x_j-ary alphabet also not containing $\star$ for all $1 \leq j \leq k$. Define an $N \times k$ array A in which the ith row of A contains symbols from $\Sigma_i \cup \{\star\}$, and the jth column of A contains symbols from $\Delta_j \cup \{\star\}$. If there exist i, j for which $\Sigma_i \cap \Delta_j = \emptyset$, the entry in the (i, j)-cell must be $\star$. Let $\Delta = \bigcup_{j=1}^{k} \Delta_j$. A *t-restriction* is a χ-tuple $\mathcal{T} = ((\mathcal{P}_1, T_1), \cdots, (\mathcal{P}_\chi, T_\chi))$, where $\mathcal{P}_i \subseteq \Delta^t$ and $T_i \in \{\exists, \forall\}$. Each set $\mathcal{P}_i$ is a *demand*. For each $\mathcal{P}_i$, if $T_i = \exists$, then at least λ rows of A contains some element of $\mathcal{P}_i$; if $T_i = \forall$, then for each element of $\mathcal{P}_i$, at least λ rows contain that element. For a collection $\mathcal{S}$ of t-tuples, define $\partial^i(\mathcal{S})$ to be the set of all $(t - i)$-tuples obtained by deleting any i columns from a t-tuple of $\mathcal{S}$. An array $\mathsf{A} = (a_{ij})$ λ*-satisfies* a given $(\mathcal{P}_i, T_i) \in \mathcal{T}$ if and only if for all $0 \leq j \leq t$ and any set $S = \{i_1, \ldots, i_{|S|}\} \in \mathcal{H}$,

1. if $T_i = \exists$, then for each $\mathcal{P} \in \partial^j(\mathcal{P}_i)$, there exist λ rows $1 \leq r_1 < \cdots < r_\lambda \leq N$ such that $(a_{r_\ell, c_{i_1}}, \cdots, a_{r_\ell, c_{i_{|S|}}}) \in \mathcal{P}$ for all $1 \leq \ell \leq \lambda$ and $S \in \mathcal{H}$ when $|S| = t - j$; or

2. if $T_i = \forall$, then for each $\mathcal{P} \in \partial^j(\mathcal{P}_i)$, and for all $(\sigma_1, \cdots, \sigma_{t-j}) \in \mathcal{P} \cap \prod_{m=1}^{t-j} \Delta_{c_m}$, there exist λ rows $1 \le r_1 < \cdots < r_\lambda \le N$ such that $(a_{r_\ell, c_{i_1}}, \cdots, a_{r_\ell, c_{i_{|S|}}}) = (\sigma_1, \cdots, \sigma_{|S|})$ for all $1 \le \ell \le \lambda$ and $S \in \mathcal{H}$ when $|S| = t - j$.

If the array λ-satisfies each of the $(\mathcal{P}_i, T_i) \in \mathcal{T}$, the array *$\lambda$-satisfies* $\mathcal{T}$. If an array A on N rows and k columns (and corresponding symbol set cardinalities for rows and columns) λ-satisfies a t-restriction $\mathcal{T}$, denote it by $\mathsf{TRA}_\lambda(N, k, \mathcal{H}, (v_1, \cdots, v_N), (x_1, \cdots, x_k), \mathcal{T})$. If $v_1 = \cdots = v_N$, A is *homogeneous*; otherwise, it is *heterogeneous*. If $x_1 = \cdots = x_k$, A is *uniform*; otherwise, it is *mixed*. When $\lambda = 1$, we omit it from the notation. When $T_1 = \cdots = T_\chi$, $\mathcal{T}$ is a *monotone* t-restriction. Most literature has concentrated on monotone t-restrictions with $\mathcal{H}$ being the hypergraph containing all possible hyperedges of size at most t on k vertices, and $\lambda = 1$.

This framework is very general, and it encompasses a number of well-studied combinatorial arrays. However, because it is also quite technical, we establish more restrictive notation for some specific arrays next. Choose an integer t, and form the set $\mathcal{M}_t$ of multisets whose elements are nonnegative integers and for which the sums of the elements of a multiset is t. Let $\mathcal{W} \subseteq \mathcal{M}_t$. A *$\mathcal{W}$-separating hash family* meets the following condition: When $C = \{c_1, \cdots, c_t\} \subseteq \binom{[k]}{t}$ and $W_1, \cdots, W_s$ is a partition of C with $\{|W_1|, \cdots, |W_s|\} \in \mathcal{W}$, define $\mathcal{D} = \{(y_1, \ldots, y_t) \in \Delta_{c_1} \times \cdots \times \Delta_{c_t} : y_c = y_{c'}$ only if c, c' belong to the same class of $W\}$. Then the demand $(\mathcal{D}, \exists)$ is met. When $\mathcal{W}$ consists of all partitions in $\mathcal{M}_t$ containing s parts, a $\mathcal{W}$-separating hash family is *(t, s)-distributing*. When $\mathcal{W}$ contains a single set $W = \{w_1, \ldots, w_s\}$, the family is separating of *type* $\{w_1, \ldots, w_s\}$.

Of primary concern here are the (t, s)-distributing hash families with $s = t$. Such a family is a *perfect hash family*. In order to refer to objects of this type, we employ standard notation. A perfect heterogeneous hash family is denoted as a $\mathsf{PHHF}_\lambda(N; k, (v_1, \cdots, v_N), t)$, and a homogeneous one is written as a $\mathsf{PHF}_\lambda(N; k, v, t)$. When a PHHF_λ A contains at least λ rows that contain distinct. symbols in each column of a t-set T of columns, it *λ-separates T*.

An example of a (homogeneous) $\mathsf{PHF}_1(6; 12, 3, 3)$ is given in Figure 1. It is a 6×12 array (6 rows, 12 columns) on the three symbols $\{0, 1, 2\}$, in which every 3-set of columns is 1-separated. For the 6×3 subarray involving columns 8, 9, and 10, only the last row consists of distinct symbols. Also, 148 of the 3-sets of columns are exactly 1-separated; 44 are exactly 2-separated; 19 are exactly 3-separated; 4 are exactly 4-separated; and none is 5 or 6-separated. There is no $\mathsf{PHF}(5; 12, 3, 3)$ [9], so this array has the fewest possible rows.

Fig. 1 A perfect hash family $\mathsf{PHF}_1(6; 12, 3, 3)$

								↓	↓	↓		
	0	1	2	2	1	2	2	0	1	1	0	0
	0	2	1	0	2	2	2	1	0	1	2	1
	1	0	0	2	2	2	1	1	2	1	0	2
	2	0	1	1	2	0	2	0	1	1	2	1
	2	0	2	1	2	1	0	2	2	1	1	0
→	2	0	1	2	1	1	2	[2]	[0]	[1]	2	1

Fig. 2 A SHF(3; 16, 4, {1, 2})

											↓				↓	↓
	1	1	1	1	2	2	2	2	3	3	3	3	4	4	4	4
	1	2	3	4	1	2	3	4	1	2	3	4	1	2	3	4
→	1	2	3	4	2	1	4	3	3	4	1	2	4	3	2	1

Fig. 3 A DHHF(10; 13, **v**, 5, 2) with $\mathbf{v} = (9^3 3^4 4^1 5^1 2^1)$

					↓	↓				↓		↓	↓
	6	7	8	3	4	0	2	2	3	0	5	1	1
	3	1	1	7	2	6	8	4	3	0	2	0	5
	8	5	1	4	2	3	2	6	7	0	1	3	0
	0	2	0	2	2	0	0	1	1	1	1	2	0
	0	0	2	1	1	1	2	0	0	2	2	0	1
→	1	1	2	2	2	0	1	0	0	2	1	0	0
	1	0	1	2	0	0	2	0	0	1	2	2	1
	1	1	0	1	0	3	2	0	2	0	1	0	2
	0	0	3	0	1	0	0	2	4	0	0	1	0
	0	⋆	⋆	⋆	⋆	1	⋆	⋆	1	⋆	⋆	0	1

The notation $\mathsf{SHHF}(N; k, (v_1, \cdots, v_N), \{w_1, \cdots, w_s\})$ is used for a separating hash family. More simply $\mathsf{SHF}(N; k, v, \{w_1, \cdots, w_s\})$ is used when it is homogeneous. Figure 2 gives an example of a (homogeneous) $\mathsf{SHF}(3; 16, 4, \{1, 2\})$. It is a 3×16 array on the four symbols $\{1, 2, 3, 4\}$ that is not a perfect hash family, because columns 11, 15, and 16 are separated by none of the three rows. However, in the 3×3 subarray consisting of these three columns, each of the three $\{1, 2\}$-separations is accomplished by a row.

A distributing hash family is denoted by $\mathsf{DHHF}(N; k, (v_1, \cdots, v_N), t, s)$; as before, a homogeneous DHHF is a $\mathsf{DHF}(N; k, v, t, s)$. Figure 3 gives a (heterogeneous) $\mathsf{DHHF}(10; 13, \mathbf{v}, 5, 2)$ with $\mathbf{v} = (9, 9, 9, 3, 3, 3, 3, 4, 5, 2)$. Often one uses an exponential notation that indicates the repetition in the exponent: $\mathbf{v} = (9^3 3^4 4^1 5^1 2^1)$.

Hash families in general, and perfect hash families in particular, play a central role in the construction of arrays that satisfy various t-restrictions. Indeed, they form the essential ingredients in a general technique known as *composition* or *column replacement*, which we describe next. The parameter t of the PHF is often referred to as its *strength*.

Construction 1 *Suppose there exist:*

1. *A, a $\mathsf{PHF}_\chi(M; \ell, k, t)$; and*
2. *B, a $\mathsf{TRA}_\lambda(N; k, \mathcal{H}, (v_1^{s_1} \cdots, v_\rho^{s_\rho}), x^k, \mathcal{T})$ with $N = \sum_{i=1}^{\rho} s_i v_i$ and $\mathcal{H} = \binom{[k]}{t}$.*

Construct an $NM \times \ell$ array, C, by replacing each symbol γ in A by the column indexed by γ in B. Then C is a

$$\mathsf{TRA}_{\chi\lambda}(NM; \ell, \mathcal{H}', ((M \cdot v_1)^{s_1} \cdots, (M \cdot v_\rho)^{s_\rho}), x^\ell, \mathcal{T})$$

with $\mathcal{H}' = \binom{[\ell]}{t}$.

Fig. 4 A CA(13; 3, 10, 2)

0	0	0	0	0	0	0	0	0	0
1	1	1	1	1	1	1	1	1	1
1	1	1	0	1	0	0	0	0	1
1	0	1	1	0	1	0	1	0	0
1	0	0	0	1	1	1	0	0	0
0	1	1	0	0	1	0	0	1	0
0	0	1	0	1	0	1	1	1	0
1	1	0	1	0	0	1	0	1	0
0	0	0	1	1	1	0	0	1	1
0	0	1	1	0	0	1	0	0	1
0	1	0	1	1	0	0	1	0	0
1	0	0	0	0	0	0	1	1	1
0	1	0	0	0	1	1	1	0	1

Construction 1 provides strong motivation for the study of perfect hash families, as it underlies the easy generation of 'large' arrays meeting t-restrictions. We outline one example of this, introducing a well-studied t-restriction that employs universal quantification. When for every t-set $\{c_1, \ldots, c_t\}$ of columns, the demand $(\Delta_{c_1} \times \cdots \times \Delta_{c_t}, \forall)$ is to be met, the array is a *mixed-level covering array*, denoted by $\mathsf{MCA}(N; t, (v_1, \cdots, v_k))$; when the array is homogeneous, it is a *covering array* of *strength* t, denoted by $\mathsf{CA}(N; t, k, v)$. In any $\mathsf{CA}(N; k, k, v)$, symbols can be permuted *in each column independently* so that the first row consists entirely of a single symbol. This yields a *constant row*, and when the CA has been modified in this way, it is *standardized*. Figure 4 gives an example of a standardized CA(13; 3, 10, 2).

Because a PHF with M rows leads to a TRA with NM rows, one wants the PHF ingredient to have as few rows as possible. The *perfect hash family (row) number*, $\mathsf{PHFN}_\lambda(k, v, t)$, is the minimum N for which a $\mathsf{PHF}_\lambda(N; k, v, t)$ exists. This notation does not extend naturally to heterogeneous hash families, because the number of rows is to be determined. To circumvent this notational issue, we often instead consider maximizing the number of columns rather than minimizing the number of rows. More formally, the *perfect hash family (column) number* $\mathsf{PHHFK}_\lambda(N, \mathbf{v}, t)$ is defined to be the maximum k for which a $\mathsf{PHHF}_\lambda(N; k, \mathbf{v}, t)$ exists. For homogeneous hash families, the notation $\mathsf{PHFK}_\lambda(N, v, t)$ is used. For homogeneous families, one can easily change between row and column numbers:

$$\mathsf{PHFN}_\lambda(k, v, t) = \min(N : \mathsf{PHFK}_\lambda(N, v, t) \geq k)$$

$$\mathsf{PHFK}_\lambda(N, v, t) = \max(k : \mathsf{PHFN}_\lambda(k, v, t) \leq N).$$

A $\mathsf{PHF}_\lambda(N; k, v, t)$ is *optimal* if $N = \mathsf{PHFN}_\lambda(k, v, t)$. Much study has been devoted to determining perfect hash family numbers for as many parameters as possible, as well as what structure underlies optimal PHFs. Moreover, one would hope to provide an explicit representation of the PHF with those parameters, particularly for constructing other combinatorial objects and t-restrictions. If this is not possible, then knowing asymptotics on this quantity is important in helping to determine asymptotics for other objects.

2 The Basics

First we state elementary relationships among perfect hash family numbers. In order to treat heterogeneous situations as well, we employ perfect hash family column numbers.

Additional rows cannot reduce the number of columns that can be achieved:

Fact 1 $\mathsf{PHHFK}_\lambda(N, (v_1, \ldots, v_N), t) \leq \mathsf{PHHFK}_\lambda(N + 1, (v_1, \ldots, v_{N+1}), t)$ whenever $v_{N+1} \geq 0$.

Reducing the size of column sets to be separated also cannot reduce the number of columns.

Fact 2 $\mathsf{PHHFK}_\lambda(N, (v_1, \ldots, v_N), t) \leq \mathsf{PHHFK}_\lambda(N, (v_1, \ldots, v_N), t-1)$ if $t \geq 2$.

Reducing λ enables one to remove rows without reducing the number of columns.

Fact 3 Let $\mathbf{v} = (v_1, \ldots, v_N)$. Let i be an integer with $1 \leq i \leq N$, and let $\mathbf{w} = (v_1, \ldots, v_{i-1}, v_{i+1}, \ldots v_N)$. Then $\mathsf{PHHFK}_\lambda(N, \mathbf{v}, t) \leq \mathsf{PHHFK}_{\lambda-1}(N-1, \mathbf{w}, t)$ if $\lambda \geq 2$.

Increasing the number of symbols in a row cannot reduce the number of columns.

Fact 4 Let $\mathbf{v} = (v_1, \ldots, v_N)$. Let i be an integer with $1 \leq i \leq N$, and let $\mathbf{w} = (v_1, \ldots, v_{i-1}, v_i + 1, v_{i+1}, \ldots v_N)$. Then $\mathsf{PHHFK}_\lambda(N, \mathbf{v}, t) \leq \mathsf{PHHFK}_\lambda(N, \mathbf{w}, t)$.

Changing the number of columns is also of interest. Removing a column is straightforward, but adding a column can leave $\binom{k}{t-1}$ t-sets of columns unseparated. Naively one could add λ rows for each to obtain

Fact 5 $\mathsf{PHFN}_\lambda(k, v, t) \leq \mathsf{PHFN}_\lambda(k + 1, v, t) \leq \mathsf{PHFN}_\lambda(k, v, t) + \lambda\binom{k}{t-1}$.

Walker and Colbourn [63] show a better bound, later generalized by Martirosyan and van Trung [46].

In order to avoid situations in which a row does not have enough symbols to separate any t-set of columns, we have

Fact 6 Let $\mathbf{v} = (v_1, \ldots, v_N)$. Let i be an integer with $1 \leq i \leq N$, and let $\mathbf{w} = (v_1, \ldots, v_{i-1}, v_{i+1}, \ldots v_N)$. Then $\mathsf{PHHFK}_\lambda(N, \mathbf{v}, t) = \mathsf{PHHFK}_\lambda(N - 1, \mathbf{w}, t)$ if $v_i < t$.

One can also consider reducing the number of symbols in a row.

Fact 7 Let $\mathbf{v} = (v_1, \ldots, v_N)$. Let i be an integer with $1 \leq i \leq N$, and let $\mathbf{w} = (v_1, \ldots, v_{i-1}, v_i - 1, v_{i+1}, \ldots v_N)$. Then $\mathsf{PHHFK}_\lambda(N, \mathbf{v}, t) \leq \left\lceil \frac{v_i-1}{v_i} \mathsf{PHHFK}_\lambda(N, \mathbf{w}, t) \right\rceil$.

Iterating Fact 7 until Fact 6 applies, one obtains

Theorem 1 (Martirosyan and van Trung [46, Theorem 7.2]) $\mathsf{PHFN}(\lceil \frac{k(t-1)}{v} \rceil, v, t) \leq \mathsf{PHFN}(k, v, t)-1$. *Equivalently,* $\mathsf{PHFK}(N-1, v, t) \geq \lceil \frac{t-1}{v} \mathsf{PHFK}(N, v, t) \rceil$.

Finally we describe a *row amalgamation* method for reducing the number of rows, which essentially comes from [15, 57]. In a $\mathsf{PHHF}_\lambda(N; k, (v_1, \ldots, v_N), t)$, select two rows i and j with $1 \le i < j \le N$. From these two, form a single row whose entries are ordered pairs, with the first coordinate being the entry from row i and the second from row j. Delete rows i and j (with v_i and v_j symbols), and add the new row with $v_i v_j$ symbols. This method can reduce the number of times a t-set of columns is separated, but this number cannot be reduced to 0.

Fact 8 Let $\mathbf{v} = (v_1, \ldots, v_N)$ and $\mathbf{w} = (w_1, \ldots, w_{N-1})$. Let i and j be integers with $1 \le i < j \le N$ so that $\{v_1, \ldots, v_N\} \setminus \{v_i, v_j\} = \{w_1, \ldots, w_{N-2}\}$, and $w_{N-1} = v_i v_j$. Then $\mathsf{PHHFK}_\lambda(N, \mathbf{v}, t) \le \mathsf{PHHFK}_{\max(1,\lambda-1)}(N-1, \mathbf{w}, t)$.

Now let us dispense with some easier parameter sets. If $N < \lambda$, there are insufficient rows to λ-separate any t-set; so we assume that $N \ge \lambda$. Now $\mathsf{PHFN}_\lambda(k, v, 1) = \lambda$ for all $k, v \ge 1$, and $\lambda \ge 1$, because any row separates all 1-sets of columns. Henceforth we only consider cases with $t \ge 2$. Fact 3 underlies the following:

Fact 9 $\mathsf{PHHFK}_\lambda(\lambda, (v_1, \ldots, v_\lambda), t) = \min(v_i : 1 \le i \le \lambda)$.

Because of this, we concentrate on cases in which no λ rows are each permitted to contain k or more distinct symbols. It is natural to ask whether one can obtain larger values of k when the number of rows is allowed to exceed λ.

In general, *recursive* constructions combine ingredient PHFs to make 'larger' ones. Many of the facts given provide easy examples of recursive constructions. Of course, because being a perfect hash family of index λ is a t-restriction, so column replacement or composition (Construction 1) is a recursive construction. In Section 3, Theorem 3 also provides a recursive construction.

3 Few Rows

No PHHF_λ with fewer than λ rows exists; when there are λ rows, Fact 9 applies. Suppose that $v_1 \ge \cdots \ge v_N \ge t$ and that a $\mathsf{PHHF}_\lambda(N; v_\lambda + 1, (v_1, \ldots, v_N), t)$ exists. Using Fact 3 we remove the first $\lambda - 1$ rows to obtain a $\mathsf{PHHF}_1(N - \lambda + 1; v_\lambda + 1, (v_\lambda, \ldots, v_N), t)$. Each row contains at least one pair of columns in which a symbol is repeated. Let $\{\gamma_i, \gamma_i'\}$ be such a pair of column indices with a repeated symbol in row i for $\lambda \le i \le N$. Then $\bigcup_{i=\lambda}^{N} \{\gamma_i, \gamma_i'\}$ is a set of at most $2(N - \lambda + 1)$ columns that is separated by no row of the PHHF_1. When $N \le \frac{t}{2} + \lambda - 1$, this is a contradiction. Hence we conclude

Fact 10 When $v_1 \ge \cdots \ge v_N \ge t$, $\mathsf{PHHFK}_\lambda(N, (v_1, \ldots, v_N), t) > v_\lambda$ only if $N \ge \lceil \frac{t+1}{2} \rceil + \lambda - 1$.

Although this condition is not sufficient whenever $v_1 \geq \cdots \geq v_N \geq t$, it does establish, for example, that $\mathsf{PHFK}_1(N, v, t) = v$ whenever $1 \leq N \leq \frac{t}{2}$. In order to increase the number of columns, one therefore requires further rows.

We describe a $\mathsf{PHF}_\lambda(s+\lambda; m(s+\lambda), m(s+\lambda-1)+1, 2s+1)$ whenever $m \geq 2$ and $s \geq 1$, generalizing a result of Walker and Colbourn [63] when $\lambda = 1$.

Construction 2 *Let $s \geq 1$, $m \geq 2$, and $\lambda \geq 1$. A $\mathsf{PHF}_\lambda(s+\lambda; m(s+\lambda), m(s+\lambda-1)+1, 2s+1)$ is constructed as follows. Form a set of $m(s+\lambda-1)$ elements X, and let ∞ be an element not in X. Then the desired PHF_λ contains exactly one occurrence of ∞ in each column, and contains each element of X exactly once in each row.*

Now Construction 2 yields more columns than symbols, and by Fact 10 it has the fewest rows for which this is possible with strength $t = 2s+1$. As a function of v, k grows linearly. This linear relationship is not restricted to the minimum number of rows, Blackburn [13] explored this phenomenon when $\lambda = 1$, and explicit computations, again for $\lambda = 1$, are pursued in [29]. We apply Blackburn's techniques to treat all λ.

To begin, we suppose that $k > v_1 \geq \cdots \geq v_N$, for otherwise we can either reduce λ by Fact 3 or conclude that one row suffices when $\lambda = 1$. Then every row contains at least one element that is repeated. The key idea is to classify the entries in each row; an entry is a *singleton* for this row when it appears exactly once in the row, and a *replicate* otherwise. Now suppose that a $\mathsf{PHHF}_\lambda(N; k, (v_1, \ldots, v_N), N-(\lambda-2)+s)$ with $0 \leq s \leq \frac{t-1}{2}$ has a column γ with at most $s+\lambda-1$ singletons, and hence at least $N-s-\lambda+1 = t-2s-1$ replicates. Form a set C of at most $t-2s$ columns by including γ, and a column index from each of the $t-2s-1$ rows that contains the same symbol as in column γ. Now choose any s further rows, and for each add a pair of column indices for columns containing the same symbol in this row to C. In total, C now contains at most $t-2s+2s = t$ column indices, and C is not separated in any of $t-2s-1+s = N-\lambda+1$ rows. But then C is not λ-separated, because at most $\lambda-1$ rows remain.

So $\lambda + s = t - N + 2(\lambda-1)$ is the minimum number of singletons in each column. In a row having v_i symbols, at most $v_i - 1$ can be singletons. However, there must be at least $k(t-N+2(\lambda-1))$ singletons in total. It follows that

$$k(t-N+2(\lambda-1)) \leq \sum_{i=1}^{N}(v_i - 1).$$

Hence we obtain

Lemma 1 *A $\mathsf{PHHF}_\lambda(N; k, (v_1, \ldots, v_N), t)$ with $N-(\lambda-2) \leq t \leq 2(N-(\lambda-2))-1$ satisfies*

$$k \leq \max\left(t, v_1, \ldots, v_N, \frac{\sum_{i=1}^{N}(v_i-1)}{t-N+2(\lambda-1)}\right).$$

Lemma 1 ensures that for a $\mathsf{PHF}_\lambda(N; k, v, t)$ with $N - (\lambda - 2) \le t \le 2(N - (\lambda - 2)) - 1$, (or, equivalently, $\frac{t+1}{2} + \lambda - 2 \le N \le t + \lambda - 2$), k grows at most linearly as a function of v.

For $\lambda = 1$, Blackburn [13] establishes that when $N = t + \lambda - 1$, k may grow superlinearly in v. We extend this construction to treat all values of λ next.

Construction 3 *Let $t \ge 2$, $\lambda \ge 1$, and $a \ge 2$. Then there exists a* $\mathsf{PHF}_\lambda(t + \lambda - 1; a^{t+\lambda-1}, a^{t-\lambda-2}, t)$. *The set of all vectors from $\{1, \dots, a\}^{t+\lambda-1}$ index the columns. In each column, in the ith row place the vector from $\{1, \dots, a\}^{t+\lambda-2}$ obtained by deleting the entry in the ith coordinate of the column index.*

The verification that this is a $\mathsf{PHF}_\lambda(t + \lambda - 1; a^{t+\lambda-1}, a^{t-\lambda-2}, t)$ comes essentially from [13]. Suppose to the contrary that there are t rows $\rho_1, \dots, \rho_t$ in which t columns $\gamma_1, \dots, \gamma_t$ are not separated. Form a graph G on vertex set $\{\gamma_1, \dots, \gamma_t\}$; for each $\rho \in \{\rho_1, \dots, \rho_t\}$, place an edge in G between some two vertices whose columns share a symbol in row ρ, and colour the edge with ρ. Now G has t vertices and t edges (of t different colours), and hence contains a cycle, say on vertices $\{v_0, \dots, v_\ell\}$. Let $e_i = \{v_i, v_{i+1}\}$ for $0 \le i < \ell$ have colour c_i, and let $e_\ell = \{v_\ell, v_0\}$ have colour c_ℓ. For $0 \le i \le \ell$, the two columns indexed by e_i agree in coordinate c_i *and in no other*. Because all edge colours in the cycle are distinct, the columns indexed by each of $\{e_0, \dots, e_{\ell-1}\}$ agree in coordinate c_ℓ, but e_ℓ requires that they disagree, which yields the contradiction.

Fact 1 now guarantees that k can grow superlinearly in v whenever $N \ge t+\lambda-1$, in contrast with the requirement that $\mathsf{PHFK}_\lambda(t + \lambda - 2, v, t) \le \max(v, \frac{1}{\lambda}(t + \lambda - 2)(v - 1))$ from Lemma 1.

Of course, practical interest is in obtaining a large number of columns, but understanding the situation with few rows has an important consequence.

Theorem 2 *Let $N = \alpha(t - 1) + \beta$ with $1 \le \beta \le t - 1$. Then* $\mathsf{PHFK}_\lambda(N + \lambda - 1, v, t) \le \mathsf{PHFK}_1(N, v, t) \le v^\alpha(t - 1 + \beta(v - 1)) \le (t - 1)v^{\lceil \frac{N}{t-1} \rceil}$.

Proof For the first inequality, apply Fact 3. For the second, consider a $\mathsf{PHF}_1(N; k, v, t)$. Repeatedly amalgamate rows (Fact 8) to form a $\mathsf{PHHF}_1(t - 1; k, ((v^{\alpha+1})^\beta (v^\alpha)^{t-1-\beta}), t)$. Apply Lemma 1 to conclude that $k \le v^\alpha(t - 1 + \beta(v - 1))$.

Row amalgamation can reduce λ when it exceeds 1, and hence Theorem 2 employs amalgamation only when $\lambda = 1$. Consequently, it yields a useful upper bound when $\lambda = 1$, but we anticipate that the bound is weak when $\lambda > 1$.

In the 'linear' range when $\frac{t+1}{2} + \lambda - 2 \le N \le t + \lambda - 2$, Lemma 1 establishes that for some constant $c(N - (\lambda - 1), t - N - (\lambda - 1), \lambda)$, the existence of a $\mathsf{PHF}_\lambda(N; k, v, t)$ requires that $k \le c(N - (\lambda - 1), t - N - (\lambda - 1), \lambda)v$. Blackburn [13] devises a linear programming formulation to explicitly determine the constant $d(N, t - N)$ so that $k = d(N, t - N)v(1 + o(1))$ when $\lambda = 1$. In order to establish the lower bound asymptotically, he develops a construction technique using coverings. In [29], the method is extended to produce explicit constructions for small values of v, and to treat the generalization to distributing hash families (with $\lambda = 1$).

The key idea is to employ hash families with an additional property. A $\mathsf{DHHF}(t; k, (v_1, \ldots, v_t), t, p)$ is *fractal* if $t \leq 2$, or if, for each row j, deleting row j yields a fractal $\mathsf{DHHF}(t-1; k, (v_1, \ldots, v_{j-1}, v_{j+1}, \ldots, v_t), t-1, \min(p, t-1))$. Fractal PHHF s are fractal DHHFs with $p = t$. A $\mathsf{DHHF}(t; k, (v_1, \ldots, v_n), t, p)$ is α-*fractal* if it is fractal and at least α rows of the DHHF contain all distinct symbols. In addition to fractal DHHF s, we require a second ingredient. An (n, m, d)-*covering* of *type* $(\rho_0, \ldots, \rho_{m-1})$ is a collection of n subsets $\{P_0, \ldots, P_{n-1}\}$ of $\{0, \ldots, m-1\}$ satisfying:

1. $|\{P_r : 0 \leq r < n, P_r \ni c\}| = \rho_c$ for $0 \leq c < m$ and
2. For every $S \subseteq \{0, \ldots, m-1\}$ with $|S| = d$, S is a subset of some set in $\{P_1, \ldots, P_{n-1}\}$.

Extending the methods in [13] to use fractal and heterogeneous ingredients, and distributing hash families rather than just perfect ones, we have the following.

Theorem 3 ([29]) *Suppose that there exist*

- *an* (n, m, d)*-covering* $\mathcal{P} = \{P_0, \ldots, P_{n-1}\}$ *of type* $(\rho_0, \ldots, \rho_{m-1})$ *and*
- *for each* $0 \leq c < m$*, a* ρ_c*-fractal* $\mathsf{DHHF}(n; k_c, (v_{0,c}, \ldots, v_{n-1,c}), n, p)$ *in which, for* $0 \leq r < n$*, row* r *contains all distinct symbols when* $c \in P_r$.

Then there exists a $\mathsf{DHHF}(n; \sum_{c=0}^{m-1} k_c, (w_0, \ldots, w_{n-1}), n+d, p')$ *where*

$$w_r = \sum_{c=0}^{m-1} v_{r,c} \text{ for } 0 \leq r < n, \text{ and}$$
$$p' = \begin{cases} p & \text{if } p < n-d \\ n+d & \text{if } p \geq n-d. \end{cases}$$

One easy result uses the trivial $(\binom{m}{d}, m, d)$-covering:

Corollary 1 *Let* $m > d \geq 1$ *be integers. Suppose that a fractal*

$$\mathsf{PHHF}\left(\binom{m-1}{d}; \kappa, (w_0, \ldots, w_{\binom{m-1}{d}-1}), \binom{m-1}{d}\right)$$

exists. Let σ *be the sum of the* $m-d$ *largest elements in* $\{w_i : 0 \leq i \leq \binom{m-1}{d} - 1\}$*. Then a* $\mathsf{PHF}(\binom{m}{d}; m\kappa, d\kappa + \sigma, \binom{m}{d} + d)$ *exists.*

A sufficient condition for a PHHF to be fractal establishes that fractal PHHFs are common; nevertheless, not all PHHF s are fractal.

Lemma 2 ([29]) *If a* $\mathsf{PHHF}(t; k, (v_1, \ldots, v_t), t)$ *has at most one singleton in each row, it is fractal.*

It is not clear whether there is a useful generalization of the fractal definition to cases when the strength is not the same as the number of rows, although a variant of this idea is used in [25] in a recursive construction. Theorem 3 gives a powerful tool for the construction of DHHF s with few rows, but one needs many ingredients in order to apply it effectively. Therefore we now turn to construction techniques.

4 The Connection with Codes

One direct method for constructing a variety of hash families relies on the existence of error-correcting codes. A *code* with parameters $(n, k, d)_q$ is a set of k distinct vectors (*codewords*) of length n over an alphabet of size q, so that every two distinct codewords are at Hamming distance at least d. Then $A_q(n, d)$ denotes the largest k for which there is a $(n, k, d)_q$ code, and $A(n, d, w)$ denotes the largest k for which there is a $(n, k, d)_2$ code in which each codeword has weight w (i.e., has w 1's). See [19] for some bounds on $A(n, d, w)$. Determining exact values for $A_q(n, d)$ and $A(n, d, w)$ in general remains a major challenge.

Alon [2] shows a connection between codes and PHF_1s; see also Atici et al. [10]. We give the easy generalization for higher index:

Theorem 4 *If there is an $(n, k, d)_q$ code, then for any t, λ such that $\binom{t}{2} < \frac{n-\lambda+1}{n-d}$, there is a $\mathsf{PHF}_\lambda(n; k, q, t)$.*

Proof Let C be an $(n, k, d)_q$ code. Construct an array A that has the codewords of C as its columns. Let $L = \{c_1, \cdots, c_t\}$ be a set of t columns of A. Two distinct columns of L can agree in at most $n - d$ rows, so the number of rows in which not all columns of L disagree is at most $(n-d)\binom{t}{2}$. Provided that $(n-d)\binom{t}{2} < n-\lambda+1$, A has at least λ rows that separate L.

In general, Theorem 4 yields a PHF from a code, but not every PHF need arise in this way. However, when $t = 2$ the correspondence is exact (see Mehlhorn [47] and Atici et al. [10] when $\lambda = 1$):

Theorem 5 *An $(n, k, \lambda)_q$ code is equivalent to a $\mathsf{PHF}_\lambda(n; k, q, 2)$.*

It follows that $\mathsf{PHFK}_1(N, v, 2) = v^N$ and hence $\mathsf{PHFN}_1(k, v, 2) = \lceil \frac{\log k}{\log v} \rceil$. By considering all codewords in $\{0, \ldots, v-1\}^N$ whose entries sum to 0 (mod v), one has $\mathsf{PHFK}_2(N, v, 2) = v^{N-1}$ (the upper bound is a consequence of Theorem 1). When $\lambda \geq 3$, one wants $(n, k, d)_q$ codes with $d \geq 3$. Numerous constructions and bounds are known [45], but in general exact values are not.

There is a $\mathsf{PHHF}_1(N; \prod_{i=1}^{N} v_i, (v_1, \cdots, v_N), 2)$ when $v_1, \cdots, v_N \geq 2$ for any $N \geq 1$, obtained by taking all possible column vectors. In the heterogeneous case, one has a correspondence with codewords in which each coordinate has its own alphabet, but such codes have not been much studied.

Turning to cases with $t \geq 3$, Theorem 4 has been extensively employed, particularly to Reed–Solomon codes to make PHFs with $\lambda = 1$ [10]. We formulate a generalization using design-theoretic terminology. A *transversal design*, $\mathsf{TD}(s, k, n)$ is a triple $(V, \mathcal{G}, \mathcal{B})$ where V is a set of kn points, partitioned into k groups $\mathcal{G} = \{G_1, \cdots, G_k\}$, and $|G_i| = n$ for all i. Furthermore, $\mathcal{B}$ contains n^s blocks

$\{B_1, \ldots, B_{n^s}\}$ of size k with $|B_i \cap G_j| = 1$ for all i, j, and $|B_i \cap B_j| \leq s$ for all $i \neq j$. A standard construction of transversal designs over the finite field $\mathbb{F}_q$ (so that q is a prime power) follows.

Construction 4 *A* $\mathsf{TD}(s, k, q)$ *with* $k \leq q + 1$ *exists. Let* $X = \{x_1, \ldots, x_k\} \subseteq \mathbb{F}_q \cup \{\infty\}$. *The elements of the* TD *are* $\mathbb{F}_q \times X$. *The groups are* $\{\mathbb{F}_q \times \{x_i\} : 1 \leq i \leq k\}$. *For each polynomial* $a_0 + a_1 y + \cdots + a_{s-1} y^{s-1}$ *of degree* $s - 1$ *with coefficients from* $\mathbb{F}_q$, *form a block that contains element* (b, z) *whenever* $z \in X$ *and (1)* $b = a_{s-1}$ *when* $z = \infty$, *or (2)* $b = a_0 + a_1 z + \cdots + a_{s-1} z^{s-1}$ *otherwise (all arithmetic performed in* $\mathbb{F}_q$*).*

A transversal design constructed in this manner is called *linear*. Treating the blocks of the $\mathsf{TD}(s, k, q)$ from Construction 4 as columns so that block $\{(b_1, x_1), \ldots, (b_k, x_k)\}$ corresponds to a column having entry b_i in row x_i, we obtain a $k \times q^s$ array C on q symbols. In fact, because the difference between two polynomials of degree at most $s - 1$ is also a polynomial of degree at most $s - 1$, and such a polynomial has at most $s - 1$ roots, the columns of C form a $(k, q^s, k - s + 1)_q$ code, so Theorem 4 applies. When constructed in this way, the PHF_λ is called *linear* as well. However, because the code has a natural algebraic interpretation, often much more can be said. Suppose that there is a set X for which every set of t polynomials of degree $s-1$ disagree on some value of X. This can arise when $|X| < (s-1)\binom{t}{2}$. For example, Blackburn [14] shows that a $\mathsf{PHF}(3; r^3, r^2, 3)$ exists for all $r \geq 2$, and that a $\mathsf{PHF}(6; p^2, p, 4)$ exists when $p = 11$ and whenever $p \geq 17$ is prime. This phenomenon has been extensively examined when $\lambda = 1$. A PHF is *optimal linear* if it is linear and no linear PHF exists having fewer rows. Blackburn [14] provides explicit constructions of PHFs, some of which have the fewest possible rows of any PHF with the same number of columns, same number of symbols, and same strength.

Blackburn and Wild [16] showed that if q is a sufficiently large prime power, there is an optimal linear $\mathsf{PHF}(s(t-1); q^s, q, t)$ for $s, t \geq 2$. For specific choices of s and t, characterizations of the number of rows that suffice for small prime powers q have been carried out by Barwick and Jackson [11, 12], and Colbourn and Ling [24]. These provide numerous explicit examples of PHF_1s that are easily constructed. The extension of these ideas to larger values of λ is straightforward.

It remains an open question whether an optimal linear PHF always has the fewest rows for any PHF with these parameters [14, 16]. The linear perfect hash families always consist of rows in which the numbers of occurrences of each symbol are as equal as possible. Of course, this equireplication cannot be required for all parameter sets; consider, for example, Construction 2. Nevertheless, it appears plausible that once the number of rows is large enough, there is a PHF in which every row is nearly equireplicated. If true, this constraint could simplify the development of further constructions.

However, we do not expect that linear PHFs lead to the largest number of columns. Consider, for example, $\mathsf{PHFK}(3, v, 3)$. By [14], $\mathsf{PHFK}(3, v, 3) = \Omega(v^{1.5})$; however, Walker and Colbourn [63] found solutions for small v that suggest a larger growth rate, and asked whether $\mathsf{PHFK}(3, v, 3) = o(v^2)$. Fuji-Hara [37] constructs $\mathsf{PHF}(3; v^5, v^3, 3)$ for a prime power $v \geq 3$, to establish that $\mathsf{PHFK}(3, v, 3) = \Omega(v^{5/3})$. Shangguan and Ge [56] solved the question of Walker and Colbourn: For sufficiently large v and arbitrary $\varepsilon > 0$, $v^{2-\varepsilon} < \mathsf{PHFK}(3, v, 3) = o(v^2)$. A similar result for $\mathsf{PHFK}(4, v, 4)$ is also proved.

One does not need transversal designs constructed over the field $\mathbb{F}_q$ in order to produce a code. It is well known that a $\mathsf{TD}(2, k, v)$ is equivalent to $k - 2$ mutually orthogonal Latin squares of side v (see [23], for example). Via this connection, one can generalize a result of Stinson, Wei, and Zhu [61] to $\lambda \geq 1$, by also employing Theorem 4:

Theorem 6 ([61]) *If there are at least* $s = \binom{t}{2} + \lambda - 2$ *MOLS of order* n, *there exists a* $\mathsf{PHF}_\lambda(s + 2 + \lambda; n^2, n, t)$.

The same authors generalize this statement to mutually orthogonal $n \times m$ latin rectangles on $\max(m, n)$ symbols, obtaining a PHF_1 with mn columns. Dinitz et al. [32] establish in some cases that the number of rows employed by Theorem 6 can be reduced by ensuring that the corresponding TD avoids certain forbidden configurations.

Another generalization of transversal designs is to block designs. Let X be a set of v *points*, and $\mathcal{B}$ be a set of b subsets of X, called *blocks* of X each with k points. Then $(X, \mathcal{B})$ is a *balanced incomplete block design* (BIBD) if every point occurs in r blocks, and every pair of points occurs in λ blocks. We denote this by $\mathsf{BIBD}(v, b, r, k, \lambda)$. By a simple counting argument, $vr = bk$ and $\lambda(v - 1) = r(k - 1)$. A BIBD is *resolvable* if $\mathcal{B}$ can be partitioned into r *parallel classes*, where each class contains $\frac{v}{k}$ disjoint blocks. We denote this by $\mathsf{RBIBD}(v, b, r, k, \lambda)$. Brickell [18] and Atici et al. [10] proved that if there is an $\mathsf{RBIBD}(v, b, r, k, \lambda)$ and $r > \lambda\binom{w}{2}$, there is a $\mathsf{PHF}_1(r; v, \frac{v}{k}, w)$. For results on the existence and asymptotics of RBIBDs, see [23].

5 Asymptotic Bounds and Algorithms

The probabilistic method yields bounds on PHFN_λ. A basic bound is obtained as follows. Let A be a random $N \times k$ array on v symbols; by this, each entry is selected uniformly at random and independent of all other entries. Choose t distinct column indices $C = \{c_1, \cdots, c_t\}$. The probability that C is separated in a single row A is the probability that in this row, all columns of C contain distinct entries. This probability is $\phi_{t,v} = \frac{\binom{v}{t}t!}{v^t}$, and the probability that C is not separated in this row is $1 - \phi_{t,v}$. Often we write ϕ and $1 - \phi$ when t and v are fixed in this context.

Because all rows are chosen independently, the number of rows in which C is not separated is equal to ρ with probability $\binom{N}{\rho}(1 - \phi)^\rho \phi^{N-\rho}$. Define the probability $\psi_{\lambda,N,t,v}$ (or, more simply, $\psi_{\lambda,N}$) to be $\sum_{\rho=0}^{\lambda-1} \binom{N}{\rho}(1 - \phi)^\rho \phi^{N-\rho}$. By linearity of expectation, the expected number of t-sets of columns that are not λ-separated in

A is $\binom{k}{t}\psi_{\lambda,N}$. When this expected number is less than 1, a $\mathsf{PHF}_\lambda(N; k, v, t)$ surely exists. When $\lambda = 1$, $\psi_{1,N} = (1-\phi)^N$. Then taking logarithms of $\binom{k}{t}(1-\phi_{t,v})^N < 1$, we obtain an easily stated bound, first established by Mehlhorn:

Theorem 7 ([47]) $\mathsf{PHFN}_1(k, v, t) \leq \left\lceil \frac{\log\binom{k}{t}}{f(v,t)} \right\rceil$, *where* $f(v, t) = \log(v^t) - \log(v^t - t!\binom{v}{t})$.

Stinson et al. [60] improved this bound with an expurgation method. We generalize their technique (for $\lambda = 1$) using the technique of post-processing [62] or oversampling [28].

Theorem 8 $\mathsf{PHFN}_1(k, v, t) \leq \min_{x \geq 0}\left(N : \binom{k+x}{t}(1 - \phi_{t,v})^N < x + 1\right)$.

Proof Start with an array with $k + x$ columns, and let E be the expected number of unseparated t-sets of columns in this array if each entry is chosen uniformly and independently. If $E < x + 1$, then by deleting one column from each of the (at most) x unseparated t-sets, we must have a $\mathsf{PHF}_1(N; k, v, t)$. So we desire that $\binom{k+x}{t}(1 - p)^N < x + 1$; find the minimum value of N such that there exists an $x \geq 0$ for which this inequality holds.

In a related context [28], the choice $x = \frac{k}{t-1}$ was shown to be the best; the same value works for PHF_1s as well. Of course, this method extends in the natural way to higher values of λ.

Stein [58], Lovász [44], and Johnson [40] devise a greedy strategy for explicitly producing solutions to certain covering problems of the size guaranteed by the basic probabilistic method. Their method encompasses t-restrictions. In the context of perfect hash families with $\lambda = 1$, the method constructs the family one row at a time. As long as there remains at least one t-set to separate, the method chooses a next row to maximize the number of t-sets separated for the first time by this row. Let $\mathcal{U}_i$ denote the set of t-sets of columns that are not yet separated after i rows have been selected. If row $i + 1$ is selected at random, linearity of expectations determines that the expected size of $\mathcal{U}_{i+1}$ is $(1 - \phi)|\mathcal{U}_i|$. Hence the greedy selection guarantees that $|\mathcal{U}_{i+1}| \leq \lfloor(1 - \phi)|\mathcal{U}_i|\rfloor$. Our goal is to find the smallest N such that $|\mathcal{U}_N| < 1$; this ensures that we have found an array with N rows that separates all t-sets. Because $|\mathcal{U}_0| = \binom{k}{t}$, we have $|\mathcal{U}_N| \leq \binom{k}{t}(1 - \phi)^N$. Consequently this greedy strategy leads to explicit constructions meeting the bounds of Theorem 7 (and, with small modifications, Theorem 8). In fact, it typically yields smaller values of N than are guaranteed by these results, because (1) expectations can be replaced by their integer floors and (2) the greedy strategy may choose a row that separates many more t-sets than the expectation.

Unfortunately, even when t and v are fixed, a naive implementation of this greedy strategy is not efficient, because it asks for the examination of v^k rows in order to select a maximum. No efficient algorithm is known for computing a row that separates the maximum number of t-sets among a given set $\mathcal{U}$ of t-sets. One could, however, be content with a row that separates at least the expected number rather than the maximum, without affecting the analysis. A randomized approach to select row $i+1$ could first calculate $\phi|\mathcal{U}_i|$, the expected number separated, and generate random rows until one separates at least this expected number from $\mathcal{U}_i$. Such a method guarantees to produce a solution meeting the bound on N, but does not guarantee to run in polynomial time.

Using conditional expectations to derandomize this approach, Colbourn [21] devised the density method, a deterministic method for constructing perfect hash families that is efficient when t and v are fixed. We wish to produce the row $i+1$ efficiently to separate at least $\phi|\mathcal{U}_i|$ t-sets from $\mathcal{U}_i$. The key idea is to consider partial rows, in which some entries have been *fixed* to one of the v values, and some are *free* to take on any value. For such a partial row, one can efficiently calculate the expected number of t-sets in $\mathcal{U}_i$ that are separated when the free entries are chosen uniformly at random. Then, as long as free entries remain, we can choose one and tentatively fix it to each of the v possible values, computing the expectation for each. Then we fix the free entry permanently to a value that leads to a largest expectation. At each stage, the expected number of t-sets of $\mathcal{U}_i$ separated by this (partial) row cannot decrease. Hence, once all entries are fixed, a row has been efficiently found that separates at least the expected number for a random row.

In practice, the density method often completes with a number of rows that is significantly smaller than guaranteed by Theorem 7 (see [21]). We expect that improvements in a similar method for covering arrays [52, 53] can be applied to hash families as well.

Unfortunately, the extension of these algorithmic methods to cases with $\lambda > 1$ is not immediate. One row cannot, by itself, λ-separate a t-set. Hence rather than simply tracking t-sets that are 0-separated, progress towards constructing a perfect hash family of index $\lambda > 1$ is measured by the number of ℓ-separated t-sets for each $0 \leq \ell < \lambda$. Nevertheless, we outline one method to follow the Stein–Lovász–Johnson paradigm. Suppose that i rows have been selected, and that the t-sets that are not yet separated at least λ times are in one of $\mathcal{U}_{i,0}, \ldots, \mathcal{U}_{i,\lambda-1}$ where $\mathcal{U}_{i,\ell}$ contains all t-sets that are ℓ-separated in the first i rows. Extending the earlier analysis, when t, v, and λ are fixed, one can efficiently calculate the expected number $\psi((\mathcal{U}_{i,0}, \ldots, \mathcal{U}_{i,\lambda-1}), N-i)$ of t-sets that are not separated at least λ times if $N-i$ further rows are selected uniformly at random. Then one (greedily) chooses row $i+1$ so that

$$\psi((\mathcal{U}_{i+1,0}, \ldots, \mathcal{U}_{i+1,\lambda-1}), N-i-1) \leq \psi((\mathcal{U}_{i,0}, \ldots, \mathcal{U}_{i,\lambda-1}), N-i).$$

Set $\mathcal{U}_{0,0}$ to contain all $\binom{k}{t}$ t-sets, and $\mathcal{U}_{0,\ell} = \emptyset$ for $1 \leq \ell < \lambda$. Choose N so that $\psi((\mathcal{U}_{0,0}, \ldots, \mathcal{U}_{0,\lambda-1}), N) < 1$. The key difference when $\lambda > 1$ is that one presupposes the determination of N, the target number of rows, a step that is not needed when $\lambda = 1$.

The bound of Theorem 7 can be improved in a different manner using the Lovász local lemma, a tool extensively used in combinatorics [6, 35]. We employ the symmetric version here:

Theorem 9 *Let $E_1, \cdots, E_n$ be events in a probability space, E_i is mutually dependent of at most d other events for all i, and $\Pr[E_i] \leq p$ for all i. If $ep(d+1) \leq 1$, then with nonzero probability all of the events simultaneously do not occur.*

In the setting of PHF_1s, an event E_i is that the ith t-set of columns is not separated, so that $\Pr[E_i] = (1-\phi)^N$. Events E_i and E_j are mutually dependent if and only if the corresponding t-sets have non-empty intersection, so $d = \binom{k}{t} - \binom{k-t}{t} - 1$. When $e\,(1-p)^N \left(\binom{k}{t} - \binom{k-t}{t}\right) \leq 1$, there is a PHF_1 with those parameters:

Theorem 10 ([31]) $\mathsf{PHFN}_1(k, v, t) \leq \left\lceil \frac{\log\left(\binom{k}{t}-\binom{k-t}{t}\right)+1}{f(v,t)} \right\rceil$, *where* $f(v,t) = \log(v^t) - \log(v^t - t!\binom{v}{t})$.

The same argument shows that when $e\psi_{\lambda,N,t,v}\left(\binom{k}{t} - \binom{k-t}{t}\right) \leq 1$, there is a $\mathsf{PHF}_\lambda(N; k, v, t)$.

An improvement on Theorem 10 for many parameter sets was found by Procacci and Sanchis [51] using the algorithmic cluster expansion local lemma [54].

In landmark work, Moser and Tardos [48] develop a constructive method for objects whose sizes match the bound provided by the Lovász Local Lemma, and their techniques have been shown to extend to the cluster expansion local lemma in [8, 50]. The method is randomized and runs in expected polynomial time when the number of rows is as dictated by the (original or cluster expansion) LLL bound. In the context of hash families, one first computes the value of N that the LLL bound asserts is sufficient, and generates a random $N \times k$ array on v symbols. Now order the $\binom{k}{t}$ t-sets of columns arbitrarily, and fix this ordering throughout. To check the array, consider each t-set in order until one is not λ-separated or all are checked. If all are checked, the current array is the desired PHF_λ. Otherwise, for the first t-set of columns that fails, randomly resample the entries in each row in each column of the t-set, and start checking again from the first t-set. Remarkably, this *column resampling* method terminates with a solution in expected polynomial time when t is fixed [8, 48, 50].

Because the value for N used is an upper bound, but typically not the exact value, one can apply the same column resampling approach for smaller values of N. When N is smaller, there is no guarantee of expected polynomial running time, and indeed no guarantee that the method will terminate at all. (This is discussed in a different context in [20].) Nevertheless, column resampling provides a practical algorithm for producing perfect hash families that are substantially smaller than the bounds. Of course, when resampling an $N \times t$ subarray, the number of t-sets that are separated fewer than λ times may increase. Moreover, the likelihood of this occurring increases as N decreases, as one would expect. Therefore one finds that when N is well below the bound, most column resamplings make the situation worse. To avoid wholesale changes in the array resulting in a significant increase in the number of unseparated t-sets, one could replace a single column within an unseparated t-set rather than all t. In general, one could consider partial resamplings of the type in [39].

Starting with a $\mathsf{PHF}_\lambda(N; k, v, t)$ and adding a randomly chosen column, one can choose always to replace the new column when there is an unseparated t-set. This *random extension* method [28] amounts to choosing possible random columns to adjoin until one is found that λ-separates all $\binom{k}{t-1}$ t-sets of columns containing the new one. In practical computations, often many additional columns can be adjoined. We do not report computational results for the methods discussed, instead referring the reader to [33].

Density-based and column resampling algorithms are conceptually simple, and while they can be useful for making perfect hash families with moderately large parameters, one should not expect either to produce optimal PHFs in general; indeed, the optimal linear perfect hash families have dramatically fewer rows than are guaranteed by the probabilistic methods. Unfortunately, known direct constructions address quite limited parameter sets. Nevertheless, we have seen that recursive constructions rely on finding suitable ingredients with small values of the parameters. For this reason, it is natural to consider *metaheuristics* that identify a set of changes allowed and a rule for determining when one applies the change to the array, and when one rejects the change and keeps the current array. Evidently the long-term objective of any such method is to reduce the number of t-sets that are unseparated, eventually to 0.

Arguably, the Moser–Tardos algorithm is such a method, in which the possible changes are the resamplings of all entries of an unseparated t-set of columns, whose rule accepts every change. Applying the method of [39] would choose from a smaller set of possible changes but still accept all. One might prefer changes that reduce, or at least do not increase, the number of unseparated t-sets. *Local optimization* or *hill-climbing* accepts changes only when the change does not increase the number of t-sets. As one might expect, it can happen that none of the offered changes is accepted, and the method is stuck at a local optimum with no escape. For this reason, more sophisticated metaheuristic methods (simulated annealing [41], tabu search [38], the great deluge algorithm [34], constraint or answer-set programming [43], for example) can be applied. In general, these methods require either that the separation status of each t-set be stored or frequently recalculated; whichever is done, one small change can affect the status of many t-sets. As a result these methods are (at least at present) limited to 'small' values of the parameters. Many such perfect hash families that are the smallest known have been found using metaheuristic methods [33], although none guarantees an improvement on column resampling or the density-based methods in general.

A different strategy, *post-optimization*, starts with a $\mathsf{PHF}_\lambda(N; k, v, t)$. The method takes advantage of situations in which certain entries are superfluous to ensuring separation, replacing such entries by a 'don't care' marker $\star$. Through a series of alterations of individual entries without making any t-set unseparated, the method attempts to make an entire row contain only $\star$ entries. When this can be done, the row can be deleted to form a $\mathsf{PHF}_\lambda(N-1; k, v, t)$; again, some computational results are reported in [33]. This strategy is discussed for general t-restrictions in [26].

6 Concluding Remarks

In this selective survey of combinatorial aspects of perfect hash families, we have found that many methods generalize in a natural manner to treat existence for fixed $\lambda > 1$. Perfect hash families are central in the construction of a wide variety of other combinatorial objects, particularly because of Construction 1. The extension of existence results for $\lambda = 1$ to higher λ can therefore provide a valuable tool in the construction of arrays for a variety of practical applications. An example arises in the construction of locating arrays (a type of t-restriction) with separation greater than 1 [55]; in that context, the increase in separation entails a substantial increase in the number of rows, and hence also in computation time. We argue that a viable alternative is to instead construct a perfect hash family with index $\lambda > 1$, and employ column replacement (as exemplified by Construction 1) to address the t-restriction. We anticipate that this framework can prove useful in similar applications in which the effects of a single row are prone to error or mismeasurement.

In addition to ensuring that every t-set of columns be separated at least λ times, one might address the more stringent requirement that every t-set be separated at least $\underline{\lambda}$ and at most $\overline{\lambda}$ times. When $\underline{\lambda} = \overline{\lambda}$, such a PHF is *perfectly balanced* [3]. Alon and Gutner [3] establish that a perfectly balanced $\mathsf{PHF}_\lambda(N; k, v, t)$ can exist only when $N = \Omega(k^{\lfloor t/2 \rfloor})$ for t fixed. Contrast this with the $\Theta(\log k)$ growth rate for PHF_1s to understand why perfectly balanced PHF s are not frequently used. On the other hand, a $\mathsf{PHF}(N; k, v, t)$ is δ*-balanced* for some $\delta \geq 1$ if there is a value $T > 0$ so that every t-set of columns is separated at least $\frac{T}{\delta}$ and at most δT times [4]. Alon and Gutner [4] show that for any fixed $\delta > 1$, there is a δ-balanced $\mathsf{PHF}(N; k, v, t)$ with N close to $2^{O(t \log\log t)} \log k$; so, for fixed t, the growth rate is the same as for PHF_1s. Their approach relies (in small part) on the binomial distribution of the number of times a t-set is separated and the application of Chernoff bounds. Moreover, their techniques yield an explicit construction method in principle; its practical effectiveness for intermediate values of k has not been explored.

When δ-balanced PHF s are used in Construction 1 with different t-restrictions, the array constructed inherits from the balanced PHF a lower bound on the number of rows in which the t-restriction is met. However, the t-restriction may be met in a row arising from a row of the PHF despite failure of the PHF to separate in this row; hence balanced PHFs need not result in balanced t-restrictions through Construction 1. For these reasons, it is reasonable to focus on extending known methods, and finding new methods, for constructing perfect hash families of index $\lambda > 1$.

Acknowledgments This work is supported in part by the U.S. National Science Foundation grants #1421058 and #1813729. Thanks to Randy Compton, Stephanie Forrest, Daniel Horsley, Erin Lanus, Kaushik Sarkar, and Violet Syrotiuk for helpful discussions, and to an anonymous reviewer for very useful comments.

References

1. Y. Akhtar, F.K.H. Phoa, construction of cost-efficient designs with guaranteed repeated measurements on interaction effects (2019). Preprint
2. N. Alon, Explicit construction of exponential sized families of k-independent sets. Discrete Math. **58**(2), 191–193 (1986)
3. N. Alon, S. Gutner, Balanced hashing, color coding and approximate counting, in *Parameterized and Exact Computation, 4th International Workshop, IWPEC 2009*, Copenhagen (2009), pp. 1–16
4. N. Alon, S. Gutner, Balanced families of perfect hash functions and their applications. ACM Trans. Algorithms **6**(3), 54:1–54:12 (2010)
5. N. Alon, M. Naor, Derandomization, witnesses for Boolean matrix multiplication and construction of perfect hash functions. Algorithmica **16**(4–5), 434–449 (1996)
6. N. Alon, J.H. Spencer, *The Probabilistic Method* (Wiley, New York, 2004)
7. N. Alon, D. Moshkovitz, S. Safra, Algorithmic construction of sets for k-restrictions. ACM Trans. Algorithms **2**, 153–177 (2006)
8. R.G. Alves, A. Procacci, Witness trees in the Moser-Tardos algorithmic Lovász local lemma and Penrose trees in the hard-core lattice gas. J. Stat. Phys. **156**(5), 877–895 (2014)
9. M. Atici, Hash families: recursive constructions and applications to cryptography. PhD dissertation, University of Nebraska (1996)
10. M. Atici, S.S. Magliveras, D.R. Stinson, W.D. Wei, Some recursive constructions for perfect hash families. J. Comb. Des. **4**(5), 353–363 (1996)
11. S.G. Barwick, W.A. Jackson, A sequence approach to linear perfect hash families. Des. Codes Cryptogr. **45**(1), 95–121 (2007)
12. S.G. Barwick, W.A. Jackson, Geometric constructions of optimal linear perfect hash families. Finite Fields Appl. **14**(1), 1–13 (2008)
13. S.R. Blackburn, Perfect hash families with few functions. Unpublished manuscript
14. S.R. Blackburn, Perfect hash families: Probabilistic methods and explicit constructions. J. Comb. Theory Ser. A **92**(1), 54–60 (2000)
15. S.R. Blackburn, Frameproof codes. SIAM J. Discrete Math. **16**(3), 499–510 (2003)
16. S.R. Blackburn, P.R. Wild, Optimal linear perfect hash families. J. Comb. Theory Ser. A **83**(2), 233–250 (1998)
17. S.R. Blackburn, M. Burmester, Y. Desmedt, P.R. Wild, Efficient multiplicative sharing schemes, in *Advances in Cryptology - EUROCRYPT '96, International Conference on the Theory and Application of Cryptographic Techniques* (1996), pp. 107–118
18. E.F. Brickell, A problem in broadcast encryption, in *5th Vermont Summer Workshop on Combinatorics and Graph Theory* (1991)
19. A.E. Brouwer, T.Etzion, Bounds for binary constant weight codes. IEEE Trans. Inf. Theory **36**, 1334–1380 (1990)
20. J.D. Catarata, S. Corbett, H. Stern, M. Szegedy, T. Vyskocil, Z. Zhang, The Moser-Tardos resample algorithm: Where is the limit? (an experimental inquiry), in *Proceedings of the Nineteenth Workshop on Algorithm Engineering and Experiments ALENEX* (2017), pp. 159–171
21. C.J. Colbourn, Constructing perfect hash families using a greedy algorithm, in *Coding and Cryptology* (2008). https://doi.org/10.1142/6915

22. C.J. Colbourn, Covering arrays and hash families, in *NATO Science for Peace and Security Series, D: Information and Communication Security*. Information Security, Coding Theory and Related Combinatorics, vol. 29 (2011), pp. 99–135
23. C.J. Colbourn, J.H. Dinitz, *Handbook of Combinatorial Designs* (CRC Press, Boca Raton, 2007)
24. C.J. Colbourn, A.C.H. Ling, Linear hash families and forbidden configurations. Designs Codes Cryptogr. **52**(1), 25–55 (2009)
25. C.J. Colbourn, A.C.H. Ling, A recursive construction for perfect hash families. J. Math. Cryptol. **3**(4), 291–306 (2009)
26. C.J. Colbourn, P. Nayeri, Randomized post-optimization for t-restrictions, in *Information Theory, Combinatorics, and Search Theory*. Lecture Notes in Computer Science, vol. 7777 (Springer, Heidelberg, 2013), pp. 597–608
27. C.J. Colbourn, D. Horsley, V.R. Syrotiuk, Strengthening hash families and compressive sensing. J. Discrete Algorithms **16**, 170–186 (2012)
28. C.J. Colbourn, E. Lanus, K. Sarkar, Asymptotic and constructive methods for covering perfect hash families and covering arrays. Designs Codes Cryptogr. **86**, 1–31 (2017)
29. C.J. Colbourn, R.E. Dougherty, D. Horsley, Distributing hash families with few rows. Theor. Comput. Sci. **800**, 31–41 (2019)
30. Z.J. Czech, G. Havas, B.S. Majewski, Perfect hashing. Theor. Comput. Sci. **182**, 1–143 (1997)
31. D. Deng, D.R. Stinson, R. Wei, The Lovász local lemma and its applications to some combinatorial arrays. Designs Codes Cryptogr. **32**(1–3), 121–134 (2004)
32. J.H. Dinitz, A.C.H. Ling, D.R. Stinson, Perfect hash families from transversal designs. Aust. J. Comb. **37**, 233–242 (2007)
33. R.E. Dougherty, Perfect hash family tables for t=3 to 11 (2017). http://www.public.asu.edu/~redoughe/phf_pages/phf_tables.html
34. G. Dueck, New optimization heuristics: the great deluge algorithm and the record-to-record travel. J. Comput. Phys. **104**(1), 86–92 (1993)
35. P. Erdős, L. Lovász, Problems and results on 3-chromatic hypergraphs and some related questions, in *Infinite and Finite Sets* (North-Holland, Amsterdam, 1975), pp. 609–627
36. A. Fiat, M. Naor, Broadcast encryption, in *Proceedings of the 13th Annual International Cryptology Conference on Advances in Cryptology, CRYPTO '93* (1994), pp. 480–491
37. R. Fuji-Hara, Perfect hash families of strength three with three rows from varieties on finite projective geometries. Designs Codes Cryptogr. **77**(2–3), 351–356 (2015)
38. F. Glover, M. Laguna, Tabu search, in *Handbook of Combinatorial Optimization*, vol. 3 (Kluwer Academic Publishers, Boston, MA, 1998), pp. 621–757
39. D.G. Harris, A. Srinivasan, The Moser-Tardos framework with partial resampling, in *2013 IEEE 54th Annual Symposium on Foundations of Computer Science—FOCS 2013* (IEEE Computer Society, Los Alamitos, CA, 2013), pp. 469–478
40. D.S. Johnson, Approximation algorithms for combinatorial problems. J. Comput. Syst. Sci. **9**, 256–278 (1974)
41. S. Kirkpatrick, C.D. Gelatt, M.P. Vecchi, Optimization by simulated annealing. Science **220**(4598), 671–680 (1983)
42. E. Knill, W.J. Bruno, D.C. Torney, Non-adaptive group testing in the presence of errors. Discrete Appl. Math. **88**(1), 261–290 (1998)
43. V. Lifschitz, Answer set programming and plan generation. Artif. Intell. **138**(1), 39–54 (2002)
44. L. Lovász, On the ratio of optimal integral and fractional covers. Discrete Math. **13**(4), 383–390 (1975)
45. F.J. MacWilliams, N.J.A. Sloane, *The Theory of Error-Correcting Codes* (North-Holland Publishing Co., Amsterdam/New York/Oxford, 1977)
46. S. Martirosyan, T. Van Trung, Explicit constructions for perfect hash families. Designs Codes Cryptogr. **46**(1), 97–112 (2008)
47. K. Mehlhorn, On the program size of perfect and universal hash functions, in *23rd Annual Symposium on Foundations of Computer Science (SFCS 1982)* (1982), pp. 170–175

48. R.A. Moser, G. Tardos, A constructive proof of the general Lovász local lemma. J. ACM **57**(2), 11 (2010)
49. I. Newman, A. Wigderson, Lower bounds on formula size of Boolean functions using hypergraph entropy. SIAM J. Discrete Math. **8**(4), 536–542 (1995)
50. W. Pegden, An extension of the Moser-Tardos algorithmic local lemma. SIAM J. Discrete Math. **28**(2), 911–917 (2014)
51. A. Procacci, R. Sanchis, Perfect and separating hash families: new bounds via the algorithmic cluster expansion local lemma. Annales de l'Institut Henri Poincaré D. Comb. Phys. Interact. **5**(2), 153–171 (2018)
52. K. Sarkar, C.J. Colbourn, Upper bounds on the size of covering arrays. SIAM J. Discrete Math. **31**(2), 1277–1293 (2017)
53. K. Sarkar, C.J. Colbourn, Two-stage algorithms for covering array construction. J. Comb. Des. **27**, 475–505 (2019)
54. A.D. Scott, A.D. Sokal, The repulsive lattice gas, the independent-set polynomial, and the Lovász local lemma. J. Stat. Phys. **118**(5–6), 1151–1261 (2005)
55. S.A. Seidel, K. Sarkar, C.J. Colbourn, V.R. Syrotiuk, Separating interaction effects using locating and detecting arrays, in *International Workshop on Combinatorial Algorithms* (2018), pp. 349–360
56. C. Shangguan, G. Ge, Separating hash families: a Johnson-type bound and new constructions. SIAM J. Discrete Math. **30**(4), 2243–2264 (2016)
57. J.N. Staddon, D.R. Stinson, R. Wei, Combinatorial properties of frameproof and traceability codes. IEEE Trans. Inf. Theory **47**, 1042–1049 (2001)
58. S.K. Stein, Two combinatorial covering theorems. J. Comb. Theory Ser. A **16**, 391–397 (1974)
59. D.R. Stinson, On some methods for unconditionally secure key distribution and broadcast encryption. Designs Codes Cryptogr. **12**(3), 215–243 (1997)
60. D.R. Stinson, T. Van Trung, R. Wei, Secure frameproof codes, key distribution patterns, group testing algorithms and related structures. J. Stat. Plann. Inference **86**(2), 595–617 (2000)
61. D.R. Stinson, R. Wei, L. Zhu, New constructions for perfect hash families and related structures using combinatorial designs and codes. J. Comb. Des. **8**(3), 189–200 (2000)
62. E. Van Den Berg, E. Candès, G. Chinn, C. Levin, P.D. Olcott, C. Sing-Long, Single-photon sampling architecture for solid-state imaging sensors. Proc. Natl. Acad. Sci. **110**(30), E2752–E2761 (2013)
63. R.A. Walker II, C.J. Colbourn, Perfect hash families: constructions and existence. J. Math. Cryptol. **1**(2), 125–150 (2007)

A Note on Randomly Colored Matchings in Random Bipartite Graphs

Alan Frieze

Abstract We are given a bipartite graph that contains at least one perfect matching and where each edge is colored from a set $Q = \{c_1, c_2, \ldots, c_q\}$. Let $Q_i = \{e \in E(G) : c(e) = c_i\}$, where $c(e)$ denotes the color of e. The perfect matching color profile $mcp(G)$ is defined to be the set of vectors $(m_1, m_2, \ldots, m_q) \in [n]^q$ such that there exists a perfect matching M such that $|M \cap Q_i| = m_i$. We give bounds on the matching color profile for a randomly colored random bipartite graph.

1 Introduction

We consider the following problem: we are given a random bipartite graph G in which each edge is given a random color from a set $Q = \{c_1, c_2, \ldots, c_q\}$. An edge e is colored $c(e) = c_i$ with probability α_i where $\alpha_i > 0$ is a constant. Let $Q_i = \{e \in E(G) : c(e) = c_i\}$, where $c(e)$ denotes the color of e. The perfect matching color profile $mcp(G)$ is defined to be the set of vectors $(m_1, m_2, \ldots, m_q) \in [n]^q$ such that there exists a perfect matching M such that $|M \cap Q_i| = m_i$. We give bounds on the matching color profile for a randomly colored random bipartite graph.

Randomly colored random graphs have been studied recently in the context of (1) rainbow matchings and Hamilton cycles, see, for example, [2, 3, 7, 10]; (2) rainbow connection see, for example, [4, 8, 9, 11, 12]; (3) pattern colored Hamilton cycles, see, for example, [1, 6]. This paper can be considered to be a contribution in the same genre. One can imagine a possible interest in the color profile via the following scenario: suppose that A is a set of tools and B is a set of jobs where edge $\{a, b\}$ indicates that b can be completed using a. If colors represent people, then one might be interested in equitably distributing jobs. I.e. determining whether $(n/q, n/q, \ldots, n/q) \in mcp(G)$. In any case, we find the problem interesting.

A. Frieze (✉)
Department of Mathematical Sciences, Carnegie Mellon University, Pittsburgh, PA, USA
e-mail: alan@random.math.cmu.edu

A. M. Raigorodskii, M. Th. Rassias (eds.), *Discrete Mathematics and Applications*, Springer Optimization and Its Applications 165,
https://doi.org/10.1007/978-3-030-55857-4_8

We will consider G to be the random bipartite graph $G_{n,n,p}$ where $p = \frac{\log n+\omega}{n}$, $\omega = \omega(n) \to \infty$ where $\omega = o(\log n)$. Erdő and Rényi [5] proved that G has a perfect matching w.h.p. We will prove the following theorem: let $\alpha_1, \alpha_2, \ldots, \alpha_q, \beta$ be positive constants such that $\alpha_1 + \alpha_2 + \cdots + \alpha_q = 1$ and $\beta < 1/q$. Let

$$\alpha_{\min} = \min\{\alpha_i : i \in [q]\}.$$

Theorem 1 *Let G be the random bipartite graph $G_{n,n,p}$ where $p = \frac{\log n+\omega}{n}$, $\omega = \omega(n) \to \infty$ where $\omega = o(\log n)$. Suppose that the edges of G are independently colored with colors from $C = \{c_1, c_2, \ldots, c_q\}$ where $\mathbf{Pr}(c(e) = c_i) = \alpha_i$ for $e \in E(G), i \in [q]$. Let $m_1, m_2, \ldots, m_q$ satisfy: (1) $m_1 + \cdots + m_q = n$ and (2) $m_i \geq \beta n, i \in [q]$. Then w.h.p., there exists a perfect matching M in which exactly m_i edges are colored with c_i, $i = 1, 2, \ldots, q$.*

It is clear that w.h.p. $(n, 0, \ldots, 0) \notin mcp(G)$. This is because the bipartite graph induced by edges of color c_1 is distributed as $G_{n,n,\alpha_1 p}$ and this contains isolated vertices w.h.p. On the other hand, if $p \geq \frac{q(\log n+\omega)}{\alpha_{\min} n}$, then w.h.p. $mcp(G) = [n]^q$. To see this, suppose that $m_1 \leq m_2 \leq \cdots m_q \leq n$. Suppose we have found a matching that uses m_i edges of color c_i for $i \geq 0$. Let $n' = n - m_1 - \cdots - m_i$. Then the random bipartite graph induced by vertices not in M and having edges of color c_i has density at least $\frac{q\alpha_i n'}{\alpha_{\min} n} \cdot \frac{\log n+\omega}{n'} \geq \frac{\log n'+\omega/2}{n'}$ and so has a perfect matching w.h.p.

Open Question What is the threshold for $mcp(G) = [0, n]^q$?

2 Structural Lemma

Suppose that the bipartition of $V(G)$ is denoted A, B. For sets $S \subseteq A, T \subseteq B$ we let $e_i(S, T)$ denote the number of $S : T$ edges of color c_i. We say that vertex u is c_i-adjacent to vertex v if the edge $\{u, v\}$ exists and has color c_i.

Lemma 1 *Let $p = \frac{\log n+\omega}{n}$, $\omega = \omega(n) \to \infty$ where $\omega = o(\log n)$. Then w.h.p.*

(a) $S \subseteq A, T \subseteq B$ and $\gamma_a \log n \leq |S| \leq n_0 = \gamma_a n/\log n$ and $|T| \leq \alpha_i \eta |S| \log n$ where $\gamma_a = \eta/(20\alpha_i)$ implies that $e_i(S : T) \leq 2\alpha_i \eta |S| \log n$ for $i = 1, 2, \ldots, q$.
(b) There do not exist sets $X \subseteq S \subseteq A, T \subseteq B$ and $i \in [q]$ such that $|S|, |T| \geq \beta n$ and $|X| = \gamma_b |S|/\log n$, $\gamma_b = 10\log(e/\beta)/\alpha_i$ and such that each $x \in X$ is c_i-adjacent to fewer than $\alpha_i \beta \log n/10$ vertices in T.
(c) There do not exist sets $X \subseteq S \subseteq A, T \subseteq B$ and $i \in [q]$ such that $|S|, |T| \geq \beta n$ and $|X| = |S|/\log n$ and a set $Z \subseteq T, |Z| = \gamma_b n/\log n$ such that each $x \in X$ is c_i-adjacent to $k = \frac{10\log n}{\log\log n}$ vetices in Z.
(d) There do not exist sets $S \subseteq A, T \subseteq B$ and $i \in [q]$ such that $|S|, |T| \geq \beta n$ such that there are more than $\gamma_d n/\log n$, $\gamma_d = \frac{4}{\alpha_i}\log\left(\frac{e}{\beta}\right)$ vertices in T that not c_i-adjacent to a vertex in S.

(e) *Fix $\gamma, \delta > 0$ constants. Then w.h.p. there do not exist sets S, T with $|S| = |T| = \gamma n/\log n$ such that $e_i(S,T) \geq \delta |S| \log n/\log\log n$.*
(f) *There do not exist sets $S \subseteq A$, $T \subseteq B$ and $i \in [q]$ such that $|S|, |T| \geq \beta n/10$ such that $e_i(S,T) = 0$.*

Proof

(a) The probability that the condition is violated can be bounded by

$$\sum_{s=\gamma_a \log n}^{n_0} \sum_{t=1}^{\alpha_i \eta s \log n} \binom{n}{s}\binom{n}{t}\binom{st}{2\alpha_i \eta s \log n}(\alpha_i p)^{2\alpha_i \eta s \log n}$$

$$\leq \sum_{s=\gamma_a \log n}^{n_0} \sum_{t=1}^{\alpha_i \eta s \log n} \left(\frac{ne}{s}\right)^s \left(\frac{ne}{t}\right)^t \left(\frac{est\alpha_i p}{2\alpha_i \eta s \log n}\right)^{2\alpha_i \eta s \log n}$$

$$\leq \sum_{s=\gamma_a \log n}^{n_0} \sum_{t=1}^{\alpha_i \eta s \log n} \left(\frac{ne}{s}\right)^s \left(\frac{ne}{\alpha_i \eta s \log n}\right)^{\alpha_i \eta s \log n} \left(\frac{etp}{2\eta \log n}\right)^{2\alpha_i \eta s \log n}$$

$$\leq \sum_{s=\gamma_a \log n}^{n_0} \sum_{t=1}^{\alpha_i \eta s \log n} \left(\left(\frac{ne}{s}\right)^{1/2\alpha_i \eta \log n} \left(\frac{ne}{\alpha_i \eta s \log n}\right)^{1/2} \cdot \frac{e^{1+o(1)}\alpha_i s \log n}{2n}\right)^{2\alpha_i \eta s \log n}$$

$$\leq \sum_{s=\gamma_a \log n}^{n_0} \sum_{t=1}^{\alpha_i \eta s \log n} \left(\frac{s \log n}{n}\right)^{\alpha_i \eta s \log n - s} (\log n)^s \left(\frac{e^{3/2+o(1)}\alpha_i^{1/2}}{2\eta^{1/2}}\right)^{2\alpha_i \eta s \log n}$$

$$= o(1).$$

(b) The probability that the condition is violated can be bounded by

$$\binom{n}{\beta n}^2 \binom{\beta n}{\gamma_b n/\log n} \left(e^{-\alpha_i \beta/4}\right)^{\gamma_b n} \leq \left(\left(\frac{e}{\beta}\right)^{(2\beta+o(1))} e^{-\alpha_i \beta \gamma_b/4}\right)^n = o(1).$$

The factor $e^{-\alpha_i \beta \gamma_b/4}$ comes from applying a Chernoff bound.

(c) We can assume w.l.o.g. that $|S| = |T| = \beta n$. The probability that the condition is violated can be bounded by

$$\binom{n}{\beta n}^2 \binom{\beta n}{n/\log n}\binom{\beta n}{\gamma_b n/\log n}\left(\binom{\gamma_b n/\log n}{k}(\alpha_i p)^k\right)^{n/\log n}$$

$$\leq \left(\frac{e}{\beta}\right)^{(2\beta+o(1))n} \left(\frac{e\gamma_b \alpha_i}{k}\right)^{kn/\log n} = o(1).$$

(d) The probability that the condition is violated can be bounded by

$$\binom{n}{\beta n}^2 \binom{\beta n}{\gamma_d n/\log n}(1-\alpha_i p)^{\beta n \gamma_d n/\log n} \leq \left(\left(\frac{e}{\beta}\right)^{2+o(1)} e^{-\alpha_i \gamma_d}\right)^{\beta n} = o(1).$$

(e) The probability that the condition is violated can be bounded by

$$\binom{n}{\gamma n/\log n}^2 \binom{\gamma^2 n^2/(\log n)^2}{\delta n/\log\log n} p^{\delta n/\log\log n}$$
$$\leq \left(\frac{e\log n}{\gamma}\right)^{2\gamma n/\log n} \left(\frac{\gamma^2 e \log\log n}{\delta \log n}\right)^{\delta n/\log\log n} = o(1).$$

(f) The probability that the condition is violated can be bounded by

$$2^{2n}(1-p)^{\beta^2 n^2/100} = o(1).$$

□

3 Proof of Theorem 1

Proof Assume from now on that the high probability conditions of Lemma 1 are in force. Let M be a perfect matching and let $\mu_i = |M \cap Q_i|$ for $i \in [q]$. Suppose that $\mu_1 > m_1 \geq \beta n$ and $\beta n \leq \mu_2 < m_2$. We show that we can find another matching M' such that $|M' \cap Q_1| = \mu_1 - 1$ and $|M' \cap Q_2| = \mu_2 + 1$. We do this by finding an alternating cycle with edge sequence $C = (e_1, f_1, \ldots, e_\ell, f_\ell)$ and vertex sequence $(x_1 \in A, y_1 \in B, x_2, \ldots, x_\ell, y_\ell, x_1)$ such that (1) $e_i = \{x_i, y_i\} \in M$, (2) $f_i = \{y_i, x_{i+1}\} \notin M, i \in [\ell]$, (3) $e_1 \in Q_1$, and (4) $E(C) \setminus \{e_1\} \subseteq Q_2$. Repeating this for pairs of colors, one over-subscribed and one under-subscribed we eventually achieve our goal. It is sufficient to consider this case, seeing as we can always w.h.p. find a matching that has been randomly colored with $\approx \alpha_i n$ edges of color c_i, $i = 1, 2, \ldots, q$.

Next let $A_i = V(M \cap Q_i) \cap A$ and $B_i = V(M \cap Q_i) \cap B$ for $i \in [q]$ and for $S \subseteq A$ let $N_i(S) = \{b \in B : \exists a \in S \ s.t..\ \{a,b\} \in Q_i\}$ and $N_i(a) = N_i(\{a\})$. Then let

$$D_0' = \left\{a \in A_2 : |N_2(a) \cap B_2| \geq \frac{\alpha_2 \beta \log n}{10}\right\}.$$

$$D_0 = \left\{a \in A_1 : |N_2(a) \cap M(A_2 \setminus D_0')| \leq k_0 = \frac{10\log n}{\log\log n}\right\}.$$

It follows from Lemma 1(b) that

$$|M(A_2 \setminus D_0')| \leq \frac{\gamma_b n}{\log n}.$$

It then follows from Lemma 1(c) that if $W_0 = A_1 \setminus D_0$, then

$$|W_0| \leq \frac{n}{\log n}. \tag{1}$$

We now define a sequence of sets $W_0, W_1, \dots$ where W_{j+1} is obtained from W_j by adding a vertex of $A_2 \setminus W_j$ for which $|N_2(a) \cap M(W_j)| \geq k_0$. Now consider $S = W_t, T = M(W_t)$ for some $t \geq 1$. Then we have

$$|S| = |T| \leq t + \frac{n}{\log n} \text{ and } e_2(S, T) \geq tk_0.$$

Given Lemma 1(e) with $\delta = 5, \gamma = 2$, we see that this sequence stops with $t = t^* \leq 4n/\log n$. So we now let $R_0 = A_2 \setminus W_{t^*}$. We note that

$$\begin{aligned} &|R_0| \geq \beta n - \frac{5n}{\log n} \\ &a \in R_0 \text{ implies } |N_2(a) \cap M(R_0)| \geq \frac{\alpha_2 \beta \log n}{10} - k_0. \end{aligned} \tag{2}$$

We now fix some $a_0 \in R_0$ and define a sequence of sets $X_0, Y_0, X_1, Y_1, \dots$ where $X_j \subseteq R_0$ and $Y_j \subseteq B_2$. We let $X_0 = \{a_0\}$ and then having defined $X_i, i \geq 0$ we let

$$Y_i = N_2(X_i) \text{ and } X_{i+1} = \left(M^{-1}(Y_i) \setminus \bigcup_{j \leq i} X_j \right) \cap R_0.$$

We claim that for $i \geq 0$,

$$|X_i| \leq \frac{n}{200 \log n} \text{ implies that } |X_{i+1}| \geq \frac{\alpha_2 \beta \log n}{25} |X_i|. \tag{3}$$

We verify (3) below. Assuming its truth, there exists a smallest k such that

$$|X_k| \geq \frac{\alpha_2 \beta n}{5000}. \tag{4}$$

Starting with $\widehat{Y}_0 = \{b_0\}$ where $b_0 = M(a_0) \in \hat{R}_0$, we can similarly construct a sequence of sets $\widehat{Y}_1, \widehat{X}_1, \dots$ where $\widehat{X}_j \subseteq M^{-1}(\hat{R}_0)$ and $\widehat{Y}_j \subseteq \hat{R}_0$. Here $\hat{R}_0$ is the equivalently defined set to R_0 in B_2. We can assume that $b_0 \in \hat{R}_0$, because of the sizes of the sets $R_0, \hat{R}_0$. More precisely, by (1), there will be $o(n)$ choices for a_0 for which $b_0 \notin \hat{R}_0$. Having defined $\widehat{Y}_i$ we let

$$\widehat{X}_i = N_2(\widehat{Y}_i) \text{ and } \widehat{Y}_{i+1} = \left(M(\widehat{X}_i) \setminus \bigcup_{j \le i} \widehat{Y}_j \right) \cap \hat{R}_0$$

and then let $\widehat{Y}_{i+1} = M(\widehat{X}_i)$. The equivalent of (3) will be

$$|\widehat{Y}_i| \le \frac{n}{200 \log n} \text{ implies that } |\widehat{Y}_{i+1}| \ge \frac{\alpha_2 \beta \log n}{25} |\widehat{Y}_i|. \tag{5}$$

Assuming its truth, there exists ℓ such that

$$|\widehat{Y}_\ell| \ge \frac{\alpha_2 \beta n}{5000}. \tag{6}$$

It follows from Lemma 1(f) that at least 9/10 of the vertices of A_1 have a c_2-neighbor in $\hat{R}_0$ and at least 9/10 of the vertices of B_1 have a c_2-neighbor in R_0. We deduce from this that there is a pair $x_0 \in A_1, y_0 = M(x_0) \in B_1$ such that $N_2(x_0) \cap \hat{R}_0 \neq \emptyset$ and $N_2(y_0) \cap R_0 \neq \emptyset$. This defines an alternating cycle $x_0, u_0, P_1, b_0, a_0, P_2, v_0, y_0, x_0$. Here u_0 is a c_2-neighbor of x_0 in $\hat{R}_0$ and P_1 is (the reversal of) a path from u_0 to b_0 and P_2 is the path from a_0 to $v_0 \in X_k, v_0 \in N_2(y_0)$. This completes the proof of Theorem 1.

Verification of (3)**,** (5) We have by the assumption $a_0 \in R_0$ that

$$|X_1| = |Y_1| \ge \frac{\alpha_2 \beta \log n}{10} - o(\log n).$$

Now suppose that $1 \le |X_i| \le n/(200 \log n)$. Then, by (2),

$$e_2(X_i : (N_2(X_i) \setminus M(A_2 \setminus R_0))) \ge \frac{(\alpha_2 \beta \log n)|X_i|}{10 + o(1)}.$$

Applying Lemma 1(a) we see that

$$|N_2(X_i) \setminus M(A_2 \setminus R_0)| \ge \frac{(\alpha_2 \beta \log n)|X_i|}{20 + o(1)}. \tag{7}$$

Because the sets $X_1, X_2, \ldots$ expand rapidly, the total size of $\bigcup_{j \le i} X_j$ is small compared with the R.H.S of (7) and (3) follows. The argument for (5) is similar.

□

4 Concluding Remarks

We have established that w.h.p. $mcp(G)$ is almost all of $[0, n]^q$ and posed the question of finding the exact threshold for $mcp(G) = [0, n]^q$. It seems technically feasible to extend our results to randomly colored $G_{n,p}$. We leave this for future research. It would be of some interest to analyze other spanning subgraphs from this point of view, e.g. Hamilton cycles.

Acknowledgment Research supported in part by NSF grant DMS1661063.

References

1. M. Anastos, A.M. Frieze, Pattern colored Hamilton cycles in random graphs. SIAM J. Discrete Math. **33**, 528–545 (2019)
2. D. Bal, A.M. Frieze, Rainbow matchings and Hamilton cycles in random graphs. Random Struct. Algorithms **48**, 503–523 (2016)
3. C. Cooper, A.M. Frieze, Multi-coloured Hamilton cycles in random edge-coloured graphs. Comb. Probab. Comput. **11**, 129–134 (2002)
4. A. Dudek, A.M. Frieze, C. Tsourakakis, Rainbow connection of random regular graphs. SIAM J. Discrete Math. **29**, 2255–2266 (2015)
5. P. Erdős, A. Rényi, On random matrices. Publ. Math. Inst. Hungar. Acad. Sci. **8**, 455–461 (1964)
6. L. Espig, A.M. Frieze, M. Krivelevich, Elegantly colored paths and cycles in edge colored random graphs. SIAM J. Discrete Math. **32**, 1585–1618 (2018)
7. A. Ferber, M. Krivelevich, Rainbow Hamilton cycles in random graphs and hypergraphs, in *Recent Trends in Combinatorics*, ed. by A. Beveridge, J.R. Griggs, L. Hogben, G. Musiker, P. Tetali. IMA Volumes in Mathematics and Its Applications (Springer, Cham, 2016), pp. 167–189
8. A.M. Frieze, C.E. Tsourakakis, Rainbow connectivity of sparse random graphs. Electron. J. Comb. **19** (2012). https://doi.org/10.37236/2784
9. A. Heckel, O. Riordan, The hitting time of rainbow connection number two. Electron. J. Comb. **19** (2012). https://doi.org/10.37236/2708
10. S. Janson, N. Wormald, Rainbow Hamilton cycles in random regular graphs. Random Struct. Algorithms **30**, 35–49 (2007)
11. N. Kamcev, M. Krivelevich, B. Sudakov, Some remarks on rainbow connectivity. J. Graph Theory **83**, 372–383 (2016)
12. M. Molloy, The rainbow connection number for random 3-regular graphs. Electron. J. Comb. **24**, P3.49 (2017)

Prime Difference Champions

S. Funkhouser, D. A. Goldston, D. Sengupta, and J. Sengupta

Abstract A Prime Difference Champion (PDC) for primes up to x is defined to be any element of the set of one or more differences that occur most frequently among all positive differences between primes $\leq x$. Assuming an appropriate form of the Hardy–Littlewood Prime Pair Conjecture we can prove that for sufficiently large x the PDCs run through the primorials. Numerical results also provide evidence for this conjecture as well as other interesting phenomena associated with prime differences. Unconditionally we prove that the PDCs go to infinity and further have asymptotically the same number of prime factors when counted logarithmically as the primorials.

2000 Mathematics Subject Classification Primary 11N05; Secondary 11P32, 11N36

1 Introduction

The study of prime gaps has been historically one of the most insightful probes of the nature of the primes. The gaps, referring here strictly to the positive differences between consecutive primes, are a particularly natural subject of investigation because they exhibit global characteristics that are amenable to rigorous analytical treatment. Some of the most important theorems related to primes are those

S. Funkhouser
Department of Natural Sciences, Trident Technical College, North Charleston, SC, USA

D. A. Goldston (✉)
Department of Mathematics and Statistics, San José State University, San José, CA, USA
e-mail: daniel.goldston@sjsu.edu

D. Sengupta · J. Sengupta
Department of Mathematics and Computer Science, Elizabeth City State University, Elizabeth City, NC, USA
e-mail: dcsengupta@ecsu.edu; jdsengupta@ecsu.edu

A. M. Raigorodskii, M. Th. Rassias (eds.), *Discrete Mathematics and Applications*, Springer Optimization and Its Applications 165,
https://doi.org/10.1007/978-3-030-55857-4_9

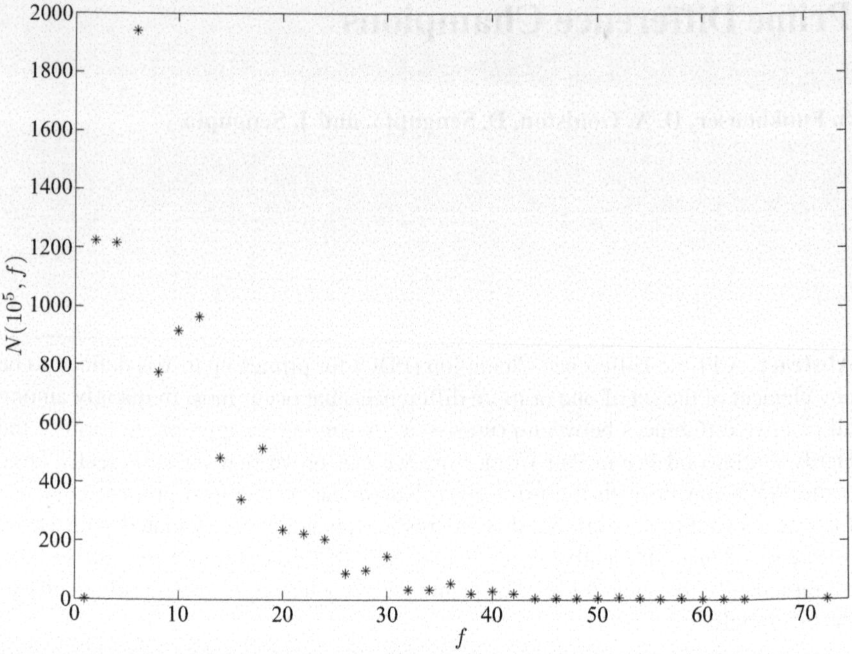

Fig. 1 Gap count for $x = 10^5$, showing only non-zero $N(x, d)$

concerning the gaps. Most prominently, we may state the Prime Number Theorem equivalently in terms of the asymptotic behavior of the average local gap.

An elemental component of the study of prime gaps is the counting function,

$$N(x, d) := \sum_{\substack{p_{n+1} \le x \\ p_{n+1} - p_n = d}} 1\,, \tag{1}$$

giving the number of gaps of size d within the sequence of primes no greater than some x, where p_n is the n-th prime. As an example Figure 1 is a plot of $N(10^5, d)$. From $N(x, d)$ we may readily obtain the average gap among primes not exceeding x. The most commonly occurring gap (or gaps) among primes no greater than some x is the value (or values) of d for which $N(x, d)$ is maximal. With

$$N^*(x) := \max_d N(x, d) \tag{2}$$

the set

$$J^*(x) := \{d \colon N(x, d) = N^*(x)\} \tag{3}$$

formally describes the most common gap(s), known as the *prime jumping champion(s)* (PJCs), for a given x. Figure 2 shows the PJCs for all prime $x \le 1800$.

Several analytical results concerning the PJCs are relevant here. In 1999, based on heuristic arguments and extensive numerical studies, Odlyzko, Rubinstein, and

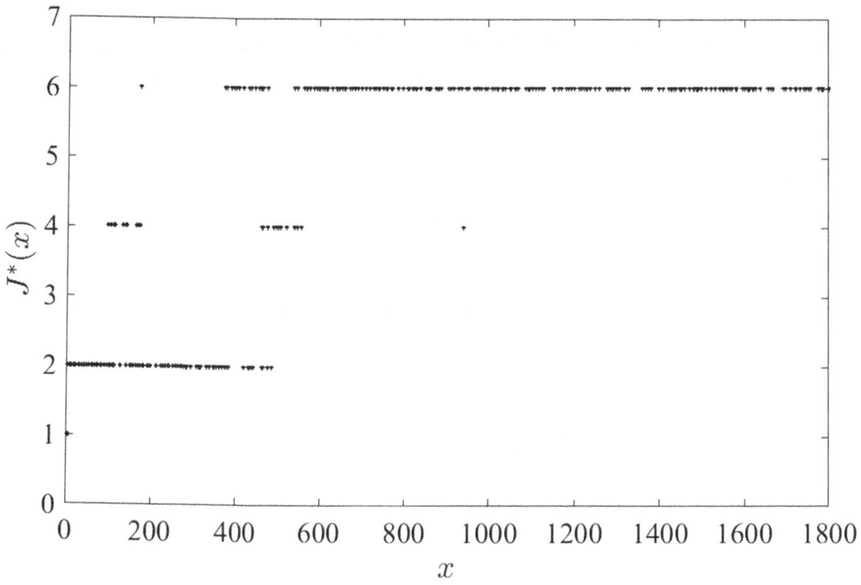

Fig. 2 Prime jumping champions for all odd, prime $x \leq 1800$

Wolf (ORW) [8] conjectured that the PJCs greater than unity are 4 and the primorials,

$$\left\{p_k^\sharp\right\}_{k=1}^{\infty} = 2, 6, 30, 210, 2310, \dots , \tag{4}$$

where

$$p_k^\sharp := \prod_{j=1}^{k} p_j \tag{5}$$

is the k-th primorial. They also advanced a weaker implication of this conjecture, namely that the PJCs are asymptotically infinite and that any given prime divides all sufficiently large PJCs. In 1980 Erdos and Straus [1] proved, assuming the Hardy–Littlewood prime pair conjecture (HLPPC), that the PJCs tend to infinity. In 2011 Goldston and Ledoan extended the method described in [1] to give a proof that any given prime will divide all sufficiently large jumping champions [3]. Soon thereafter they gave also a proof that sufficiently large PJCs run through the primorials assuming a sufficiently strong form of the Hardy–Littlewood prime pair and prime triple conjectures [4].

It is important to note that, despite the body of work devoted to the prime gaps, we know virtually nothing unconditionally about the PJCs. Numerically 6 is the PJC for $947 \leq x \leq 10^{15}$, and it unlikely that numerical work will ever find any

PJC other than 6 for ranges where computations are feasible. Unconditionally we cannot even disprove that the PJC is 2 for all sufficiently large x, or eliminate any given even number as sometimes being a PJC. The study of differences between primes, however, need not be restricted to differences between consecutive primes. In fact, the HLPPC is for the generalized (positive) differences between primes not exceeding a given x, and it is natural to first study pair differences in a given sequence before examining more complicated gap questions.

The purpose of this work is to present a broad investigation of the prime differences and their so-called champions. We find that, in contrast to the situation with the gaps, rich behaviors are evident in computationally accessible ranges and, more importantly, both conditional and unconditional theorems may be formulated to explain many prominent characteristics of the PDCs. The work is organized as follows. Section 2 contains definitions of the central terms and some basic numerical studies, in which significant regularities are immediately evident. Section 3 contains an overview of the HLPPC, with appropriate adaptations necessary for this present analysis. Section 4 contains the results of several numerical studies in support of the HLPPC. Section 5 outlines how the HLPPC may be used to probe the nature of the prime differences. In Section 6 it is proven under condition of the HLPPC that the PDCs run through the primorials for sufficiently large x. Section 7 reviews some relevant properties of logarithmic sums. In Section 8 it is proven unconditionally that the PDCs tend to infinity and have many prime factors.

2 Counting Prime Differences

Let p and p' denote primes and let d be a positive integer. Analogously to $N(x, d)$ we define the counting function

$$G(x, d) := \sum_{\substack{p, p' \leq x \\ p' - p = d}} 1 \tag{6}$$

to give the number of prime differences equal to d among primes no greater than x. Figure 3 shows the count of prime differences for $x = 10^5$. For convenience we have excluded all odd differences, being those associated with the anomalously even $p_1 = 2$. Figure 4 shows a zoomed-in view of Figure 3. Figures 3 and 4 are representative for all sufficiently large x within the computationally explored range, which so far extends to $x = 2 \times 10^8$.

A prime difference champion for a given x is any d for which $G(x, d)$ attains its maximum,

$$G^*(x) := \max_d G(x, d) . \tag{7}$$

Analogously to $J^*(x)$ we define formally the PDC (or PDCs) for a given x by the set

$$D^*(x) := \{d \colon G(x, d) = G^*(x)\} . \tag{8}$$

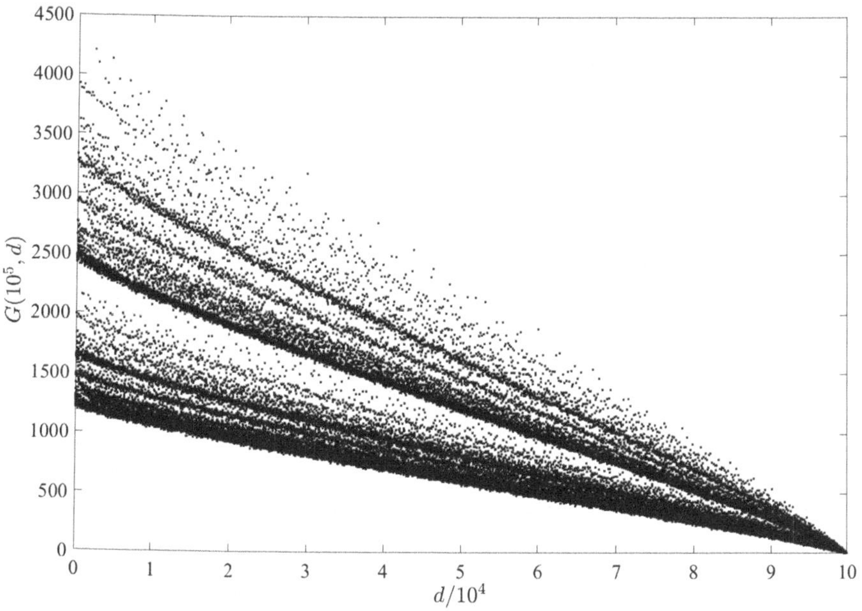

Fig. 3 Difference count for $x = 10^5$

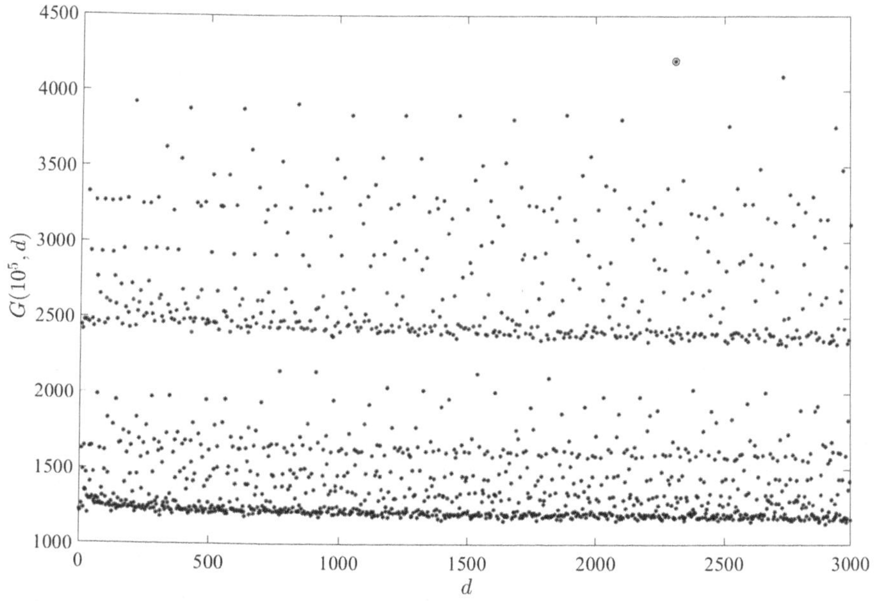

Fig. 4 Detail of difference count for $x = 10^5$; the PDC is 2310 and the corresponding count is circled

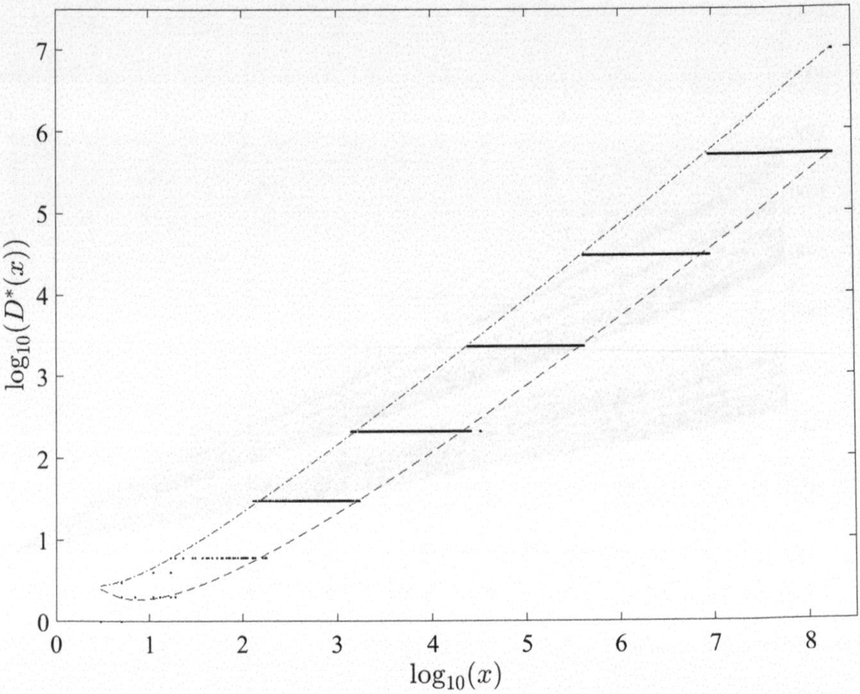

Fig. 5 Prime difference champions for all odd, prime x no greater than 2×10^8. The dot-dashed line is a plot of $\log_{10}(x/\log(x))$ and the dashed line is a plot of $\log_{10}(x/\log^2(x))$, in association with Theorem 1

As $D^*(x)$ is a step function of x with possible steps only when x is a prime, we only need consider values of $x = p \geq 3$ in examining the behavior of the PDCs. For example, we have $D^*(3) = \{1\}$, $D^*(5) = \{1, 2, 3\}$, $D^*(7) = \{2\}$, $D^*(11) = \{2, 4\}$, $D^*(13) = \{2\}$, $D^*(17) = \{2, 4, 6\}$. Figure 5 is a plot of the PDCs for all prime $x \leq 2 \times 10^8$. The dashed and dot-dashed lines in Figure 5 are associated with Theorem 1 as introduced in Section 5.

The basic computational results exemplified in Figures 3, 4, and 5 intimate several new global properties of the prime differences and the PDCs. For sufficiently large x, $G(x, d)$ exhibits a rich structure bearing a statistically strong signature of periodicity [2]. As shown in Section 3 we may attribute the periodic structure to behavior encapsulated in the HLPPC. Aside from 1, 3, and 4 all PDCs in the numerically explored range of x are primorials. Additionally the PDCs run through the primorials in a broadly step-wise manner as x increases.

It is instructive to review some of the details of the observed transitions in the PDCs. Table 1 shows the smallest x and the largest known prime x for which $D^*(x)$ includes each respective primorial within the observed range. Because the last observed instance of 9699690 represents only the arbitrarily chosen end of calculations it does not represent any meaningful transition point and is omitted

Table 1 Transition points for PDCs

Primorial	Smallest x for occurrence as PDC	Largest known prime x for occurrence as PDC
6	$p_7 = 17$	$p_{41} = 179$
30	$p_{32} = 131$	$p_{269} = 1723$
210	$p_{224} = 1423$	$p_{3523} = 32843$
2310	$p_{2718} = 24499$	$p_{35000} = 414977$
30030	$p_{34903} = 413863$	$p_{609928} = 9120277$
510510	$p_{607867} = 9087131$	$p_{10657197} = 191945597$
9699690	$p_{10561154} = 190107653$	–

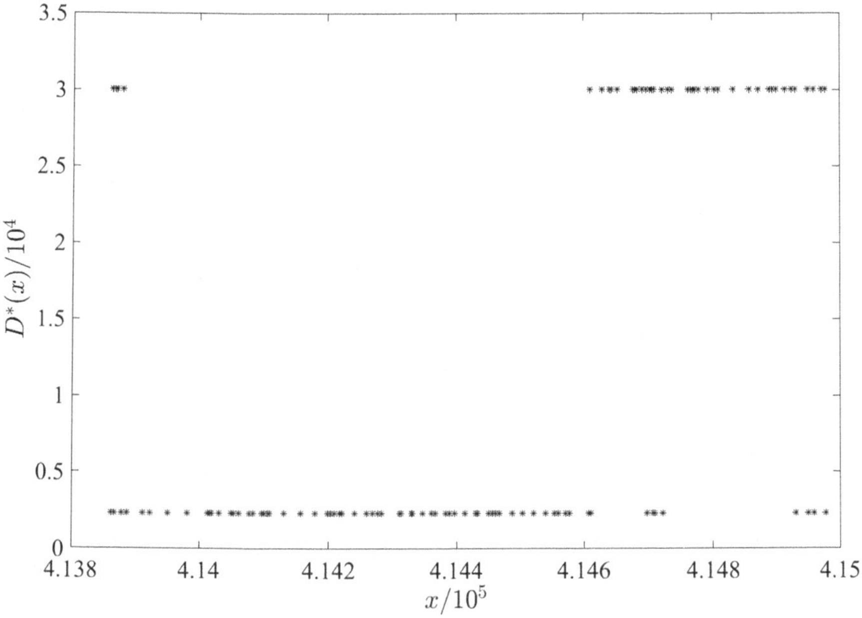

Fig. 6 Zoom-in view of $D^*(x)$ over region where transition from 210 to 2310 occurs

from the table. Note that the step-wise transitions occur for x in the vicinity of p_n such that n is roughly equal to the next largest primorial. For example, the transition from 210 to 2310 begins at $n = 2718$, and so on [2]. This empirical trend may be understood as a condition of the HLPPC, as formalized at the end of Section 5.

Within the transition regions the PDC is observed to exhibit some oscillatory behavior between the two primorials involved in the transition. For example, Figure 6 is a plot of $D^*(x)$ beginning with the first occurrence of 30030 and ending at the last occurrence of 2310. Note that during the transitions the PDC is observed only to be one or both of the two respective primorials involved.

3 The Hardy–Littlewood Prime Pair Conjecture

Hardy–Littlewood in their famous paper Partitio Numerioum III [6] conjectured an asymptotic formula for $G(x, d)$. This is often called the prime pair conjecture, and states that

$$G(x, d) \sim \mathfrak{S}(d) \frac{x}{(\log x)^2}, \quad \text{as } x \to \infty. \tag{9}$$

The singular series $\mathfrak{S}(d)$ is defined for all integers $d \neq 0$ by

$$\mathfrak{S}(d) = \begin{cases} 2C_2 \prod\limits_{\substack{p \mid d \\ p>2}} \left(\frac{p-1}{p-2} \right), & \text{if } d \text{ is even, } d \neq 0; \\ 0, & \text{if } d \text{ is odd;} \end{cases} \tag{10}$$

where

$$C_2 = \prod_{p>2} \left(1 - \frac{1}{(p-1)^2} \right) = 0.66016\ldots \tag{11}$$

with the product extending over all primes $p > 2$. We see that for d a positive even integer the singular series may be written as

$$\mathfrak{S}(d) = 2C_2 \prod_{\substack{p \mid d \\ p>2}} \left(1 + \frac{1}{p-2} \right), \tag{12}$$

and therefore $\mathfrak{S}(d)$ has local maximums when d is a primorial.

In the case of the prime number theorem we know that we obtain a better approximation by assuming that a prime p has a density or probability of occurring of $\frac{1}{\log p}$ rather than taking the constant density $\frac{1}{\log x}$ for all the primes up to x. We expect the same is true for prime pairs, and Hardy and Littlewood conjectured that one should replace

$$\frac{x}{(\log x)^2} \quad \text{by} \quad \mathrm{li}_2(x) = \int_2^x \frac{dt}{(\log t)^2}, \tag{13}$$

in which case the conjecture should hold with a much smaller error term.

Although Hardy and Littlewood in this case did not specifically consider the situation where $d = d(x) \to \infty$, it is reasonable to suppose that a form of the HLPPC will hold in this situation. We see in the definition of $G(x, d)$ that if $d > 0$, then the conditions $p, p' \leq x$ and $p' = p + d$ implies that $p \leq x - d$. Hence we see for $d > 0$ that

$$G(x, d) = \sum_{\substack{p \le x-d \\ p+d \text{ is prime}}} 1, \tag{14}$$

and thus we may conjecture for $d > 0$ that since p has density $\frac{1}{\log p}$ and $p + d$ has density $\frac{1}{\log(p+d)}$

$$G(x, d) = \mathfrak{S}(d) \int_2^{x-d} \frac{dt}{\log t \log(t + d)} + E(x, d), \quad \text{as } x \to \infty, \tag{15}$$

where $E(x, d)$ represents an error term. For convenience let us define

$$I(x, d) := \int_2^{x-d} \frac{dt}{\log t \, \log(t + d)} \tag{16}$$

and

$$\widetilde{G}(x, d) := \mathfrak{S}(d) I(x, d) \tag{17}$$

such that

$$G(x, d) = \widetilde{G}(x, d) + E(x, d), \tag{18}$$

where $\widetilde{G}(x, d)$ represents presumably the asymptotic difference count.

A strong conjecture is that for $2 \le d \le x - x^{\epsilon}$

$$E(x, d) \ll (x - d)^{\frac{1}{2}+\epsilon}. \tag{19}$$

The actual conjecture we need to resolve the present problems concerning the PDCs is much weaker than this, and is stated appropriately as follows.

Conjecture 1 *We have that* (15) *holds with* $E(x, d) = o(\frac{x}{(\log x)^4})$ *uniformly for* $2 \le d \le \frac{8}{9}x$.

4 Numerical Tests of the Hardy–Littlewood Conjecture

Before proceeding to the proofs it is worthwhile to examine some numerical results related to the behaviors expected of the HLPPC, as expressed in Conjecture 1. The studies described here provide novel, detailed tests of the validity of the HLPPC.

First let us examine the representative behaviors of the terms on the right-hand side of (15). Figure 7 is a plot of $\widetilde{G}(10^5, d)$, for comparison to Figure 3. Figure 8

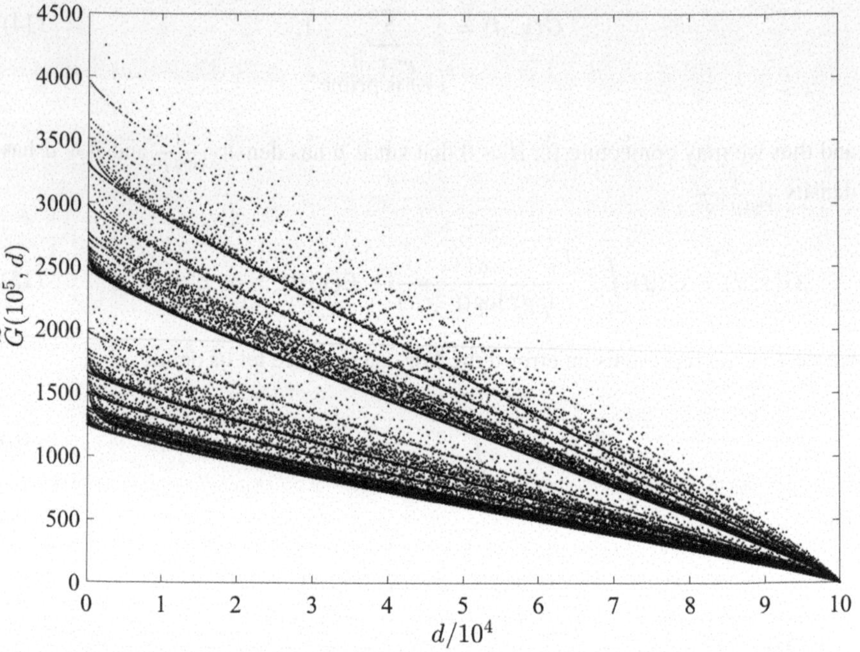

Fig. 7 Asymptotic difference count, $\widetilde{G}(x, d)$, for $x = 10^5$. Compare to Figure 3

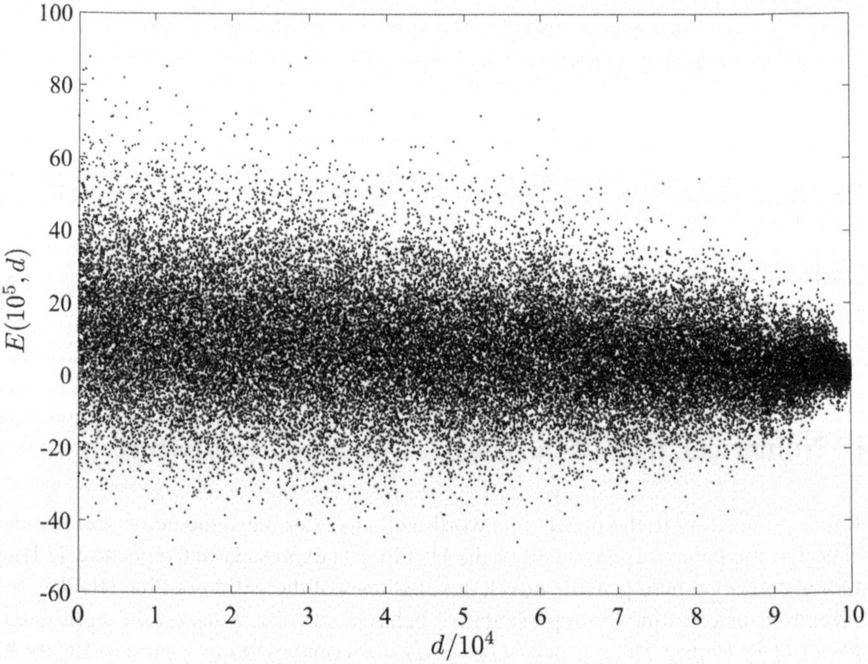

Fig. 8 Error term associated with asymptotic count for $x = 10^5$

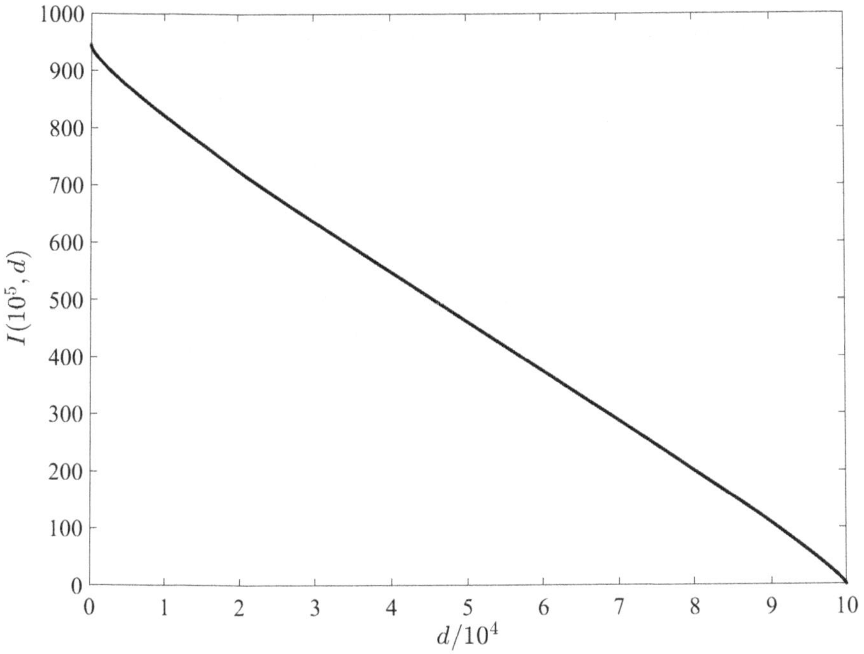

Fig. 9 Numerically computed integral term for $x = 10^5$

shows the corresponding error term, $E(10^5, d)$. Finally Figure 9 shows $I(10^5, d)$. The behaviors shown in these figures are representative for sufficiently large x, and the precision of the numerical integrations is such that the associated uncertainties are negligible on the scales of all of the figures.

Next, note that we have

$$\sum_d G(p_n, d) = \frac{n(n-1)}{2} \tag{20}$$

by construction, where summations over d are taken for $d = 1, 2, 4, 6, 8, \ldots$. In accordance with the HLPPC we therefore expect

$$\sum_d \widetilde{G}(p_n, d) \sim \frac{n^2}{2}. \tag{21}$$

In order to test (21) it is appropriate to define the relative error

$$\mu(x) := \frac{\sum_d \left(\widetilde{G}(x, d) - G(x, d)\right)}{\sum_d G(x, d)}. \tag{22}$$

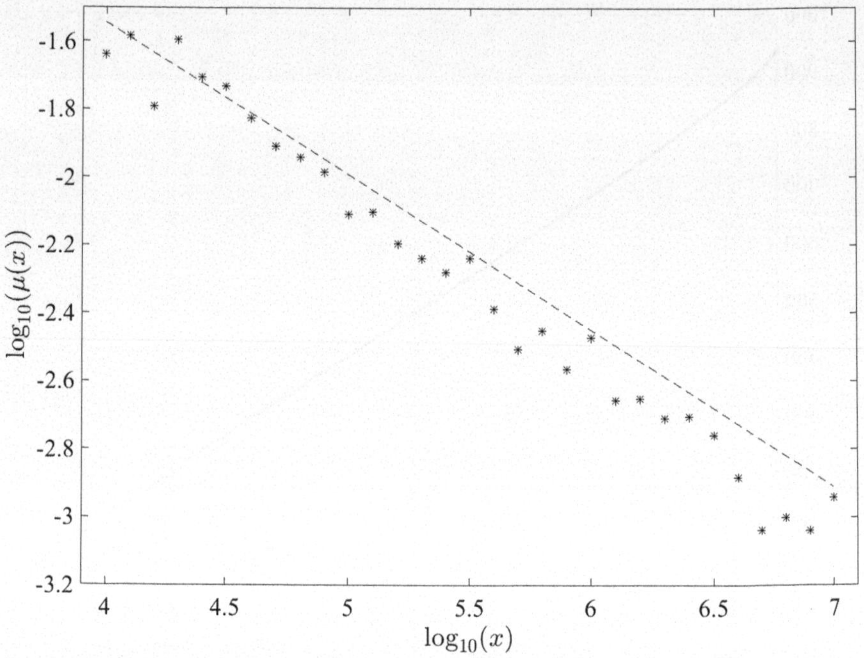

Fig. 10 Relative error in the sum of $\widetilde{G}(x, d)$, for certain $x = p_n$, plotted with asterisks. The dashed line is a plot of $1/\sqrt{\pi(x)}$, in corresponding logarithmic fashion

Figure 10 is a plot of $\mu(x)$ for roughly logarithmically spaced, prime $x \in [10^4, 10^7]$. The dashed line is $1/\sqrt{\pi(x)}$, where $\pi(x)$ is the usual prime counting function, hence $\pi(p_n) = n$. An analysis of the possible relationship between $\mu(x)$ and a function of the form $c_\mu/\sqrt{\pi(x)}$, for some constant c_μ, is reserved for future work. It is sufficient here to note that $\mu(x)$ would vanish asymptotically if the observed trends should persist *ad infinitum*.

It is also instructive to consider the variance,

$$\nu(x) := \sum_d \left(\widetilde{G}(x, d) - G(x, d)\right)^2. \tag{23}$$

Figure 11 is a plot $\nu(x)/\pi(x)^2$ for the same particular $x = p_n$ as in Figure 10. Note that the plotted points are all in the vicinity of 0.16. The asymptotic behavior of $\nu(x)$ and the apparent broad proportionality between $\nu(x)$ and $\pi(x)^2$ are reserved for future studies.

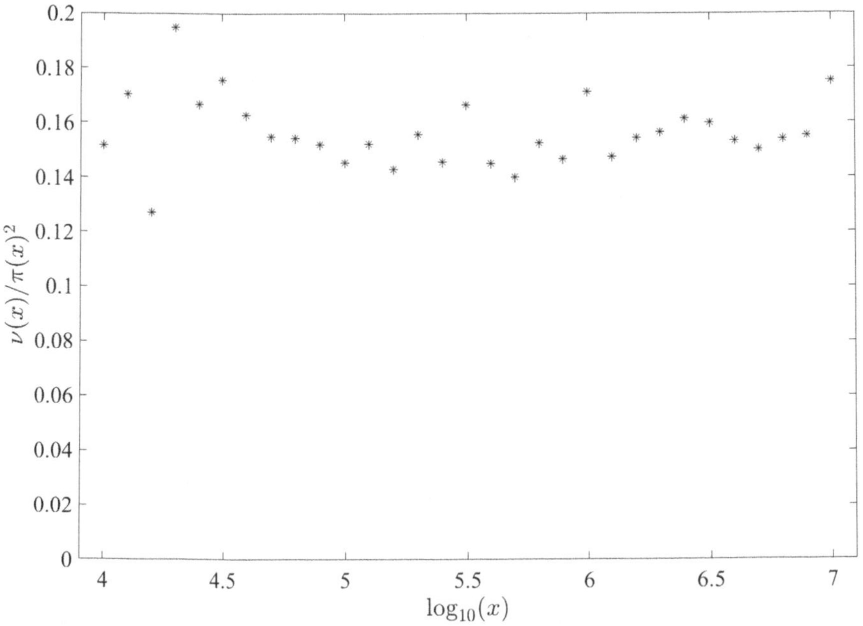

Fig. 11 Plot of variance, $\nu(x)$, divided by $\pi(x)^2$ for certain $x = p_n$

5 Sketch of Solution of the PDC Problem Using Conjecture 1

To determine the PDCs, we use Conjecture 1 to obtain formulas for $G(x, d)$ in various ranges of d. We will assume here $2 \le d \le \frac{8}{9}x$. We want to evaluate $I(x, d)$ asymptotically, but we will never need to be more accurate than terms with size smaller than $o\left(\frac{x}{(\log x)^4}\right)$. Hence, we take

$$I(x,d) = \int_{\frac{x}{(\log x)^5}}^{x-d} \frac{dt}{\log t \, \log(t+d)} + O\left(\frac{x}{(\log x)^5}\right), \tag{24}$$

and use integration by parts to obtain

$$I(x,d) = \frac{x-d}{\log x \, \log(x-d)} + \int_{\frac{x}{(\log x)^5}}^{x-d} \frac{\left(\log(t+d) + \frac{t}{t+d}\log t\right)dt}{(\log t \, \log(t+d))^2} + O\left(\frac{x}{(\log x)^5}\right). \tag{25}$$

Since $\frac{t}{t+d} = 1 - \frac{d}{t+d}$, we obtain

$$I(x,d) = \frac{x-d}{\log(x-d)\,\log x} + \int_{\frac{x}{(\log x)^5}}^{x-d} \frac{dt}{(\log t)^2 \log(t+d)} + \int_{\frac{x}{(\log x)^5}}^{x-d} \frac{dt}{\log t(\log(t+d))^2}$$
$$+ O\left(\frac{d}{(\log x)^2}\right) + O\left(\frac{x}{(\log x)^5}\right). \tag{26}$$

We now define, for fixed integers $m, n \geq 1$,

$$I_{m,n} := \int_{\frac{x}{(\log x)^5}}^{x-d} \frac{dt}{(\log t)^m (\log(t+d))^n}, \tag{27}$$

and then obtain by the same integration by parts argument as above

$$I_{m,n} = \frac{x-d}{(\log x)^n\,(\log(x-d))^m} + mI_{m+1,n} + nI_{m,n+1} + O\left(\frac{d}{(\log x)^2}\right) + O\left(\frac{x}{(\log x)^5}\right). \tag{28}$$

Furthermore we have the trivial estimate $I_{m,n} \ll \frac{x}{(\log x)^{m+n}}$. Hence

$$\begin{aligned} I(x,d) &= \frac{x-d}{\log x\,\log(x-d)} + I_{2,1} + I_{1,2} + O\left(\frac{d}{(\log x)^2}\right) + O\left(\frac{x}{(\log x)^5}\right) \\ &= \frac{x-d}{\log x\,\log(x-d)} + \frac{x-d}{\log x(\log(x-d))^2} + \frac{x-d}{(\log x)^2 \log(x-d)} \\ &\quad + 2I_{3,1} + I_{2,2} + I_{2,2} + 2I_{1,3} + O\left(\frac{d}{(\log x)^2}\right) + O\left(\frac{x}{(\log x)^5}\right) \\ &= (x-d)\Bigg(\frac{1}{\log x\,\log(x-d)} + \frac{1}{(\log x)^2 \log(x-d)} + \frac{1}{\log x(\log(x-d))^2} \\ &\quad + \frac{2}{\log x(\log(x-d))^3} + \frac{2}{(\log x)^2(\log(x-d))^2} + \frac{2}{(\log x)^3 \log(x-d)} \\ &\quad + O\left(\frac{d}{x(\log x)^2}\right) + O\left(\frac{1}{(\log x)^5}\right)\Bigg). \end{aligned} \tag{29}$$

Since for $2 \le d \le \frac{8}{9}x$,

$$\log(x-d) = \log x + \log\left(1 - \frac{d}{x}\right) = \log x + O\left(\frac{d}{x}\right), \tag{30}$$

we conclude that

$$\begin{aligned} I(x,h) &= (x-d)\left(\frac{1}{(\log x)^2} + \frac{2}{(\log x)^3} + \frac{6}{(\log x)^4} + O\left(\frac{d}{x(\log x)^2}\right) + O\left(\frac{1}{(\log x)^5}\right)\right) \\ &\quad \times \left(1 + O\left(\frac{d}{x\log x}\right)\right) = \frac{x-d}{(\log x)^2} \\ &\quad \times \left(1 + \frac{2}{\log x} + \frac{6}{(\log x)^2} + O\left(\frac{1}{(\log x)^3}\right) + O\left(\frac{d}{x\log x}\right)\right). \end{aligned} \tag{31}$$

Letting

$$H(x,d) := \frac{I(x,d)}{x-d}(\log x)^2 = 1 + \frac{2}{\log x} + \frac{6}{(\log x)^2} + O\left(\frac{1}{(\log x)^3}\right) + O\left(\frac{d}{x\log x}\right). \tag{32}$$

Conjecture 1 takes the form that for $2 \le d \le \frac{8}{9}x$,

$$G(x,d) = \mathfrak{S}(d)\frac{x-d}{(\log x)^2}H(x,d)\left(1 + o\left(\frac{1}{(\log x)^2}\right)\right). \tag{33}$$

This will be the formula we use in what follows.

Before proceeding to the proof, we explain how this formula implies our theorem. The factor $\frac{H(x,d)}{(\log x)^2}$ is essentially constant and can be ignored. From (12) the singular series $\mathfrak{S}(d)$ increases on the sequence of primiorials and therefore if $d < p_k^\sharp$, then $\mathfrak{S}(d) < \mathfrak{S}(p_k^\sharp)$. Thus $G(x,d)$ will also grow on the sequence of primorials as long as d is small enough that the linear decreasing factor $(x-d)$ does not overwhelm the increase from the singular series. From (5) and the Prime Number Theorem we have

$$\log p_k^\sharp = \sum_{p \le p_k} \log p \sim p_k, \qquad \text{as } k \to \infty, \tag{34}$$

which relates the primorial to the primes that constitute it. The critical range is when $x^{1-o(1)} \le p_k^\sharp \le x$ and thus here $p_k \sim \log x$. The maximum of $G(x,d)$ will occur at the primorial where

$$\mathfrak{S}(p_k^\sharp)(x - p_k^\sharp) > \mathfrak{S}(p_{k+1}^\sharp)(x - p_{k+1}^\sharp) \tag{35}$$

for the first time, and this occurs when

$$\frac{\mathfrak{S}(p_{k+1}^\sharp)}{\mathfrak{S}(p_k^\sharp)} \sim \frac{x - p_k^\sharp}{x - p_{k+1}^\sharp}. \tag{36}$$

The left-hand side is

$$= 1 + \frac{1}{p_{k+1} - 2} \sim 1 + \frac{1}{\log x}, \tag{37}$$

while the right-hand side is

$$= 1 + \frac{p_{k+1}^\sharp - p_k^\sharp}{x - p_k^\sharp} \sim 1 + \frac{p_{k+1}^\sharp}{x}. \tag{38}$$

Hence these two expressions match when $p_{k+1}^\sharp \sim \frac{x}{\log x}$ and we expect that the PDC will usually be $\lfloor \frac{x}{\log x} \rfloor^\sharp$, where $\lfloor x \rfloor^\sharp$ is a floor function with respect to the primorials, defined by

$$\lfloor x \rfloor^\sharp = p_k^\sharp \quad \text{if} \quad p_k^\sharp \leq x < p_{k+1}^\sharp. \tag{39}$$

The exception and most delicate situation is where the primorial $p_{k+1}^\sharp$ is very close to $\frac{x}{\log x}$, in which case either it or the previous primorial $p_k^\sharp$ may be the PDC. Here $p_k^\sharp$ will be very close to $\frac{x}{(\log x)^2}$. Notice the singular series has the same value at $p_k^\sharp$ and $2p_k^\sharp$, so in order to show this latter value is not the PDC we need to use the inequality $x - p_k^\sharp > x - 2p_k^\sharp$ in Conjecture 1. These terms differ in Conjecture 1 in the $\frac{x}{(\log x)^4}$ term, which we can only distinguish by taking $E(x, d) = o\left(\frac{x}{(\log x)^4}\right)$ for the error in Conjecture 1.

The result we prove is the following.

Theorem 1 *Assume Conjecture 1. Let* $0 < \delta \leq \frac{1}{4}$ *be a given number. Then for* x *sufficiently large, if the interval* $[(1 + \frac{\delta}{2})\frac{x}{(\log x)^2}, (1 - \frac{\delta}{2})\frac{x}{\log x}]$ *contains a primorial, then that primorial is the PDC. If this interval does not contain a primorial, then both of the intervals* $[(1 - \delta)\frac{x}{(\log x)^2}, (1 + \frac{\delta}{2})\frac{x}{(\log x)^2})$ *and* $((1 - \frac{\delta}{2})\frac{x}{\log x}, (1 + \delta)\frac{x}{\log x}]$ *will contain primorials and one or the other or sometimes both will be the PDCs.*

Corollary 1 *Assume Conjecture 1. Then all sufficiently large PDCs are primorials, and every sufficiently large primorial will be the PDC for some* x.

Before proceeding to the proof of Theorem 1 it is instructive to refer again to the numerical calculations presented in Section 2. The dot-dashed line in Figure 5 is a logarithmic plot of $x/\log(x)$ and represents an asymptotic upper bound on the PDCs in accordance with Theorem 1. The dashed line is a logarithmic plot of

$x/\log^2(x)$ and represents the corresponding lower bound. The analytical bounds are in excellent agreement with the numerical data. Furthermore, in accordance with Corollary 1, we find that the PDCs in the observed range are all primorials for $x \geq 19$. Note also that the upper bound on the PDCs is essentially the prime counting function, which explains the observed tendency for the steps in the PDCs to occur where $x = p_n$ is roughly equal to the next largest primorial.

6 Proof of Theorem 1

Theorem 1 follows from the following two lemmas. First it is convenient to define a ceiling function with respect to the sequence of primorials, analogously to $\lfloor x \rfloor^\sharp$, such that

$$\lceil x \rceil^\sharp = p_k^\sharp \quad \text{if} \quad p_{k-1}^\sharp < x \leq p_k^\sharp. \tag{40}$$

Lemma 1 *Assume Conjecture 1, and let* $0 < \delta \leq \frac{1}{4}$ *be a given number. Define*

$$p_a^\sharp := \left\lfloor \left(1 - \frac{\delta}{2}\right) \frac{x}{\log x} \right\rfloor^\sharp. \tag{41}$$

Then, for x sufficiently large and $2 \leq d < p_a^\sharp$, we have $G(x, d) < G(x, p_a^\sharp)$.

Lemma 2 *Assume Conjecture 1, and let* $0 < \delta \leq \frac{1}{4}$ *be a given number. Define*

$$p_b^\sharp := \left\lceil \left(1 + \frac{\delta}{2}\right) \frac{x}{(\log x)^2} \right\rceil^\sharp. \tag{42}$$

Then, for x sufficiently large and $x \geq d > p_b^\sharp$, we have $G(x, p_b^\sharp) > G(x, d)$.

Proof of Theorem 1 There can be at most one primorial in the interval $[(1 + \frac{\delta}{2})\frac{x}{(\log x)^2}, (1 - \frac{\delta}{2})\frac{x}{\log x}]$ since if $p_k^\sharp = x^{1+o(1)}$ we have $p_{k+1}^\sharp \sim p_k^\sharp \log x$ and $\log x$ times the left endpoint of this interval is larger than the right endpoint. If there is a primorial in this interval, then clearly it is $p_a^\sharp = p_b^\sharp$, and by Lemmas 1 and 2 it is the PDC.

Next both of the intervals $[(1 - \delta)\frac{x}{(\log x)^2}, (1 - \frac{\delta}{2})\frac{x}{\log x})$ and $((1 + \frac{\delta}{2})\frac{x}{(\log x)^2}, (1 + \delta)\frac{x}{\log x}]$ will contain at least one primorial, but if $[(1 + \frac{\delta}{2})\frac{x}{(\log x)^2}, (1 - \frac{\delta}{2})\frac{x}{\log x}]$ does not contain a primorial, then both of the intervals $[(1 - \delta)\frac{x}{(\log x)^2}, (1 + \frac{\delta}{2})\frac{x}{(\log x)^2})$ and $((1 - \frac{\delta}{2})\frac{x}{\log x}, (1 + \delta)\frac{x}{\log x}]$ must contain consecutive primorials, say $p_j^\sharp$ and $p_{j+1}^\sharp$, and $p_j^\sharp = p_a^\sharp$ and $p_{j+1}^\sharp = p_b^\sharp$. At least one of these must be the PDC, and since, for fixed d, $N(x, d)$ increases by steps of 1 as x increases, the PDC will transition from

$p_a^\sharp$ to $p_b^\sharp$ and there must be at least one value of x where $N(x, p_a^\sharp) = N(x, p_b^\sharp)$ and the PDC is $\{p_a^\sharp, p_b^\sharp\}$. □

In proving Lemmas 1 and 2 we use (33) to examine the ratio

$$\frac{G(x, p_k^\sharp)}{G(x, d)} = \frac{\mathfrak{S}(p_k^\sharp)}{\mathfrak{S}(d)} \frac{x - p_k^\sharp}{x - d} \frac{H(x, p_k^\sharp)}{H(x, d)} \left(1 + o(\frac{1}{(\log x)^2})\right). \tag{43}$$

For $2 \le d < p_k^\sharp$ we have $\mathfrak{S}(p_{k-1}^\sharp) \ge \mathfrak{S}(d)$ and therefore

$$\frac{\mathfrak{S}(p_k^\sharp)}{\mathfrak{S}(d)} = \left(1 + \frac{1}{p_k - 2}\right) \frac{\mathfrak{S}(p_{k-1}^\sharp)}{\mathfrak{S}(d)} \ge 1 + \frac{1}{p_k - 2}. \tag{44}$$

Also, for $2 \le d_1, d_2 \le \frac{8}{9}x$

$$\begin{aligned} \frac{x - d_2}{x - d_1} = \frac{1 - \frac{d_2}{x}}{1 - \frac{d_1}{x}} &= \left(1 - \frac{d_2}{x}\right)\left(1 + \frac{d_1}{x} + \left(\frac{d_1}{x}\right)^2 + \left(\frac{d_1}{x}\right)^3 + \cdots\right) \\ &\ge \left(1 - \frac{d_2}{x}\right)\left(1 + \frac{d_1}{x}\right) \\ &= 1 + \frac{d_1 - d_2}{x} - \frac{d_1 d_2}{x^2}, \end{aligned} \tag{45}$$

and

$$\frac{H(x, d_2)}{H(x, d_1)} = 1 + O\left(\frac{d_1 + d_2}{x \log x}\right) + O\left(\frac{1}{(\log x)^3}\right). \tag{46}$$

Proof of Lemma 1 In Lemma1 we have $2 \le d < p_a^\sharp \le (1 - \frac{\delta}{2}) \frac{x}{\log x}$ and $\mathfrak{S}(p_{a-1}^\sharp) \ge \mathfrak{S}(d)$. Therefore by (44) we have

$$\frac{\mathfrak{S}(p_a^\sharp)}{\mathfrak{S}(d)} \ge 1 + \frac{1}{p_a - 2} = 1 + \frac{1}{\log x}(1 + o(1)); \tag{47}$$

by (45) we have

$$\frac{x - p_a^\sharp}{x - d} = 1 - \frac{p_a^\sharp}{x} + \frac{d}{x} + O\left(\frac{1}{(\log x)^2}\right) \ge 1 - \frac{1 - \frac{\delta}{2}}{\log x} + O\left(\frac{1}{(\log x)^2}\right), \tag{48}$$

and by (46) we have

$$\frac{H(x, p_a^\sharp)}{H(x, d)} = 1 + O\left(\frac{1}{(\log x)^2}\right). \tag{49}$$

Therefore by (43) we have

$$\begin{aligned}\frac{G(x,p_a^\sharp)}{G(x,d)} &\geq \left(1+\frac{1}{\log x}(1+o(1))\right)\left(1-\frac{1-\frac{\delta}{2}}{\log x}+O\left(\frac{1}{(\log x)^2}\right)\right)\\ &\quad\times\left(1+O\left(\frac{1}{(\log x)^2}\right)\right)\left(1+o\left(\frac{1}{(\log x)^2}\right)\right)\\ &\geq 1+\frac{\frac{\delta}{2}}{\log x}-o\left(\frac{1}{\log x}\right)\\ &> 1+\frac{\frac{\delta}{4}}{\log x}\\ &> 1,\end{aligned} \tag{50}$$

for all sufficiently large x. This proves Lemma 1. □

Proof of Lemma 2 We now assume $d > p_b^\sharp$. Here $(1+\frac{\delta}{2})\frac{x}{(\log x)^2} \leq p_b^\sharp \leq (1+\delta)\frac{x}{\log x}$ for x sufficiently large. We need to divide the proof of Lemma 2 into cases depending on the size of d.

Case 1. Suppose $p_b^\sharp < d < 2p_b^\sharp$. In this range $\mathfrak{S}(p_{b-1}^\sharp) \geq \mathfrak{S}(d)$ so that just as before in (44) we have

$$\frac{\mathfrak{S}(p_b^\sharp)}{\mathfrak{S}(d)} \geq 1+\frac{1}{p_b-2} = 1+\frac{1}{\log x}(1+o(1)). \tag{51}$$

By (45)

$$\frac{x-p_b^\sharp}{x-d} = 1-\frac{p_b^\sharp}{x}+\frac{d}{x}+O\left(\frac{(p_b^\sharp)^2}{x^2}\right) \geq 1+O\left(\frac{1}{(\log x)^2}\right), \tag{52}$$

and by (46)

$$\frac{H(x,p_b^\sharp)}{H(x,d)} = 1+O\left(\frac{p_b^\sharp}{x\log x}\right)+O\left(\frac{1}{(\log x)^3}\right) = 1+O\left(\frac{1}{(\log x)^2}\right). \tag{53}$$

Hence by (43)

$$\begin{aligned}\frac{G(x,p_b^\sharp)}{G(x,d)} &\geq \left(1+\frac{1}{\log x}(1+o(1))\right)\left(1+O\left(\frac{1}{(\log x)^2}\right)\right)^2\left(1+o\left(\frac{1}{(\log x)^2}\right)\right)\\ &\geq 1+\frac{\frac{1}{2}}{\log x}\\ &> 1\end{aligned} \tag{54}$$

for all sufficiently large x, which proves Lemma 2 in this range.

Case 2. Suppose $2p_b^\sharp \le d < \min(p_{b+1}^\sharp, \frac{8}{9}x)$. Since $d < p_{b+1}^\sharp$, we have $\mathfrak{S}(p_b^\sharp) \ge \mathfrak{S}(d)$, and thus

$$\frac{\mathfrak{S}(p_b^\sharp)}{\mathfrak{S}(d)} \ge 1. \tag{55}$$

Now $d \le \frac{8}{9}x$ is needed to stay in the range where (33), (45), and (46) are valid. Using the inequality $d - p_b^\sharp \ge \frac{d}{2}$ which is valid when $2p_b^\sharp \le d$, we have by (45) for $d \le \frac{8}{9}x$,

$$\begin{aligned} \frac{x - p_b^\sharp}{x - d} &\ge 1 + \frac{d - p_b^\sharp}{x} - \frac{dp_b^\sharp}{x^2} \\ &\ge 1 + \frac{d}{2x} - \frac{d(1+\delta)}{x \log x} \\ &> 1 + \frac{d}{3x}, \end{aligned} \tag{56}$$

and by (46) for $d \le \frac{8}{9}x$,

$$\frac{H(x, p_b^\sharp)}{H(x,d)} = 1 + O\left(\frac{d}{x \log x}\right) + O\left(\frac{1}{(\log x)^3}\right) = 1 + O\left(\frac{d}{x \log x}\right). \tag{57}$$

Hence by (43)

$$\begin{aligned} \frac{G(x, p_b^\sharp)}{G(x,d)} &\ge \left(1 + \frac{d}{3x}\right)\left(1 + O\left(\frac{d}{x \log x}\right)\right)\left(1 + o\left(\frac{1}{(\log x)^2}\right)\right) \\ &> 1 + \frac{d}{4x} \\ &> 1 \end{aligned} \tag{58}$$

for all sufficiently large x, which proves Lemma 2 in this range. Notice we needed $E(x,d) = o\left(\frac{x}{(\log x)^4}\right)$ in Conjecture 1 for this last step.

Case 3. Suppose $p_{b+1}^\sharp \le d < \frac{8}{9}x$. We see that $p_{b+2}^\sharp \sim (\log x) p_{b+1}^\sharp \sim (\log x)^2 p_b^\sharp > (1 + \frac{\delta}{3})x$ for x sufficiently large, and hence $p_{b+2}^\sharp$ is larger than the range of d here. Hence in this range $\mathfrak{S}(p_{b+1}^\sharp) \ge \mathfrak{S}(d)$. Further we have shown $p_{b+1}^\sharp > \frac{x}{\log x}$ and therefore $\frac{x}{\log x} \le d < \frac{8}{9}x$. Hence

$$\begin{aligned}\frac{\mathfrak{S}(p_b^\sharp)}{\mathfrak{S}(d)} &= \frac{\mathfrak{S}(p_{b+1}^\sharp)}{\mathfrak{S}(d)}\left(1+\frac{1}{p_{b+1}-2}\right)^{-1}\\ &\geq \left(1+\frac{1}{\log x}(1+o(1))\right)^{-1}\\ &\geq 1-\frac{1}{\log x}(1+o(1)).\end{aligned} \tag{59}$$

By (45), and using $p_b^\sharp \ll \frac{d}{\log x}$ in this range,

$$\begin{aligned}\frac{x-p_b^\sharp}{x-d} &\geq 1+\frac{d-p_b^\sharp}{x}-\frac{dp_b^\sharp}{x^2}\\ &\geq 1+\frac{d}{x}-\left(\frac{d}{x}+1\right)\frac{p_b^\sharp}{x}\\ &\geq 1+\frac{d}{x}-\frac{2p_b^\sharp}{x}\\ &\geq 1+\frac{d}{x}+O\left(\frac{d}{x\log x}\right)\\ &\geq 1+\frac{d}{x}(1+o(1)).\end{aligned} \tag{60}$$

By (46) for $\frac{x}{\log x} \leq d < \frac{8}{9}x$,

$$\frac{H(x,p_b^\sharp)}{H(x,d)} = 1+O\left(\frac{d}{x\log x}\right)+O\left(\frac{1}{(\log x)^3}\right) = 1+o\left(\frac{d}{x}\right), \tag{61}$$

and hence by (43)

$$\begin{aligned}\frac{G(x,p_b^\sharp)}{G(x,d)} &> \left(1-\frac{1}{\log x}(1+o(1))\right)\left(1+\frac{d}{x}(1+o(1))\right)\left(1+o\left(\frac{d}{x}\right)\right)\left(1+o\left(\frac{1}{(\log x)^2}\right)\right)\\ &= 1+\left(\frac{d}{x}-\frac{1}{\log x}\right)(1+o(1)).\end{aligned} \tag{62}$$

Since here

$$\begin{aligned}\frac{d}{x} \geq \frac{p_{b+1}^\sharp}{x} &= \frac{p_b^\sharp \log x(1+o(1))}{x}\\ &\geq \frac{(1+\frac{\delta}{2})\frac{x}{(\log x)^2}\log x(1+o(1))}{x}\\ &> \frac{1+\frac{\delta}{3}}{\log x}\end{aligned} \tag{63}$$

for all sufficiently large x, we conclude

$$\frac{G(x, p_b^\sharp)}{G(x, d)} > 1 + \frac{\frac{\delta}{4}}{\log x} > 1 \tag{64}$$

for all sufficiently large x, which proves Lemma 2 in this range.

Case 4. Suppose $\frac{8}{9}x \leq d \leq x - \frac{x}{(\log x)^3}$. We use the sieve upper bound, for $1 < y \leq x$

$$G(x, d) - G(x - y, d) \leq 8\mathfrak{S}(d)\frac{y}{(\log y)^2}\left(1 + O\left(\frac{\log\log 3y}{\log y}\right)\right), \tag{65}$$

see Theorem 5.3 or Corollary 5.8.2 of [5]. Since trivially $G(d, d) = 0$, we see that $G(x, d) = G(x, d) - G(d, d) = G(x, d) - G(x - (x - d), d)$, and therefore taking $y = x - d \leq \frac{x}{9}$ in the sieve estimate above for $\frac{8}{9}x \leq d \leq x - \frac{x}{(\log x)^3}$, we have

$$\begin{aligned} G(x, d) &\leq 8\mathfrak{S}(d)\frac{x - d}{(\log(x - d))^2}\left(1 + O\left(\frac{\log\log 3(x - d)}{\log(x - d)}\right)\right) \\ &\leq \frac{8}{9}\mathfrak{S}(p_{b+1}^\sharp)\frac{x}{(\log x)^2}(1 + o(1)) \\ &< \frac{9}{10}\mathfrak{S}(p_b^\sharp)\frac{x}{(\log x)^2}. \end{aligned} \tag{66}$$

On the other hand, by Conjecture 1 in the simple form in (9)

$$G(x, p_b^\sharp) \sim \mathfrak{S}(p_b^\sharp)\frac{x}{(\log x)^2}, \tag{67}$$

and thus $G(x, p_b^\sharp) > G(x, d)$.

Case 5. Suppose $x - \frac{x}{(\log x)^3} \leq d \leq x$. Then if $p' - p = d$, then $p' \geq p + x - \frac{x}{(\log x)^3} > x - \frac{x}{(\log x)^3}$. Then

$$G(x, d) = \sum_{\substack{p, p' \leq x \\ p' - p = d}} 1 \leq \sum_{x - \frac{x}{(\log x)^3} < p' \leq x} 1 \leq \frac{x}{(\log x)^3} \tag{68}$$

and Lemma 2 follows since this is smaller than the size of $G(x, p_b^\sharp)$ we just used in Case 4.

This completes the proof of Lemma 2. □

7 Logarithmically Weighted Sums and Products of Primes

Starting with (12), we take logarithms and obtain

$$\mathfrak{S}(d) = 2C_2 \exp\left(\sum_{\substack{p|d \\ p>2}} \log\left(1 + \frac{1}{p-2}\right)\right). \tag{69}$$

These types of products and sums over primes with *logarithmic weighting* are much easier to evaluate than the unweighted primes in the prime number theorem and can be estimated using some elementary results of Merten, see [7] or [9].

Lemma 3 (Merten) *We have*

$$\prod_{p\leq y}\left(1 - \frac{1}{p}\right)^{-1} = e^{\gamma} \log y + O(1), \tag{70}$$

and

$$\sum_{p\leq y} \frac{1}{p} = \log\log y + b + O\left(\frac{1}{\log y}\right), \tag{71}$$

where γ is Euler's constant and b is a constant.

We often will make use of the estimate, for fixed constants $0 < a < b$,

$$\sum_{a\log x\leq p\leq b\log x} \frac{1}{p} \ll \frac{1}{\log\log x}, \tag{72}$$

which follows immediately on differencing in (71). This estimate in turn implies that

$$\begin{aligned}
\prod_{a\log x\leq p\leq b\log x}\left(1 + O\left(\frac{1}{p}\right)\right) &= \exp\left(\sum_{a\log x\leq p\leq b\log x} \log\left(1 + O\left(\frac{1}{p}\right)\right)\right) \\
&= \exp\left(\sum_{a\log x\leq p\leq b\log x} O\left(\frac{1}{p}\right)\right) \\
&= \exp\left(O\left(\frac{1}{\log\log x}\right)\right) \\
&= 1 + O\left(\frac{1}{\log\log x}\right).
\end{aligned} \tag{73}$$

Now consider the divisor sum

$$M(d) := \sum_{p|d} \frac{1}{p}, \tag{74}$$

and let

$$M^*(x) := \max_{1 \le d \le x} M(d). \tag{75}$$

It is clear that

$$M^*(x) = M(\lfloor x \rfloor^\sharp) = \sum_{p \le p_k} \frac{1}{p}, \tag{76}$$

where $\lfloor x \rfloor^\sharp = p_k^\sharp$. We have by (34) that $p_k \sim \log x$ and therefore by Lemma 3

$$M^*(x) = \log\log\log x + O(1). \tag{77}$$

Finally, returning to (69), we have by (77)

$$\mathfrak{S}(d) \ll \exp\left(M(d) + O(1)\right) \le \exp\left(M^*(d) + O(1)\right) \ll \log\log d. \tag{78}$$

8 The Prime Difference Champions Go to Infinity

We now prove the PDCs go to infinity and have many prime factors.

Theorem 2 *Suppose $d^* = d^*(x)$ is a prime difference champion for primes $\le x$. Then $d^* \to \infty$ as $x \to \infty$ and the number of distinct prime factors of d^* also goes to infinity as $x \to \infty$.*

First, we need a simplier version of the same sieve bound for prime pairs we used in the proof of Case 4 of Lemma 2.

Lemma 4 *We have for d positive and even*

$$G(x, d) \le C\,\mathfrak{S}(d) \frac{x}{(\log x)^2} (1 + o(1)), \tag{79}$$

uniformly for $1 \le d \le x$, where C is a constant.

The value $C = 4$ follows from the Bombieri–Vinogradov theorem, see [5]. (Slightly smaller values of C are known to hold.)

Our next lemma finds values of d for which $G(x, d)$ is large and thus gives a lower bound on $G^*(x)$.

Lemma 5 *Let* $1 \leq q \leq x$. *Then*

$$\sum_{1\leq m\leq \frac{x}{q}} G(x, mq) \geq \frac{1}{2}\left(\frac{x^2}{\phi(q)(\log x)^2} - \frac{x}{\log x}\right)(1 + o(1)). \tag{80}$$

Proof Let $\pi(x; q, a)$ denote the number of primes $\leq x$ which are congruent to a modulo q. Since

$$\sum_{\substack{1\leq a\leq q\\(a,q)=1}} \left(\pi(x; q, a) - \frac{\pi(x)}{\phi(q)}\right)^2 \geq 0, \tag{81}$$

we have on multiplying out that

$$\sum_{\substack{1\leq a\leq q\\(a,q)=1}} \pi(x; q, a)^2 \geq 2\frac{\pi(x)}{\phi(q)} \sum_{\substack{1\leq a\leq q\\(a,q)=1}} \pi(x; q, a) - \frac{\pi(x)^2}{\phi(q)}. \tag{82}$$

When $(a, q) > 1$, we have $\pi(x; q, a) = 0$ or 1 Therefore

$$\sum_{\substack{1\leq a\leq q\\(a,q)>1}} \pi(x; q, a) = \sum_{\substack{1\leq a\leq q\\(a,q)>1}} \pi(x; q, a)^2 \leq \sum_{p|q} 1 \ll \log q, \tag{83}$$

and therefore we drop the condition $(a, q) = 1$ in both sums above and obtain

$$\sum_{1\leq a\leq q} \pi(x; q, a)^2 \geq 2\frac{\pi(x)}{\phi(q)} \sum_{1\leq a\leq q} \pi(x; q, a) - \frac{\pi(x)^2}{\phi(q)} + O\left(\frac{\pi(x)\log q}{\phi(q)}\right) + O(\log q). \tag{84}$$

Now

$$\sum_{1\leq a\leq q} \pi(x; q, a) = \pi(x), \tag{85}$$

and

$$\begin{aligned}\sum_{1\le a\le q}\pi(x;q,a)^2 &= \sum_{1\le a\le q}\left(\sum_{\substack{p,p'\le x\\ p'\equiv p\equiv a\ (\mathrm{mod}\ q)}} 1\right)\\ &= \sum_{\substack{p,p'\le x\\ p'\equiv p(\mathrm{mod}\ q)}} 1\\ &= \sum_{\substack{-x\le d\le x\\ q|d}}\ \sum_{\substack{p,p'\le x\\ p'-p=d}} 1\\ &= \pi(x)+2\sum_{\substack{1\le d\le x\\ q|d}} G(x,d),\end{aligned}\tag{86}$$

and therefore we obtain

$$\begin{aligned}2\sum_{\substack{1\le d\le x\\ q|d}} G(x,d) &\ge \frac{\pi(x)^2}{\phi(q)}-\pi(x)+O\left(\frac{\pi(x)\log q}{\phi(q)}\right)+O(\log q)\\ &\ge \left(\frac{x^2}{\phi(q)(\log x)^2}-\frac{x}{\log x}\right)(1+o(1)),\end{aligned}\tag{87}$$

where in the last line we used the prime number theorem in the form

$$\pi(x)=\frac{x}{\log x}(1+o(1)).\tag{88}$$

□

Proof of Theorem 2 Since

$$G^*(x)\frac{x}{q}\ge\sum_{1\le m\le\frac{x}{q}} G(x,mq),\tag{89}$$

by Lemma 5 we conclude that, recalling $\phi(q)=q\prod_{p|q}(1-\frac{1}{p})$,

$$\begin{aligned}G^*(x) &\ge \frac{1}{2}\frac{q}{\phi(q)}\left(\frac{x}{(\log x)^2}-\frac{q}{\log x}\right)(1+o(1))\\ &= \frac{1}{2}\prod_{p|q}\left(1-\frac{1}{p}\right)^{-1}\left(\frac{x}{(\log x)^2}-\frac{q}{\log x}\right)(1+o(1)).\end{aligned}\tag{90}$$

We now choose $q = \lfloor \frac{x}{(\log x)^2} \rfloor^\sharp = p_\ell^\sharp$, and thus $p_\ell \sim \log x$. Hence

$$G^*(x) \geq \frac{1}{2} \prod_{p \leq p_\ell} \left(1 - \frac{1}{p}\right)^{-1} \frac{x}{(\log x)^2}(1 + o(1)). \tag{91}$$

By Lemma 4, we have that if d^* is a PDC for primes $\leq x$, then

$$G^*(x) \leq \mathcal{C}\,\mathfrak{S}(d^*) \frac{x}{(\log x)^2} (1 + o(1)). \tag{92}$$

Combining (91) and (92) we have

$$\mathcal{C}\,\mathfrak{S}(d^*) \geq \frac{1}{2} \prod_{p \leq p_\ell} \left(1 - \frac{1}{p}\right)^{-1} (1 + o(1)), \tag{93}$$

which by Lemma 3 gives

$$\mathfrak{S}(d^*) \geq \frac{e^\gamma}{2\mathcal{C}} \log p_\ell(1 + o(1)) \gg \log\log x, \tag{94}$$

and hence by the first part of (78) we obtain

$$\sum_{p | d^*} \frac{1}{p} \gg \log\log\log x. \tag{95}$$

Therefore $d^* \to \infty$ and the number of distinct prime factors of d^* also go to infinity, as $x \to \infty$, which proves Theorem 2. □

We can obtain a slightly more precise result with a little more effort.

Theorem 3 *We have*

$$\sum_{\substack{p \leq 2\log x \\ p \nmid d^*(x)}} \frac{1}{p} \leq \log(2\mathcal{C})(1 + o(1)) < 2.08. \tag{96}$$

Corollary 2 *We have*

$$\sum_{p | d^*(x)} \frac{1}{p} \sim M^*(x) \sim \log\log\log x. \tag{97}$$

Hence we see that the PDCs must have asymptotically the maximal number of prime factors when the factors are weighted logarithmically. However, this does not imply that the PDCs must have many small prime factors, but we can prove they must have a few "small" prime factors.

Corollary 3 *For sufficiently large x, every PDC is divisible by an odd prime $\leq$ 25, 583.*

Proof of Theorem 3 Continuing from (93) and using (10) we obtain

$$2C_2 \prod_{\substack{p|d^* \\ p>2}} \left(\frac{p-1}{p-2}\right) \prod_{\substack{p\leq p_\ell \\ p>2}} \left(1-\frac{1}{p}\right) \geq \frac{1}{C}(1+o(1)). \tag{98}$$

We write the left-hand side of this equation as

$$2C_2 \prod_{\substack{p|d^* \\ p>p_\ell}} \left(\frac{p-1}{p-2}\right) \prod_{\substack{p|d^* \\ 2<p\leq p_\ell}} \left(\frac{p-1}{p-2}\right)\left(1-\frac{1}{p}\right) \prod_{\substack{p\leq p_\ell \\ p\nmid d^*}} \left(1-\frac{1}{p}\right) = 2C_2 P_1 P_2 P_3. \tag{99}$$

First, just as in (72),

$$\begin{aligned} 1 \leq P_1 &\leq \prod_{\substack{p|d^* \\ p>p_\ell}} \left(1+\frac{1}{p-2}\right) \\ &\leq \prod_{\substack{p|\lfloor x\rfloor^\sharp \\ p>p_\ell}} \left(1+\frac{1}{p-2}\right) \\ &\leq \prod_{\frac{1}{2}\log x\leq p\leq 2\log x} \left(1+\frac{1}{p-2}\right) \\ &= 1+O\left(\frac{1}{\log\log x}\right). \end{aligned} \tag{100}$$

Thus $P_1 = 1+o(1)$. Next, since

$$\left(\frac{p-1}{p-2}\right)\left(1-\frac{1}{p}\right) = \frac{(p-1)^2}{p(p-2)} = \left(1-\frac{1}{(p-1)^2}\right)^{-1}, \tag{101}$$

we see

$$P_2 \leq \prod_{p>2}\left(1-\frac{1}{(p-1)^2}\right)^{-1} = \frac{1}{C_2}. \tag{102}$$

Finally, by (73)

$$P_3 = \prod_{\substack{p \le 2\log x \\ p \nmid d^*}} \left(1 - \frac{1}{p}\right)\left(1 + O\left(\frac{1}{\log\log x}\right)\right). \tag{103}$$

We conclude from (98) and (99) that

$$\prod_{\substack{p \le 2\log x \\ p \nmid d^*}} \left(1 - \frac{1}{p}\right) \ge \frac{1}{2C}(1 + o(1)). \tag{104}$$

Taking logarithms we have

$$\sum_{\substack{p \le 2\log x \\ p \nmid d^*}} -\log\left(1 - \frac{1}{p}\right) \le \log(2C)(1 + o(1)). \tag{105}$$

Since

$$-\log(1 - x) = x + \frac{x^2}{2} + \frac{x^3}{3} + \cdots \qquad \text{for } |x| < 1, \tag{106}$$

we see $-\log(1 - x) > x$ for $0 < x < 1$ and hence taking $C = 4$,

$$\sum_{\substack{p \le 2\log x \\ p \nmid d^*}} \frac{1}{p} \le \log(8)(1 + o(1)) < 2.08, \tag{107}$$

which proves (96). □

Corollary 2 follows from (77) and (96). Corollary 3 follows from (96) because

$$\sum_{3 \le p \le p_{2817}} \frac{1}{p} > \log 8 \tag{108}$$

and $p_{2817} = 25583$. Hence every sufficiently large PDC is divided by both 2 and an odd prime ≤ 25583.

References

1. P. Erdős, E.G. Straus, Remarks on the differences between consecutive primes. Elem. Math. **35**(5), 115–118 (1980)
2. S. Funkhouser, C. Cartwright, D. Sengupta, B. Williams, Periodicity in the intervals between primes, National Technical Information Service, Accession number ADA580795, 2012

3. D.A. Goldston, A.H. Ledoan, Jumping champions and gaps between consecutive primes. Int. J. Number Theory **7**(6), 1–9 (2011)
4. D.A. Goldston, A.H. Ledoan, The jumping champion conjecture. Mathematika **61**(3), 719–740 (2015)
5. H. Halberstam, H.-E. Richert, *Sieve Methods*. London Mathematical Society Monographs, vol. 4 (Academic, London, New York, San Francisco, 1974)
6. G.H. Hardy, J.E. Littlewood, Some problems of 'Partitio numerorum'; III: on the expression of a number as a sum of primes. Acta Math. **44**(1), 1–70 (1923). Reprinted as pp. 561–630 in *Collected Papers of G. H. Hardy*, vol. I (Edited by a committee appointed the London Mathematical Society) (Clarendon Press, Oxford, 1966)
7. A.E. Ingham, *The Distribution of Prime Numbers*. Cambridge Tracts in Mathematics and Mathematical Physics, vol. 30 (Cambridge University Press, Cambridge, 1932)
8. A. Odlyzko, M. Rubinstein, M. Wolf, Jumping champions. Exp. Math. **8**(2), 107–118 (1999)
9. H.L. Montgomery, R.C. Vaughan, *Multiplicative Number Theory*. Cambridge Studies in Advanced Mathematics, vol. 97 (Cambridge University Press, Cambridge, 2007)

Exponential Variational Integrators Using Constant or Adaptive Time Step

Odysseas Kosmas and Dimitrios Vlachos

Abstract In this book article, at first we survey some recent advances in variational integrators focusing on the class of them known as exponential variational integrators, applicable in finite dimensional mechanical systems. Since these integrators are based on the space and time discretization, we start with a brief summary of the general development of the discrete mechanics and its application in describing mechanical systems with space-time integration algorithms. We, then, make an attempt to treat briefly in depth only the particular topic of adaptive time step exponential variational integrators. To this aim, the action integral along any curve segment is defined using a discrete Lagrangian that depends on the endpoints of the segment and on a number of intermediate points of interpolation. This Lagrangian is then, at any time interval, written as a weighted sum of the Lagrangians corresponding to a set of the chosen intermediate points to obtain high order integrators. The positions and velocities are interpolated here using special exponential functions. Finally, we derive exponential higher order variational integration methods for the numerical integration of systems with oscillatory solutions. The obtained exponential variational integrators using constant or adaptive time step are tested for the numerical solution of several problems showing their good behavior to track oscillatory solutions. Furthermore, we use the space-time geodesic approach of classical mechanics to explore whether the new methodology may be effective in adaptive time schemes.

O. Kosmas
Modelling and Simulation Centre, MACE, University of Manchester, Manchester, UK
e-mail: odysseas.kosmas@manchester.ac.uk

D. Vlachos (✉)
Department of Informatics & Telecommunications, University of Peloponnese, Tripoli, Greece
e-mail: dvlachos@uop.gr

A. M. Raigorodskii, M. Th. Rassias (eds.), *Discrete Mathematics and Applications*,
Springer Optimization and Its Applications 165,
https://doi.org/10.1007/978-3-030-55857-4_10

1 Introduction

During the last decades, there have been developed several numerical integration methods for Lagrangian systems, where the integrator is derived by discretizing the Hamilton's principle. This class of integration methods is known as discrete variational integrators and have specific advantages that make them attractive for many applications in mechanical systems. They are appropriate for both conservative and nearly dissipative (forced) systems. The conservative nature of variational integrators can allow substantially more accurate simulations at lower cost [1].

By understanding the geometry of space one may choose better discretizations while by understanding the geometric viewpoint one may recover the symmetries and invariants of the physical system (conservation of energy, conservation of linear and angular momentum, variational principles, etc.) [2–4].

In numerical solution of ordinary differential equations, one of the most difficult problems is related to the development of integrators for highly oscillatory systems [1]. As is well known, standard numerical schemes may require a huge number of time steps to track the oscillations. But, even with small size steps they may alter the dynamics, unless the chosen method has specific advantages. A useful category of them is that of geometric integrators, numerical schemes that preserve some geometric features of the dynamical system. These integrators provide to simulations longer time running without spurious effects (like bad energy behavior of conservative systems) than the traditional ones [5–7].

These methods are automatically symplectic thanks to the resulting good energy behavior. Also the symmetries of the discrete Lagrangian lead to conservation of orbital and angular momenta by the integrator.

In the class of asynchronous variational integrators (AVI) a refinement was developed that uses different time steps at different points in space and particularly in regions where the specific problem requires more (or less) accuracy (see [8]). The AVI are based on space-time discretizations which allow different time steps for different elements in a finite element mesh.

So far, in order to improve the numerical integration of highly oscillatory problems, e.g. [1], and derive methods as well as error bounds for families of quadrature methods that use for the required derivatives the finite difference approximations, special techniques have been developed. Alternatively, the use of the phase-lag property of [9], trigonometric fitting, and phase-fitting techniques lead to methods based on variable coefficients that depend on the characteristic frequency of the problem [10]. The latter, is known as exponential (or trigonometric) fitting and has been formulated long ago [9, 11, 12]. Exponentially fitting algorithms are considered as natural extensions of the classical polynomial fitting due to their characteristic property to approach the classical ones because the involved dominant frequencies tend to zero. The important problem of convergence of exponentially fitted methods, especially of the known as multi-step ones, has been investigated within Lyche's theory [12] (for a comprehensive discussion the reader is referred to [9]).

The main benefits of the variational integrators and phase fitting are exploited in previous works [13–15] for Lagrangian problems similar to that employed for testing ordinary differential equations (the harmonic oscillator with a given frequency ω). Furthermore, the exponential variational integrators, which solve exactly the test system, have been applied to general Lagrangian problem of single particle motion (the planar two-body problems) by determining the frequency ω at every step of the integration.

In recent years, a great number of phenomena are investigated with remarkably complex computer models and codes. Computational experiments, i.e., runs of these codes with various input data covering a wide range, lead to predictions (through the provided output) of several physical observables and parameters. In most of the cases the runs are computationally expensive and often our objective is the required computer experiments to be less time-consuming predictors of the output for the given data.

In solving special problems based on ordinary differential equations (ODEs) using numerical integration schemes, computational cost may be appreciably reduced by time adaptivity or using time adaptive steps [16, 17]. Admittedly, this tool possesses significant advantages with respect to the efficiency, the computational accuracy, and the ease in the implementation. The use, however, of symplectic integrators performs remarkably well in problems involving Hamiltonian integrations [7, 18, 19].

Many authors have, so far, addressed various derivations and have adopted symplectic integrators with variable time steps, despite the fact that the early results were not really promising [20–22]. Essentially, two main types of time variation steps have been utilized. In the first, the time step was explicitly varied in the flow of the time, a mostly problematic choice, while in the second, the time step was adopted while using the dynamical variables of the system (particle positions q, corresponding momenta p, etc.). For the case of the variable time, the derived equations are no longer in canonical Hamiltonian form leading to rather unreliable results.

Adaptive time step integrators may reduce some of the aforementioned shortcomings with high order non-symplectic schemes. Those are recently adopted [17, 23].

In improving the Galerkin type high order integrators [13–15], in such a way that adaptive time stepping to be used, the combination of space-time [16, 18] and geodesic view point of [24, 25] approaches are considered. These provide the possibility to overcome various problems that appear when symplectic integrators with variable time steps are employed. One can derive an optimal time step adaptation method computationally cheaper as much as possible.

Towards this end, a general formulation to derive high order variational integrators is presented (Section 2). They are tested on the numerical solution of some examples of the general N-body problem. In connection to those integrators we formulate the combined space-time and geodesic ideas of adaptive time stepping (Section 4) which we use to derive the proposed schemes (Section 5) and test (Section 6) in a couple of numerical applications. Finally, the advantages of the derived method are summarized in Section 7.

2 The Advantages of Variational Integrators

High order variational integrators that are applicable to physical systems where the Lagrangian is of separable form, are derived by following similar steps to those followed in the discrete variational calculus, see e.g. [4]. Thus, for a smooth and finite dimensional configuration manifold Q, one defines the discrete Lagrangian L_d through the mapping

$$L_d : Q \times Q \to \mathbb{R}. \tag{1}$$

This Lagrangian may be considered as an approximation of a continuous action obtained as

$$L_d(q_k, q_{k+1}, h_k) \approx \int_{t_k}^{t_{k+1}} L(q, \dot{q})dt. \tag{2}$$

Then, one defines also the action sum S_d as

$$S_d : Q^{N+1} \to \mathbb{R}, \tag{3}$$

($N \in \mathbb{N}$), that corresponds to the above Lagrangian as

$$S_d(\gamma_d) = \sum_{k=0}^{N-1} h_k L_d(q_k, q_{k+1}, h_k), \tag{4}$$

where $\gamma_d = (q_0, \ldots, q_N)$ denotes the discrete trajectory of the studied system. Following the procedure of the continuous Mechanics we can further compute the derivative of L_d as

$$dL_d(q_0, q_1) = D_1 L_d(q_0, q_1) + D_2 L_d(q_0, q_1), \tag{5}$$

interpreting $D_i L_d$ as the derivative with respect to the i-argument of L_d. According to the discrete variational principle, the solutions of the discrete system are determined from the L_d. Thus, in order to obtain the equation describing the motion of the system, we extremize the action sum S_d, over all the intermediate points of the trajectory γ_d by keeping the endpoints q_0 and q_N fixed. The resulting system of difference equations are

$$h_{k-1} D_2 L_d(q_{k-1}, q_k, h_{k-1}) + h_k D_1 L_d(q_k, q_{k+1}, h_k) = 0, \tag{6}$$

where $k = 1, \ldots, N-1$. These equations are known as discrete Euler–Lagrange equations [4, 13].

To derive high order methods addressed in this work, we approximate the action integral along the curve segment with endpoints q_k and q_{k+1} (see Figure 1), we use a discrete Lagrangian that depends only on the chosen endpoints, see Equation (2).

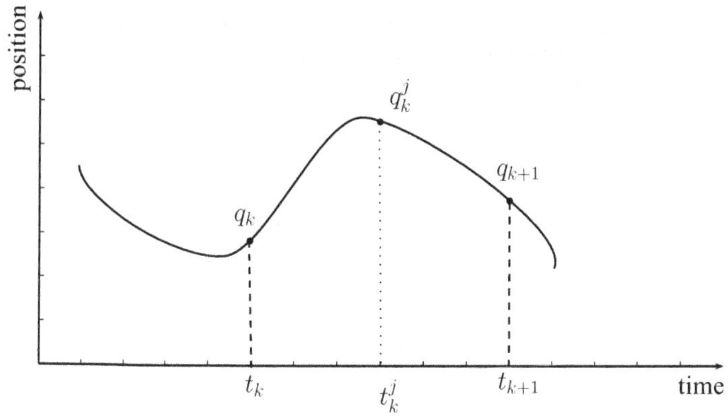

Fig. 1 Intermediate time nodes $t_k^j \in [t_k, t_{k+1}]$ and corresponding configurations q_k and q_{k+1}

This way, we obtain expressions for the configurations q_k^j and velocities $\dot{q}_k^j$ at time $t_k^j \in [t_k, t_{k+1}]$ where $j = 0, \ldots, S-1$, $S \in \mathbb{N}$. Then, by expressing the t_k^j as

$$t_k^j = t_k + C_k^j h_k \qquad \text{for} \qquad C_k^j \in [0, 1] \tag{7}$$

such that

$$C_k^0 = 0, \qquad C_k^{S-1} = 1, \tag{8}$$

where $h_k \in \mathbb{R}$ denotes the time step, we write [13]

$$\begin{aligned} q_k^j &= g_1(t_k^j) q_k + g_2(t_k^j) q_{k+1}, \\ \dot{q}_k^j &= \dot{g}_1(t_k^j) q_k + \dot{g}_2(t_k^j) q_{k+1}. \end{aligned} \tag{9}$$

Next, for the representation of the oscillatory behavior of the solution [26–29], we choose functions of the form

$$\begin{aligned} g_1(t_k^j) &= \sin\left(u - \frac{t_k^j - t_k}{h_k} u\right) (\sin u)^{-1}, \\ g_2(t_k^j) &= \sin\left(\frac{t_k^j - t_k}{h_k} u\right) (\sin u)^{-1}. \end{aligned} \tag{10}$$

For the sake of continuity, the conditions

$$g_1(t_{k+1}) = g_2(t_k) = 0 \tag{11}$$

$$g_1(t_k) = g_2(t_{k+1}) = 1 \tag{12}$$

must be fulfilled.

It should be mentioned that, for any different choice of interpolation, we define the discrete Lagrangian L_d by a weighted sum of the form [13]

$$L_d(q_k, q_{k+1}, h_k) = \sum_{j=0}^{S-1} h_k w^j L(q(t_k^j), \dot{q}(t_k^j)), \tag{13}$$

where, as can be readily proved, it holds [13, 26]

$$\sum_{j=0}^{S-1} w^j (C_k^j)^m = \frac{1}{m+1}, \tag{14}$$

with $m = 0, 1, \ldots, S-1$ and $k = 0, 1, \ldots, N-1$.

From the above equations it becomes clear that, if the time step, h_k, is equal to $h_k = h$ at every time interval, the resulting integrator is of constant time step.

By applying the above interpolation technique in combination with the trigonometric expressions of (10) and following the phase-lag analysis of [13, 26], the parameter u entering equations (10) must be determined as $u = \omega h$. For problems involving a definite frequency ω (such as the harmonic oscillator), the parameter u can be easily computed. However, for the solution of periodic orbit problems (orbital problems) of the general N-body problem, where no unique frequency of the motion can, in general, be determined, a new parameter u must be computed by estimating the frequency of the motion for any individual moving mass of the system [14, 15].

3 Exponential Integrators

When trying to solve numerically Hamiltonian systems of the form

$$\ddot{q} + \Omega q = g(q), \qquad g(q) = -\nabla U(q), \tag{15}$$

where Ω is a diagonal matrix (it may contain diagonal entries ω with large modulus) and $U(q)$ is a smooth potential function, we mostly are interested in the long time behavior of the numerical solutions. In such cases, application of the above methods imposes ωh to be rather large. Then, because an exact discretization of Equation (15) satisfies the equation $q_{n+1} - 2\cos(h\omega)q_n + q_{n-1} = 0$, we may write

$$q_{n+1} - 2\cos(h\omega)q_n + q_{n-1} = h^2 \psi(\omega h) g(\phi(\omega h) q_n), \tag{16}$$

where the functions $\psi(\omega h)$ and $\phi(\omega h)$ are even, real-valued functions satisfying the conditions $\psi(0) = \phi(0) = 1$ [5]. The latter equations represent exponential integrators (see Appendix for some typical examples).

3.1 High Order Exponential Variational Integrators

If we apply the steps of deriving high order variational integrators (see Section 2) to the Hamiltonian system (15), the discrete Euler–Lagrange equations (6) lead to the expressions

$$q_{n+1} + \Lambda(u, \omega, h, S)q_n + q_{n-1} = h^2 \Psi(\omega h) g(\Phi(\omega h) q_n), \tag{17}$$

where

$$\Lambda(u, \omega, h, S) = \frac{\sum_{j=0}^{S-1} w^j \left[\dot{g}_1(t_k^j)^2 + \dot{g}_2(t_k^j)^2 - \omega^2 \left(g_1(t_k^j)^2 + g_2(t_k^j)^2 \right) \right]}{\sum_{j=0}^{S-1} w^j \left[\dot{g}_1(t_k^j) \dot{g}_2(t_k^j) - \omega^2 g_1(t_k^j) g_2(t_k^j) \right]}. \tag{18}$$

Based on the latter two expressions, we derive exponential variational integrators that use the configurations q_k^j and velocities $\dot{q}_k^j$ of (9). We then get

$$\Lambda(u, \omega, h, S) = -2\cos(\omega h). \tag{19}$$

Whenever the latter equation holds, exponentially fitted methods relying on phase fitted variational integrators can be derived [13] which means that high order variational integrators can be considered as exponential integrators. Several numerical applications of this type have been carried out [13–15] and, due to their importance, in Section 3.3 we examine them in more detail through some representative examples.

3.2 Estimation of Frequency in Three Dimensional Particle Motions

Recently, exponential variational integrator techniques have been used [13] by estimating the required frequency on the basis of the frequency ω of a harmonic oscillator. In solving the orbital N-body problem by using a constant time step, a new way of frequency estimation is necessary to find it for each body (1) at the initial time t_0 and (2) at a time t_k, $k = 1, \ldots, N-1$.

Obviously, by applying the trigonometric interpolation (10), the parameter u can be chosen as $u = \omega h$, but for problems where the domain frequency ω is fixed and known (such as the harmonic oscillator) the parameter u can be easily computed. For the orbital N-body problem, where no global frequency is determined, u must be found by estimating the individual frequency of the motion of each moving point particle.

In the case of the 3-dimensional particle motion where N masses are moving in three dimensions, assuming that $q_i(t)$, $i = 1, \ldots, N$, denotes the trajectory and $\dot{q}_i(t)$ the velocity (with magnitude $|\dot{q}_i(t)|$) of the i-th particle, the corresponding curvatures can be computed from the known expressions

$$k_i(t) = \frac{|\dot{q}_i(t) \times \ddot{q}_i(t)|}{|\dot{q}_i(t)|^3}. \tag{20}$$

After a short time h, the angular displacement of the i-th mass is

$$\frac{h|\dot{q}_i(t) \times \ddot{q}_i(t)|}{|\dot{q}_i(t)|^2}, \tag{21}$$

which for the actual frequency gives

$$\omega_i(t) = \frac{|\dot{q}_i(t) \times \ddot{q}_i(t)|}{|\dot{q}_i(t)|^2}. \tag{22}$$

From (20) and (22) the well-known relation

$$\omega_i(t) = k_i(t)|\dot{q}_i(t)| \tag{23}$$

is satisfied (see also [13]).

Focusing on the many-body physical problem described via the Lagrangian

$$L(q, \dot{q}) = \frac{1}{2}\dot{q}^T M(q)\dot{q} - V(q), \tag{24}$$

($M(q)$ represents a symmetric positive definite mass matrix and V is the potential function), we write the continuous Euler–Lagrange equations as

$$M(q)\ddot{q} = -\nabla V(q). \tag{25}$$

In this system, the frequency $\omega_i(t_k)$ for the i-th body at time t_k, $k = 1, \ldots, N-1$ given by Equation (22), takes the form

$$\omega_i(t_k) = h^{-1}\frac{\left|M^{-1}(q_k)p_k \times \left(M^{-1}(q_k)p_k - M^{-1}(q_{k-1})p_{k-1}\right)\right|}{\left|M^{-1}(q_k)p_k\right|^2}. \tag{26}$$

The quantities on the right-hand side in the latter equation are the mass matrix, the configuration, and the momentum of the i-th body. The frequency $\omega_i(t_k)$, at an initial time instant t_0 (at which the initial positions are $\bar{q}_0$ and initial momenta are $\bar{p}_0$), is

$$\omega_i(t_0) = \frac{\left|M^{-1}(\bar{q}_0)\bar{p}_0 \times \left(-M^{-1}(\bar{q}_0)\nabla V(\bar{q}_0)\right)\right|}{\left|M^{-1}(\bar{q}_0)\bar{p}_0\right|^2}. \tag{27}$$

Equations (26) and (27) provide an "estimated frequency" for each mass in the general periodic motion of the N-body problem and allow the derivation of high order variational integrators based on trigonometric interpolation in which the frequency is estimated at every time step of the integration procedure. Compared to methods which employ constant frequency, the latter integrators show better energy behavior, i.e., smaller oscillation amplitude of the total energy is obtained [13, 14]. Before closing this section, it should be mentioned that, the linear stability of the above method is comprehensively analyzed in previous works [13, 14, 30].

3.3 Examples of Constant Time Step Exponential Integrators

Focusing on the numerical solution of the orbital problem of N-bodies moving in the gravitational field, we write the Lagrangian function as [5, 6]

$$L(q,\dot{q}) = \frac{1}{2}\sum_{i=1}^{N} m_i \dot{q}_i^2 + \sum_{i=1,j=1,i\neq j}^{N} G\frac{m_i m_j}{||q_i - q_j||}t \tag{28}$$

(G denotes the gravitational constant). In this section we study the motion of (1) the planar two-body problem and (2) that of the multi-body solar system and test the performance of the above mentioned novelties on these two systems.

3.3.1 Planar Two-Body Problem

As a first test of the above technique, we study the motion of the simple system of two objects that interact with each other through a central potential. The most famous example of this system is the planar Kepler's problem in which two masses attract each other with the gravitational force. In the solar system, such an interaction leads to elliptic orbits for the Sun–planet system and hyperbolic orbits for the Sun–comet system.

By choosing the heavier body as the center of the coordinate system, the motion remains planar. Denoting the position of the second body by $q = (q_1, q_2) \in \mathbb{R}^2$, the system's Lagrangian (28) takes the simple form

$$L(q,\dot{q}) = \frac{1}{2}\dot{q}^2 + \frac{1}{|q|}. \tag{29}$$

(for simplicity, the masses of the bodies and the gravitational constant are considered equal to 1). Also, as initial conditions we assume

$$q = (1-\epsilon, 0) \qquad \text{and} \qquad \dot{q} = \left(0, \sqrt{\frac{1+\epsilon}{1-\epsilon}}\right), \tag{30}$$

where $\epsilon \in \mathbb{R}$ is the eccentricity of the orbit [5].

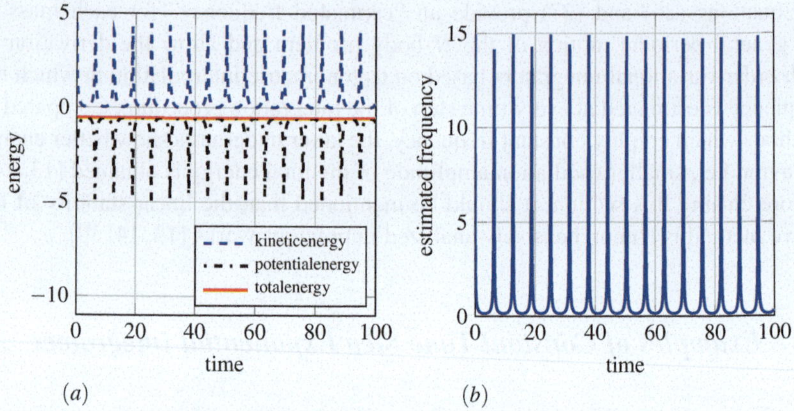

Fig. 2 Planar two-body problem with eccentricity $\epsilon = 0.8$, time step $h = 0.01$, and order of the method $S = 5$ for 10^4 steps. (**a**) Energy evolution using trigonometric interpolation. (**b**) The estimated frequency for the moving mass using the expression (22)

In trying to solve numerically the above problem, a difficulty arises when the eccentricity of the elliptical orbit is high. For example, some periodic comets have eccentricities between $0.7 \leq \epsilon \leq$ just below 1 (e.g., for Neptune's third largest moon Nereid $\epsilon = 0.750$, Halley's comet has $\epsilon = 0.967$, etc.). For this reason, we have chosen to test the above methods in the description of elliptical orbital problems with very high eccentricities for the two cases described below.

(i) In the first computational experiment, we consider the eccentricity $\epsilon = 0.8$ and we choose time step $h = 0.01$. We use trigonometric interpolation with $u = \omega h$ and $S = 5$. The good energy behavior of the method, obtained for a simulation of 10^4 steps, is illustrated in Figure 2a. The frequency is estimated at every integration step through the application of (22)–(26). The time variation is shown in Figure 2b where the peaks represent the estimated frequency at points close to the perihelion (the point where the moving mass is nearest to the central body). A comparison of the total energy evolution using constant (dashed line) and estimated (solid line) frequency is shown in Figure 3. As can be seen, even for small eccentricities ($\epsilon = 0.2$) the amplitude of the energy oscillation is smaller when the frequency is estimated at every time step of the integration process. Similar results are obtained for higher eccentricities.

(ii) In the next simulation experiment, we integrate the two-body problem in the cases of elongated orbits with the high eccentricities: $\epsilon = 0.6$, $\epsilon = 0.7$, and $\epsilon = 0.8$ and 10^3 time steps in order to explore the need of using high order variational schemes. For this reason, we test the long term behavior of two methods, the trigonometric and the Störmer–Verlet method [5], which is also variational, but of second order accuracy. For both methods, the results for the configuration q_k of the body's orbit are illustrated in Figure 4. These results demonstrate the excellent behavior of the higher accuracy method, even for

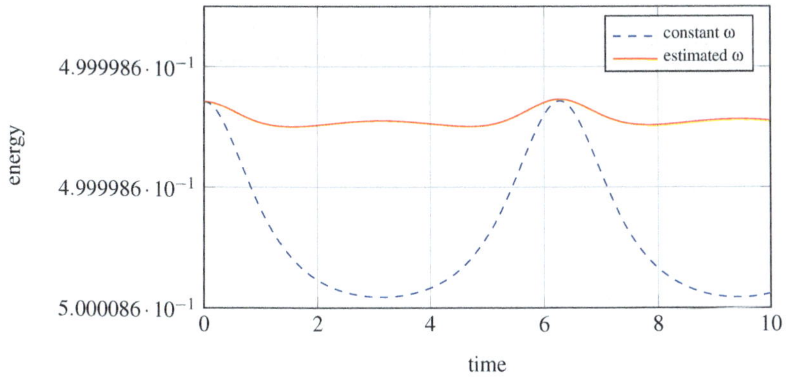

Fig. 3 Planar two-body problem with $\epsilon = 0.2$ using trigonometric interpolation for $S = 5$, $h = 0.01$. Evolution of the total energy for constant and estimated frequency using (22)

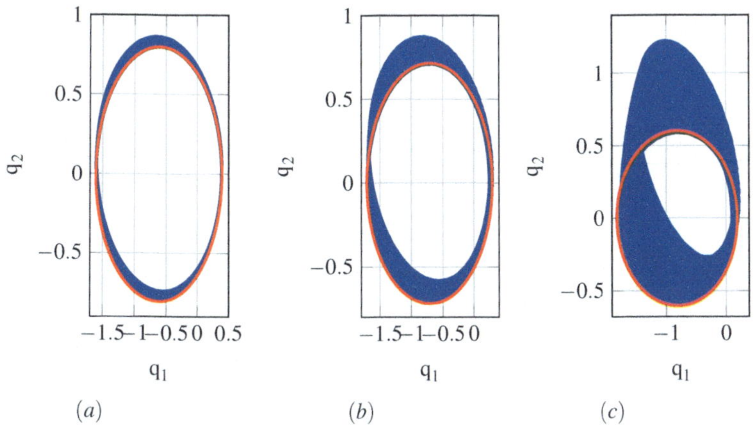

Fig. 4 Planar two-body problem for $h = 0.01$ for 10^3 steps for eccentricities (**a**) $\epsilon = 0.6$, (**b**) $\epsilon = 0.7$ and (**c**) $\epsilon = 0.8$. Long term behavior of the Störmer–Verlet method of [5] and the one that uses trigonometric interpolation with $S = 5$

orbits with extremely high eccentricity and large number of periods. On the other hand, the Störmer–Verlet method gives perturbed orbits even for small eccentricities, showing the necessity of employing higher order schemes.

3.3.2 The Modified Solar System

For multi-body systems, the advantages of choosing the parameter u via the frequency estimation of Equation (22) are illustrated by adopting the modified solar system with two planets [5]. For this system, which is described by the Lagrangian

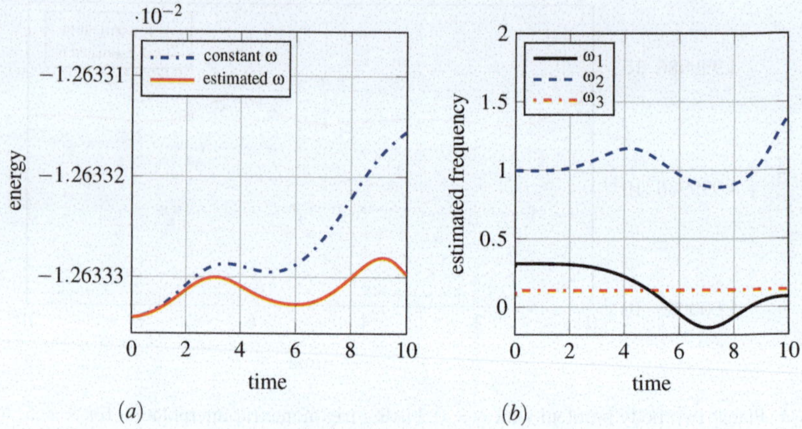

Fig. 5 Modified solar system with $h = 0.01$ and $S = 3$. (**a**) Total energy evolution using trigonometric interpolation for constant ω_i (blue line) and estimated ω_i at every time step (black line). (**b**) Estimated frequency for the three bodies of the modified solar system using the expression (22)

function (28) with $N = 3$, we choose $m_1 = 1$, $m_2 = m_3 = 10^{-2}$ and assume initial configurations and velocities given by

$$q_1 = (0, 0),\ q_2 = (1, 0),\ q_3 = (4, 0) \tag{31}$$

$$\dot{q}_1 = (0, 0),\ \dot{q}_2 = (0, 1),\ \dot{q}_3 = (0, 0.5). \tag{32}$$

The resulted motion of the two planets is nearly circular with periods equal to $T_1 \approx 2\pi$ and $T_2 \approx 14\pi$, respectively [5].

At first, we compare the results of (i) a variational integrator using trigonometric interpolation with constant frequencies, ω_i, $i = 1, 2, 3$, during the integration procedure, with those of (ii) a variational integrator derived by estimating the parameter u at every time step using Equation (22). In Figure 5a, we plot the total energy resulting from these two methods. The advantage of the second method at every step is obvious. In Figure 5b the evolution of the estimated angular velocities for each body (ω_1, ω_2, and ω_3, respectively) is shown.

For both numerical tests, the number of intermediate points is $S = 3$ while the time step is $h = T_1/365 = 2\pi$, i.e., equal to the period of the first planet.

4 Derivation of Time Adaptive Integrators Through the Geodesic Approach

When one studies the discrete Euler–Lagrange equation and geodesics, the main concern is how to find the equations of motion of a particle restricted to a particular curved surface (a sphere, a torus, etc.). As is well known, if a particle is constrained to move on a particular surface, it would follow the path of a geodesic on that surface. For example, on a sphere, it would follow a great circle in its motion. The more general problem related to this issue is how to model the trajectory of a particle constrained to move on a manifold M (or simply surface) in the discrete mechanics on the basis of geodesic approach. In general, to illustrate the algorithm performance on manifolds one, for example, may compute minimal geodesics, shortest paths on a sphere, a torus, etc. and also test the computation of equidistance curves, as well as shortest paths and geodesic distances on synthetic (complicated) objects.

In many physical applications, for the numerical solutions of the governing ordinary differential equations, the class of integration schemes based on adaptive time integrators perform remarkably well. The derivation of time adaptive integrators starts from the continuous Lagrangian formulation. Here, as an example, we consider physical problems described through the simple Lagrangian

$$L(x, \dot{x}) = \frac{1}{2}\dot{x}^2 - V(x), \qquad x \in \mathbb{R} \tag{33}$$

and the corresponding second order Euler–Lagrange differential equation

$$\ddot{x} = -\frac{\partial V}{\partial x}. \tag{34}$$

By choosing the initial conditions as $x_0 = x(0)$ and $\dot{x}_0 = \dot{x}(0)$, an expression of $x(t)$ can be determined and adopted for some time interval $t \in [0, T]$, as a solution of (34).

We then write down the generalized Lagrangian

$$\tilde{L} = \frac{1}{2}x'^2 + \frac{1}{2V}t'^2, \tag{35}$$

where, in order to disentangle from dots representing time derivatives, the primes denote differentiation with respect to some parameter λ [25] assuming that $t = t(\lambda)$ and, thus, $x = [t(\lambda)]$. For the latter Lagrangian the corresponding Euler–Lagrange equations and the relevant initial conditions take the form

$$x'' = -\frac{1}{2V^2}\frac{\partial V}{\partial x}t'^2, \quad x_0 = x(0), \quad x_0' = \dot{x}_0 t_0' \tag{36a}$$

$$t'' = \frac{1}{V}\frac{\partial V}{\partial x}t'x', \qquad t_0 = 0, \qquad t'(0) = \alpha V(x_0). \tag{36b}$$

It is worth mentioning that, even though $\tilde{L}$ depends upon V and couples the space and time variables in a non-trivial manner, the evolution equations for x depend only on $\partial V/\partial x$. Furthermore, we note that one could add on V any constant without changing the x-dynamics [24, 25].

We now consider two functions of the parameter λ, namely $\tilde{x}(\lambda)$ and $t(\lambda)$, that are further assumed to be solutions of equations (36) for some time interval $\lambda \in [0, \tilde{T}]$. For these solutions we can write $\tilde{x}(\lambda) = x(2t/\sqrt{\alpha})$ as long as both sides of (36) are explicitly defined, that is, as long as the solutions for x and $\tilde{x}$ differ only by an arbitrary constant. This constant, in essence, operates just as a time rescaling [24, 25].

In exploring for appropriate expressions for $\tilde{x}(\lambda)$ and $t(\lambda)$, we adopt the two Lagrangians

$$L_1 = \sqrt{\acute{x}^2 + f(x)\acute{t}^2}, \qquad L_2 = \frac{1}{2}\left(\acute{x}^2 + f(x)\acute{t}^2\right). \tag{37}$$

The action corresponding to L_1 is invariant under arbitrary reparametrization of λ, whereas the L_2 action is only affine reparametrization invariant. This leads to Euler–Lagrange equations corresponding to L_2 and hence they are affine time reparametrization invariants.

The Euler–Lagrange equations that come out of L_1 are

$$\frac{d}{d\lambda}\left(\frac{\acute{x}}{\sqrt{\acute{x}^2 + f(x)\acute{t}^2}}\right) = \frac{\acute{t}^2}{2\sqrt{\acute{x}^2 + f(x)\acute{t}^2}}\frac{\partial f}{\partial x} \tag{38a}$$

$$\frac{d}{d\lambda}\left(\frac{f(x)\acute{x}}{\sqrt{\acute{x}^2 + f(x)\acute{t}^2}}\right) = 0. \tag{38b}$$

The later equations are also reparametrization invariants with respect to λ, i.e., they are invariant under the replacements $\lambda = \lambda(\mu)$ and $d\lambda/d\mu \neq 0$. This means that, a solution of (38) defines a curve in the space (x, t). Furthermore, this solution gives us information on which curve does it belong, but it does not show us the exact point at that curve. The curve in question acts as a geodesic information for the system of equations as well as for its solution. The later equations are then considered to be evolution equations, which provide us with, not only the shape of the curve, but also with its parametrization [25].

5 Time Adaptive Exponential Variational Integrators

In this section we apply the steps followed in Section 2, in the Lagrangians L_1 and L_2 of equations (37). Using (13), for the length action given by L_1, the corresponding discrete Lagrangian reads [24, 29]

$$L_{1d}(q_k, q_{k+1}, h_k) = \sum_{j=0}^{S-1} h_k w^j \sqrt{\left(\acute{x}_k^j\right)^2 + f\left(x_k^j\right)\left(\acute{t}_k^j\right)^2}, \tag{39}$$

where the x_k^j are defined using (9) and $\acute{x}_k^j$, $\acute{t}_k^j$ using the expression [29]

$$\acute{q}_k^j = \frac{\partial q_k^j}{\partial \lambda} = \frac{\partial t}{\partial \lambda}\left(\dot{g}_1(t_k^j)q_k + \dot{g}_2(t_k^j)q_{k+1}\right) = \dot{g}_1(t_k^j)q_k + \dot{g}_2(t_k^j)q_{k+1}. \quad (40)$$

For the Lagrangian (39), the discrete Euler–Lagrange equations (6) give the discrete analogues of (38a) as

$$\begin{aligned}
&\sum_{j=0}^{S-1} w^j \frac{h_k}{2d_{k,k-1}}\Big[2\dot{g}_2(t_k^j)\left(\dot{g}_1(t_k^j)x_{k-1} + \dot{g}_2(t_k^j)x_k\right) \\
&\quad + \frac{\partial}{\partial x_k} f\left(g_1(t_k^j)x_{k-1} + g_2(t_k^j)x_k\right)\left(\dot{g}_1(t_k^j)t_{k-1} + \dot{g}_2(t_k^j)t_k\right)^2\Big] \\
&\quad + \sum_{j=0}^{S-1} w^j \frac{h_{k+1}}{2d_{k+1,k}}\Big[2\dot{g}_1(t_k^j)\left(\dot{g}_1(t_k^j)x_k + \dot{g}_2(t_k^j)x_{k+1}\right) \\
&\quad + \frac{\partial}{\partial x_k} f\left(g_1(t_k^j)x_k + g_2(t_k^j)x_{k+1}\right)\left(\dot{g}_1(t_k^j)t_k + \dot{g}_2(t_k^j)t_{k+1}\right)^2\Big] = 0, \quad (41)
\end{aligned}$$

and of (38b) as

$$\begin{aligned}
&\sum_{j=0}^{S-1} w^j \frac{h_k \dot{g}_2(t_k^j)}{d_{k,k-1}}\left[f\left(g_1(t_k^j)x_{k-1} + g_2(t_k^j)x_k\right)\left(\dot{g}_1(t_k^j)t_{k-1} + \dot{g}_2(t_k^j)t_k\right)^2\right] \\
&\quad + \sum_{j=0}^{S-1} w^j \frac{h_{k+1}\dot{g}_1(t_k^j)}{d_{k+1,k}}\left[f\left(g_1(t_k^j)x_k + g_2(t_k^j)x_{k+1}\right)\left(\dot{g}_1(t_k^j)t_k + \dot{g}_2(t_k^j)t_{k+1}\right)^2\right] = 0.
\end{aligned} \quad (42)$$

In the latter equation $d_{k+1,k}$ is given by [29]

$$\begin{aligned}
d_{k+1,k} = &\Big\{\left[\dot{g}_1(t_k^j)x_k + \dot{g}_2(t_k^j)x_{k+1}\right]^2 \\
&+ f\left(g_1(t_k^j)x_{k-1} + g_2(t_k^j)x_k\right)\left[\dot{g}_1(t_k^j)t_{k-1} + \dot{g}_2(t_k^j)t_k\right]^2\Big\}^{\frac{1}{2}} \quad (43)
\end{aligned}$$

and $d_{k,k-1}$ by

$$\begin{aligned}
d_{k,k-1} = &\Big\{\left[\dot{g}_1(t_{k_1}^j)x_{k-1} + \dot{g}_2(t_{k-1}^j)x_k\right]^2 + \\
&f\left(g_1(t_{k-1}^j)x_{k-2} + g_2(t_{k-1}^j)x_{k-1}\right)\left[\dot{g}_1(t_{k-1}^j)t_{k-2} + \dot{g}_2(t_{k-1}^j)t_{k-1}\right]^2\Big\}^{\frac{1}{2}}. \quad (44)
\end{aligned}$$

In accordance with the continuous formulation, Equations (41) and (42) are not independent. To solve the above system, we can choose arbitrary step sizes in either time t or space x direction and solve these equations for the x or t, respectively.

Once the discrete Euler–Lagrange equations (41) and (42) are solved, we get a sequence of points $(x_0, t_0), \ldots, (x_N, t_N)$, where $t_0, \ldots, t_N$ does not necessarily represent the physical time. Using this sequence of points, for the discrete Hamiltonian we may write [24, 25]

$$H_d(x_0, x_1, h_0) = -h_0 D_3 L_d(x_0, x_1, h_0) - L_d(q_0, q_1, h_0)\,. \tag{45}$$

Recalling that the energy of the system, expressed by its Hamiltonian, is the conjugate variable of the physical time, i.e.,

$$H_d(x_0, x_1, h_0) = H_d(x_1, x_2, h_1), \tag{46}$$

we can restore the physical time.

6 Numerical Results

In this section, we apply the above time adaptive exponential variational integrators in the following systems: (1) the simple pendulum and (2) the orbital two-body problems with extremely high eccentricities of their periodic orbits.

6.1 Harmonic Oscillator

The numerical scheme derived in Section 5 is tested below in the case of a simple pendulum described through the (approximate) Lagrangian

$$L(q, \dot{q}) = \frac{1}{2}\dot{q}^2 - \frac{1}{2}\omega^2 q^2, \tag{47}$$

leading to the equation of motion

$$\ddot{q} = -\omega^2 q. \tag{48}$$

Using the interpolation of (9), the discrete Lagrangian that provides the system's equations of motion takes the form

$$
\begin{aligned}
L_d(q_k, q_{k+1}) =& \frac{h}{2}\Bigg[\sum_{j=0}^{S-1} w^j \big(\dot{g}_1(t_k^j)q_k + \dot{g}_2(t_k^j)q_{k+1}\big)^2 \\
&\qquad -\omega^2 \sum_{j=0}^{S-1} w^j \big(g_1(t_k^j)q_k + g_2(t_k^j)q_{k+1}\big)^2\Bigg].
\end{aligned}
\tag{49}
$$

For the latter Lagrangian, following Section 5, the discrete Euler–Lagrange equations provide the two-step variational integrator [15, 28]

$$
q_{k+1} + \frac{\sum_{j=0}^{S-1} w^j \Big[\dot{g}_1(t_k^j)^2 + \dot{g}_2(t_k^j)^2 - \omega^2\big(g_1(t_k^j)^2 + g_2(t_k^j)^2\big)\Big]}{\sum_{j=0}^{S-1} w^j \Big[\dot{g}_1(t_k^j)\dot{g}_2(t_k^j) - \omega^2 g_1(t_k^j)g_2(t_k^j)\Big]} q_k + q_{k-1} = 0.
\tag{50}
$$

In order to demonstrate the benefits of the latter integrator on the numerical accuracy of the obtained methods, we compare its performance in the following two cases: (1) in the methods adopting constant time step (see Section 2) and (2) in the methods proposed in Section 5 (adaptive time step methods). We check the energy error at a specific integration time $t = 3$ (arbitrary taken) for five different frequencies $\omega \in \{1, 5, 10, 15, 20\}$ and initial conditions $(q_0, p_0) = (2, 1)$, see Figure 6. As can be seen both methods increase the energy error as the frequency of the problem increases. Secondly, even though for relatively small values of $\omega < 5$ both methods lead to energy error smaller than about 10^{-11}, for high frequency values, constant time step schemes lead to clearly larger energy error.

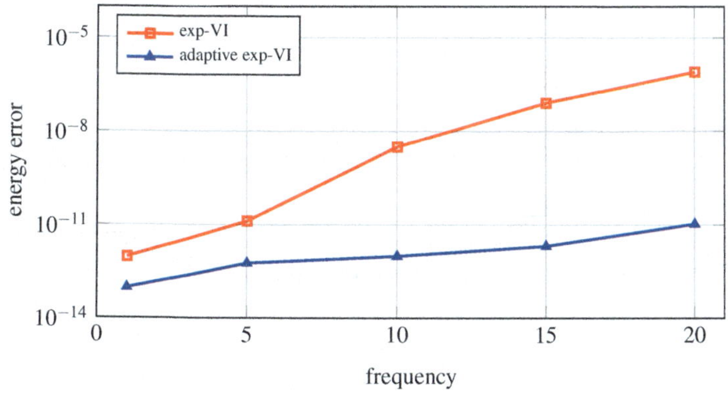

Fig. 6 Energy error for the harmonic oscillator using trigonometric interpolation (Section 2) versus the time adaptive one (Section 5) for the frequencies $\omega = 1$, $\omega = 5$, $\omega = 10$, $\omega = 15$, and $\omega = 20$

In computing the above results, both methods were considered to be third order methods, i.e., $S = 4$, while similar results have been obtained for other choices of S. We should also note that, we have chosen the same initial time step $h = 0.05$ for all results of Figure 6.

In order to illustrate how the specific choice of the time step affects significantly the computational cost, a prominent concept for our present work, we consider below some more complicated examples.

6.2 Orbits of the Two-Body Problem with Extremely High Eccentricities

To check the efficiency of the proposed technique, we consider again the Kepler's two-body problem discussed in Section 3.3.1 but now in the case of periodic orbits with remarkably high eccentricities (just below unity, $\epsilon = 0.99$). We compare the performance for long term integrations (10^6 periods) of the methods of Section 2 with that of the methods of Section 5. Figure 7 shows the exact orbit obtained with the method of Section 2 (solid line), the calculated points for the first period (points labeled with ∘) and the calculated points for the last period (points labeled with □). While most of the standard symplectic schemes (among them the one discussed in Section 2) fail to track the periodic orbit for such a high eccentricity, see [17], when adaptive time step is utilized, the obtained integrator is extremely stable, keeping the orbit close to the exact one. For this numerical experiment the observed energy error is oscillating around much smaller amplitude values (around 10^{-7}).

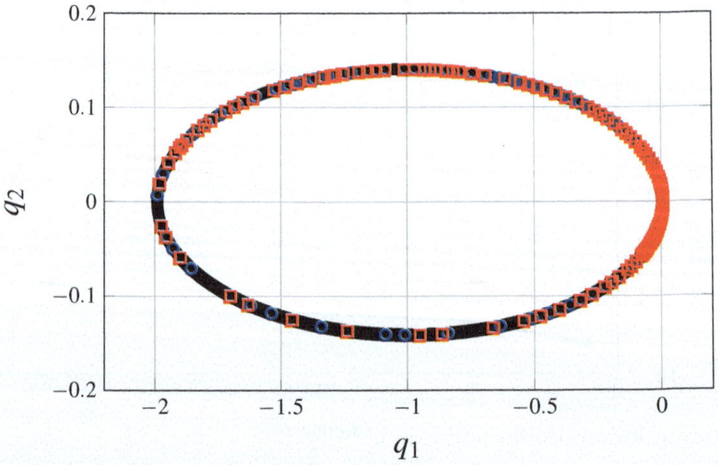

Fig. 7 Periodic orbits of the 2-body problem with eccentricity 0.99 for 10^6 periods. (1) Exact solution (solid line) and (2) calculated points for the first (open circle) and last period (open square) using the exponential variational integrators of Section 5

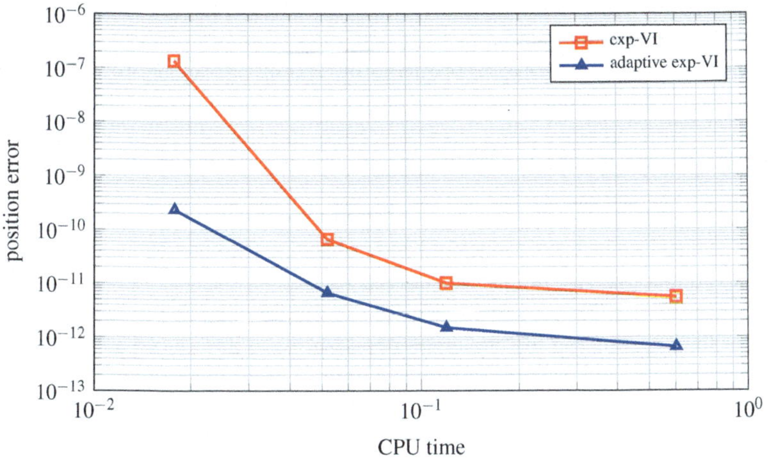

Fig. 8 Position error versus CPU time, for a 2-body problem moving in periodic orbit with eccentricity $\epsilon = 0.99$, taken at an arbitrary time $t = 3$ through numerical integration with an exponential variational integrator that uses constant time step (red line) compared to the one that uses adaptive time step (blue line)

As a final benefit of the proposed method, we explore the numerical convergence. To this aim, we choose as initial conditions the $(q_0, p_0) = (2, 2)$ and the time interval $[0, 3]$. We first calculate the global errors for the position $q(t)$ at $t = 3$ (arbitrary taken, but following [15, 31]) while using constant time steps $h \in \{0.01, 0.05, 0.1, 0.5, 1\}$. Figure 8 shows the resulting errors versus the computational time needed to obtain them (red line). It is obvious that smaller position errors are obtained for short time steps, which leads to longer computational time. When the adaptive time step exponential integrator of Section 5 is applied (blue line), the position error obtained is remarkably smaller. It should be mentioned that, in obtaining these results, we forced the proposed schemes to take the same computational time with that taken when constant time step is applied.

7 Conclusions

In this article, at first we reviewed briefly the concepts and relevant literature in discrete Lagrangian Mechanics and specifically in the discrete variational integrators developed the last decades for solving numerically the discrete Euler–Lagrangian equations. A great number of books and research articles published in this field is incorporated in the reference list below. These works have appreciably contributed in the development, enhancement, and enrichment of the topic of discrete variational integrators.

We have also presented a brief review on the extensions and refinements of the class of discrete exponential variational integrators with a particular regard to time adaptive exponential integrators. Focusing on systems of which the Lagrangian is of separable form, a methodology for deriving high order exponential variational integrators with adaptive time step has been developed. After the above, the following concluded remarks are extracted.

1. The procedure and methodology developed unfold the standard Euler–Lagrange character to its space-time manifold and translate it through the geodesic (shortest route) connecting two points on a curved surface.
2. In contrast to all the previous extensions, from the adaptive time step methods, rather than optimizing the choice of step sizing, we introduced an artificial time step parameter, and used the energy behavior in order to calculate the actual one.
3. Specifically, the proposed methods do not need to optimize the step size and, instead, one can employ the space-time geodesic formulation to generate an adaptive scheme that still preserves general underlying geometric structure properties of the system.

Finally, it is noteworthy to mention that, simulation tests showed that, this technique integrates efficiently stiff systems (like the two-body problem with very high eccentricity up to $\epsilon = 0.99$) while conserving at the same time all the benefits of the classical variational integrators.

Acknowledgement Dr. Odysseas Kosmas wishes to acknowledge the support of EPSRC via grant EP/N026136/1 "Geometric Mechanics of Solids."

Appendix

By denoting $\mathrm{sinc}(\xi) = \sin(\xi)/\xi$, special cases of the exponential integrators described using (16) can be obtained, i.e.,

- Gautschi type exponential integrators [11] for

$$\psi(\Omega h) = \mathrm{sinc}^2\left(\frac{\Omega h}{2}\right), \qquad \phi(\Omega h) = 1$$

- Deuflhard type exponential integrators [32] for

$$\psi(\Omega h) = \mathrm{sinc}(\Omega h), \qquad \phi(\Omega h) = 1$$

- García-Archilla et al. type exponential integrators [33] for

$$\psi(\Omega h) = \mathrm{sinc}^2(\Omega h), \qquad \phi(\Omega h) = \mathrm{sinc}(\Omega h)$$

Finally, in [5] a way to write the Störmer–Verlet algorithm as an exponential integrators is presenting.

References

1. B. Engquist, A. Fokas, E. Hairer, A. Iserles, *Highly Oscillatory Problems* (Cambridge University Press, Cambridge, 2009)
2. J. Wendlandt, J.E. Marsden, Mechanical integrators derived from a discrete variational principle. Physica D **106**, 223–246 (1997)
3. C. Kane, J.E. Marsden, M. Ortiz, Symplectic-energy-momentum preserving variational integrators. J. Math. Phys. **40**, 3353–3371 (2001)
4. J.E. Marsden, M. West, Discrete mechanics and variational integrators. Acta Numer. **10**, 357–514 (2001)
5. E. Hairer, C. Lubich, G. Wanner, Geometric numerical integration illustrated by the Störmer-Verlet method. Acta Numer. **12**, 399–450 (2003)
6. B. Leimkuhler, S. Reich, *Simulating Hamiltonian Dynamics*. Cambridge Monographs on Applied and Computational Mathematics (Cambridge University Press, Cambridge, 2004)
7. S. Ober-Blöbaum, Galerkin variational integrators and modified symplectic Runge–Kutta methods. IMA J. Numer. Anal. **37**, 375–406 (2017)
8. A. Lew, J.E. Marsden, M. Ortiz, M. West, Asynchronous variational integrators. Arch. Ration. Mech. Anal. **167**, 85–156 (2003)
9. L. Brusca, L. Nigro, A one-step method for direct integration of structural dynamic equations. Int. J. Numer. Methods Eng. **15**, 685–699 (1980)
10. H. Van de Vyver, A fourth-order symplectic exponentially fitted integrator. Comput. Phys. Commun. **174**, 255–262 (2006)
11. W. Gautschi, Numerical integration of ordinary differential equations based on trigonometric polynomials. Numer. Math. **3**, 381–397 (1961)
12. T. Lyche, Chebyshevian multistep methods for ordinary differential equations. Numer. Math. **19**, 65–75 (1972)
13. O.T. Kosmas, D.S. Vlachos, Phase-fitted discrete Lagrangian integrators. Comput. Phys. Commun. **181**, 562–568 (2010)
14. O.T. Kosmas, D.S. Leyendecker, Analysis of higher order phase fitted variational integrators. Adv. Comput. Math. **42**, 605–619 (2016)
15. O.T. Kosmas, S. Leyendecker, Variational integrators for orbital problems using frequency estimation. Adv. Comput. Math. (2018). https://doi.org/10.1007/s10444-018-9603-y
16. J.E. Marsden, G.W. Patrick, S. Shkoller, Multisymplectic geometry, variational integrators, and nonlinear PDEs. Commun. Math. Phys. **199**, 351–395 (1998)
17. E. Hairer, Variable time step integration with symplectic methods. Appl. Numer. Math. **25**, 219–227 (1997)
18. C. Kane, J.E. Marsden, M. Ortiz, Symplectic energy-momentum preserving variational integrators. J. Math. Phys. **40**, 3353 (1999)
19. M. Leok, J. Zhang, Discrete Hamiltonian variational integrators. IMA J. Numer. Anal. **31**, 1497–1532 (2011)
20. R.D. Skeel, Variable step size destabilizes the Störmer/leapfrog/Verlet method. BIT Numer. Math. **33**, 172–175 (1993)
21. M.P. Calvo, J.M. Sanz-Serna, The development of variable-step symplectic integrators, with application to the two-body problem. SIAM J. Sci. Comput. **14**, 936–952 (1993)
22. J.P. Wright, Numerical instability due to varying time steps in explicit wave propagation and mechanics calculations. J. Comput. Phys. **140**, 421–431 (1998)

23. S. Reich, Backward error analysis for numerical integrators. SIAM J. Numer. Anal. **36**, 1549–1570 (1999)
24. S. Nair, Time adaptive variational integrators: a space-time geodesic approach. Phys. D Nonlinear Phenom. **241**, 315–325 (2012)
25. O.T. Kosmas, D.S. Vlachos, A space-time geodesic approach for phase fitted variational integrators. J. Phys. Conf. Ser. **738**, 012133 (2016)
26. O.T. Kosmas, S. Leyendecker, Phase lag analysis of variational integrators using interpolation techniques. Proc. Appl. Math. Mech. **12**, 677–678 (2012)
27. O.T. Kosmas, Charged particle in an electromagnetic field using variational integrators. ICNAAM Numer. Anal. Appl. Math. **1389**, 1927 (2011)
28. O.T. Kosmas, D.S. Vlachos, Local path fitting: a new approach to variational integrators. J. Comput. Appl. Math. **236**, 2632–2642 (2012)
29. O.T. Kosmas, D. Papadopoulos, Multisymplectic structure of numerical methods derived using nonstandard finite difference schemes. J. Phys. Conf. Ser. **490**, 012205 (2014)
30. O.T. Kosmas, S. Leyendecker, Stability analysis of high order phase fitted variational integrators. Proc. WCCM XI - ECCM V - ECFD VI **1389**, 865–866 (2014)
31. A. Stern, E. Grinspun, Implicit-explicit integration of highly oscillatory problems. SIAM Multiscale Model. Simul. **7**, 1779–1794 (2009)
32. P. Deuflhard, A study of extrapolation methods based on multistep schemes without parasitic solutions. Z. Angew. Math. Phys. **30**, 177–189 (1979)
33. B. García-Archilla, M.J. Sanz-Serna, R.D. Skeel, Long-time-step methods for oscillatory differential equations. SIAM J. Sci. Comput. **20**, 930–963 (1999)

Disjoint Chorded Cycles in Graphs with High Ore-Degree

Alexandr Kostochka, Derrek Yager, and Gexin Yu

Abstract In 1963, Corrádi and Hajnal proved that for all $k \geq 1$, every graph with at least $3k$ vertices and minimum degree at least $2k$ has k vertex-disjoint chorded cycles. In 2010, Chiba, Fujita, Gao, and Li proved that for all $k \geq 1$, every graph with $|G| \geq 4k$ and minimum Ore-degree at least $6k - 1$ contains k (vertex-)disjoint chorded cycles. In 2016, Molla, Santana, and Yeager refined this to characterize all graphs with at least $4k$ vertices and minimum Ore-degree at least $6k - 2$ that do not have k disjoint chorded cycles. We further strengthen this to characterize the graphs with Ore-degree at least $6k - 3$ that do not have k disjoint chorded cycles.

Mathematics Subject Classification 05C35, 05C38, 05C70, 05D99

1 Introduction

For a graph G, we will use $V(G)$ and $E(G)$ to denote the vertex set and edge set, respectively, with $|G| = |V(G)|$ and $||G|| = |E(G)|$. Also, we will use $\delta(G)$ to denote the minimum degree of G. The *minimum Ore-degree* is

$$\sigma_2(G) = \min\{d_G(x) + d_G(y) : xy \notin E(G)\}.$$

A. Kostochka (✉)
University of Illinois at Urbana–Champaign, Urbana, IL, USA

Sobolev Institute of Mathematics, Novosibirsk, Russia
e-mail: kostochk@math.uiuc.edu

D. Yager
University of Illinois at Urbana–Champaign, Urbana, IL, USA
e-mail: yager2@illinois.edu

G. Yu
College of William & Mary, Williamsburg, VA, USA
e-mail: gyu@wm.edu

A. M. Raigorodskii, M. Th. Rassias (eds.), *Discrete Mathematics and Applications*, Springer Optimization and Its Applications 165,
https://doi.org/10.1007/978-3-030-55857-4_11

There are various results presenting sufficient conditions for a (multi)graph to have a given number of disjoint cycles. In this paper, by *disjoint* we mean *vertex-disjoint*. In 1963, Corrádi and Hajnal verified a conjecture of Erdős:

Theorem 1 (Corrádi–Hajnal [3]) *Let $n, k \geq 1$ be integers such that $n \geq 3k$. If $\delta(G) \geq 2k$, then G contains k disjoint cycles.*

Enomoto [5] and Wang [16] independently strengthened the result by replacing the minimum degree condition with a minimum Ore-degree condition.

Theorem 2 (Enomoto [5], Wang [16]) *Every graph G on $|G| \geq 3k$ vertices with $\sigma_2(G) \geq 4k - 1$ contains k vertex-disjoint cycles.*

Kierstead, Kostochka, and Yeager [13] described all extremal graphs for Theorem 1, and Kierstead, Kostochka, Molla, and Yeager [11] described all extremal graphs for Theorem 2. Other results in this vein, including the results on multigraphs with multiple edges and/or loops allowed the reader can find in [4, 10, 12].

A *chord* of a cycle H in a graph G is an edge $e \in E(G) \setminus E(H)$ with both endpoints in $V(H)$. Hence, a *chorded cycle* is a cycle with a chord. Several interesting extensions of the results on disjoint cycles to disjoint chorded cycles were obtained in the last decade, see, e.g., [1, 7–9, 15]. In particular, Finkel [6] proved the chorded cycle analog of Theorem 1.

Theorem 3 (Finkel [6]) *Every graph G on $|G| \geq 4k$ vertices with $\delta(G) \geq 3k$ contains k vertex-disjoint chorded cycles.*

This was then generalized by Chiba, Fujita, Gao, and Li [2]. In particular, the following appears as a corollary in their paper.

Theorem 4 (Chiba–Fujita–Gao–Li [2]) *Every graph G with at least 4k vertices and $\sigma_2(G) \geq 6k - 1$ contains a collection of k vertex-disjoint chorded cycles.*

More recently, Molla, Santana, and Yeager [14] characterized what occurs in the extremal case.

Theorem 5 (Molla–Santana–Yeager [14]) *For $k \geq 2$, let G be a graph with $n = |G| \geq 4k$ and $\sigma_2(G) \geq 6k - 2$. Then G does not contain k vertex-disjoint chorded cycles if and only if $G \in \{G_1(n,k), G_2(k)\}$, where $G_1(n,k) = K_{3k-1,n-3k+1}$ for $n \geq 6k - 2$ and $G_2(k) = K_{3k-2,3k-2,1}$ for $k \geq 2$ (Figure 1).*

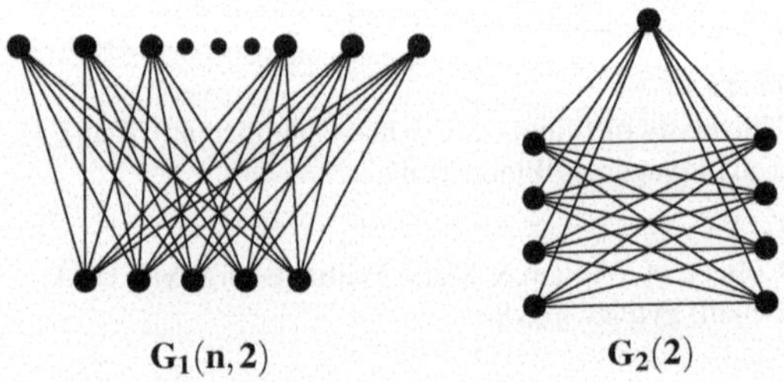

Fig. 1 Graphs for Theorem 5 with $k = 2$

Our main theorem sharpens Theorem 5: we relax the Ore-degree restriction, and show that this relaxation does not much affect the exceptional graphs. We also observe that further relaxation, even by 1, results in much more complicated exceptions.

1.1 Notation

For $n \in \mathbb{N}$, let $[n] = \{1, 2, 3, \ldots, n\}$. The neighborhood of $v \in V(G)$ in a subgraph $H \subseteq G$ is denoted by $N_H(v)$, and $d_H(v) = |N_H(v)|$. We will use $||S, T||$ to denote the number of edges from S to T regardless of whether they are vertex sets or subgraphs.

A cycle of length t will be called a *t-cycle*. For $H \subseteq G$, we will use $G[H]$ to denote the graph G induced by $V(H)$, i.e. $G[V(H)]$. Likewise, we will use = to denote graph isomorphism. Hence, we will frequently write $G[P] = K_4$ instead of $G[V(P)] \cong K_4$. Also, when there is no ambiguity, H^- will denote the isomorphism class of H with one removed edge and H^+ will denote the isomorphism class of H with a single added edge. We will make frequent use of the graphs K_4^- and C_5^+. Also, the *paw* is the 4-vertex graph obtained from a copy K of K_3 by adding one vertex adjacent to exactly one vertex in K, i.e. $K_{1,3}^+$.

Let K_n denote the complete graph on n vertices, and let $K_{n_1,n_2,\ldots,n_t}$ denote the complete t-partite graph with parts of size $n_1, n_2, \ldots, n_t$.

1.2 Main Result

We sharpen Theorem 5, by relaxing the Ore-degree constraint by 1. We prove that the new graphs with no k disjoint cycles differ very little from the ones in Theorem 5. Our exceptional graphs include $G_1(n,k) = K_{3k-1,n-3k+1}$, $G_2(k) = K_{3k-2,3k-2,1}$, and G_3 which is produced from a K_7 after removing the edges of a triangle T and adding a vertex whose neighborhood is $V(T)$ (see Figure 2). Now, if $G_2(k)$ has partite sets $\{v\}, A, B$, then we define $G_2^*(k) = G_2(k) - vx$ for any $x \in A \cup B$ and $G_2^{**}(k) = G_2(k) - vx - vy$ for any $x \in A$, $y \in B$. Our exceptional graphs also include $G_1^-(n,k)$, $G_2^*(k)$, and $G_2^{**}(k)$.

Theorem 6 *Let $k \geq 2$, G be an n-vertex graph with $n \geq 4k$, and $\sigma_2(G) \geq 6k-3$. Then G does not contain k vertex-disjoint chorded cycles if and only if*

$$G_1^-(n,k) \subseteq G \subseteq G_1(n,k),\ G_2^{**}(k) \subseteq G \subseteq G_2(k),\ \textit{or}\ G = G_3.$$

Observe that if we relax the bound by 1 more to $\sigma_2(G) \geq 6k-4$, then we run into significantly more exceptional graphs. For example, we would have to consider all graphs $G_1(n,k) - M$, where M is a matching or exactly two edges incident to the same vertex in the independent set of size $n-(3k-1)$. Or, we could have $G_2(k) - M$, where M is a matching, or M consists of up to $3k-1$ edges incident to the dominating vertex or M consists of two edges incident to the same vertex. Or,

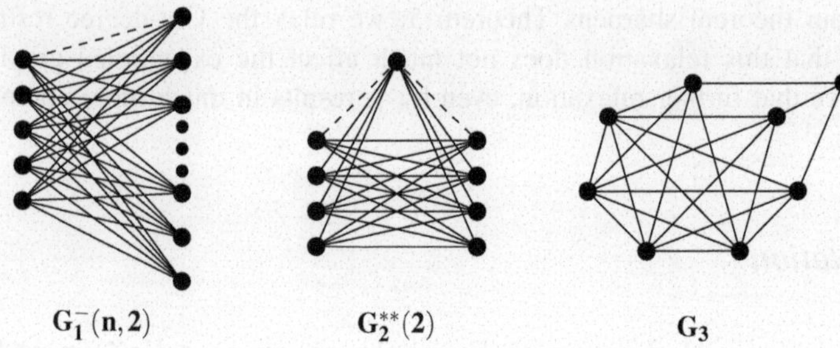

Fig. 2 Graphs for Theorem 6 with $k = 2$. Dashed lines indicate missing edges

we could even have G consisting of a K_3 and a K_7 with one vertex in each identified together, i.e. K_7 with a pendant triangle. This graph has $\sigma_2(G) = 6(2) - 4 = 8$, but every chorded cycle must exist in the K_7. This results in at most one chorded cycle. This list is by no means exhaustive, but it does show the significant jump in complexity beyond the few graphs in Theorem 6.

1.3 Outline

In Section 2, we set up our proof and present some known facts to aid in the main proof. We then break up our main proof into Sections 3–5. The strategy for our proof is to take an optimal collection of chorded cycles, and then consider the set R of the remaining vertices. In Section 3, we handle the case where $G[R]$ does not have a spanning cycle. Then, we address the case when $G[R]$ does have a spanning cycle but $k \geq 3$ in Section 4. Lastly, we handle the pesky case when $G[R]$ has a spanning cycle and $k = 2$ in Section 5. We tie all this together and conclude the proof in Section 6.

2 Setup and Preliminaries

2.1 Setup

Consider the minimum k such that Theorem 6 fails. For a given $n \geq 4k$, let G be an edge-maximal n-vertex graph satisfying the Ore condition

$$d(y) + d(z) \geq 6k - 3 \text{ for every distinct } y, z \in V(G) \text{ with } yz \notin E(G) \qquad (1)$$

such that G has at most $k - 1$ disjoint chorded cycles.

For such a graph G, let $\mathcal{F}$ be a collection of disjoint chorded cycles chosen by the following rules:

(O1) the number of chorded 4-cycles is maximum,
(O2) subject to the preceding, the number of K_4 is maximum,
(O3) subject to the preceding, the k–tuple $(F_1, \ldots, F_k)$ has $(|F_1|, \ldots, |F_k|)$ least lexicographically, where $|F_i| = \infty$ for nonexistent F_i.
(O4) subject to the preceding, the k-tuple $(||F_1||, \ldots, ||F_k)||$ is greatest lexicographically, where we use a similar convention of $||F_i|| = 0$ for nonexistent F_i.
(O5) subject to the preceding, the total number of chords in the cycles of $\mathscr{F}$ is maximum,
(O6) subject to the preceding, the length of a longest path P in $R := V(G) - V(\mathscr{F}) = \{v_1, \ldots, v_r\}$ is maximum. If $|P| = |R|$, then the number of Hamiltonian cycles in $G[R]$ is maximum, unless $|R| = 4$ in which case we maximize the copies of $K_{1,3}^+$ in $G[R]$,
(O7) subject to the preceding, $|E(G[R])|$ is maximum, and
(O8) subject to the preceding, $\sum_{v \in R} d_G(v)$ is maximum.

Let $R = \{v_1, \ldots, v_r\}$ and $P = v_1 v_2 \cdots v_p$. Let $\mathscr{F} = \{F_1, F_2, \ldots\}$, where F_i is the chorded cycle $x_{i,1} x_{i,2} \cdots x_{i,s_i}$. When the cycle F is unambiguous, we will use $\{x_1, x_2, \ldots\}$ to denote $V(F)$.

A vertex v is *low* if $d(v) \leq 3k - 2$ and *high* otherwise. By (1), the set of low vertices in G forms a clique. We will heavily use the following lemma from [14].

Lemma 7 ([14]) *Let G be a graph, $\mathscr{F}$ an optimal collection, $v \in R$, and $F \in \mathscr{F}$.*

(1) If $||v, F|| \geq 4$, then $||v, F|| = 4 = |F|$, and $F = K_4$.
(2) If $||v, F|| = 3$, then $|F| \in \{4, 5, 6\}$. Moreover,

(a) if $|F| = 4$, then F has a chord incident to the nonneighbor of v;
(b) if $|F| = 5$, then F is singly chorded, and the endpoints of the chord are disjoint from the neighbors of v;
(c) if $|F| = 6$, then F has three chords, with $F = K_{3,3}$, and $G[F + v] = K_{3,4}$.

Since we will often refer to Lemma 7, unless noted otherwise, we will use the following conventions:

- if $F = K_4$, then $V(F) = \{x_1, x_2, x_3, x_4\}$,
- if $F = K_4^-$, then $V(F) = \{x_1, x_2, x_3, x_4\}$ with missing edge $x_1 x_3$,
- if $F = C_5^+$, then $V(F) = \{x_1, x_2, x_3, x_4, x_5\}$ with unique chord $x_3 x_5$, and
- if $F = K_{3,3}$, then $V(F) = \{x_1, x_2, x_3, x_4, x_5, x_6\}$ with parts $A = \{x_1, x_3, x_5\}$ and $B = \{x_2, x_4, x_6\}$.

2.2 Preliminaries

In view of Lemma 7, the following observations will be useful.

Lemma 8 *If $k \geq 2$ and $F = K_{3,3}$, then*

(i) for every $v \in R$*, either* $N(v) \cap V(F) \subseteq A$ *or* $N(v) \cap V(F) \subseteq B$*, and*
(ii) for every $u, v \in R$ *with* $uv \in E(G)$*, if there is* $x \in N(v) \cap N(u) \cap V(F)$*, then* $N(v) \cap V(F) = N(u) \cap V(F) = \{x\}$.

Lemma 9 *For all* $x \in V(F)$*,* $d_F(x) \leq 3$.

An easy observation is that

If a vertex v in a graph G has 3 *neighbors on a path* P' *disjoint from* v*, then* $G[P' + v]$ *has a chorded cycle.*

(2)

This yields:

Lemma 10 *For every path* $P' \subseteq G[R]$ *and a vertex* $v \in R - V(P')$*,* $||v, P'|| \leq 2$*. In particular, if* $v_1 v_2 \dots v_s$ *is a path in* $G[R]$*, then* $||v_1, \{v_2, \dots, v_s\}|| \leq 2$.

We first prove some lemmas that apply to all k.

Lemma 11 *For* $k \geq 2$*,* $|R| \geq 4$.

Proof Suppose that $|R| \leq 3$ (possibly, $R = \emptyset$). Let $F_t = x_1 x_2 \dots x_s x_1$ be the last existing chorded cycle in the k-tuple from Rule (O3) and $|F_t| = s$. Since $t < k$ and $n \geq 4k$, $s \geq 5$.

Claim 11.1 For all $1 \leq i \leq t-1$, $||F_t, F_i|| \leq 3s$.

Proof Suppose that for some $i < t$, $||F_t, F_i|| \geq 3s + 1$. Let F_i be the chorded cycle $y_1 \dots y_{s_i} y_1$. Then there is an $x_k \in V(F_t)$ with $||x_k, F_i|| \geq 4$.

If $s_i \geq 5$, then there is $1 \leq j \leq s_i - 1$ such that x_k has at least 3 neighbors on a path in $F_i - y_j - y_{j+1}$. In this case, $G[F_i - y_j - y_{j+1} + x_k]$ contains a chorded cycle F_i' shorter than F_i, and hence the family $\mathscr{F}' = \{F_1, \dots, F_{i-1}, F_i'\}$ is better than $\mathscr{F}$ by (O3), a contradiction. Thus, we only need to consider the case $s_i = 4$. In this case, $V(F_i) \subseteq N(x_k)$.

If $F_i = K_4^-$, then there is $y \in F_i$ such that $G[F_i - y + x_k] = K_4$. Since $s \geq 5$, this means that the family $\mathscr{F}' = \mathscr{F} - F_i - F_t + (F_i - y + x_k)$ has more K_4 than $\mathscr{F}$, contradicting (O2). Thus, $F_i = K_4$.

If there exists a vertex $y \in V(F_i)$ with $||y, F_t|| \geq 5$, then $G[F_t - x_{k-1} - x_k + y]$ contains a chorded cycle and $G[F_i - y + x_k] = K_4$, contradicting (O3). Thus,

$$3s + 1 \leq ||F_i, F_t|| \leq 4|F_i| = 16. \tag{3}$$

This means $s = 5$ and each $y \in V(F_i)$ has exactly 4 neighbors in F_t. So, if any $y \in F_i$ is not adjacent to x_{k+1}, then $G[F_i - y + x_k] = K_4$ and $G[F_t - x_k - x_{k+1} + y]$ contains a chorded cycle F_t' that is shorter than F_t. This contradicts (O3). Thus each $y \in V(F_i)$ is adjacent to x_{k+1}. Considering x_{k+1} in place of x_k, we get that each $y \in F_i$ is adjacent to x_{k+2}, and so on. Then each $y \in F_i$ is adjacent to each $x \in F_t$, contradicting (3). This proves the claim.

Claim 11.2 $s = 5$.

Proof Suppose that $s \geq 6$. First, recall that for all $v \in V(F_t)$, $d_{F_t}(v) \leq 3$ by Lemma 9. Then, by the previous claim,

$$\sum_{v \in V(F_t)} d_{G-R}(v) \leq 3s(t-1) + 3s = 3st \leq 3s(k-1).$$

Also, since $s \geq 6$, the biggest clique of F_t is size at most 3. Since low vertices appear in a clique, each low vertex of F_t can be paired with a nonadjacent high vertex in F_t, which implies that

$$\sum_{v \in V(F_t)} d_G(v) \geq (6k-3) \cdot \frac{s}{2}. \tag{4}$$

Hence,

$$||V(F_t), R|| \geq \left(3k - \frac{3}{2}\right) s - 3(k-1)s = \frac{3}{2}s \geq 9.$$

By the assumption $|R| \leq 3$, if $||V(F_t), R|| \geq 10$, then some vertex in R has at least 4 neighbors in F_t which implies $F_t = K_4$ by Lemma 7, a contradiction to $s \geq 6$. Therefore,

$$||F_t, R|| = 9, s = 6, \text{ and } F_t = K_{3,3}. \tag{5}$$

This implies that F_t has at most 2 low vertices, each of which we can pair with a nonadjacent high vertex, and there will remain at least two unpaired high vertices. So, we can improve (4) to

$$\sum_{v \in V(F_t)} d_G(v) \geq 2(6k-3) + 2(3k-1) = 18k - 8,$$

and so $||F_t, R|| \geq (18k - 8) - 3(k-1) \cdot 6 = 10$, contradicting (5).

Claim 11.3 F_t has at least 2 low vertices.

Proof Suppose that F_t has at most one low vertex, and a has the smallest among the vertices in F_t. Let a' be a nonneighbor of a in F_t. Then all vertices in $F_t - a - a'$ are high vertices, and

$$\sum_{v \in V(F_t)} d_G(v) \geq (d_G(a) + d_G(a')) + 3(3k-1) \leq (6k-3) + 9k - 3 = 15k - 6.$$

Also, by Claim 11.1,

$$\sum_{v \in V(F_t)} d_{G-R}(v) \leq 3st \leq 15t \leq 15(k-1).g$$

So, $||F_t, R|| \geq (15k-6)-(15k-15) = 9$. Since $|F_t| = 5$, if $v \in R$ has $||v, F_t|| \geq 4$, then $G[F_t+v]$ contains a shorter chorded cycle, a contradiction. Then by pigeonhole and $|R| \leq 3$, we have $|R| = 3$, say $R = \{z_1, z_2, z_3\}$, and each $z_i \in R$ has exactly 3 neighbors in F_t. By Lemma 7, if we consider F_t as a 5-cycle $x_1x_2x_3x_4x_5x_1$ with the unique chord x_3x_5, then $N(z_1) \cap N(z_2) \cap V(F_t) = \{x_1, x_2, x_4\}$. Hence $G[R \cup F_t]$ has a 4-cycle $z_1x_1z_2x_2z_1$ with chord x_1x_2, contradicting (O1).

Now, if we were to have two chorded 5-cycles $F, F' \in \mathscr{F}$, then we can choose $x, y \in V(F)$ and $u, v \in V(F')$. Then, $G[\{x, y, u, v\}] = K_4$ since the set of low vertices forms a clique. This contradicts (O1). Thus F_t is the only 5-cycle in $\mathscr{F}$, and hence $|V(\mathscr{F})| = 4(t-1)+5 = 4t+1 \leq 4k-3$. Since $n \geq 4k$, we have $|R| \geq 4$ unless $n = 4k$ and $\mathscr{F}$ consists of $(k-2)$ 4-cycles and one 5-cycle.

We now handle the case $n = 4k$. Since K_{4k} contains k disjoint K_4, G has a non-edge xy. Since G is an edge-maximal counterexample, $G + xy$ contains k disjoint chorded cycles $F_1, \ldots, F_k$. Since $n = 4k$, these are all 4-cycles. By the choice of G, after deleting xy from $G + xy$, we ruin one of the F_j and thus produce $|R| = 4$.

Lemma 12 *If $k = 2$, then $\delta(G) \geq 3$.*

Proof Let $v \in V(G)$ be a vertex with $d_G(v) = \delta(G)$ and $G' := G - N[v]$. If $d_G(v) \leq 1$, then there are $n-2$ nonneighbors, all of degree at least 8. Then G' is a graph on $n-2$ vertices with $\delta(G') \geq 7$; in particular, $|G'| \geq 8$. So by Theorem 3, G' has 2 disjoint chorded cycles.

Now, suppose v is a vertex with $d_G(v) = 2$, say $N(v) = \{v_1, v_2\}$. For each $u \in V(G')$, $uv \notin E(G)$ so $d_G(u) \geq (6k-3)-2 = 7$ and hence $d_{G'}(u) \geq 5$.

Case 1 $v_1v_2 \in E(G)$. If there exists $u \in N_G(v_1) \cap N_G(v_2) - v$, then uv_1vv_2u is a 4-cycle with chord v_1v_2. Since $\delta(G'-u) \geq 5-1 = 4$, Theorem 3 provides a second chorded cycle unless $|G'-u| = n-4 < 4$, which is not the case. Therefore, we have two disjoint chorded cycles.

Hence, for every $u \in V(G')$, $||u, \{v, v_1, v_2\}|| \leq 1$ and so $\delta(G') \geq 6$. Again, Theorem 3 provides that G' has two disjoint chorded cycles unless $|G'| < 8$. But then, $G' \cong K_7$. Thus, $d_{G'}(u) = 6$ but $d_G(u) \geq 7$ for each $u \in V(G')$. Then, $||\{v_1, v_2\}, G'|| \geq 7$ and so $||v_i, G'|| \geq 4$ for some $i \in \{1, 2\}$. Hence, for each $u_1, u_2, u_3 \in N_{G'}(v_i)$, $G[\{v_iu_1u_2u_3v_i\}] = K+4$ and $G' - \{u_1, u_2, u_3\} = K_4$, a contradiction.

Case 2 $v_1v_2 \notin E(G)$. By symmetry, we may assume $d_G(v_1) \geq d_G(v_2)$ so that the Ore condition (1) yields $d_G(v_1) \geq 5$. Consider $G'' := G - v - v_2$. Since v is not adjacent to the vertices in $V(G'') - v_1$,

$$\text{for all } u \in V(G'' - v_1),\ d_G(u) \geq 7 \text{ and hence, } d_{G''}(u) \geq 6. \tag{6}$$

Since $d_{G''}(v_1) \geq 4$, $\sigma_2(G'') \geq 10$ and so by Theorem 5, either $n'' := |G''| = 7$, or $G'' = G_1(n'', 2)$, or $G'' = G_2(2)$.

Subcase 2.1 $n'' = 7$. Then $G'' = K_7$ and since each $u \in V(G'') - v_1$ has $d_G(u) \geq 7$, $uv_2 \in E(G)$ for all such u. Hence, $\delta(G - v) \geq 6$ and $|V(G) - v| = 8$ so that $G - v$ has 2 disjoint chorded cycles by Theorem 3.

Subcase 2.2 $G'' = G_1(n'', 2)$. Call the partite sets A, B with $|A| = 5$ and $|B| = n'' - 5$. For each $u \in B$, $d_{G''}(u) = 5$ so $d_G(u) \leq 6$. For each $u \in A$, $d_G(u) = d_{G''}(u) + ||u, \{v, v_2\}|| \leq n'' - 5 + 1$. By the Ore condition, we have $n'' \geq 9$ so that there exists $u \in B - v_1$. Then, we contradict the Ore condition of G since $uv \notin E(G)$ but $d_G(u) + d_G(v) \leq 6 + 2 = 8 < 9$.

Subcase 2.3 $G'' = G_2(2)$. Then, only one vertex in $u \in V(G'')$ has $d_{G''}(u) \geq 6$ and so u is the only vertex in G with degree at least 7 since $d_G(v), d_G(v_2) \leq 6$. This contradicts (6).

Lemma 13 *If $k = 2$, then G is 3-connected.*

Proof If G is disconnected, then by Lemma 12 we can use Theorem 4 component-wise to yield two disjoint chorded cycles, a contradiction.

Suppose G has a cut-vertex x. Let Y be a component of $G - x$ disjoint from $\mathscr{F}$. By Lemma 12, $\delta(G) \geq 3$ so that $\delta(G[Y]) \geq 2$. But the set $S := \{v \in Y : d_Y(v) = 2\}$ induces a clique since for all $s, t \in S$, $d_G(s) + d_G(t) \leq 6$. Hence, $\sigma_2(G[Y]) \geq 5$ and so $G[Y]$ contains a chorded cycle by Theorem 4, so we have 2 disjoint chorded cycles, a contradiction.

Now, suppose that there exists a separating set S with $|S| = 2$. Say $S = \{u, v\}$ and $G - S$ has components A and B.

Case 1 For all $w \in V(G) - S$, $d_G(w) \geq 4$. Let $G_A = G[A + u]$ and $G_B = G[B + v]$. If $d_{G_A}(u) \geq 2$, then $\sigma_2(G_A) \geq 5$, otherwise $\sigma_2(G_A - u) \geq 5$. In both cases, G_A contains a chorded cycle. Similarly, G_B has a chorded cycle, a contradiction.

Case 2 There exists a vertex $y \in V - S$ such that $d_G(y) = 3$, say $y \in A$. By (1),

$$d_G(w) \geq 6 \quad \text{for all} \quad w \in B. \tag{7}$$

Let P' be a shortest u, v–path in $G[B \cup S]$.

Subcase 04 For some $w \in B - V(P')$, $||w, P'|| \geq 4$. Then $||w, P' - v|| \geq 3$ and so $G[P' - v + w] \subseteq G[B + u]$ contains a chorded cycle. Take a maximal path $Q := v_0 v_1 \cdots v_s$ in $G[A + v]$ with $v_0 = v$ and $v_i \in A$ for $i > 0$. If $d_{G[A+v]}(v_s) \geq 3$, then $||v_s, Q|| \geq 3$ and so $G[Q]$ contains a chorded cycle. Otherwise, by Lemma 12, $d_{G[A+v]}(v_s) = 2$ and $d_G(v_s) = 3$. Since Q is maximal, $N_{G[A+v]}(v_s) = \{v_{s-1}, v_i\}$ for some $0 \leq i \leq s - 2$ and $\{u, v\}$ being a separating set means that $N_G(v_s) = \{v_{s-1}, v_i, u\}$.

Consider v_{i+1}. Path $v_0 v_1 \cdots v_i v_s v_{s-1} \cdots v_{i+1}$ has the same length as Q, so

$d_{G[A+v]}(v_{i+1}) = 2$ and $d_G(v_{i+1}) = 3$. Hence by (1), $v_{i+1}v_s \in E(G)$ and so $i+1 = s-1$. Also, $N_G(v_{s-1}) = \{v_{s-2}, v_s, u\}$ so that $uv_{s-1}v_{s-2}v_su$ is a 4-cycle with chord $v_{s-1}v_s$. By the subcase, $||w, P' - u|| \geq 3$ so that $G[P' - u + w] \subseteq G[B + v]$ contains a chorded cycle. Thus we found 2 disjoint chorded cycles. This completes the subcase.

Subcase 05 For all $w \in B - P'$, $||w, P'|| \leq 3$. First, $B - P' \neq \emptyset$ as otherwise $x \in V(P')$ and $xy \notin E(G)$ implies that $d_G(x) \geq 9 - d_G(y) = 6$ so that P' has a chord, contradicting it being a shortest u, v–path. Then, $\delta(G[B - P']) \geq 6 - 3 = 3$ and so $G[B - P']$ contains a chorded cycle by Theorem 4. Now consider $G[A + P'] \supseteq G[A + S]$. All maximum paths Q' in $G[A + P']$ have endpoints of degree 2 in $G[A + P']$ or else we have our chorded cycle. Therefore, by Lemma 12 all such maximum paths are u, v–paths. Consider one such path $Q' := v'_0v'_1 \cdots v'_t$ where $u = v'_0, v = v'_t$. Note, $t = 1$ implies A is already disconnected in G and $t = 2$ implies $N(v'_1) = \{u, v\}$, contradicting $\delta(G) \geq 3$, so we assume $t \geq 3$. By Lemma 12, $d_{G[A+P']}(v'_{t-1}) \geq 3$.

If xv'_{t-1} for some $x \in A - Q'$, then $v'_0 \cdots v'_{t-1}x$ is a part of a different maximum path, hence must have v as an endpoint, i.e. there exists $0 < i < t - 1$ and vertices $x_{i+1}, \cdots, x_{t-2} \in A - Q'$, where $x = x_{t-2}$ such that $Q'' := v'_0 \cdots v'_ix_{i+1} \cdots x_{t-2}v'_{t-1}v'_t$ is maximum. In other words, $G' := G[P' \cup Q' \cup \{x_{i+1}, \ldots, x_{t-2}\}]$ is a Θ-graph with 3-vertices v'_i and v'_{t-1}, and so $x = x_{t-2}$ cannot have a third neighbor in G'. So, there exists $y \in A - V(Q') - V(Q'')$ such that $xy \in E(G)$. But then $v'_0 \cdots v'_{t-1}xy$ is longer than Q' contradicting Q' being maximum.

Thus, $N_G(v'_{t-1}) \subseteq Q'$ and so $v'_jv'_{t-1}$ for some $0 \leq j \leq t - 3$. But then we have a cycle $vP'uQ'v$ with chord $v'_jv'_{t-1}$. Hence, each of $G[A + P']$ and $G[B - P']$ contains a chorded cycle, thus providing 2 disjoint chorded cycles, a contradiction.

Lemma 14 *Suppose $k \geq 2$, $R = \{v_1, v_2, v_3, v_4\}$, $E(G[R]) = \{v_1v_2, v_2v_3, v_3v_4, v_4v_1\}$, and $F \in \mathscr{F}$ is such that $F = K_4$. If $||R, F|| \geq 11$, then*

(a) $||R, F|| = 11$ *and*
(b) $\sum_{j=1}^{4} d_G(v_j) \geq 12k - 5$.

Proof Suppose $||R, F|| \geq 11$. We proceed in a series of claims. First, we show that

$$||x_i, R|| \leq 3 \quad \text{for each } 1 \leq i \leq 4. \tag{8}$$

Indeed, assume $x_1v_j \in E(G)$ for each $1 \leq j \leq 4$. Since $||R, F|| \geq 11$, we may assume $||v_1, F|| \geq \lceil 11/4 \rceil = 3$. Then $G[F - x_1 + v_1] \supseteq K_4^-$ and $G[R - v_1 + x_1] = K_4^-$, a contradiction to Rule (O1). This proves (8).

By (8), we may assume that $||x_i, R|| = 3$ for $1 \leq i \leq 3$ and $||x_4, R|| \geq 2$. For $1 \leq i \leq 3$, let $j(i)$ be such that $v_{j(i)}x_i \notin E(G)$. Our next claim is:

$$||v_{j(i)}, F|| \leq 1 \text{ for each } 1 \leq i \leq 3. \tag{9}$$

Indeed, suppose that, for example, $j(1) = 4$ and $||v_4, F|| \geq 2$. Since $v_4x_1 \notin E(G)$, this means $G[F - x_1 + v_4]$ contains K_4^-. But by the choice of x_1, also $G[R - v_4 + x_1] = K_4^-$, a contradiction to Rule (O1). This proves (9).

If there exist i_1, i_2 such that $j(i_1) \neq j(i_2)$, then by (9), $||R, F|| = ||\{v_{j(i_1)}, v_{j(i_2)}\}, F|| + ||R - \{v_{j(i_1)}, v_{j(i_2)}\}, F|| \leq 2 \cdot 1 + 2 \cdot 4 = 10$, a contradiction. Thus, we may assume

$$j(1) = j(2) = j(3) = 4. \tag{10}$$

We can now say more:

$$x_4v_1, x_4v_3 \notin E(G). \tag{11}$$

If (11) does not hold, then by symmetry we may assume $x_4v_1 \in E(G)$. In this case, by (10), $G[F - x_1 + v_1] = K_4$ and $G[R - v_1 + x_1] = K_{1,3}^+$. But a $K_{1,3}^+$ is "better" than a C_4 by (O6), a contradiction.

Since $||R, F|| \geq 11$, (8) and (11) together imply $||R, F|| = 11$ and $N(x_4) \cap R = \{v_2, v_4\}$. So, if one of v_1 or v_3, say v_3, is low, then x_4 is high, and replacing F with $G[F - x_4 + v_3]$ yields another K_4, and $R - v_3 + x_4$ again induces C_4. But it contradicts Rule (O8), since $d(x_4) > d(v_3)$. Hence, both v_1 and v_3 are high, and $\sum_{j=1}^{4} d_G(v_j) = d_G(v_2) + d_G(v_4) + 2(3k - 1) \geq (6k - 3) + 2(3k - 1) = 12k - 5$. This proves the lemma.

Lemma 15 *For $k \geq 3$, $K_4 \in \mathscr{F}$ or $\{K_4^-, K_4^-\} \subseteq \mathscr{F}$.*

Proof Suppose that $K_4 \notin \mathscr{F}$ and $\{K_4^-, K_4^-\} \nsubseteq \mathscr{F}$. Let $G' = G - V(F_1)$. We first show that

$$\sigma(G') \geq 6(k - 1) - 3. \tag{12}$$

Otherwise, there exist $x, y \in V(G')$ such that $d_{G'}(x) + d_{G'}(y) \leq 6(k - 1) - 4 = 6k - 10$. Then, $||\{x, y\}, F_1|| \geq 7$ and so by symmetry we may assume $||x, F_1|| \geq 4$. If $|F_1| \geq 5$, then $G[F_1 + x]$ contains a chorded cycle shorter than F_1, contradicting (O3). Otherwise, $F_1 = K_4^-$ and $G[F_1 + x] \supseteq K_4$, contradicting (O2). This proves (12).

By the minimality of G, either $n - |F_1| < 4(k - 1)$ or G' has $k - 2$ disjoint cycles, and G' is an excluded graph. Note that $G' \neq G_3$ since G_3 contains K_4. If $||F_1, x|| \leq 2$ for all $x \in V(G')$, then $\sigma(G') \geq 6k - 3 - 4 = 6(k - 1) - 1$. By Theorem 4, G' has $k - 1$ disjoint cycles, so G has k disjoint chorded cycles, a contradiction. Then, either $n - |F_1| < 4(k - 1)$, or $G' \supseteq G_1^-(n - |F_1|, k - 1)$, or $G' \supseteq G_2^{**}(k - 1)$, and $||x, F_1|| = 3$ for some $x \in V(G')$ so that by Lemma 7, we have the following cases:

Case 1 $F_1 = K_4^-$. In this case, $n - |F_1| \geq 4(k - 1)$, so G' is an excluded graph. Since $G' \supseteq G_2^{**}(k-1)$ contains K_4^-, we have $G' \supseteq G_1^-(n-4, k-1)$ with partite sets A and B, where $|A| = (n-4)-3(k-1)+1 = n-3k$ and $|B| = 3(k-1)-1 = 3k-4$.

There exists a vertex $y_0 \in A$ such that for each $y \in A - y_0$, $||y, F_1|| = 3$, $N_{G'}(y) = B$, and $||y_0, F_1|| \geq 2$ or else we violate (12). For the same reason, $|A| \geq |B|$ and so $|A| \geq |B| \geq 5$. Note, for each $y \in A - y_0$, we have $x_1, x_3 \in N(y)$. So, take some $y_1 \in A - y_0$ and suppose by symmetry that $N(y_1) \cap F_1 = \{x_1, x_2, x_3\}$. We have the chorded cycle F_1' comprised of $y_1x_2x_4x_3y_1$ with chord x_2x_3. Then, the remaining graph $G'' := G - \{y_1, x_2, x_3, x_4\}$ contains $K_{|A|-1,|B|+1} - x_1y_0$ since x_1 is adjacent to all of $A - y_0$. G'' has $k-1$ disjoint chorded cycles unless $|A - y_1| = n - 3k < 3k - 3$. But then, (12) implies $n - 3k = 3k - 4$ and so A and B are interchangeable. Namely, there exists $z_0 \in B$ such that each $z \in B - z_0$ has $||z, F_1|| = 3$ and $N_{G'}(z) = A$. Moreover, if any $y \in A - y_0$ and $z \in B - z_0$ share the same neighborhood in F_1, say $N(y) \cap N(z) \cap V(F_1) = \{x_1, x_3, x_2\}$, then $G[\{x_1, x_2, y, z\}] = K_4$, a contradiction. So, we can take any $y_1, y_2 \in A - y_0$ and $z_1, z_2 \in B - z_0$ where $N(y_1) \cap V(F_1) = N(y_2) \cap V(F_1) = \{x_1, x_3, x_2\}$ and $N(z_1) \cap V(F_1) = N(z_2) \cap V(F_1) = \{x_1, x_3, x_4\}$ to get 4-cycles $y_1z_2x_1x_2y_1$ and $y_2z_1x_4x_3y_2$ with chords y_1x_1 and z_1x_3, respectively, a contradiction.

Case 2 $F_1 = C_5^+$ with unique chord x_3x_5. If $n - 5 < 4(k - 1)$, then $n \geq 4k$ implies $n = 4k$. Now, $\{x_1, x_3\}$ and $\{x_2, x_4\}$ form independent sets, so for $W := \{x_1, x_2, x_3, x_4\}$, we have $||W, V(G')|| \geq 2(6k - 3 - 5) = 12k - 16$. For each $y \in V(G')$, $||y, F_1|| \leq 3$ and so we have $||W, V(G')|| \leq 12k - 15$. Hence, there exists a $y_0 \in V(G')$ such that for each $y \in V(G') - y_0$, $||y, W|| = 3$. For such a y, if $yx_4 \notin E(G)$, then $G[\{y, x_1, x_2, x_3\}] = K_4$, a contradiction.

Thus, $n-5 < 4(k-1)$ and hence G' is an excluded graph. Whether $G' \supseteq G_1^-(n-5, k-1)$ or $G' \supseteq G_2^{**}(k-1)$, there exists an independent set A of size $3k - 5 \geq 3$. Moreover, there is a vertex $y_0 \in A$ such that for all vertices $y_1, y_2 \in A - y_0$, we have $d_{G'}(y_1), d_{G'}(y_2) = 3k - 4$, $y_1y_2 \notin E(G)$, and so $||y_1, F_1||, ||y_2, F_1|| = 3$ by the Ore condition. Hence, $N(y_1) \cap N(y_2) \supseteq \{x_1, x_2\}$ in which case we have the 4-cycle $y_1x_1y_2x_2y_1$ with chord x_1x_2, contradicting (O1).

Case 3 $F_1 = K_{3,3}$ with partite sets $A' = \{x_1, x_3, x_5\}$ and $B' = \{x_2, x_4, x_6\}$. Then $G_2^{**}(k-1), G_3 \nsubseteq G'$ as each of them has smaller chorded cycles than $K_{3,3}$. Suppose $n-6 < 4(k-1)$. Then, since each partite set of F_1 has at most one low vertex, we can by symmetry assume x_1, x_2 are high vertices, in which case $||\{x_1, x_2\}, V(G')|| \geq 2(3k-1-3) = 6k-8 \geq 4(k-1) > n-6$. So, there is a vertex $y \in N(x_1) \cap N(x_2) \cap V(G')$, and hence $yx_1x_4x_3x_2y$ is a 5-cycle with chord x_1x_2, contradicting (O3).

So, $n - 6 \geq 4(k - 1)$ and $G' \supseteq G_1^-(n - 6, k - 1)$ with partite sets A and B, where $|A| = (n - 6) - 3(k - 1) + 1 = n - 3k - 2$ and $|B| = 3(k - 1) - 1 = 3k - 4$.

By (1), there exists a vertex $y_0 \in A$ such that for each $y \in A - y_0$, $||y, F_1|| = 3$, $N_{G'}(y) = B$, and $||y_0, F_1|| \geq 2$. By Lemma 7, for each such y, either $N(y) \cap V(F_1) = A'$ or $N(y) \cap V(F_1) = B'$.

Subcase 3.1 All $y \in A - y_0$ have neighbors in one partite set, say A'. Then no vertex $z \in V(B)$ has a neighbor in A' or else we have a chorded 5-cycle, contradicting (O3). Then, for each pair $x, x' \in B'$, we have $||\{x, x'\}, B|| \geq 6k - 9$. This implies that there exists some vertex $z_0 \in B$, such that for all $z \in B - z_0$,

$N(z) \cap V(F_1) = B'$. Hence, our graph G is $K_{n-3k+1,3k-1} - y_0z_0$ or $K_{n-3k+1,3k-1}$, i.e. $G_1^-(n, k-1)$ or $G_1(n, k-1)$, a contradiction.

Subcase 3.2 $A - y_0$ is partitioned into two non-empty sets A_1, A_2 where for all $y \in A_1$, $N(y) \cap V(F_1) = A'$ and for all $y \in A_2$, $N(y) \cap V(F_1) = B'$. Again, we cannot have any $z \in V(B)$ with $||z, F_1|| \neq 0$, as otherwise we contain a C_5^+. Hence, for each $x \in B'$, $N(x) \subseteq A' \cup A_2 + y_0$ and for each $y \in A'$, $N(y) \subseteq B' \cup A_1 + y_0$. If say $|A_1| \leq |A_2|$, then for each $x, x' \in A'$,

$$d(x)+d(x') \leq 2\cdot 3+2|A_1|+||y_0, \{x, x'\}|| \leq 8+2\left\lfloor \frac{|A|-1}{2} \right\rfloor \leq n-3k+5. \quad (13)$$

Therefore, $n \geq 9k - 8$. If $|A_1|, |A_2| \geq 3$, then take any $A_1' \subseteq A_1, A_2' \subseteq A_2$ each with $|A_1'|, |A_2'| = 3$ and so $G[A_1', A'], G[A_2', B'] = K_{3,3}$. Then, $G - \{A_1', A', A_2', B'\} \supseteq K_{n-3k-8,3k-4}$. If $n > 9k - 8$ or $k \geq 4$, then this remaining graph contains $k - 2$ disjoint chorded cycles because $n - 3k - 8 \geq 3(k - 2)$, a contradiction. But if $k = 3$ and $n = 9k - 8 = 19$, then $n - 3k + 5 = 6k - 3$, and so we need equality in (13). However, $|A| = n - 3k - 2 = 8$ is even and so this is impossible.

Otherwise, we may assume $|A_2| \leq 2$. Since B' is independent, it contains two high vertices, say $x, x' \in B'$. For $k \geq 3$, $d_G(x) + d_G(x') \leq 2|A_2 + y_0| + 2 \cdot 3 \leq 12$. Hence, $||\{x, x'\}, B|| \geq 6k - 3 - 12 \geq 3$ and so there exist $z, z' \in B \cap (N(x) \cup N(x'))$. Since $|B \cup B'| = 4k - 1$, y_0 can have at most one nonneighbor in $B \cup B'$. So we can assume $y_0z, y_0x \in E(G)$. But then for each $z'' \in N(y_0) \cap (B - z)$ and $y \in A - y_0$, $xy_0z''yzx$ is a 5-cycle with chord y_0z, contradicting (O3).

3 Case: *G*[*R*] Does Not Have a Hamiltonian Path

Suppose that our collection $\mathscr{F}$ utilizing Rules (O1)–(O8) yields a remainder vertex set R where $G[R]$ does not contain a Hamiltonian path, i.e. $G[R]$ does not contain a spanning path of length $|R|$.

Lemma 16 *For $k \geq 2$ and all $z \in R - V(P)$, $d_{G[R]}(z) \geq 2$.*

Proof Suppose there is a vertex $z \in R - V(P)$ such that $d_{G[R]}(z) \leq 1$. By Lemma 10 and since P is maximum, $d_P(v_1) \leq 2$ and $||v_1, R - V(P)|| = 0$ and symmetrically for v_p. Therefore $d_{G[R]}(v_1), d_{G[R]}(v_p) \leq 2$. Since $v_1z, v_pz \notin E(G)$, $2d(z) + d(v_1) + d(v_p) \geq 12k - 6$. By the case and Lemma 10, $2d_{G[R]}(z) + d_{G[R]}(v_1) + d_{G[R]}(v_p) \leq 6$, hence

$$||\{v_1, v_p\}, \mathscr{F}|| + 2||z, \mathscr{F}|| \geq 12k - 6 - 6 = 12(k - 1). \quad (14)$$

Claim 16.4 For all $F \in \mathscr{F}$, if $||\{v_1, v_p\}, F|| + 2||z, F|| \geq 12$, then

(a) $||\{v_1, v_p\}, F|| + 2||z, F|| = 12$,
(b) $|\mathscr{F}| = k - 1$,
(c) $||v, F|| = 3$ for all $v \in \{v_1, v_p, z\}$, and
(d) $d_{G[R]}(z) = 1, d_{G[R]}(v_1) = 2$, and $d_{G[R]}(v_p) = 2$.

Proof Suppose that there exists $F \in \mathscr{F}$ such that $||\{v_1, v_p\}, F|| + 2||z, F|| \geq 13$. Since $||\{v_1, v_p\}, F|| \leq 8$, $||z, F|| \geq 3$. Moreover, there is some $v \in \{v_1, v_p, z\}$ such that $||v, F|| = 4$, so $F = K_4$ by Lemma 7. Since some vertex has 4 neighbors in F, if possible choose $x \in [(N(v_1) \cup N(v_p)) \setminus N(z)] \cap V(F)$, otherwise just choose $x \in [N(v_1) \cup N(v_p)] \cap V(F)$. Then $G[F - x + z] = K_4$ and $G[V(P) + x]$ contains a path longer than P, contradicting (O6). This proves 1.

Hence, by (14), $||\{v_1, v_p\}, F|| + 2||z, F|| = 12$ for all $F \in \mathscr{F}$ and thus $|\mathscr{F}| = k - 1$, $d_{G[R]}(v_1) = d_{G[R]}(v_p) = 2$, and $d_{G[R]}(z) = 1$. This proves 2 and 4. If $||v, F|| = 4$ for some $v \in \{v_1, v_p, z\}$, then again $F = K_4$. If possible choose $x \in [(N(v_1) \cup N(v_p)) \setminus N(z)] \cap V(F)$, otherwise just choose $x \in [N(v_1) \cup N(v_p)] \cap V(F)$. Either $G[F - x + z] \supseteq K_4$ and $G[P + x]$ contains a path longer than P, or $||z, F|| = 2$ in which case $G[F - x + z] = K_4^-$ and $xv_1 \cdots v_p x$ is a cycle with chord incident to v_1 since $d_{G[R]}(v_1) = d_P(v_1) = 2$. In any case, we have a contradiction, thus proving 3, and this completes the claim.

By Lemma 7, we have four cases:

Case 1 $F = K_4$. There exists $x \in N(v_1) \cap N(v_p) \cap V(F)$ and so $G[F - x + z] \supseteq K_4^-$ and $G[P + x]$ has a cycle with chord incident to v_1 since $d_P(v_1) = 2$. Thus, we have k disjoint chorded cycles, a contradiction.

Case 2 $F = K_4^-$. By Lemma 7, $\{x_1, x_3\} \subseteq N(z) \cap N(v_1) \cap N(v_p) \cap V(F)$. Then $G[F - x_1 + z] = K_4^-$ and $G[P + x_1]$ has a cycle with chord incident to v_1 since $d_P(v_1) = 2$. Thus, we have k disjoint chorded cycles, a contradiction.

Case 3 $F = C_5^+$. By Lemma 7, $\{x_1, x_2, x_4\} \subseteq N(z) \cap N(v_1) \cap N(v_p) \cap V(F)$. Then $G[\{z, x_1, x_2, v_1\}] \supseteq K_4^-$, contradicting (O1).

Case 4 $F = K_{3,3}$. We may assume $N(z) \cap V(F) = A = \{x_1, x_3, x_5\}$. If there exists $v \in \{v_1, v_p\}$ with $N(v) \cap V(F) = B = \{x_2, x_4, x_6\}$, then $G[F - x_2 + z] = K_{3,3}$ and $G[P + x_2]$ contains a path longer than P, contradicting (O6). Hence, $N(z) \cap V(F) = N(v_1) \cap V(F) = N(v_p) \cap V(F) = A$. Moreover, since this is the last case, each $F_i \in \mathscr{F}$ also is a $K_{3,3}$. Then, $d_G(z) = 1 + 3(k-1) = 3k - 2$, so z is a low vertex. Since $zx_2 \notin E(G)$, x_2 is not low. However, $G[F - x_2 + z] = K_{3,3}$ and P is still a path in $G[R - z + x_2]$ so that either there is a longer path and we contradict (O6), or the degree sum in R has increased, contradicting (O8).

Lemma 17 *Let $k \geq 2$ and $P_1 = z_1 z_2 \cdots z_s$ be a maximal path in $G[R - V(P)]$. Then, $d_{G[R]}(z_1) = 2$ or $d_{G[R]}(z_s) = 2$.*

Proof By Lemma 16, $d_{G[R]}(z_1) \geq 2$ and $d_{G[R]}(z_s) \geq 2$ so suppose that $d_{G[R]}(z_1)$, $d_{G[R]}(z_s) \geq 3$. If $s = 1$, then $||z_1, P|| \geq 3$, contradicting Lemma 10. Therefore

$z_1 \neq z_s$. If $d_{P_1}(z_1) = d_{P_1}(z_s) = 1$, then each vertex has at least two neighbors on P, say $z_1v_i, z_1v_j, z_sv_m, z_sv_\ell \in E(G)$ with $i < j$ and $m < \ell$. By symmetry assume $i < m$. Then $z_1 \cdots z_s v_\ell v_{\ell-1} \cdots v_i z_1$ is a cycle with chord $z_s v_m$, so we have k disjoint chorded cycles, a contradiction. Otherwise, some endpoint of P_1 has a chord on its own maximal path, say $z_1z_i \in E(G)$ for some $3 \leq i \leq s$. Still $z_1v_m, z_sv_j \in E(G)$ for some $1 \leq m, j \leq p$. Then $z_1 \cdots z_s v_j \cdots v_m z_1$ is a cycle with chord z_1z_i, so we have k disjoint chorded cycles, a contradiction.

Lemma 18 *If $k \geq 2$, then $G[V(P)]$ is not a cycle.*

Proof Suppose $G[V(P)]$ is the cycle $v_1v_2 \ldots v_pv_1$. For notation, let $P_1 = P$, and let $P_2 = w_1w_2 \ldots w_q$ be a longest path in $G[R] - V(P_1)$. By the maximality of P_1 and P_2, neither of w_1 nor w_q has a neighbor in $R - V(P_1) - V(P_2)$ and by Lemma 16, $d_{G[R]}(w_1) \geq 2$ and $d_{G[R]}(w_q) \geq 2$, so we conclude that

$$q \geq 3, d_{G[R]}(w_1){=}d_{G[R]}(w_q){=}2, \textit{and } w_1 \textit{and } w_q \textit{are incident to chord(s) of } P_2. \tag{15}$$

Let $W_1 = \{v_1, v_2\}$, $W_2 = \{w_1, w_q\}$, and $W = W_1 \cup W_2$. Since $v_1w_1, v_2w_q \notin E(G)$,

$$||W, V(G){-}R|| \geq \sum_{w \in W} d_G(w) - \sum_{w \in W} d_{G[R]}(w) \geq 2(6k{-}3){-}4(2) = 12(k{-}1){-}2. \tag{16}$$

Hence if $|\mathscr{F}| \leq k - 2$, then there is $F \in \mathscr{F}$ with $||W, F|| \geq 13$. So, there is $w' \in W$ with $||w', F|| \geq 4$. Then by Lemma 7, $F = K_4$. If $w' \in W_j$, then to have $||W, F|| \geq 13$, there is $w'' \in W_{3-j}$ with $||w'', F|| \geq 3$. Therefore, we have $w \in W_2$ and $v \in W_1$ with $||w, F||, ||v, F|| \geq 3$. Let $x \in [N(v) \setminus N(w)] \cap V(F)$ if possible, otherwise simply choose $x \in N(v) \cap V(F)$. Thus $G[F - x + w] = K_4$ and $G[P + x]$ has a path longer than P. This contradicts (O6). Therefore,

$$|\mathscr{F}| = k - 1. \tag{17}$$

By (16), there is $F \in \mathscr{F}$ with

$$||W, F|| \geq 10. \tag{18}$$

By symmetry, we may assume

$$||v_1, F|| \geq ||v_2, F|| \text{ and } ||w_1, F|| \geq ||w_q, F||. \tag{19}$$

Let $||W_i, F|| \geq 5$ with $W_i := \{v_1', v_2'\}$ and $W_{3-i} := \{w_1', w_q'\}$. Assume first that $F = K_{3,3}$. By Lemma 7, we may assume that $N(v_1') \cap V(F) = A$ and $\{x_2, x_4\} \subseteq N(v_2') \cap V(F)$. Since $||W_{3-i}, F|| \geq 4$, both w_1', w_q' have neighbors on F, say $x \in N(w_1') \cap V(F)$ and $x' \in N(w_q') \cap V(F)$, respectively, where we may assume $N(w_1') \cap V(F) \subseteq A$ and $N(w_q') \cap V(F) \subseteq B$ by Lemma 8. Then both

$G[V(P_{3-i}) \cup \{x, x'\}]$ and $G[V(P_i) \cup (F - \{x, x'\})]$ contain chorded cycles, thus producing k disjoint chorded cycles.

Now assume that $F \neq K_{3,3}$. Then there exists $x \in N_G(v_1') \cap N_G(v_2') \cap V(F)$. Note, this holds even if $F = C_5^+$, as if $v' \in W$ has $N(v') \cap V(F) = \{x_3, x_5\}$, i.e. the endpoints of the chord, then we have $G[\{v', x_3, x_4, x_5\}] = K_4^-$, contradicting (O1). Note that $G[V(P_i) \cup \{x\}]$ contains a chorded cycle. It follows that each of w_1', w_q' has at most two neighbors on $F - x$, and w_1' and w_q' cannot both have neighbors on $F - x$, for otherwise, $G[V(P_{3-i}) \cup (F - x)]$ contains a chorded cycle and we obtain k disjoint chorded cycles. Therefore, $||W_{3-i}, F - x|| \leq 2$.

If x is adjacent to both w_1' and w_q', then by the same argument, $||W_i, F - x|| \leq 2$, a contradiction to $||W_i, F|| \geq 5$. So $||W_{3-i}, F|| \leq 3$. It follows that $||W_i, F|| \geq 7$, and thus for some $v' \in W_i$, $||v', F|| \geq 4$, so $F = K_4$, and v_1', v_2' have at least three common neighbors on F.

Since $||W_i, F|| \leq 8$, $||W_{3-i}, F|| \geq 2$. So we may assume that either w_1' has two neighbors on F, or w_1', w_q' both have neighbors on F. In the former case, let $x \in N_G(v_1') \cap N_G(v_2') \cap V(F) \setminus N_G(w_1')$, then $G[\{w_1'\} \cup (F - x)] \supseteq K_4^-$ and $G[V(P_i) \cup \{x\}]$ contains a chorded cycle. In the latter case, let $x_1, x_q \in V(F)$ where $x_1 w_1', x_q w_q' \in E(G)$, respectively and $x \in N_G(v_1') \cap N_G(v_2') - x_1 - x_q$. Then $G[V(P_i) \cup \{x\}]$ and $G[V(P_{3-i}) \cup \{x_1, x_q\}]$ contain chorded cycles.

Lemma 19 *Suppose $k \geq 2$, $V(P) \neq R$, and $z \in R - V(P)$ is the endpoint of a maximal path in $R - V(P)$ with $d_{G[R]}(z) = 2$. Let $W = \{v_1, v_p, z\}$. Then,*

(a) For all $F \in \mathscr{F}$, $||W, F|| \leq 9$,
(b) $|\mathscr{F}| = k - 1$, and
(c) P has a chord with an endpoint in $\{v_1, v_p\}$.

Proof Suppose that $||W, F|| \geq 10$ for some $F \in \mathscr{F}$. Then, either $||z, F|| = 4$ or $||v, F|| = 4$ for some $v \in \{v_1, v_p\}$. If $||z, F|| = 4$, then some $x \in V(F)$ has $xv_1 \in E(G)$. So $G[F - x + z] = K_4$ but $xv_1 \cdots v_p$ is longer than P, contradicting (O6). If say $||v_1, F|| = 4$, then $||\{z, v_p\}, F|| \geq 6$ and so there exists some $x \in N(z) \cap N(v_p) \cap V(F)$. So $G[F - x + v_1] = K_4$ and $v_2 \cdots v_p x z$ is longer than P contradicting (O6). This proves Lemma 1.

Due to P being maximum and $G[P]$ not a cycle by Lemma 18, W is an independent set so $d(v_1) + d(v_p) + d(z) \geq 9k - 4$. Note, at most one of these vertices is low. Then,

$$||W, \mathscr{F}|| \geq 9k - 4 - 6 = 9(k - 1) - 1 \tag{20}$$

By Lemma 1, if $|\mathscr{F}| \leq k - 2$, then $||W, \mathscr{F}|| \leq 9(k - 2) < 9(k - 1) - 1$, a contradiction. This proves Lemma 2.

For Lemma 3, if P had no chord at an endpoint, then $d_{G[R]}(v_1) = d_{G[R]}(v_p) = 1$ and so we could improve (20) to $||W, \mathscr{F}|| \geq 9k - 4 - 4 = 9(k - 1) + 1$, a contradiction to Lemma 1 or Lemma 2. This completes the proof.

Below we continue to use the special set $W = \{v_1, v_p, z\}$, where $z \in R - V(P)$ is the endpoint of a maximal path in $R - V(P)$ with $d_{G[R]}(z) = 2$.

Lemma 20 *If $k \geq 2$ and $F \in \mathscr{F}$ with $F = K_4^-$ or $F = C_5^+$, then $||W, F|| \leq 7$. Hence, $K_4^-, C_5^+ \notin \mathscr{F}$.*

Proof Suppose that there is some $F \in \mathscr{F}$ such that $||W, F|| \geq 8$. If $F = C_5^+$, then by Lemma 7, there are two vertices $w, w' \in W$ such that $N(w) \cap V(F) = N(w') \cap V(F)$. Namely, $G[\{w, w'\} \cup (N(w) \cap V(F))] \supseteq K_4^-$, a contradiction to (O1). Hence, $F = K_4^-$.

If $||z, F|| = 3$, then for each $x \in \{x_1, x_3\}$, $G[F - x + z] = K_4^-$ and $G[P + x]$ contains a longer path than P since there is some $v \in \{v_1, v_p\}$ with $||v, F|| = 3$ as well.

Otherwise, $||z, F|| = 2$ and $||v, F|| = 3$ for every $v \in \{v_1, v_p\}$. If there is $x \in N(z) \cap \{x_1, x_3\}$, then again $G[F - x + v_1] = K_4^-$ and $G[P + x]$ contains a path longer than P, contradicting (O6). Hence, $N(z) \cap V(F) = \{x_2, x_4\}$. But then, $G[F - x_1 + z] = K_4^-$ and $x_1 v_1 \cdots v_p x_1$ is a cycle with chord from Lemma 3. This gives us k disjoint chorded cycles, a contradiction. This proves the first part of the claim.

Since W is an independent set, we have $||W, \mathscr{F}|| \geq 9k - 4 - 6 = 9(k-1) - 1$. By Lemma 1 and the first claim, if there exists a single cycle of type K_4^- or C_5^+, then we have $||W, \mathscr{F}|| \leq 7 + 9(k-2) < 9(k-1) - 1$, a contradiction. This completes the proof.

Lemma 21 *Suppose that z is the same from Lemma 19 and $||\{v_1, v_p, z\}, F|| \geq 8$. If $k = 2$, then $F = K_{3,3}$. If $k \geq 3$, then $F = K_{3,3}$, or $F = K_4$ and $||z, F|| = 0$.*

Proof Suppose that $F \neq K_{3,3}$, then by Lemma 20, $F = K_4$. Then some $x \in V(F)$ is adjacent to v_1 or v_p.

If $||z, F|| = 4$, then $G[F - x + z] = K_4$ and $G[P + x]$ contains a path longer than P, a contradiction.

If $||z, F|| = 3$, then $||\{v_1, v_p\}, F|| \geq 5$, and some $x' \in F$ is adjacent to both v_1 and v_p. In this case, $G[F - x' + z]$ contains K_4^- and $G[P + x']$ contains a cycle with chord due to Lemma 19.

If $||z, F|| = 1$, then $||\{v_1, v_p\}, F|| \geq 7$. We may assume that $||v_1, F|| = 4$ and $||v_p, F|| \geq 3$. If v_p and z have a common neighbor, say x, on F, then $G[F - x + v_1] = K_4$ and $G[P - v_1 + x + z]$ contains a path longer than P. If v_p and z have no common neighbor on F, let $x \in F$ be the neighbor of z, then $G[F - x + v_p] = K_4$ and $G[P - v_p + x + z]$ contains a path longer than P.

Finally, let $||z, F|| = 2$. If say $||v_1, F|| = 4$, then v_p and z have no common neighbors $x \in V(F)$ (otherwise, we get $G[F - x + v_1] = K_4$ and $G[P - v_1 + x + z]$ contains a path longer than P); hence, there exists $x \in N(v_1) \cap N(v_p) \cap V(F) \setminus N(z)$ so that $G[F - x + z] = K_4^-$ and $G[P + x]$ contains a chorded cycle by Lemma 3. So $||v_1, F|| = ||v_p, F|| = 3$. If $N(v_1) \cap V(F) = N(v_p) \cap V(F)$, then some $x \in N(v_1) \cap V(F)$ is not a neighbor of z, thus $G[P + x]$ and $G[F - x + z]$ both contain chorded cycles. If $N(v_1) \cap V(F) \neq N(v_p) \cap V(F)$, then z has no neighbors in $N(v_1) \cap F - N(v_p)$ and $N(v_p) \cap F - N(v_1)$ as we can contradict (O6) again, so $N(z) \cap F = N(v_1) \cap N(v_p) \cap F$. Let $\{x_1, x_2\} = V(F) - N(z)$ and $x_1 \in N(v_1)$ (thus $x_2 \in N(v_p)$). Let z' be the other endpoint of P_1. If $z = z'$, then $||z, P|| = 2$. Let $N(z) \cap V(P) = \{v_i, v_j\}$ with $i < j$. Then $v_1 v_2 \cdots v_j z x_3 v_1$ is a chorded cycle and $G[F - x_3 + v_p] = K_4^-$. So $z' \neq z$. Then similar to z, $N(z') \cap V(F) = N(z) \cap V(F) = \{x_3, x_4\}$. Now $G[P + x_1 + x_2]$ contains a chorded cycle and $G[\{x_3, x_4, z, z'\}] \supseteq K_4^-$.

Therefore, $||z, F|| = 0$. Now let $k = 2$. Note that $d_G(v_1) \leq 6$, and by Lemma 17, $d_G(z) = d_{G[R]}(z) = 2$, a contradiction to (1).

By Lemma 21, each copy of $F = K_{3,3} \in \mathscr{F}$ has $||w, F|| = 3$ for all $w \in W$. Moreover, each

$$V(F) \text{ has partite sets } A_F, B_F \text{ such that } ||W, A_F|| = 0.$$

Otherwise, if v_1 and v_p share a neighbor $x \in V(F)$, then $G[V(P) + x]$ contains a chorded cycle and $G[F - x + z] = K_{3,3}$. If instead z and say v_1 share a neighbor $x \in V(F)$, then $G[F - x + v_p] = K_{3,3}$ and $z x v_1 \cdots v_{p-1}$ is a path longer than P, contradicting (O6).

Now, choose $z' \in A_F$. If $||z', R|| \geq 3$, then $G[F - z' + z] = K_{3,3}$, P is still a path in $G[R - z + z']$, so we either contradict (O6) or (O7) since $d_{R'}(z') \geq 3 > 2 \geq d_{G[R]}(z)$. If $||z', R|| = 2$, then the same process yields a contradiction to (O8) since z is low but $zz' \notin E(G)$ implies z' is not low. Hence, $||z', R|| \leq 1$. So, $d(z') \leq 3(k-2) + 4 + 1 = 3k - 1$, but z' not low implies that we have $||z', F_1|| = 4$, where F_1 is still our unique copy of $K_4 \in \mathscr{F}$. Hence, for each $x \in V(F_1) \cap (N(v_1) \cup N(v_p))$, $G[F - z' + z] = K_{3,3}$ and $G[F_1 - x + z'] = K_4$ and $G[V(P) + x]$ contains a path longer than P, contradicting (O6). Thus, this completes the case if $k \geq 3$.

All that remains is to handle the base case of $k = 2$.

Lemma 22 *If $k = 2$, $G - V(P)$ forms a subgraph of $K_{3,m}$, where $m = n - 3 - |V(P)|$, such that v_1, v_p have only neighbors in the partite set of size 3.*

Proof Let $P_1 = z_1 \dots z_s$ be the longest path in $R - V(P)$. Consider $W = \{v_1, v_p, z_1\}$, where by Lemma 17 we assume that $d_R(z_1) = 2$. Clearly, W is an independent set. So at most one vertex in W is low. Then $d_G(v_1) + d_G(v_p) + d_G(z_1) \geq 9 + 5 = 14$. As $d_{G[R]}(x) \leq 2$ for each $x \in W$, we have $||W, F|| \geq 14 - 6 = 8$. By Lemma 21, $F = K_{3,3}$. Let A_F, B_F be the two parts of F. Let $A_F = \{x_1, x_3, x_5\}$ and $B_F = \{x_2, x_4, x_6\}$.

We claim that v_1, v_p, and z_1 have only neighbors in one part of F, say A_F. By symmetry, assume that $N(v_1) \cap V(F) = A_F$. Suppose $N(v_p) \cap F \subseteq A_F$. If $N(z_1) \cap V(F) \subseteq B_F$, let $x_1 \in N(v_1) \cap N(v_p) \cap V(F)$, then $G[F - x_1 + z_1] = K_{3,3}^-$ and $G[P + x_1]$ contains another chorded cycle.

Now suppose $N(v_p) \cap V(F) \subseteq B_F$. If $N(z_1) \cap F \subseteq B_F$, then v_p and z_1 have a common neighbor, say x_2, thus $G[F - x_2 + v_1] = K_{3,3}$ and $z x_2 v_p v_{p-1} \cdots v_2$ is a path longer than P, contradicting (O6). If $N(z_1) \cap V(F) \subseteq A_F$, then when $||z_1, F|| = 3$, let $x_2 \in N(v_p)$, then $G[F - x_2 + z_1] = K_{3,3}$ and $G[P + x_2]$ contains a path longer than P, and when $||v_p, F|| = 3$, let $x_1 \in N(z_1)$, then $G[F - x_1 + v_p] = K_{3,3}$ and $G[P - v_p + x_1 + z]$ contains a path longer than P.

By symmetry, z_s has only neighbors in A_F as well. Let $x_1, x_3 \in N(z_1) \cap V(F)$ and $z_s x_3 \in E(G)$ and $x_5 \in N(v_1) \cap N(v_p) \cap V(F)$. Then $G[P + x_5]$ and $G[P_1 \cup \{x_1, x_3, x_2\}]$ both contain chorded cycles since P has a chord by Lemma 19. So $P_1 = \{z_1\}$. It follows that $R - P$ consists of isolated vertices and each of the vertices in $R - P$ has neighbors in A_F.

Lemma 23 *If $k = 2$ and $G[R]$ has no Hamiltonian path, then G contains 2 disjoint chorded cycles.*

Proof By Lemma 22, we may assume that G consists of a path P and a subgraph H of $K_{3,m}$ with parts $A = \{x_1, x_3, x_5\}$ and $B = \{y_1, \ldots, y_m\}$ such that v_1, v_p are all not adjacent to B. As B is an independent set, at most one vertex in B is low, so we may assume that $d_G(y_i) \geq 5$ for each $1 \leq i \leq m - 1$.

Also, at most one of v_1, v_p is low, and $d(v_1), d(v_p) \leq 5$. So we may assume that $d(v_1) = 5, d(v_p) \geq 4$, and $d(y_m) \geq 4$. It follows that v_1 is adjacent to all vertices of A and v_p is adjacent to at least two vertices in A. Likewise, for each $i \in [m - 1]$, since y_i has only three neighbors in H, it has at least two neighbors in P. As $R - P \neq \emptyset, m - 1 \geq 3$. We may assume that $v_p x_3 \in E(G)$ and $y_m x_3, y_m x_5 \in E(G)$.

Case 1 There exists $1 \leq i < j \leq p$ and a vertex in $B - y_m$, say y_1, with $y_1 v_i, y_1 v_j \in E(G)$, and a vertex, say y_2 with $y_2 v_s \in E(G)$, such that $i < j < s$ or $s < i < j$. If $i < j < s$, then $v_1 P v_j y_1 x_1 v_1$ contains chord $y_1 v_i$, and $v_s P v_p x_3 y_3 x_5 y_2 v_s$ contains chord $x_3 y_2$. Similarly, if $s < i < j$, then $v_p P v_i y_1 x_3 v_p$ contains chord $y_1 v_j$, and $v_1 P v_s y_2 x_1 y_3 x_5 v_1$ contains chord $x_5 y_2$. This proves the case.

For Case 1 to not apply for all vertices in $B - y_m$, we need $N(y_s) \cap V(P) = \{v_i, v_j\}$. We now handle that case.

Case 2 For each $s \in [m - 1]$, $N(y_s) \cap V(P) = \{v_i, v_j\}$, where $i < j$. Let $v_1 v_t \in E(G)$. If $t \leq i$, then $v_1 P v_i y_1 x_1 v_1$ contains chord $v_1 v_t$, and $v_j P v_p x_3 y_2 x_5 y_3 v_j$ contains chord $x_3 y_3$.

If $i < j < t$, then $v_1 v_t P v_p x_3 y_4 x_5 v_1$ contains chord $v_1 x_3$ and $G[x_1, y_1, y_2, y_3, v_i, v_j] = K_{3,3}$.

So $v_1 v_t \in E(G)$ with $i < t \leq j$. Since $y_m v_p \notin E(G)$, either $d_{G[R]}(v_p) = 2$ or $N(y_m) = \{x_1, x_3, x_5, v_i, v_j\}$.

Subcase 2.1 We do not have a chord incident to v_p and $N(y_m) = \{x_1, x_3, x_5, v_i, v_j\}$. Now, $j < p - 1$ as otherwise $P' = v_1 \cdots v_{p-1} y_1$ is a maximum path and $G[V(H) - y_1 + v_p] \supseteq K_{3,3}$, contradicting (O7) since $||y_1, P' - y_1|| = 2$ but $||v_p, P - v_p|| = 1$. Note that also $d_{G[V(P)]}(v_{p-1}) \leq 3$ so $||v_{p-1}, H|| \geq 1$. By the case, a neighbor must be in $\{x_1, x_3, x_5\}$, say $v_{p-1} x_1 \in E(G)$. Hence,

$x_1v_1v_tv_{t+1}\cdots v_px_1$ is a cycle with chord x_1v_{p-1} and $G[v_i, x_3, x_5, y_1, y_2, y_3] = K_{3,3}$. This produces two disjoint chorded cycles and ends the subcase.

Subcase 2.2 $d_{G[R]}(v_p) = 2$. By symmetry with v_1, $v_pv_q \in E(G)$ for some $i \leq q < j$. Now $G[x_1, x_5, v_i, y_1, y_2, y_3] = K_{3,3}$.

If $i < t \leq q \leq j$, then $v_1v_tPv_px_3v_1$ contains chord v_pv_q. Otherwise $q < t$, and so $G[V(P)]$ is a Θ-graph.

If $i < q < t < j$, then $v_1v_tv_{t-1}\cdots v_qv_pv_{p-1}v_jy_4x_3v_1$ contains chord v_px_3 unless $m = 4$ and $v_jy_4 \notin E(G)$. But then $N(y_m) \supseteq A$ by (1) and so we can swap the roles of y_1 and y_4, i.e. $G[x_1, x_5, v_i, y_2, y_3, y_4] = K_{3,3}$ and $v_1v_tv_{t-1}\cdots v_qv_pv_{p-1}v_jy_1x_3v_1$ contains chord v_px_3.

If $i < q < t = j$, then $v_1v_q \notin E(G)$ and $d_{G[P]}(v_q) = 3$ implies $||v_q, A|| \geq 9 - 5 - 3 = 1$, say $x_1v_q \in E(G)$. Then $x_1v_1v_tv_{t-1}\cdots v_qv_px_1$ is a cycle with chord x_1v_q.

If $i = q < t < j$, a similar argument holds with $G[x_1, x_5, v_j, y_2, y_3, y_4] = K_{3,3}$ and $x_1v_pv_qv_{q+1}\cdots v_tv_1x_1$ is a cycle with chord x_1v_t.

Now it remains to handle when $i = q$ and $j = t$ so $i = q < t = j$. First, we claim

$$i = j - 2.$$

Otherwise, $||v, H|| \geq 2$ for $v \in \{v_{i+1}, v_{i+2}\}$. By the case, we have $N(v) \cap V(H) \subseteq A$. Hence, there exists $A \cap N(v_{i+1}) \cap N(v_{i+2}) \neq \emptyset$, say x_1. Then, $x_1v_1\cdots v_{i+1}x_1$ is a cycle with chord x_1v_i and $G[V(H) - x_1 + v_j] \supseteq K_{3,3}$, producing two disjoint chorded cycles.

Next, we claim $i = 2$. Note, $G[P]$ is a Θ-graph so $d_{G[P]}(v_2) = 2$, as otherwise $G[P]$ contains a chorded cycle. So, $||v_2, H|| \geq 2$ since $v_2v_p \notin E(G)$. If $v_2y_k \in E(G)$ for some $k \in [m-1]$, then $y_kv_2\cdots v_jy_k$ is a cycle with chord y_kv_i. Since $N(v_2) \cap V(H)$ is contained in either A or B, by symmetry assume $x_1v_2, x_3v_2 \in E(G)$. Then $v_1v_2x_1v_px_3v_1$ is a 5-cycle with chord v_1x_1, contradicting (O3).

By symmetry, we also have $j = p - 1$. This together with $i = 2$ and $i = j - 2$ yields $i = q = 2$, $j = t = 4$, and $p = 5$ so that $G[V(P)] = K_{2,3}$. By the symmetry of $K_{2,3}$, $\{v_1, v_3, v_5, y_m\}$ is an independent set so at most one vertex has degree 4. Thus, $G \in \{G_1(n,2), G_1^-(n,2)\}$ with parts $\{x_1, x_3, x_5, v_2, v_4\}$ and $B \cup \{v_1, v_3, v_5\}$, a contradiction.

This contradicts our initial assumption that G is a counterexample to Theorem 6. This completes the $k = 2$ case, and hence the case where $G[R]$ does not have a Hamiltonian path.

4 Case: $G[R]$ Has a Hamiltonian Path and $k \geq 3$

In this section we consider the case that $G[R]$ does have a Hamiltonian path and $k \geq 3$. We first prove a stronger version of Lemma 11.

Lemma 24 *For $k \geq 3$, $|R| \geq 5$.*

Proof Suppose that $|R| \leq 4$. Then by Lemma 11, $|R| = 4$, say $R = \{v_1, v_2, v_3, v_4\}$.

Case 1 $G[R] \subseteq C_4$. First we show that

$$||R, F|| \leq 12 \text{ for every } F \in \mathscr{F}. \tag{21}$$

Indeed, suppose that $||R, F|| > 12$ for some $F \in \mathscr{F}$. By Lemma 7, $F = K_4$. Since $||R, F|| \geq 13$, there is $v_i \in R$ with $||v_i, F|| = 4$ and there is $x \in V(F)$ with $||x, R|| = 4$. So, $G[V(F) - x + v_i] = K_4$ and $G[R - v_i + x] \supset G[R]$. This contradicts either (O1) or (O8). This shows (21).

Our next claim is

$$\text{if } F \in \mathscr{F} \text{ and } F = C_5^+, \text{ then } ||R, F|| \leq 9. \tag{22}$$

Indeed, suppose $F \in \mathscr{F}$ is the cycle $x_1x_2x_3x_4x_5x_1$ with chord x_3x_5 and $||R, F|| \geq 10$. By Lemma 7, for each $1 \leq i \leq 4$, $||v_i, F|| \leq 3$ and if $||v_i, F|| = 3$, then $N(v_i) \cap F = \{x_1, x_2, x_4\}$. So, there are $v, v' \in R$ with $N(v) \cap F = N(v') \cap F = \{x_1, x_2, x_4\}$. In this case, replacing F in $\mathscr{F}$ by the chorded 4-cycle $G[\{v, v', x_1, x_2\}]$ we obtain a family $\mathscr{F}'$ that is better than $\mathscr{F}$ by (O1). This proves (22).

Next we show that

$$\text{if } F \in \mathscr{F} \text{ and } F = K_4^-, \text{ then } ||R, F|| \leq 10. \tag{23}$$

Indeed, suppose $F \in \mathscr{F}$ is the cycle $x_1x_2x_3x_4x_1$ with chord x_2x_4 and $||R, F|| \geq 11$. By Lemma 7, for each $1 \leq i \leq 4$, $||v_i, F|| \leq 3$ and if $||v_i, F|| = 3$, then $N(v_i) \supseteq \{x_1, x_3\}$. This means that there is $1 \leq j \leq 4$ such that each $v \in R - v_j$ has 3 neighbors in F and v_j has at least 2 neighbors in F. If $x_1v_j \in E(G)$, then for each $j' \in [4] - j$, $G[F - x_1 + v_{j'}] = K_4^-$ and $G[R - v_{j'} + x_1] \supset G[R]$, which contradicts either (O1) or (O8). Thus, $x_1v_j \notin E(G)$ and symmetrically $x_3v_j \notin E(G)$, implying $N(v_j) \cap V(F) = \{x_2, x_4\}$. Then, $G[V(F) - x_1 + v_j] = K_4^-$ and $G[R - v_j + x_1] \supset G[R]$, again contradicting either (O1) or (O8). This proves (23).

Since R has two disjoint pairs of nonadjacent vertices,

$$||R, \mathscr{F}|| \geq 2(6k - 3) - 2||G[R]|| \geq 12(k - 1) - 2 = 12(k - 2) + 10. \tag{24}$$

So, by (21),

$$\begin{aligned} &\mathscr{F} \text{ has } k - 1 \text{ chorded cycles and} \\ &||G[R]|| \geq 3, \text{ i.e. } G[R] \text{ is a cycle } v_1v_2v_3v_4v_1 \text{ or a path } v_1v_2v_3v_4. \end{aligned} \tag{25}$$

Now we show

$$K_4 \in \mathscr{F}. \tag{26}$$

Indeed, otherwise by Lemma 15, $\mathscr{F}$ contains two cycles F_1 and F_2 inducing K_4^- in G. Hence by (21) and (23), $||R, \mathscr{F}|| \leq 2 \cdot 10 + 12(t-2) = 12(k-1) - 4$, a contradiction to (24).

We also need

$$\text{if } F \in \mathscr{F} \text{ and } F = K_4, \text{ then } ||R, F|| \leq 11. \tag{27}$$

If $G[R] = C_4$, then this follows from Lemma 14; so by (25) we may assume $G[R]$ is a path $v_1v_2v_3v_4$. Suppose $||R, F|| \geq 12$. If $||v_1, F|| = 4$, then for each $x \in V(F)$, $||x, R - v_1|| \leq 1$, since otherwise $G[F - x + v_1] = K_4$ and $G[R - v_1 + x]$ has a Hamiltonian path and more edges than $G[R]$, which contradicts (O7). But then $||R, F|| \leq 8$. If $||v_1, F|| = 3$, then each $x \in F$ has at most 2 neighbors in $R - v_1$, since otherwise $G[F - x + v_1] \supseteq K_4^-$ and $G[R - v_1 + x] = K_4^-$, which contradicts (O1). But in this case $||R, F|| \leq 3 + 2 \cdot 4 = 11$, as claimed. Thus, $||v_1, F|| \leq 2$ and symmetrically $||v_4, F|| \leq 2$. Since $||R, F|| \geq 12$, this yields $||v_1, F|| = ||v_4, F|| = 2$ and $||v_2, F|| = ||v_3, F|| = 4$. Then, let $x_1 \in N(v_1) \cap V(F)$, $x_4 \in N(v_4) \cap V(F) - x_1$, and x_2, x_3 be the remaining vertices in F. In this notation, $G[\{v_1, v_2, x_1, x_2\}] \supseteq K_4^-$ and $G[\{v_3, v_4, x_3, x_4\}] \supseteq K_4^-$, a contradiction to (O1).

Our last claim is:

$$\text{if } F \in \mathscr{F}, F = K_{3,3} \text{ and } R \text{ contains } \ell \text{ low vertices, then } ||R, F|| \leq 12 - \ell. \tag{28}$$

Indeed, since low vertices form a clique in G and $G[R]$ is bipartite, $\ell \leq 2$. So, there is nothing to prove if $||R, F|| \leq 10$. Also by (21) we may assume $\ell \geq 1$. Suppose $||R, F|| \geq 13 - \ell \geq 11$. By Lemma 7, there is $1 \leq j \leq 4$ such that each vertex in $R - v_j$ has 3 neighbors in F and v_j has at least $||R, F|| - 9 \geq 2$ neighbors in F. By the same lemma, if the partite sets of F are A and B, then we may assume that $(N(v_1) \cup N(v_3)) \cap V(F) \subseteq A$ and $(N(v_2) \cup N(v_4)) \cap V(F) \subseteq B$. By above, at least $||R, F|| + \ell - 12 \geq 1$ low vertices in R have 3 neighbors in F; so let v_i be such a vertex. By symmetry, we may assume i is odd. Since $||\{v_2, v_4\}, B|| \geq 5$, there is $b \in B$ adjacent to both, v_2 and v_4. As $bv_i \notin E(G)$, b is high. Then $G[F - b + v_i] = K_{3,3}$, $G[R - v_i + b] \subseteq G[R]$, and $\sum_{w \in R - v_i + b} d(w) > \sum_{v \in R} d(v)$. This is a contradiction to (O8) which proves (28).

By (26), we may assume $F_1 = K_4$. So by (24), (21), (23), (22), and (27), $\mathscr{F}$ does not contain copies of K_4^- and C_5^+. In other words,

$$F_1 = K_4\, r \text{ and each chorded cycle in } \mathscr{F} \text{ is a } K_4 \text{ or a } K_{3,3}. \tag{29}$$

By (25), we have two subcases.

Subcase 1.1 $G[R]$ is a path $v_1v_2v_3v_4$. Since $||G[R]|| = 3$, (24) together with (21) imply that $||R, F|| = 12$ for all $F \in \mathscr{F}$. But this is impossible by (26) and (27).

Subcase 1.2 $G[R]$ is a cycle $v_1v_2v_3v_4v_1$. By Lemma 14, either $||R, F_1|| \le 10$ or $||R, F_1|| = 11$ and $\sum_{j=1}^{4} d_G(v_j) \ge 12k - 5$. In both cases, by (24),

$$||R, \mathscr{F} - F_1|| \ge 12(k-2).$$

In view of (21), this means for each $F \in \mathscr{F} - F_1$, $||F, R|| = 12$. Hence by (29) and (27), each such chorded cycle, and in particular, F_2 induces $K_{3,3}$. Then by (28), all vertices in R are high. So, instead of (24) we have

$$||R, \mathscr{F}|| \ge 4(3k-1) - 2||G[R]|| = 12(k-1).$$

This together with (27) and (29) contradicts (21). This proves the case.

Case 2 $G[R] \not\subseteq C_4$. If $G[R]$ does not have a degree 3 vertex, then $G[R] = K_3 + K_1$, say v_1 is the isolated vertex, and $v_2v_3v_4v_2$ is a 3-cycle. Then, $d_G(v_1)+d_G(v_2) \ge 6k-3$ and so $||\{v_1, v_2\}, \mathscr{F}|| \ge 6k - 5 = 6(k-1) + 1$. Hence, there exists some $F \in \mathscr{F}$ such that $||\{v_1, v_2\}, F|| \ge 7$. By Lemma 7, $F = K_4$. Whether $||v_1, F||$ is 3 or 4, there exists some $x \in V(F)$ such that $G[F - x + v_1] = K_4$ and $xv_2 \in E(G)$. But then $G[R - v_1 + x] \supseteq K_{1,3}^+$, contradicting (O6).

So suppose $G[R]$ does have degree 3 vertex v_1. If the set $\{v_2, v_3, v_4\}$ is independent, then it has at most one low vertex and so $||\{v_2, v_3, v_4\}, \mathscr{F}|| \ge 9k-4-3 = 9(k-1)+2$. Hence, there exists some $F \in \mathscr{F}$ such that $||\{v_2, v_3, v_4\}, F|| \ge 10$ and so a vertex, say v_2, has 4 neighbors in F. By Lemma 7, $F = K_4$. By pigeonhole, there exists $x \in N(v_3) \cap N(v_4) \cap V(F)$. Then $G[F - x + v_2] = K_4$ and $G[R - v_2 + x] \supseteq C_4$, contradicting (O6).

The last possibility is that $G[R] = K_{1,3}^+$, say v_2v_3 is an edge. Consider

$$f(R, F) := 2 \cdot ||v_4, F|| + ||v_2, F|| + ||v_3, F||. \tag{30}$$

Our main claim is that for each $F \in \mathscr{F}$

$$f(R, F) \le 12 \text{ and if } f(R, F) = 12, \text{ then } F \text{ is a } K_4. \tag{31}$$

Indeed, suppose $f(R, F) \ge 13$. Then for some $2 \le j \le 4$, $||v_j, F|| = 4$. So by Lemma 7, $F = K_4$. Since $|F| = 4$, to have $f(R, F) \ge 13$, we need $||\{v_2, v_3\}, F|| \ge 13 - 8 = 5$ and $||v_4, F|| \ge \lceil (13-8)/2 \rceil = 3$. Then there is $x \in N(v_2) \cap N(v_3) \cap V(F)$. Hence $G[F - x + v_4] \supseteq K_4^-$ and $G[R - v_4 + x] \supseteq K_4^-$, contradicting (O1).

Suppose now $f(R, F) = 12$ and $F \ne K_4$. By Lemma 7, to have $f(R, F) = 12$ we need

$$||v_2, F|| = ||v_3, F|| = ||v_4, F|| = 3. \tag{32}$$

If F induces a cycle $x_1x_2x_3x_4x_1$ with chord x_2x_4, then $\{x_1, x_3\} \subseteq N(v_2) \cap N(v_3) \cap N(v_4) \cap V(F)$. Hence $G[R - v_4 + x_1] \supseteq K_4^-$ and $G[F - x_1 + v_4] \supseteq K_4^-$, contradicting (O1).

If F induces a cycle $x_1x_2x_3x_4x_5x_1$ with chord x_3x_5, then by (32) and Lemma 7, $\{x_1, x_2, x_4\} \subseteq N(v_2) \cap N(v_3) \cap N(v_4) \cap V(F)$. So, $G[\{v_2, v_3, x_2, x_1\}] = K_4$, contradicting (O1).

If F induces a $K_{3,3}$ with parts A and B, then by (32) and Lemma 7, $N(v_i) \cap V(F) \in \{A, B\}$ for all $2 \leq i \leq 4$. By Lemma 8, we may assume $N(v_2) \cap V(F) = A$, $N(v_3) \cap V(F) = B$, and by symmetry, $N(v_4) \cap V(F) = A$. Then for each $a \in A$, $v_1v_2v_3av_4v_1$ is a chorded 5-cycle contradicting (O3). This proves (31).

Observe that

$$\sum_{F \in \mathcal{F}} f(R, F) = 2 \cdot d(v_4) + d(v_2) + d(v_3) - 6$$

$$\geq 2(6k - 3) - 6 = 12k - 12 = 12(k - 1).$$

By (31)), this means $\mathcal{F}$ has $k - 1$ disjoint chorded cycles and $f(R, F) = 12$ for each $F \in \mathcal{F}$. Also by (31),

$$\text{each } F \in \mathcal{F} \text{ induces a } K_4. \text{ In particular, } |V(G)| = 4k. \tag{33}$$

Now we show that

$$\text{for each } F \in \mathcal{F}, ||v_4, F|| = 4, \text{ and each } x \in \text{F has exactly one neighbor in } \{v_2, v_3\}. \tag{34}$$

Indeed, let $V(F) = \{x_1, x_2, x_3, x_4\}$. To have $f(R, F) = 12$, we need $||v_4, F|| \geq 2$. Furthermore, if $N(v_4) \cap V(F) = \{x_1, x_2\}$, then $||\{v_2, v_3\}, F|| = 8$. Then for $x \in V(F) - N(v_4)$, $G[F - x + v_4] \supseteq K_4^-$ and $G[R - v_4 + x] \supseteq K_4^-$, contradicting (O1). If there exists $x \in N(v_2) \cap N(v_3) \cap V(F)$ with $|N(v_4) \cap V(F) - x| \geq 2$, then we reach the same contradiction. So $||\{v_2, v_3\}, F|| \leq 5$. Then, for $f(R, F) = 12$, we have $||v_4, F|| = 4$ but also $||\{v_2, v_3\}, F|| = 4$. Now if there exists $x \in N(v_2) \cap N(v_3) \cap V(F)$, we again have $||N(v_4) \cap V(F) - x| \geq 2$. So we have $N(v_2) \cap N(v_3) \cap V(F) = \emptyset$. This proves (34).

Our last claim is that

$$\text{for each } F \in \mathcal{F}, ||v_1, F|| = 0. \tag{35}$$

Suppose now that for some $F \in \mathcal{F}$ and $x \in F$, $v_1x \in E(G)$. By (34), $G[F - x + v_4] = K_4$ and $G[R - v_4 + x] = K_4^-$, contradicting (O1).

By (35), $||v_1, \mathcal{F}|| = 0$ and so $d_G(v_1) = 3$. Let $x \in V(F_1)$. By (33), (34), and (35),

$$d_G(x) \leq |V(G) - x - v_1| - 1 \leq 4k - 3$$

Hence $d_G(x)+d_G(v_1) \leq 4k$, contradicting the condition $d_G(v_1)+d_G(x) \geq 6k-3$. This completes the case and the claim.

A *broken Θ-graph* is obtained from a Θ-graph by deleting a vertex of degree 2. We will view each broken Θ-graph as a triple $(F_{u,v}, P_u, P_v)$, where $F_{u,v}$ is a cycle with two special vertices u and v, a path P_u starts at u, finishes at u' with $V(F_{u,v}) \cap P_u = \{u\}$, and a path P_v starts at v, finishes at v' with $[V(F_{u,v}) \cup V(P_u)] \cap P_v = \{v\}$.

Observation 25 *If F is a broken Θ-graph $(F_{u,v}, P_u, P_v)$, then for each triple T of vertices, F contains a path P_T passing through all vertices of T.*

Proof Since the graph $F-(P_u-u)$ has a Hamiltonian path P', if $T \cap (V(P_u)-u) = \emptyset$, then we may take $P_T = P'$. Thus it is enough to consider the case that T has a vertex $t_1 \in P_u - u$, and similarly, a vertex $t_2 \in P_v - v$. Then for any choice of t_3, F has a $v'u'$-path containing t_3. Since any such path contains $P_v \cup P_u$, it contains all T.

Lemma 26 *Suppose $k \geq 3$, $R = \{v_1, \ldots, v_r\}$, and $G[R]$ is a Θ-graph with branching vertices v_0 and v_0'. Let $F \in \mathscr{F}$ with $F = K_4$. Then*

(a) $||R, F|| \leq 3r$, and
(b) if $|\mathscr{F}| = k-1$, then $||R, F|| \leq 2r+2$.

Proof By Lemma 24, $|R| \geq 5$, and $|R| = 5$ implies $G[R] = K_{2,3}$. Suppose the lemma does not hold for G so $G[R]$ is a spanning Θ_{j_1,j_2,j_3}-graph consisting of three paths $P_i = v_0 v_{i,1} v_{i,2} \cdots v_{i,j_i} v_0'$ connecting v_0 with v_0' for $i = 1, 2, 3$ where $j_i \geq 1$ since otherwise it is itself a chorded cycle. We proceed in a series of claims. Our first claim is:

$$||v_{i,j}, F|| \leq 3 \quad \text{for each } 1 \leq i \leq 3 \text{ and } 1 \leq j \leq j_i. \tag{36}$$

Indeed, assume that for some $v_{i,j} \in R - v_0 - v_0'$, $x_\ell v_{i,j} \in E(G)$ for each $1 \leq \ell \leq 4$. If for at least one ℓ, $||x_\ell, R - v_{i,j}|| \geq 3$, then we consider $F' = F - x_\ell + v_{i,j}$ and $R' = R - v_{i,j} + x_\ell$. By the case, $F' = K_4$, and (since $R - v_{i,j}$ is a broken Θ-graph) by Observation 25, $G[R']$ contains a chorded cycle. This contradicts (O3), so $||x_\ell, R - v_{i,j}|| \leq 2$ for all $1 \leq i \leq 4$. It follows that $||F, R|| \leq 4 + 2 \cdot 4 = 12$. This satisfies both (1) and (2), contradicting our assumption. This proves (36).

Next we show

$$||\{v_0, v_0'\}, F|| \leq 6. \tag{37}$$

Indeed, suppose (37) does not hold. Then by symmetry, we may assume $||v_0, F|| = 4$ and $N(v_0') \supseteq \{x_2, x_3, x_4\}$. Hence, for any two vertices $v, v' \in R - v_0'$, graph $G[R - v_0']$ has a path passing through v, v' and v_0. Thus if x_1 has two neighbors, say $v, v' \in R - v_0 - v_0'$, then $G[R - v_0' + x_1]$ contains a chorded cycle. Since $G[F - x_1 + v_r] = K_4$, this contradicts (O3). Therefore, $||x_1, R - v_0 - v_0'|| \leq 1$. Similarly, since $N(v_0') \supseteq \{x_2, x_3, x_4\}$, if $||x_i, R - v_0 - v_0'|| \geq 2$ for some $2 \leq i \leq 4$, then $G[R - v_0 + x_i]$ contains a chorded cycle and $G[F - x_i + v_0] = K_4$. Again, this contradicts (O3). This proves (37).

Together, (36) and (37) imply (1).

Suppose now $|\mathscr{F}| = k - 1$. Then we can strengthen (36) to

$$||v_{i,j}, F|| \le 2 \quad \text{for each } 1 \le i \le 3 \text{ and } 1 \le j \le j_i. \tag{38}$$

Indeed, assume there exists $v_{i,j} \in R - v_0 - v_0'$ with $N(v_{i,j}) \cap V(F) = \{x_2, x_3, x_4\}$. If for at least one ℓ, $||x_\ell, R - v_{i,j}|| \ge 3$, then $G[F - x_\ell + v_{i,j}] \supseteq K_4^-$, and by Observation 25, $G[R - v_{i,j} + x_\ell]$ contains a chorded cycle. Since $|\mathscr{F}| = k - 1$, we have obtained k disjoint chorded cycles, a contradiction. Therefore $||x_\ell, R - v_{i,j}|| \le 2$ for all $1 \le \ell \le 4$, and hence $||F, R|| \le 3 + 2 \cdot 4 = 11$, as claimed. This satisfies both (1) and (2), contradicting our assumption. This proves (38). Together with (37), this implies

$$||F, R|| \le 2 \cdot (r - 2) + 6 = 2r + 2.$$

Lemma 27 *For $k \ge 3$, $G[R]$ does not contain a spanning Θ-graph.*

Proof Suppose $G[R]$ contains a spanning Θ-subgraph $G' = \Theta_{r_1,r_2,r_3}$. Since $G[R]$ has no chorded cycles, $r_1, r_2, r_3 \ge 1$. Furthermore, since adding an edge to a Θ-graph creates a chorded cycle, $G[R] = G'$. Let ℓ be the number of low vertices in $G[R]$. Since $G[R]$ is triangle-free and low vertices form cliques, $\ell \le 2$. Also, $\sum_{v \in V(R)} d_G(v) \ge (3k - 1)r - \ell$. Then,

$$||R, \mathscr{F}|| \ge (3k - 1)r - \ell - 2(r + 1) = 3(k - 1)r - \ell - 2, \tag{39}$$

so there exists some cycle $F \in \mathscr{F}$ such that $||R, F|| \ge 3r - \frac{\ell+2}{k-1} \ge 3r - 2$.

Claim 27.1 If $F = K_{3,3}$, then $||R, F|| \le 3r - 2$ or we have the exception $G[R] = K_{2,3}$ and $\ell = 0$, in which case $||R, F|| \le 3r$.

Proof Suppose that $||R, F|| \ge 3r - 1$. By Lemma 7, there exists some vertex $w \in V(R)$ such that for all $v \in V(R) - v_0$, $||v, F|| = 3$ and $||v_0, F|| \ge 2$. Also by Lemma 7, for all $v \in V(R) - w$, $N(v) \cap V(F) \in \{A, B\}$. Moreover, by Lemma 8 if $v, w \in V(R)$ with $vw \in E(G)$, then $N(v) \cap N(w) \cap V(F) = \emptyset$. Hence, $G[R \cup F]$ is bipartite with parts $A \cup R_A$ and $B \cup R_B$, where $R = R_A \cup R_B$. Note, $|A \cup R_A|, |B \cup R_B| \ge 5$ since $r \ge 5$ and each of R_A and R_B contains at most one low vertex since each is an independent set. So, we can partition $A \cup R_A$ into $A' \cup R_A'$ and $B \cup R_B$ into $B' \cup R_B'$ so that $R_A' \cup R_B'$ contain no low vertices and $G[A' \cup B'] = K_{3,3}$. This contradicts (O8) unless $\ell = 0$.

Suppose $\ell = 0$ but say $r_1 > 1$. Specifically, this means we can find three vertices $v', v, v'' \in R$ such that $v'vv''$ is a path in $G[R]$, the middle vertex v is a degree 2 vertex in the Θ-graph, and $v \ne w$. By symmetry we may assume

$N(v') \cap V(F)$, $N(v'') \cap V(F) \subseteq A$ and $N(v) \cap V(F) = B$. Also, some other $u \in R - v - v' - v''$ has $N(u) \cap V(F) \subseteq A$. Take $x \in N(u) \cap N(v') \cap N(v'') \cap V(F) \subseteq B$. But then, $G[F - x + v] = K_{3,3}$, and $G[R - v + x] \supseteq \Theta^+_{r_1, r_2, r_3}$ and thus also contains a chorded cycle. This contradicts (O3). Hence, we are left with $G[R] = K_{2,3}$, and this proves the claim.

Claim 27.2 If $F = C_5^+$, then $||R, F|| \leq 2r + 1$. Moreover, since $r \geq 5$, $||R, F|| \leq 3r - 4$.

Proof Suppose $||R, F|| \geq 2r + 2$. Then, by Lemma 7, there exist two vertices $v, v' \in V(R)$ with $||v, F|| = 3$. Specifically, Lemma 7 gives $N(v) \cap N(v') \cap V(F) = \{x_1, x_2, x_4\}$, but then $vx_1v'x_2v$ forms a cycle with chord x_1x_2, contradicting (O3). This proves the claim.

Claim 27.3 If $F = K_4^-$, then $||R, F|| \leq 3r - 3$.

Proof Suppose that $||R, F|| \geq 3r - 2$. Let v_0, v_0' be the branching vertices in $G[R]$. There exists a $v' \in V(R) - v_0 - v_0'$ such that $||v', F|| = 3$. Moreover, all but possibly two vertices in R have neighbors x_1 and x_3 such that $||x, R|| \geq r - 2 \geq 3$ for every $x \in \{x_1, x_3\}$. If there exist such a $v' \in V(R) - v_0 - v_0'$ and $x \in \{x_1, x_3\}$ with $||v', F|| = 3$ and $||x, R - v'|| \geq 3$, then by Observation 25, $G[R - v' + x]$ contains a chorded cycle and $G[F - x + v'] \supseteq K_4^-$. Hence, $||\{x_1, x_3\}, R - v'|| \leq 4$ and so

$$||\{x_2, x_4\}, R|| \geq ||F, R|| - ||\{x_1, x_3\}, R|| \geq 3r - 2 - 6 = 3r - 8.$$

Then, there exists $v'' \in R$ with $||\{x_2, x_4\}, v''|| \geq \frac{1}{r}(3r - 8) = 3 - \frac{8}{r} > 1$ and so $v''x_2, v''x_4 \in E(G)$. Hence, $G[R - v'']$ is a broken Θ-graph so $G[R - v'' + x_1]$ contains a chorded cycle by Observation 25 and $G[F - x_1 + v''] = K_4^-$.

Now, if $|\mathscr{F}| < k - 1$, then by Lemma 26,

$$||R, \mathscr{F}|| \geq 3(k-2)r + 3r - \ell - 2 \geq 3(k-2)r + 11$$

gives that any type of F would yield a contradiction to (39). Hence $|\mathscr{F}| = k - 1$, and by Lemma 15, either $\{K_4^-, K_4^-\} \subseteq \mathscr{F}$ or $K_4 \in \mathscr{F}$. If $\{K_4^-, K_4^-\} \subseteq \mathscr{F}$, then Claim 27.3 contradicts (39). If $F_1 = K_4$, then by Lemma 26, $||R, F_1|| \leq 2r + 2 \leq 3r - 3$, and so all other cycles F have $||R, F|| \geq 3r - 1$. This forces the exceptional case of Claim 27.1 in which $\ell = 0$ which again contradicts (39). This completes the proof of the lemma.

Lemma 28 *If $k \geq 3$ and $r \geq 5$, then $G[R]$ has no Hamiltonian cycle.*

Proof Suppose $G[R]$ has a Hamiltonian cycle $v_1v_2 \cdots v_rv_1$. Then, since $G[R]$ has no chorded cycles, $G[R]$ is this cycle. Since R has at most two low vertices and by the conditions on G,

$$\sum_{i=1}^{r} d_G(v_i) \geq (r-4)(3k-1) + 2(6k-3) = (3k-1)r - 2.$$

Since, $d_{G[R]}(v_i) = 2$ for each i, we get $||R, V(G) - R|| \geq (3k-3)r - 2$. Hence, if $|\mathscr{F}| \leq k-2$, then

$$\text{there is } F \in \mathscr{F} \text{ such that } ||R, F|| \geq \left\lceil \frac{(3k-3)r-2}{|\mathscr{F}|} \right\rceil \geq 3r + \frac{3r-2}{k-2} > 3r. \tag{40}$$

Then there is $v_i \in R$, say v_1 with $||v_1, F|| \geq 4$. By Lemma 7, $F = K_4$, $||v_1, F|| = 4$, and there is $x \in V(F)$ such that $||x, R|| \geq ||F, R||/4 \geq (3r+1)/4 \geq 16/4 = 4$. Then $G[V(F) - x + v_1] = K_4$ and $G[R - v_1 + x]$ has a vertex x adjacent to at least 3 vertices on the path $v_2, \ldots, v_r$, a contradiction to (O3). Thus $|\mathscr{F}| = k-1$.

Similarly to (40),

$$\text{there is } F \in \mathscr{F} \text{ such that } ||R, F|| \geq \left\lceil \frac{(3k-3)r-2}{|\mathscr{F}|} \right\rceil = 3r - \frac{2}{k-1} \geq 3r - 1 > 2r. \tag{41}$$

By Lemma 7, we have the following cases.

Case 1 $F = K_4$. By symmetry, we may assume $||v_1, F|| \geq 3$. Also, there is $x \in V(F)$ with

$$||x, R|| \geq \left\lceil \frac{3r-1}{4} \right\rceil \geq \left\lceil \frac{14}{4} \right\rceil = 4.$$

Then $G[F - x + v_1]$ contains a K_4^- and $G[R - v_1 + x]$ has a vertex x adjacent to at least 3 vertices on the path $v_2, \ldots, v_r$. Thus G has k disjoint chorded cycles, a contradiction. This completes the case.

For the remaining cases, since no vertex in R may have 4 neighbors in F, by (41) we may assume that

$$||v_i, F|| = 3 \text{ for } 1 \leq i \leq r-1, \text{ and } ||v_r, F|| \geq 2. \tag{42}$$

Case 2 $F = K_4^-$. We may assume that x_1x_3 is the missing edge. By (42), for $1 \leq i \leq r-1$, vertex v_i is adjacent to x_1 and x_3. Thus $G[F - x_1 + v_1] = K_4^-$ and $G[\{x_1, v_2, v_3, v_4\}] = K_4^-$, thus giving k chorded cycles, a contradiction.

Case 3 $F = C_5^+$. By Lemma 7 and (42), v_2 and v_3 are adjacent to x_1 and x_2, i.e. $G[\{v_2, v_3, x_1, x_2\}] = K_4$, contradicting (O1).

Case 4 $F = K_{3,3}$. By Lemma 7 and the symmetry, we may assume that for every odd $1 \leq i \leq r-1$, $N(v_i) \cap F = A$ and for every even $2 \leq i \leq r-1$, $N(v_i) \cap F = B$. In particular, r is even, $r \geq 6$, and $N(v_r) \cap F \subseteq B$. Then $G[F - x_1 + v_2] = K_{3,3}$, and x_1 has 3 neighbors on the path $v_3v_4\cdots v_rv_1$, thus producing another chorded cycle, a contradiction.

Let $W_1 := \{v_1, v_2\}$, $W_2 := \{v_{r-1}, v_r\}$, and $W := W_1 \cup W_2 = \{v_1, v_2, v_{r-1}, v_r\}$. We will use ℓ to denote the number of low vertices in W.

Recall that by Lemma 28, $v_1v_r \notin E(G)$. If $v_2v_{r-1} \in E(G)$ and any of the edges v_1v_{r-1}, v_2v_r is in $E(G)$, then $G[R]$ has a cycle with the chord v_2v_{r-1}. If both edges v_1v_{r-1} and v_2v_r are in $E(G)$, then $G[R]$ has a spanning Θ-subgraph, a contradiction to Lemma 27. Thus,

$$||W_1, W_2|| \le 1. \text{ In particular } d_G(v_1)+d_G(v_2)+d_G(v_{r-1})+d_G(v_r) \ge 2(6k-3)=12k-6. \tag{43}$$

Lemma 29 *Suppose $k \ge 3$. Then for all $F \in \mathscr{F}$, $||W, F|| \le 12$; in particular, $|\mathscr{F}| = k-1$.*

Proof Suppose that $||W, F|| \ge 13$. By symmetry we may assume $||W_1, F|| \ge 7$. Then, there exists $w \in W_1$ with $||w, F|| \ge 4$. So, by Lemma 7, $F = K_4$ and $||w, F|| = 4$. By the pigeonhole principle, there is $x \in V(F)$ with $||x, W|| = 4$. Then $G[F - x + w] = K_4$ and $||x, R - w|| \ge 3$. So if $G[R] - w$ has a Hamiltonian path (or even just a path containing $W - w$), then x has 3 neighbors on this path and hence $G[R - v_1 + x]$ contains a chorded cycle, contradicting (O3). Otherwise, by the definition of W, $w = v_2$ and $d_{G[R]}(v_1) = 1$. In particular, $||v_1, F|| = 3$ and $||v_2, F|| = 4$. Let x' be the nonneighbor on v_1 in F. If $||x', R|| \ge 2$, then $G[F - x' + v_1] = K_4$ and $G[R - v_1 + x']$ has Hamiltonian path and has more edges than $G[R]$, contradicting (O6). Thus,$N_G(x') \cap R = \{v_2\}$ and for all $y \in W - v_2$, $N_G(y) \cap R = W - v_2$. Hence, $G[W_2 \cup (V(F) - \{x, x'\})] = K_4$ and $G[W_1 \cup \{x, x'\}] = K_4^-$, contradicting (O1). Therefore, $||W, F|| \le 12$.

By (43), $\sum_{w \in W} d_G(w) \ge 12k - 6$. By the definition of W,

$$||W, \mathscr{F}|| \ge 12k - 6 - 10 = 12(k-2) + 8.$$

Since $||W, F|| \le 12$ for all F_i, this implies $|\mathscr{F}| \ge k - 1$. So $|\mathscr{F}| = k - 1$.

Lemma 30 *For all $F \in \mathscr{F}$ with $F = K_4^-$, $||W, F|| \le 10$. Moreover, if $d_{G[R]}(v_1) = d_{G[R]}(v_r) = 2$, then $||W, F|| \le 9$.*

Proof Suppose that $||W, F|| \ge 11$. Then by pigeonhole and symmetry we may assume $||W_1, F|| \ge 6$. By Lemma 7, $||v_1, F|| = ||v_2, F|| = 3$. By the same argument, we can choose $w \in W_2$ such that $||w, F|| = 3$ and for $w' \in W_2 - w$, we have $||w', F|| \ge 2$. Again by Lemma 7, $\{x_1, x_3\} \subseteq N(v_1) \cap N(v_2) \cap N(w)$. So if $\{x_1, x_3\} \cap N(w') \neq \emptyset$, say $x_1w' \in E(G)$, then $G[F - x_1 + v_1] \supseteq K_4^-$, and x_1 has 3 neighbors on the path $P - v_1$, implying $G[R - v_1 + x_1]$ contains a chorded cycle. This contradicts (O3).

Hence, $N(w') \cap F = \{x_2, x_4\}$. Then $G[F - x_1 + w'] = K_4^-$, and x_1 has 3 neighbors in $R - w'$. So if $G[R - w']$ has a Hamiltonian path, then $G[R - w' + x_1]$ contains a chorded cycle, a contradiction to (O3). Otherwise, $w = v_r$, $w' = v_{r-1}$, and $d_{G[R]}(v_r) = 1$. In this case $G[F - x_1 + v_r] = K_4^-$, $G[R - v_r + x_1]$ has a Hamiltonian path and has more edges than $G[R]$, contradicting (O6). This proves the main claim.

Now suppose $||W, F|| \geq 10$ and $d_{G[R]}(v_1) = d_{G[R]}(v_r) = 2$. There are $j \geq 3$ and $q \leq r - 2$ such that $v_1 v_j, v_q v_r \in E(G)$. By Lemma 27, $3 \leq j \leq q \leq r - 2$. First we show:

$$\text{If } x \in \{x_1, x_3\} \text{ and } i \in \{1, 2\}, \text{ then } ||x, W_i|| \leq 1. \tag{44}$$

Indeed, assume say $x_1 v_1, x_1 v_2 \in E(G)$. If $||W_2, F - x_1|| \geq 3$, then $G[W_2 \cup F - x_1]$ contains a chorded 4-cycle with exactly one vertex in W_2, and $v_2 \ldots v_j v_1 x_1 v_2$ is a cycle with chord $v_1 v_2$, contradicting (O3). Thus $||W_2, F - x_1|| \leq 2$. If $||x_1, W_2|| = 2$, then symmetrically $||W_1, F - x_1|| \leq 2$ and hence $||F, W|| \leq ||x_1, W|| + 2 \cdot 2 = 4 + 4 = 8$, contradicting the assumption $||W, F|| \geq 10$. Hence,

$$||F, W_2|| \leq 2 + 1 = 3 \text{ and so } ||F, W_1|| \geq 10 - 3 = 7.$$

But this contradicts Lemma 7. This proves (44).

Recall that by Lemma 7, inequality $||W, F|| \geq 10$ implies that at least two vertices in R have 3 neighbors in F, and each vertex in R with 3 neighbors in F is adjacent to both x_1 and x_3. Together with (44), this yields that for each of $i \in \{1, 2\}$,

(a) exactly one vertex $w_i \in W_i$ has 3 neighbors in F (hence $w_i x_1, w_i x_3 \in E(G)$), and
(b) the other vertex $w_i' \in W_i$ has exactly two neighbors in F, and these neighbors are x_2 and x_4.

We know that one of x_2 and x_4, say x_2, is adjacent to w_2. Then $G[\{x_2, x_3\} \cup W_2] = K_4^-$ and $G[\{v_1, \ldots, v_j, x_1, x_4\}]$ contains a Hamiltonian cycle that is the union of the paths $v_2 \ldots v_j v_1$ and $w_1 x_1 x_2 w_1'$ which has chord $v_1 v_2$. This contradicts (O3) and proves the lemma.

Lemma 31 *For all $F \in \mathcal{F}$ with $F = C_5^+$, $||W, F|| \leq 9$.*

Proof Suppose $||W, F|| \geq 10$. By Lemma 7, there exist $v, v' \in W$ with $N(v) = N(v') = \{x_1, x_2, x_4\}$. Then, $G[\{v, v', x_1, x_2\}] \supseteq K_4^-$, contradicting (O1).

Lemma 32 *For all $F \in \mathcal{F}$ with $F = K_{3,3}$, $||W, F|| \leq 11$. Moreover, if $||W, F|| = 11$ and $\sum_{w \in W} d_R(w) \geq 8$, then $v_2 v_r \in E(G)$ or $v_1 v_{r-1} \in E(G)$, $||\{v_1, v_r\}, F|| = 5$, and $\sum_{w \in W} d_R(w) = 8$.*

Proof By Lemma 7, for each $w \in W$, $N(w) \cap F$ is contained in one part of F. So $||W, F|| \leq 12$. Assume that $||W, F|| \geq 11$. Then by Lemma 7, we may assume that $N(v_1) \cap V(F) = A = \{x_1, x_3, x_5\}$ and $N(v_2) \cap V(F) = B = \{x_2, x_4, x_6\}$.

We claim that $N(v_r) \cap V(F) \subseteq A$. For otherwise, $N(v_r) \cap V(F) \subseteq B$. Then by Lemma 7, $N(v_{r-1}) \cap V(F) \subseteq A$. Let x_2 be a neighbor of v_r, then $G[F - x_2 + v_1] =$

$K_{3,3}$ and $G[R - v_1 + x_2]$ contains a spanning cycle $x_2 v_2 \cdots v_r x_2$, a contradiction to (O6) or Lemma 28. Thus $N(v_r) \cap V(F) \subseteq A$.

Now by Lemma 7, $N(v_{r-1}) \cap V(F) \subseteq B$. In order for $||W, F|| \geq 11$, we need $||v, F|| \geq 2$ for $v \in \{v_{r-1}, v_r\}$. Let $x_2, x_4 \in N(v_{r-1})$ and $x_1, x_3 \in N(v_r)$. Then $G[F - x_2 - x_4 + v_1 + v_r] = K_{3,3}$ and $G[P - v_1 - v_r + x_2 + x_4]$ is a spanning Θ-graph of the remainder, a contradiction to (O7) or Lemma 27, unless $d_{G[R]}(v_1) = d_{G[R]}(v_r) = 2$, where $v_1 v_j, v_r v_q \in E(G)$ with $3 \leq j \leq q \leq r-2$. But then, $G[F - x_1 - x_2 + v_1 + v_2] = K_{3,3}$ and $x_2 x_1 v_r vq v_{q+1} \cdots v_{r-1} x_2$ is a cycle with chord $v_{r-1} v_r$, producing k disjoint chorded cycles.

Moreover, we only needed the path chord $v_q v_r$ with $q > 2$. A very similar argument would work if our path chord were $v_1 v_j$ with $j < r-1$. Hence, assume for the secondary claim that $||W, F|| = 11$ and $\sum_{w \in W} d_R(w) \geq 8$. If $v_1 v_{r-1}, v_2 v_r \notin E(G)$ or if $\sum_{w \in W} d_{G[R]}(w) > 8$, then $v_2 v_i \in E(G)$ for some $3 < i \leq r-1$ (or $v_j v_{r-1} \in E(G)$ for some $2 \leq j < r-2$), but $G[F - x_2 + v_1] = K_{3,3}$ and $x_2 v_2 \cdots v_{r-1} x_2$ is a cycle with chord $v_2 v_i$ (or $v_j v_{r-1}$). So $v_2 v_r \in E(G)$ or $v_1 v_{r-1} \in E(G)$, and $\sum_{w \in W} d_R(w) = 8$. If $||\{v_1, v_r\}, F|| = 6$, then $G[F - x_2 - x_4 + v_1 + v_r] = K_{3,3}$ and $G[R - v_1 - v_r + x_2 + x_4]$ is a Θ-graph, contradicting either (O7) or Lemma 27. This proves the secondary claim.

Lemma 33 *For all $F \in \mathscr{F}$ with $F = K_4$, $||W, F|| \leq 11$. Moreover, if $||W, F|| = 11$, then $v_2 v_r$ or $v_1 v_{r-1} \in E(G)$, $||\{v_1, v_r\}, F|| \leq 5$, and $\sum_{w \in W} d_G(w) = 8$.*

Proof Suppose that $||W, F|| \geq 12$. We need some extra notation. For each $x_i \in V(F)$, let $a_{i,j} := ||x_i, W_j||$. Hence, $\sum_{i,j} a_{i,j} = ||W, F|| = 12$.

Case 1 $||W_1, F|| = 8$ and $||W_2, F|| \geq 4$. If there exist j_1, j_2 such that $a_{j_1,2} + a_{j_2,2} \geq 3$, then $G[W_2 + x_{j_1} + x_{j_2}] \supseteq K_4^-$ and $G[W_1 \cup V(F) - x_{j_1} - x_{j_2}] = K_4$, contradicting (O1). Otherwise $a_{j,2} = 1$ for all j.

If $||v_r, F|| = 4$, then $G[R] = C_r$ as otherwise, for each $x \in V(F)$, $G[V(F) - x + v_1] = K_4$ but $G[R - v_1 + x] \supseteq C_r$, contradicting (O6). By Lemma 28, we have $|R| \leq 4$ which contradicts Lemma 24.

If $||v_{r-1}, F|| = 4$, then $d_{G[R]}(v_1) = 2$ as otherwise, for each $x \in V(F)$, $G[F - x + v_1] = K_4$ but $\sum\{d_{G[R]}(v) : v \in R - v_1 + x\} \geq \sum\{d_{G[R]}(v) : v \in R\} + 1$, contradicting (O7). Moreover, if $v_1 v_{r-1} \notin E(G)$, then for each $x \in V(F)$, $G[R - W_2 + x]$ contains a chorded cycle and $G[F - x + v_{r-1}] = K_4$, so we have k disjoint chorded cycles, a contradiction. Hence $v_1 v_{r-1} \in E(G)$. By symmetry, $||v_{r-2}, F|| = 4$ which gives $G[F - x + v_1] = K_4$ and $G[R - v_1 + x]$ containing a chorded cycle unless $r - 2 = 2$, i.e. $|R| = 4$ which contradicts Lemma 24.

If $||v_r, F|| \in \{1, 2, 3\}$, say $v_r x \in E(G)$, then $G[F - x + v_1] = K_4$ and $G[R - v_1 + x] = C_r$, contradicting Lemma 28 unless $|R| = 4$, in which case, we contradict Lemma 24.

Case 2 $||W_1, F|| = 7$ and $||W_2, F|| \geq 5$. Choose j_1, j_2 such that $a_{j_1,2} + a_{j_2,2} \geq 3$. Then $G[W_2 + x_{j_1} + x_{j_2}] \supseteq K_4^-$ and $G[W_1 \cup V(F) - x_{j_1} - x_{j_2}] \supseteq K_4^-$, contradicting (O1).

Case 3 $||W_1, F|| = 6$ and $||W_2, F|| = 6$. Then for $j \in \{1, 2\}$, $\{a_{1,j}, a_{2,j}, a_{3,j}, a_{4,j}\} \in \{\{2, 2, 2, 0\}, \{2, 2, 1, 1\}\}$. If there exists an i such that $(a_{i,1}, a_{i,2}) = (0, 0)$, then for each $i' \neq i$, $G[R - v_1 + x_{i'}]$ contains a chorded cycle due to $||x_{i'}, W - v_1|| = 3$, but also $G[F - x_{i'} + v_1] = K_4^-$, so we have k disjoint chorded cycles, a contradiction. Otherwise, each pair $(a_{i,1}, a_{i,2})$ has a nonzero entry.

Let $a_{i',1} = \min_i a_{i,1}$ and $a_{i'',2} = \min_i a_{i,2}$ and $k', k'' \in \{1, 2, 3, 4\} - i' - i''$ distinct. If $a_{i',1} = a_{i'',2} = 0$ and $i' = i''$, then $G[R - v_1 + x_{k'}]$ contains a chorded cycle due to $||x_{k'}, W - v_1|| = 3$, but also $G[F - x_{k'} + v_1] = K_4^-$, so we have k disjoint chorded cycles, a contradiction.

If $i' \neq i''$ or one of the minima is nonzero, then $G[W_1 + x_{i''} + x_{k''}] \supseteq K_4^-$ and $G[W_2 + x_{i'} + x_{k'}] \supseteq K_4^-$, contradicting (O1). This proves the main claim. We now prove the secondary statement in a series of claims.

Claim 33.1 For $F = K_4$, , if $d_{G[R]}(v_1) = d_{G[R]}(v_r) = 2$, then $||W, F|| \leq 9$.

Proof Suppose that $v_1v_i, v_qv_r \in E(G)$ for some $3 \leq i \leq q \leq r - 2$. For convenience, let $R_1 = \{v_1, \ldots, v_i\}$ and $R_2 = \{v_q, \ldots, v_r\}$ where $R \supseteq R_1 \cup R_2$. Suppose that $||W, F|| \geq 10$. By symmetry we may assume $||W_1, F|| \geq 5$ and so $||W_2, F|| \geq 2$. Hence, there exists $x \in V(F)$ such that $xv_1, xv_2 \in E(G)$. Then, if $||F - x, W_2|| \geq 2$, $G[F - x + \{v_r, \ldots, v_q\}]$ contains a chorded cycle and $xv_2 \cdots v_iv_1x$ is another cycle with chord v_1v_2 unless $i = q$. We postpone handling this exception to first address the case $||F - x, W_2|| \leq 1$. Therefore, $||x, W_2|| \geq 1$ and $||W_1, F|| \geq 7$. Specifically, there exists $x' \in V(F) - x$ such that $||\{x, x'\}, W_2|| \geq 2$. If these neighbors in W_2 are distinct, then $G[R_2 + x + x']$ contains a cycle with chord $v_{r-1}v_r$ and $||W_1, F - x - x'|| \geq 7 - 4 = 3$ so that $G[W_1 \cup V(F) - x - x'] \supseteq K_4^-$. If instead these neighbors in W_2 are not distinct, we can choose $y \in V(F) - x - x'$ such that $||y, W_1|| = 2$ and so $v_1yv_2 \cdots v_iv_1$ is a cycle with chord v_1v_2 and $G[F - y \cup W_2] \supseteq K_4^-$. Thus, we have k disjoint chorded cycles.

We now handle this exception where $i = q$. Recall, $||x, W_1|| = 2$. So, if $||F - x, W_2|| \geq 3$, $G[R_1 + x]$ contains a chorded cycle and $G[W_2 + F - x]$ contains a chorded cycle. Hence, $||F - x, W_2|| \leq 2$. If $||x, W_2|| = 2$, then $G[R_2 + x]$ contains a chorded cycle and $||W_1, F - x|| \geq 10 - ||W_2, F|| - ||x, W_1|| \geq 10 - 4 - 2 = 4$. So $G[F - x + W_1]$ contains a chorded cycle, thus producing k disjoint chorded cycles. Therefore, $||x, W_2|| \leq 1$ and so $||F, W_2|| \leq 3$. Thus, $||F, W_1|| \geq 7$. Choose $x'', x''' \in V(F)$ such that $||\{x'', x'''\}, W_2|| \geq 2$. Then $G[R_2 + x'' + x''']$ contains a chorded cycle and $G[W_1 + F - \{x'', x'''\}]$ contains a chorded 4-cycle since $||W_1, F - x'' - x'''|| \geq 3$. This completes the case where $i = q$ and the claim.

Claim 33.2 For $F = K_4$, if $d_{G[R]}(v_1) = 1$, $d_{G[R]}(v_r) = 2$, and $v_2v_r \notin E(G)$, then $||W, F|| \leq 10$.

Proof Suppose $||W, F|| \geq 11$. First consider if $||v_1, F|| = 4$. Then, there exists $x \in V(F)$ such that $||x, W - v_1|| \geq 2$. If $\{v_2, v_r\} \supseteq N(x)$ or $\{v_{r-1}, v_r\} \supseteq N(x)$, then $G[R - v_1 + x]$ contains a chorded cycle with $G[F - x + v_1] = K_4$, producing k chorded cycles. If instead $\{v_2, v_{r-1}\} \supseteq N(x)$, $G[F - x + v_1] = K_4$, but $G[R - v_1 + x]$

contains a Hamiltonian path $xv_2 \cdots v_r$ with $|E(G[R - v_1 + x])| > |E(G[R])|$, contradicting (O7).

If $||v_1, F|| = 3$, say $x \in V(F) - N(v_1)$. Just as in the $||v_1, F|| = 4$ case, we have that $||x, W|| \leq 1$. Hence, $||W - v_1, F - x|| \geq 11 - 3 - 1 = 7$. Specifically, there exists $x' \in V(F) - x$ such that $||x', W - v_1|| = 3$. Therefore, $G[F - x' + v_1] = K_4^-$ and $G[R - v_1 + x']$ contains a chorded cycle since $||x', W - v_1|| = 3$. Again, this yields k chorded cycles.

Now, suppose $||v_1, F|| = 2$ where $N(v_1) \cap V(F) = \{x, x'\}$ and $\{y, y'\} = N(v_1) \cap V(F) - \{x, x'\}$. If $||W_2, F|| \leq 3$, then $||W_1, F|| = 8$ and $||W_2, F|| = 3$. Now, if $||\{y, y'\}, W_1|| \geq 1$ (say $||y, W_1|| \geq 1$), then $G[\{W_1 + x + y\}] \supseteq K_4^-$ and $G[R_2 + x']$ contains a chorded cycle, producing k chorded cycles. Hence, $||\{y, y'\}, W_1|| = 0$. If $||y, W_2|| = 2$ or $yv_{r-1}, y'v_r \in E(G)$, then $G[R_2 + y + y']$ contains a chorded cycle and $G[W_1 + x + x'] = K_4$, producing k disjoint chorded cycles. Thus, we are left with the subcase where there exists $v \in W_2$ such that $vy, vy' \in E(G)$. Then, $G[F - x + v] \supseteq K_4^-$ and $||x, W - v|| = 3$ so that $G[R - v + x]$ contains a chorded cycle since there exists a path in $G[R]$ covering $W - v$. We are now left with the subcase where $||\{y, y'\}, W|| \geq 4$. In fact, the previous argument still works unless $||y, W - v_1||, ||y', W - v_1|| = 2$. Moreover, $N(y) \cap W = N(y') \cap W = \{v_2, v_{r-1}\}$. But then $||W, \{x, x'\}|| \geq 7$, so we may assume $||x, W|| = 4$. Then, $||W_1, F - x|| \geq 3$ and so $G[W_1 + c - x]$ contains a chorded cycle with $xv_rv_q \cdots v_{r-1}x$ is a cycle with chord $v_{r-1}v_r$. This again yields k disjoint chorded cycles.

Lastly, if $||v_1, F|| \leq 1$, then $||W_2, F|| \geq 11 - 5 = 6$ and so there exist at least 2 vertices $y, y' \in V(F)$ such that $||y, W_2|| = ||y', W_2|| = 2$. Moreover, we can choose y so that $||y, W_2|| = 2$ and $||W_1, F - y|| \geq 3$. Thus, $yv_{r-1} \cdots v_q v_r y$ is a cycle with chord $v_{r-1}v_r$ and $G[F - y + W_1]$ contains a chorded cycle, producing k disjoint chorded cycles. This completes the claim.

Claim 33.3 For $F = K_4$, if $d_{G[R]}(v_1) = d_{G[R]}(v_r) = 1$, then $||W, F|| \leq 10$.

Proof Suppose that $||W, F|| = 11$ since we have that $||W, F|| \leq 11$ by the main statement of the lemma and Lemmas 30–32. If $\sum_W d_{G[R]}(w) \leq 7$ and ℓ is the number of low vertices in W, then $||W, \mathcal{F}|| \geq 4(3k - 1) - \ell - \sum_W d_{G[R]}(w) \geq 12(k - 1) - 1$, a contradiction since $||W, F|| \leq 11(k - 1) < 12(k - 1) - 1$ for $k \geq 3$. So $\sum_W d_{G[R]}(w) \geq 8$.

Case 1 There exists $x \in V(F)$ such that $||x, W|| = 4$. If $v \in \{v_1, v_r\}$ has $||v, F|| \geq 3$, then $G[F - x + v] \supseteq K_4^-$ and $G[R - v + x]$ contains a chorded cycle since $||x, R - v|| \geq 3$ and $G[R - v]$ contains a Hamiltonian path. Thus, we have k chorded cycles, a contradiction, and so $||v, F|| \leq 2$ for all $v \in \{v_1, v_r\}$. Since $||W, F|| = 11$, by symmetry we may assume $||W_1, F|| \geq 6$, and so there exist distinct $y, y' \in V(F) \cap N(v_1) \cap N(v_2)$. Since $||v_1, F|| \leq 2$, then $||v_1, F|| = 2$ and $||v_2, F|| = 4$. Hence, $||W_2, F|| = 5$ and so there exists $x \in V(F) \cap N(v_{r-1}) \cap N(v_r)$. Say $x \neq y$ and choose $x' \in V(F) - x - y$ with $||x', W_2|| \geq 1$. This choice is possible due to $||W_2, F|| = 5$ implying that $||v_{r-1}, F|| \geq 3$. From this, $G[W_2 + x + x'] \supseteq K_4^-$ and $G[W_1 \cup V(F) - x - x'] \supseteq K_4^-$, thus giving us k chorded cycles, a contradiction. This completes the case.

Case 2 For all $x \in V(F)$, $||x, W|| \leq 3$. Since $\sum_W d_{G[R]}(w) \geq 8$, we have at least one chord on our path. First, consider the case that this chord e uses vertex v_2, v_{r-1}. By the case, this chord is not v_2v_r nor $v_{r-1}v_1$. In fact, if the chord is not v_2v_{r-1}, then we have two chords: v_2v_i and v_jv_{r-1} for some i, j. Due to this symmetry, we may assume $||W_1, F|| \geq 6$. Again, we will use the notation that for each $x_i \in F$, $a_{i,j} := ||x_i, W_j||$. Hence, $\sum_{i,j} a_{i,j} = ||W, F|| = 11$. If three vertices in $V(F)$ have two neighbors on one side, then we have a contradiction. Namely, if $a_{1,j} = a_{2,j} = a_{3,j} = 2$ and $i' := \arg\max_{1 \leq i \leq 3}\{a_{i,3-j}\}$, then $a_{i',3-j} \geq 1$. So for $j = 1$, $G[F - x_{i'} + v_1] \supseteq K_4^-$ and $G[R - v_1 + x_{i'}]$ contains a cycle with chord e, giving us k chorded cycles. For $j = 2$, just replace v_1 with v_r. Since at most two vertices in $V(F)$ have two neighbors on one side, assume $a_{1,1} = a_{2,1} = 2$, $a_{3,1} = 1$, and $a_{4,1} \leq 1$. By the case, $a_{1,2}, a_{2,2} \leq 1$ and so $a_{3,2} + a_{4,2} = ||\{x_3, x_4\}, W_2|| \geq 3$. Hence, $G[W_2 + x_3 + x_4] \supseteq K_4^-$ and $G[W_1 + v_1 + v_2] = K_4$, and so we have k chorded cycles, a contradiction. This completes the case and thus the claim.

Claim 33.4 Suppose $r \geq 5$, $G[R]$ contains a Hamiltonian path $v_1v_2 \ldots v_r$, and v_2v_r or $v_1v_{r-1} \in E(G)$. Let $F \in \mathscr{F}$ and $F = K_4$. If

$$||\{v_1, v_r\}, F|| \geq 6, \tag{45}$$

then $||W, F|| \leq 10$.

Proof Suppose (45) holds but $||W, F|| \geq 11$. By (45), $||v_1, F|| \geq 2$. Thus we have three cases.

Case 1 $||v_1, F|| = 4$. Since $||W, F|| \geq 11$, there is $x \in F$ with $||x, W|| \geq 3$. Then $G[F - x + v_1] = K_4$, $G[R - v_1 + x]$ has a Hamiltonian path starting from x, and x has at least 2 neighbors in $G[R - v_1 + x]$, contradicting (O7).

Case 2 $||v_1, F|| = 3$. Then by (45), $||v_r, F|| \geq 3$. If there is $x \in V(F) \cap N(v_r)$ adjacent to some $v \in \{v_2, v_{r-1}\}$, then $G[R - v_1 + x]$ contains a cycle passing through edges xv and xv_r with chord v_rv, and $G[V(F) - x + v_1] \supseteq K_4^-$, producing k disjoint chorded cycles. Otherwise, $||\{v_2, v_{r-1}, v_r\}, F|| \leq 3 \cdot 1 + 1 \cdot 2 = 5$, and hence $||W, F|| \leq 5 + 3 = 8$, a contradiction.

Case 3 $||v_1, F|| = 2$. Then by (45), $||v_r, F|| = 4$. Suppose $N(v_1) \cap F = \{x_1, x_2\}$. If at least one $x \in \{x_3, x_4\}$ is adjacent to some $v \in \{v_2, v_{r-1}\}$, then again $G[R - v_1 + x]$ contains a cycle passing through edges xv and xv_r with chord v_rv, and $G[F - x + v_1] \supseteq K_4^-$, a contradiction since then we have k disjoint chorded cycles. Otherwise, $||\{v_2, v_{r-1}\}, F|| \leq 2 \cdot 2 = 4$, and hence $||W, F|| \leq 4 + 6 = 10$, as claimed.

By these claims, we can only have $||W, F|| = 11$ if v_2v_r or $v_1v_{r-1} \in E(G)$ but the condition of Claim 33.4 is not met, i.e. $||\{v_1, v_r\}, F|| \leq 5$, as claimed. Or, if $v_1v_i \in E(G)$ or $v_jv_r \in E(G)$ for some $3 \leq i, j \leq r - 2$, either $G[R]$ contains a chorded cycle or is a spanning Θ-graph, contradicting Lemma 27. Therefore,

$d_{G[R]}(v_1) = 1$ and $d_{G[R]}(v_{r-1}) = d_{G[R]}(v_r) = 2$, and $d_{G[R]}(v_2) = 3$ so that $\sum_{w \in W} d_G(w) = 8$.

We now finish the proof of the main result of this section. Recall, $P = v_1 \dots v_r$ is a Hamiltonian path in $G[R]$, and $W = \{v_1, v_2, v_{r-1}, v_r\}$ has ℓ low vertices. Then ,

$$||W, \mathscr{F}|| \geq 4(3k-1) - \ell - \sum_{w \in W} d_{G[R]}(w).$$

On the other hand, by Lemmas 30–33, for each $F \in \mathscr{F}$, $||W, F|| \leq 11$ and $|\mathscr{F}| = k-1$. So $||W, \mathscr{F}|| \leq 11(k-1)$. It follows that $4(3k-1) - \ell - \sum_{w \in W} d_R(w) \leq 11(k-1)$, and we obtain

$$\sum_{w \in W} d_R(w) \geq k + 7 - \ell \geq 3 + 7 - 2 = 8. \tag{46}$$

If $\sum_{w \in W} d_R(w) > 8$, then by Lemmas 32 and 33 , $||W, F|| \leq 10$ for each $F \in \mathscr{F}$. Then

$$4(3k-1) - \ell - \sum_{w \in W} d_R(w) \leq ||W, \mathscr{F}|| \leq 10(k-1),$$

and we obtain $\sum_{w \in W} d_{G[R]}(w) \geq 2k + 6 - \ell \geq 2 \cdot 3 + 6 - 2 = 10$. As $d_{G[R]}(v_1), d_{G[R]}(v_r) \leq 2$ and $d_{G[R]}(v_2), d_{G[R]}(v_{r-1}) \leq 3$, we have $\ell = 2$, and $d_{G[R]}(v_1) = d_{G[R]}(v_r) = 2$ and $d_{G[R]}(v_2) = d_{G[R]}(v_{r-1}) = 3$, and $||W, F|| = 10$ for each F. By Lemma 30 and Claim 33.1, $F \notin \{K_4^-, K_4\}$, a contradiction to Lemma 15.

So $\sum_{w \in W} d_R(w) = 8$. It follows that $k = 3$ and $\ell = 2$, and $||W, F|| = 11$ for each $F \in \mathscr{F}$. By Lemmas 30–33, we may assume $\mathscr{F} = \{K_4, K_4\}$, $v_2 v_r \in E(G)$, and $||\{v_1, v_r\}, F|| \leq 5$ and $||W, F| = 11$. Since $v_2 v_r \in E(G)$, we have $d_{G[R]}(v_1) = 1$ by Lemma 27. So,

$$||\{v_1, v_r\}, \mathscr{F}|| \geq d_G(v_1) + d_G(v_r) - (d_{G[R]}(v_1) + d_{G[R]}(v_r)) \geq 6k - 3 - 3 = 12.$$

This contradicts $||\{v_1, v_r\}, F|| \leq 5$ for both F. This proves the case where $G[R]$ has a Hamiltonian path and $k \geq 3$.

5 Case: $G[R]$ Has a Hamiltonian Path and $k = 2$

Lastly, we handle the case where $k = 2$ and $G[R]$ has a Hamiltonian path. The sequence of the cases is similar to the previous section, but the proofs are different: some things are harder because we now have only one chorded cycle in $\mathscr{F}$ with which to work, and some other things are easier because $G[R \cup C]$ is the whole G.

Lemma 34 *For $k = 2$, $G[R]$ is not a Hamiltonian cycle.*

Proof Suppose that $G[R]$ is a cycle $v_1v_2 \dots v_rv_1$. By (1), and since the set of low vertices is a clique, at least $r-2$ vertices in R have degree at least 5 but only degree 2 in $G[R]$. By Lemma 11 and since low vertices form a clique, we may assume

$$||R, F|| \geq (r-2)(3k-1)+2(3k-2)-2r=3r-2 \text{ and } d_G(v_1), \dots, d_G(v_{r-2}) \geq 5. \tag{47}$$

By Lemma 7, we have four cases.

Case 1 $F = K_4$. By (47), $3 \leq ||v_1, F|| \leq 4$. So

$$G[F - x_j + v_1] \quad \text{contains a } K_4^- \text{foreach} 1 \leq j \leq 4. \tag{48}$$

If $r \geq 5$, then by (47), some x_j has at least

$$\left\lceil \frac{3r-2-||v_1, F||}{r-1} \right\rceil \geq \left\lceil \frac{3r-6}{r-1} \right\rceil = 3 - \left\lfloor \frac{3}{r-1} \right\rfloor = 3$$

neighbors in the path $G[R - v_1]$. But then $G[R - v_1 + x_j]$ contains a chorded cycle. Together with (48), we have two disjoint chorded cycles, a contradiction.

If instead $r = 4$, then by (47), we may assume that $||x_1, R|| \geq \left\lceil \frac{10}{4} \right\rceil = 3$. If $N(x_1) \supset R - v_1$, then $G[R - v_1 + x_1] = K_4^-$ and by (47), v_1 has at least two neighbors in $V(F) - x_1$. So $G[F - x_1 + v_1]$ also is a chorded cycle, a contradiction. Thus x_1 has exactly one nonneighbor in $R - v_1$, say $v_i, i \neq 1$.

If v_i has at least two neighbors in F (and hence in $F - x_1$), then again G has disjoint chorded cycles: $G[R - v_i + x_1]$ and $G[F - x_1 + v_i]$, a contradiction. So v_i has at most one neighbor in F. Since $\delta(G) \geq 3$ by Lemma 12, we may assume that $N(v_i) \cap V(F) = \{x_4\}$. Then for each $z \in \{x_1, x_2, x_3, v_{i-2}\}, d_G(z) \geq 9 - d_G(v_i) = 6$, where we interpret v_{i-2} with modularity, e.g. $v_{-1} = v_{r-1}$. So each such z is not adjacent only to v_i, and so $G \supseteq G_3(2)$. If $v_{i-1}x_4$ or $v_{i+1}x_4 \in E(G)$, say $v_{i+1}x_4 \in E(G)$, then we have two disjoint chorded cycles $G[\{v_i, v_{i+1}, x_3, x_4\}] = K_4^-$ and $G[R - v_i - v_{i+1} + x_1 + x_2] = K_4$. Hence, $G = G_3(2)$. This proves the case.

Case 2 $F = K_4^-$. By Lemma 7 and (47), for each $1 \leq i \leq r-2$ there exists $j_i \in \{2, 4\}$ such that $N(v_i) \cap V(F) = \{x_{j_i}, x_1, x_3\}$. Note, $j_i \neq j_{i+1}$ as otherwise $G[v_i, v_{i+1}, x_1, x_{j_i}\}] = K_4$, contradicting (O2). By (1),

$$d(v_{r-1}) \geq 4 \text{ and } d(v_r) \geq 4. \tag{49}$$

We claim that

$$||\{v_{r-1}, v_r\}, \{x_1, x_3\}|| = 0. \tag{50}$$

Indeed, suppose, for example, $v_r x_3 \in E(G)$. If $r \geq 5$, then $G[F - x_3 + v_1] = K_4^-$, and $G[R - v_1 + x_3]$ contains a chorded cycle since $||R - v_1, x_3|| \geq 3$, producing two disjoint chorded cycles. So, let $r = 4$, and then $G[\{v_4, v_1, x_{j_1}, x_3\}] \supseteq K_4^-$. Moreover, if v_3 has a neighbor in $\{x_{j_2}, x_1\}$, then also $G[\{v_3, v_2, x_{j_2}, x_1\}] \supseteq K_4^-$, producing two disjoint chorded cycles. So we may assume v_3 has no neighbor in $\{x_{j_2}, x_1\}$, and hence by (49), $x_3 v_3 \in E(G)$. In this case, $G[\{v_1, v_4, v_3, x_3\}] = K_4^-$ and $G[\{v_2, x_1, x_2, x_4\}] = K_4^-$, a contradiction. This proves (50).

By (50), $N(v_r) \cap V(F) = \{x_2, x_4\} = N(v_{r-1}) \cap V(F)$. So $G[\{v_r, v_{r-1}, x_2, x_4\}] = K_4$, contradicting (O2).

Case 3 $F = C_5^+$. By (47) and Lemma 7, $N(v_1) \cap V(F) = N(v_2) \cap V(F) = \{x_1, x_2, x_4\}$. Thus, $G[\{v_1, v_2, x_1, x_2\}| = K_4$, contradicting (O1).

Case 4 $F = K_{3,3}$ with parts $A = \{x_1, x_3, x_5\}$ and $B = \{x_2, x_4, x_6\}$. By Lemma 8,

$$N(v_i) \cap F \subseteq A \text{ or } N(v_i) \cap F \subseteq B \text{ for every } 1 \leq i \leq r. \tag{51}$$

Together with (47), this implies $d_G(v_1) = d_G(v_2) = 5$ and so $d_G(v_{r-1}), d_G(v_r) \geq 4$. By symmetry, we may assume $N(v_1) \cap V(F) = A$ and then this forces $N(v_2) \cap V(F) = B$. By Lemma 8, $N(v_r) \cap V(F) \subseteq B$, say $N(v_r) \cap V(F) \supseteq \{x_2, x_4\}$. Similarly $v_2 v_3 \in E(G)$ forces $N(v_3) \cap F \subseteq A$, and so on such that r is even. If $r = 4$, then G is a bipartite graph with parts $A \cup \{v_2, v_4\}$ and $B \cup \{v_1, v_3\}$, i.e. it is a subgraph of $K_{5,5}$. But after deleting any edge yz from a $K_{5,5}$, (1) fails for y and z, i.e. $d_G(y) + d_G(z) \leq 4 + 4 < 9$. Hence $G = K_{5,5} = G_1(10, 2)$, contradicting the assumption. So $r \geq 6$. Then $N(v_4) \cap V(F) = B$ and $N(v_5) \cap V(F) \subseteq A$. So $G[A \cup \{v_1, v_3, v_5\}]$ and $G[B \cup \{v_2, v_4, v_r\}]$ induce two disjoint chorded 6-cycles, a contradiction.

Lemma 35 *For $k = 2$, $G[R] \neq K_{2,3}$.*

Proof Suppose $G[R]$ is a $K_{2,3}$ with parts $V_1 = \{v_1, v_2\}$ and $V_2 = \{v_3, v_4, v_5\}$. By (1), and since the set of low vertices is a clique, similarly to (47), we may assume

$$||V_1, F|| \geq 3, \ ||V_2, F|| \geq 8. \tag{52}$$

By (52), $||R, F|| \geq 11$. By (52) and Lemma 7, we have four cases.

Case 1 $F = K_4$. First, we show that

$$\text{foreach} x \in V(F), |E(G[N(x) \cap R])| = 0. \tag{53}$$

Suppose that the neighborhood in R does have an edge, say $N(x) \supseteq \{v_1, v_5\}$. Then, at least one of $\{v_3, v_4\}$ is high, say v_3 is high so that $||v_3, F - x|| \geq 2$. Therefore, $G[R - v_3 + x] = C_5^+$ and $G[F - x + v_3] \supseteq K_4^-$, producing two disjoint chorded cycles. This proves (53).

Hence, for all $x \in V(F)$, $N(x) \cap R \subseteq V_1$ or $N(x) \cap R \subseteq V_2$. Since $||R, F|| \geq 11$, there is at most one vertex $x' \in V(F)$ with $N(x') \cap R \subseteq V_1$ and so $d_G(v_0) + d_G(v_0') \leq 8$, a contradiction to (1).

Case 2 $F = K_4^-$. Similarly to the previous case, we show that

$$\text{foreach} x \in \{x_1, x_3\}, |E(G[N(x) \cap R])| = 0. \quad (54)$$

Suppose that the neighborhood in R does have an edge, say $N(x) \supseteq \{v_1, v_5\}$. Then, at least one of $\{v_3, v_4\}$ is high, say v_3 is high so that $||v_3, F - x|| \geq 2$. Therefore, $G[R - v_3 + x] = C_5^+$ and $G[F - x + v_3] \supseteq K_4^-$, producing two disjoint chorded cycles. This proves (54).

So by (52), we may assume that the vertices in $H := \{v_1, v_3, v_4\}$ are high, and so we can say that for every $i \in \{3, 4\}$, there exists $j_i \in \{2, 4\}$ such that $N(v_i) \cap V(F) = \{x_1, x_3, x_{j_i}\}$. Also, by (54), $N(v_1) = \{x_2, x_4\}$ and $N(v_2) \subseteq \{x_2, x_4\}$ with $||v_2, \{x_2, x_4\}|| \geq 1$ by (1). Choose $j_2 \in \{2, 4\}$ such that $v_2 x_{j_2} \in E(G)$.

If $j_3 \neq j_4$, and say $j_2 = j_3$, then $v_2 x_{j_3} x_3 v_3 v_2$ and $v_1 x_{j_4} x_1 v_4 v_1$ are cycles with chords $x_{j_3} v_3$ and $x_{j_4} v_4$, respectively. Hence $j_3 = j_4$.

Suppose $v_5 x_{6-j_3} \in E(G)$. Then, $G[R - v_3 + x_{6-j_3}] \supseteq C_5^+$ and $G[F - x_{6-j_3} + v_3] = K_4^-$, producing two disjoint chorded cycles. Hence, $v_5 x_{6-j_3} \notin E(G)$.

Now that the graph is fairly defined, we may assume $j_3 = 2$. Then $v_3 x_2, v_4 x_2 \in E(G)$ and $x_{6-j_3} = x_4$. Moreover, $A := \{v_3, v_4, v_5, x_4\}$ and $B := \{v_1, v_2, x_1, x_3\}$ are independent sets. Then, $G[A \cup B] \subseteq K_{4,4}$ so that if for some $a \in A$ and $b \in B$, $ab \notin E(G)$, then $d_G(a) + d_G(b) \leq 8$ contradicting (1). Thus $G - x_2 = K_{4,4}$. If $v_2 x_4 \notin E(G)$, then $d_G(v_2) + d_G(x_4) \leq 8$, contradicting (1). Thus, we have $G \subseteq G_2(2)$ and $G \supseteq G_2(2) - x_2 v_2 - x_2 v_5$. In other words, $G_2^{**} \subseteq G \subseteq G_2$, and so G is an exceptional graph, a contradiction.

Case 3 $F = C_5^+$. By (52) and again Lemma 7, there are two distinct vertices $v, v' \in \{v_3, v_4, v_5\}$ which are high and so are adjacent to each of x_1, x_2. Thus $G[\{v, v', x_1, x_2\}] = K_4^-$, a contradiction to (O1).

Case 4 $F = K_{3,3}$. By (O3) ,

$$N(v) \cap F \subseteq A \text{ or } N(v) \cap V(F) \subseteq B \text{ for every } v \in R. \quad (55)$$

Together with (52), we can assume that $\{v_1, v_3, v_4\}$ are high and so $d_G(v_1) = d_G(v_3) = d_G(v_4) = 5$ with $d_G(v_5), d_G(v_2) \geq 4$. By symmetry, we may assume $N(v_3) \cap V(F) = A$. Then by Lemma 8, $N(v_1) \cap V(F) \subseteq B$ and $N(v_2) \cap V(F) \subseteq B$. Since by (52), $d_G(v_1) \geq 5$, we may assume $N(v_1) \cap V(F) \supseteq \{x_2, x_4\}$. So again by Lemma 8, $N(v_4) \cap V(F) = A$. If $N(v_5) \cap V(F) \subseteq B$, then v_5 has at least two neighbors in B and hence $N(v_5) \cap B \cap N(v_1) \neq \emptyset$, say $x_2 \in N(v_5) \cap B \cap N(v_1)$. But then G has a 5-cycle $v_1 v_5 x_2 x_1 x_4 v_1$ with chord $v_1 x_2$, a contradiction to (O3). Therefore, $N(v_5) \cap V(F) \subseteq A$, and G is a bipartite graph with parts $A' := A \cup \{v_1, v_2\}$ and $B' := B \cup \{v_3, v_4, v_5\}$. Furthermore, if there exist $y, y' \in A'$ and $z, z' \in B'$ such that $yz \neq y'z'$ and $yz, y'z' \notin E(G)$, then (1) is violated. This occurs when $z \neq z'$ since $d_G(z) + d_G(z') \leq 8$, and otherwise $z = z'$ so if we let $z'' \in B - z$, then $d_G(z) + d_G(z'') \leq 3 + 5 = 8$. Thus, at most one edge can be missing so that $G_1^-(11, 2) \subseteq G \subseteq G_1(11, 2)$, a contradiction.

Lemma 36 *For $k = 2$, $G[R]$ is not a Θ-graph.*

Proof Suppose $G[R]$ is a Θ_{j_1,j_2,j_3}-graph formed by three internally disjoint paths $P_i = v_0 v_{i,1} v_{i,2} \dots v_{i,j_i} v_0'$ for $i \in \{1, 2, 3\}$. Since $G[R]$ contains no chorded cycles, $j_i \geq 1$ for all i. If $j_1 = j_2 = j_3 = 1$, then $G[R] = K_{2,3}$, a contradiction to Lemma 35. So we may assume

$$j_3 = \max\{j_1, j_2, j_3\} \geq 2. \tag{56}$$

Let $R' := \{v_0, v_0', v_{1,1}, v_{2,1}, v_{3,1}, v_{3,2}\}$. Since $G[R]$ has no triangles, R contains at most 2 low vertices and so

$$||R', F|| \geq 2 \cdot 9 + 2 \cdot 5 - 14 = 14. \tag{57}$$

Moreover, we can assume by symmetry that the low vertices are contained in $\{v_{2,1}, v_0, v_{3,1}, v_{3,2}\}$ so that we can state

$$d_G(v_0'), d_G(v_{1,1}) \geq 5. \tag{58}$$

By (58) and Lemma 7, we have four cases:

Case 1 $F = K_4$. By (57), there is $x_j \in V(F)$ with $||R', x_j|| \geq \left\lceil \frac{14}{4} \right\rceil = 4$. By (58), $G[F - x_j + v_{1,1}] \supseteq K_4^-$. By the choice, x_j has at least 3 neighbors in the path $P' := v_{1,2} v_{1,3} \dots v_0' v_{2,j_2} \dots v_0 v_{3,1} v_{3,2} \dots v_{3,j_3}$. Lemma 10 yields that $G[R - v_{1,1} + x_j]$ also contains a chorded cycle. So, we have two disjoint chorded cycles, a contradiction.

Case 2 $F = K_4^-$. Since $v_{1,1}$ is high, we may assume $N(v_{1,1}) \cap V(F) = \{x_1, x_2, x_3\}$. If some $x_j \in V(F) - x_2$ has $||x_j, R - v_{1,1}|| \geq 3$, then as in Case 1, $G[F - x_j + v_{1,1}] \supseteq K_4^-$ and $||x_j, P'|| \geq 3$ so that $G[R - v_{1,1} + x_j]$ contains a chorded cycle by Lemma 10. This is a contradiction since then we have two disjoint chorded cycles. So $||F - x_2, R|| = ||F - x_2, R - v_{1,1}|| + 2 \leq 6 + 2 = 8$. Since $|R'| = 6$, this together with (57) yields $||x_2, R'|| = 6$ and $||F - x_2, R'|| = ||F - x_2, R|| = 8$. Hence, with $N(v_{1,1}) \cap V(F) = \{x_1, x_2, x_3\}$, we have

$$N(v) \cap V(F) = \{x_1, x_2, x_3\} \text{forall} v \in R \text{with} ||v, F|| \geq 3. \tag{59}$$

If $d_G(v_{2,1}) \leq 4$, then by (1), $d_G(v_{3,1}) \geq 5$ and $d_G(v_{3,2}) \geq 5$. In this case, by (59), $G[\{v_{3,1}, v_{3,2}, x_1, x_2\}] = K_4$, contradicting (O2). So $d_G(v_{2,1}) \geq 5$ and by (59), $V(F) - N(v_{1,1}) = V(F) - N(v_{2,1}) = \{x_4\}$. Hence, G has cycle $x_1 x_4 x_3 v_{1,1} v_0 v_{2,1} x_1$ with chord $x_1 v_{1,1}$ and, because $N(x_2) \supseteq R'$, cycle $x_2 v_{3,1} v_{3,2} \dots v_0' x_2$ with chord $x_2 v_{3,2}$. So, we have two disjoint chorded cycles, a contradiction.

Case 3 $F = C_5^+$. By Lemma 7, each $v \in R$ with $||v, F|| \geq 3$ is adjacent to $\{x_1, x_2, x_4\}$. By (1), at least two vertices $v, v' \in \{v_{1,1}, v_{2,1}, v_{3,1}\}$ have this property. Thus, $G[\{v, v', x_4, x_5\}] = K_4^-$, a contradiction to (O1).

Case 4 $F = K_{3,3}$. Each $v \in R' - v_0 - v_0'$ has a nonneighbor in $\{v_{1,1}, v_{3,2}\}$ and $||v_{1,1}, F||, ||v_{3,2}, F|| \leq 3$. Hence, $d_G(v) \geq 4$ for each $v \in R - v_0 - v_0'$. In particular, $||v_{3,1}, F|| \geq 2$ and $||v_{3,2}, F|| \geq 2$. By (O3) , $N(v_{3,1}) \cap N(v_{3,2}) \cap V(F) = \emptyset$. By symmetry, we may assume that $N(v_{3,1}) \cap V(F) \subseteq A$ and $N(v_{3,2}) \cap V(F) \subseteq B$. By (57), there is $x_j \in V(F)$ with $||R', x_j|| \geq \left\lceil \frac{14}{6} \right\rceil = 3$. If $x_j \in A$, then $G[F - x_j + v_{3,2}] \supseteq K_{3,3}^-$ (and so contains a chorded 6-cycle), and x_j has 3 neighbors on the path $v_{3,1}v_0v_{2,1} \dots v_0'v_{1,j_1} \dots v_{1,1}$. Together with Lemma 10, we have two disjoint chorded cycles, a contradiction. Similarly, if $x_j \in B$, then $G[F - x_j + v_{3,1}] \supseteq K_{3,3}^-$, and x_j has 3 neighbors on the path $v_{3,2}v_{3,3} \dots v_0'v_{2,j_2} \dots v_{2,1}v_0v_{1,1}$, a contradiction again.

Lemma 37 *If* $k = 2$, *then* $v_1v_3, v_rv_{r-2} \notin E(G)$.

Proof Suppose that $v_1v_3 \in E(G)$. Let $W_1 = \{v_1, v_2\}$, $W_2 = \{v_{r-1}, v_r\}$, and $R_1 = \{v_1, v_2, v_3\}$. If $v_rv_j \in E(G)$ for $j < r - 1$, then let $R_2 = \{v_r, v_{r-1}, \dots, v_j\}$. Since $v_1v_r \notin E(G)$, $d_G(v_1) + d_G(v_r) \geq 9$ and so there exists $v \in \{v_1, v_r\}$ with $||v, F|| \geq 3$. We then have the four cases of Lemma 7.

Case 1 $F = K_4$. We first claim

$$||v_r, F|| \leq 3. \tag{60}$$

Otherwise, $||v_r, F|| = 4$. If there exists $x \in N(v_1) \cap N(v_2) \cap V(F)$, then $G[R_1 + x]$ is a chorded cycle and $G[F - x + v_r] = K_4$, producing two disjoint chorded cycles. Hence, $N(v_1) \cap N(v_2) \cap V(F) = \emptyset$ but also $||W_1, F|| \leq 4$.

Now, consider if $d_{G[R]}(v_r) = 1$. Then, $||v, F|| = 2$ for $v \in W_1$. For each $x \in V(F)$, $||x, \{v_1, v_2, v_r\}|| \leq 2$ but also $d_{G[R]}(v_{r-1}) \leq 3$ so that either $v_iv_{r-1} \in E(G)$ for some $3 \leq i \leq r - 3$ or $xv_k \in E(G)$ for some $3 \leq k \leq r - 1$. If say $x \in N(v_2) \cap V(F)$, then $xv_2 \cdots v_rx$ is a cycle either with chord v_iv_{r-1} or xv_k. Subsequently, $x \in N(v_2) \cap V(F)$ implies $x \notin N(v_1)$ so that $G[F - x + v_1] = K_4^-$, producing two disjoint chorded cycles.

So, now consider if $d_{G[R]}(v_r) = 2$. Since $||W_1, F|| \leq 4$, there exists $v \in W_1$ such that $d_G(v) \leq 2 + 2 = 4$ and so $vv_{r-1} \notin E(G)$ implies $d_G(v_{r-1}) \geq 5$ so that $||v_{r-1}, F|| \geq 2$, say $x, x' \in N(v_{r-1}) \cap N(v_r) \cap V(F)$. If there exist $x_1, x_2 \in V(F)$ where $N(v_1) \cup N(v_2) \supseteq \{x_1, x_2\}$ such that $\{x_1, x_2\} \neq \{x, x'\}$, say $x' \notin \{x_1, x_2\}$, then $G[R_2 + x']$ and $G[W_1 + F - x']$ contain a chorded cycle. Hence, $||W_1, F|| \leq 1$ so that $d_G(v) = 2$ for some $v \in W_1$, contradicting Lemma 12. This proves (60).

We again consider when $d_{G[R]}(v_r) = 1$. By (60), $||v_1, F||, ||v_2, F|| \geq 9-4-2 = 3$. So, there exists $x, x' \in N(v_1) \cap N(v_2) \cap V(F)$. If possible, choose $x \in V(F)$ such that $||v_r, F - x|| \geq 2$. Then, $G[F - x + v_r] \supseteq K_4^-$ and $G[R_1 + x] = K_4^-$. Therefore, $||v_r, F - x||, ||v_r, F - x'|| \leq 1$ so that $||v_r, F|| = 2$ with $N(v_r) \cap V(F) =$

$N(v_1) \cap N(v_2) \cap V(F)$ since $||v_r, F|| \geq 2$ by (1). By (1), $||v, F|| = 4$ for $v \in W_1$. But then, there exists $x'' \in N(v_1) \cap N(v_2) \cap V(F) - x - x'$ so that $G[R_1 + x''] = K_4^-$ and $G[F - x'' + v_r] = K_4^-$, producing two disjoint chorded cycles.

Suppose now $d_{G[R]}(v_r) = 2$. By (60) and the Ore condition, $d_G(v_1), d_G(v_2) \geq 9 - d_G(v_r) \geq 4$, say $v_1x_1, v_2x_2 \in E(G)$ with $x_1 \neq x_2$.

If $||W_2, F|| \geq 7$, then there exists $x \in N(v_{r-1}) \cap N(v_r) \cap V(F) - x_1 - x_2$. Hence, $G[R_2 + x]$ contains a chorded cycle and $G[W_1 + F - x] \supseteq C_5^+$, producing two disjoint chorded cycles.

If $||W_2, F|| = 6$, then there exist $x, x' \in N(v_{r-1}) \cap N(v_r)$ and $||W_1, F|| \geq 3$. If there exists $x'' \in \{x, x'\}$ such that $|(N(v_1) \cup N(v_2)) \cap (V(F) - x'')| \geq 2$, then $G[W_1 + F - x'']$ and $G[R_2 + x'']$ each contains a chorded cycle. Hence, $N(v_1) \cap V(F), N(v_2) \cap V(F) = \{x, x'\}$. Since $||W_2, F|| \geq 5$, choose $v \in W_2$ with $||v, F|| \geq 3$. Then, $G[R_1 + x] = K_4^-$ and $G[W_2 + F - x]$ contains chorded cycles since $||W_2, F - x|| \geq ||W_2, F|| - 2 \geq 4$.

Then, we handle the case when $||W_2, F|| = 5$ and $||W_1, F|| = 4$. Again, there exists $x \in N(v_{r-1}) \cap N(v_r) \cap V(F)$. If $|(N(v_1) \cup N(v_2)) \cap (V(F) - x)| \geq 2$, then $G[W_1 + F - x]$ and $G[R_2 + x]$ contain chorded cycles. So, $N(x) \supseteq W_1 \cup W_2$, and then $G[R_1 + x] = K_4^-$ and $G[W_2 + F - x]$ contain chorded cycles since $||W_2, F - x|| \geq 3$.

Lastly, if $||W_2, F|| \leq 4$, then some vertex in W_2 is low and so $||v, F|| \geq 3$ for all $v \in W_1$. Also, by Lemma 12, there exists $x \in N(v_r) \cap V(F)$. If say $x \in N(v_1) \cap V(F)$, then $xv_1v_3 \cdots v_rx$ is a cycle with chord v_rv_j and $G[F - x + v_2] \supseteq K_4^-$, producing two disjoint chorded cycles. Hence, $N(v_1) \cap V(F) = N(v_2) \cap V(F) = V(F) - x$. But then $d_G(v_1) + d_G(v_r) \leq (3+2) + (1+2) = 8 < 9$, contradicting (1) This completes the case.

Case 2 $F = K_4^-$. If $d_{G[R]}(v_r) = 1$, then, by (1), we immediately have $||v, F|| = 3$ for all $v \in \{v_1, v_2, v_r\}$. Hence, $G[R_1 + x_1] = K_4^-$ and $G[F - x_1 + v_r] = K_4^-$, producing two disjoint chorded cycles. Now, suppose $d_{G[R]}(v_r) = 2$. We now have either W_1 or W_2 containing all low vertices in W.

If W_2, contains the low vertices, then for all $v \in W_1$, $||v, F|| = 3$. If also $|(N(v_{r-1}) \cup N(v_r)) \cap (V(F) - x)| \geq 2$ for some $x \in \{x_1, x_3\}$, then $G[F - x + W_2]$ and $G[R_1 + x]$ contain chorded cycles. Therefore, there exists $x' \in V(F)$ such that $N(x') \supseteq W_2$ and so $\{x_1, x_3\} = \{x, x'\}$. But then $G[R_2 + x']$ and $G[W_1 + F - x']$ contain chorded cycles.

Then, we have that W_1 contains the low vertices. Hence, $||W_2, F|| \geq 5$ and there exists $x \in N(v_{r-1}) \cap N(v_r) \cap V(F)$. If $||W_1, F - x|| \geq 3$, then $G[F - x + W_1]$ and $G[R_2 + x]$ contain chorded cycles. So, we have $x \in N(v_1) \cap N(v_2) \cap V(F)$ since $d_G(v_r) \leq 5$. But then, $G[R_1 + x] = K_4^-$ and $G[W_2 + F - x]$ contains a chorded cycles since $||W_2, F - x|| \geq 5 - 2 = 3$.

Case 3 $F = C_5^+$. If $d_{G[R]}(v_r) = 1$, then, by (1), we immediately have $G[\{v_1, v_2, x_1, x_2\}] = K_4$, contradicting (O2). Now, suppose $d_{G[R]}(v_r) = 2$, then for some i, $||W_i, F|| \geq 5$. Moreover, for all $v \in W_i$, $||v, \{x_3, x_4, x_5\}|| \leq 1$ as otherwise

$G[\{v, x_3, x_4, x_5\}] \supseteq K_4^-$, contradicting (O1). But then, $||W_i, \{x_1, x_2\}|| \geq 3$ forcing $G[W_i + x_1 + x_2] \supseteq K_4^-$, again contradicting (O1).

Case 4 $F = K_{3,3}$. By Lemma 8, we may assume $N(v_1) \cap V(F) \subseteq A$ and $N(v_2) \cap V(F) \subseteq B$. Also, since $d_G(v_r) \leq 5$, $||v_1, A||, ||v_2, B|| \geq 2$, say $x_1, x_3 \in N(v_1)$ and $x_2, x_4 \in N(v_2)$. But then, $v_3v_1x_1x_2v_2v_3$ is a 5-cycle with chord v_1v_2, contradicting (O3).

This completes the proof.

Lemma 38 *For $k = 2$, $d_{G[R]}(v_1) = 1$ or $d_{G[R]}(v_r) = 1$.*

Proof Suppose that $d_{G[R]}(v_1) = d_{G[R]}(v_r) = 2$. Let $v_1v_s \in E(G)$ and $v_tv_r \in E(G)$. If $s - 1 = t$, then $G[R]$ is a chorded cycle, producing two disjoint chorded cycles. If $t < s - 1$, then $G[R]$ contains a spanning Θ-subgraph H consisting of 3 paths connecting v_t with v_s. In this case, adding any edge to this subgraph creates a chorded cycle. Thus $G[R] = H$, a contradiction to Lemma 36. Hence $t \geq s$. By Lemma 37, $s \geq 4$ and $t \leq r - 3$. So the set $S = \{v_1, v_{s-1}, v_{t+1}, v_r\}$ is independent. Therefore, by symmetry, we may assume that v_r is the only vertex in S that can be low, i.e.

$$\text{for every } v \in S - v_r, d_G(v) \geq 5 \text{ and } d_G(v) + d_G(v_r) \geq 9. \tag{61}$$

By (61), we also have

$$||S, F|| \geq 11, \text{ and if } d_G(v_r) = 3, \text{ then } ||S, F|| \geq 13. \tag{62}$$

We claim that

$$\text{for } q \in \{s, t\}, \text{ there is at most one edge } v_iv_\ell \text{ with } i < q < \ell. \tag{63}$$

Indeed, suppose there are edges $v_{i_1}v_{\ell_1}$ and $v_{i_2}v_{\ell_2}$ with $i_1, i_2 < q$ and $\ell_1, \ell_2 > q$. By symmetry, we may assume that $\ell_2 \geq \ell_1$ and $i_2 \leq i_1$. Then the cycle $v_{i_2}v_{i_2+1}\dots v_{\ell_2}v_{i_2}$ has chord $v_{i_1}v_{\ell_1}$. This proves (63).

By (61) and Lemma 7, we have four cases.

Case 1 $F = K_4$. If there is $x \in V(F)$ with $||x, S|| = 4$, then $G[F - x + v_1] \supseteq K_4^-$ and x has 3 neighbors on the path $v_2 \cdots v_r$ so that $G[\{x, v_2, \dots, v_r\}]$ contains a chorded cycle by Lemma 10, a contradiction to (O3). Otherwise, $||S, F|| \leq 12$ and by (62), $d_G(v_r) \geq 4$. In this case, again by (62), there is $x \in V(F)$ with $||x, S|| = 3$. Let $v \in S - N(x)$. Then, $G[F - x + v] \supseteq K_4^-$ and x has 3 neighbors on any Hamiltonian path in $G[R - v]$, producing two disjoint chorded cycles.

Case 2 $F = K_4^-$. Since G does not contain K_4, $||v, \{x_2, x_4\}|| = 1$ for all $v \in R$ with $||v, F|| \geq 3$. Therefore, $d_G(v_1) \leq 5$ and so $d_G(v_r) \geq 4$. In particular, $|N(v_r) \cap V(F)| \geq 2$. If $N(v_r) \cap V(F) = \{x_2, x_4\}$, then $G[F - x_1 + v_r] = K_4^-$ and x_1 is adjacent to the 3 vertices in $S - v_r$ on the path $v_1v_2 \cdots v_{r-1}$. Otherwise, by the symmetry between x_1 and x_3, we may assume $x_1v_r \in E(G)$. In this case, $G[F - x_1 + v_1] = K_4^-$ and x_1 is adjacent to the 3 vertices in $S - v_1$ on the path $v_2v_3 \cdots v_r$, producing two disjoint chorded cycles.

Case 3 $F = C_5^+$. By Lemma 7, $N(v) \cap V(F) = \{x_1, x_2, x_4\}$ for $v \in S - v_r$. Hence, $G[\{v_1, v_{s-1}, x_1, x_2\}] = K_4^-$, contradicting (O1).

Case 4 $F = K_{3,3}$. Let $V_1 := \{v_1, \ldots, v_{s-1}\}$ contain ℓ_1 low vertices and $V_2 := \{v_{t+1}, \ldots, v_r\}$ contain ℓ_2 low vertices. Also, let $W_1 = \{v_1, v_2\}$ and $W_2 = \{v_r, v_{r-1}\}$. If $\ell_1 > 1$, then by (63), $\ell_2 = 0$. So by switching the roles of V_1 and V_2 if needed, we can assume $\ell_1 \leq \ell_2$; in particular, $\ell_1 \leq 1$. For this assumption, we will no longer follow the symmetrical assumption of (61), i.e. some vertex in $S - v_r$ could be low instead of v_r.

By Lemma 7, for each vertex $v \in S + v_2$, either $N(v) \cap V(F) \subseteq A$ or $N(v) \cap V(F) \subseteq B$. Let $S_A = \{v \in S + v_2 : N(v) \cap V(F) \subseteq A\}$ and similarly for S_B. Since $S + v_2 = S_A \cup S_B$, we may assume that $|S_A| \geq \lceil 5/2 \rceil = 3$. By (61), there is $x \in A$ with $||x, S_A|| = |S_A|$. We now claim

$$\{v_1, v_2\} \not\subseteq S_A \tag{64}$$

Suppose $\{v_1, v_2\} \subseteq S_A$. If $|N(v_1) \cap N(v_2) \cap V(F)| \geq 2$, then $G[A + v_1 + v_2] \supseteq K_4^-$, contradicting (O3). So, $||W_1, A|| \leq 4$ and at least one of $\{v_1, v_2\}$ is low. Since V_1 has at most one low vertex and $d_G(v_r) \leq 5$, there is a unique $x' \in N(v_1) \cap N(v_2) \cap V(F)$. Say, $x' = x_1$. Since $||\{v_1, v_2\}, \{v_{t+1}, v_r\}|| = 0$, v_{t+1} and v_r are both high. If $v \in \{v_{t+1}, v_r\}$ is such that $v \in S_B$, then $x_1v_1v_sv_{s-1} \cdots v_2x_1$ is a cycle with chord v_1v_2 and $G[F - x_1 + v] = K_{3,3}$, producing two disjoint chorded cycles. Hence $v_{t+1}, v_r \in S_A$. Then, $x_1v_1v_s \cdots v_2x_1$ is a cycle with chord v_1v_2 and $x_3v_{t+1} \cdots v_rx_5x_4x_3$ is a cycle with chord x_3v_r, producing two disjoint chorded cycles. This proves (64).

Hence, $S_B \neq \emptyset$. If there is $v \in S_B$ such that $||v, B|| \geq 2$, then $G[F - x + v]$ has a chorded 6-cycle, and $G[R - v + x]$ contains a chorded cycle since $||x, R - v|| \geq |S_A| \geq 3$ and $G[R - v]$ contains a spanning path. Hence, $||v, B|| \leq 1$, in fact $||v, B|| = 1$ by (1), and $v \notin S$. This forces $d_G(v) \leq 4$, $v = v_2$ and so all $w \in S$ are high and $S_B = \{v_2\}$. Moreover, we have $v_2v_i \in E(G)$ for some $4 \leq i \leq r - 1$ by (1). Since $\ell_1 \leq \ell_2$, v_i is low, $i > t + 1$, and all vertices $v \in R - v_2 - v_i$ are high with $||v, B|| \leq 1$.

If $v_{s-2} \neq v_2$, then $d_{G[R]}(v_{s-2}) \leq 3$ and v_{s-2} high implies $||v_{s-2}, A|| \geq 2$. Hence, $G[A + v_{s-2} + v_{s-1}] \supseteq K_4^-$, contradicting (O1). Therefore, $v_{s-2} = v_2$. By symmetry, $v_{t+2} = v_{r-1}$ and so $s = 4$, $t = r - 3$. Hence, $i = r - 1$.

Now, since $v_i = v_{r-1} \in V_2$ and $v_2v_{r-1} \in E(G)$, $d_{G[R]}(v_{r-1}) = 3$, so $||v_{r-1}, F|| \geq 1$. So $v_{r-1}x \in E(G)$ for some $x \in V(F)$. If $x \in A$, $N(x) \supseteq \{v_{r-2}, v_{r-1}, v_r\}$ so that $G[\{x, v_{r-2}, v_{r-1}, v_r\}] = K_4$, contradicting (O2). Otherwise, $x \in B$. If $x = x_2$, then $x_2v_2v_3v_4 \cdots v_{r-1}x_2$ is a cycle with chord v_2v_i and $G[F - x_2 + v_1] = K_{3,3}$, producing two disjoint chorded cycles. Therefore $x_2v_{r-1} \notin E(G)$, but say $x_4v_{r-1} \in E(G)$. Now, $d_{G[R]}(v_4) = 3$ as otherwise

$v_4v_j \in E(G)$ for some $6 \leq j \leq r-3$, and then $v_2 \cdots v_{r-1}v_2$ is a cycle with chord v_4v_j, contradicting (O3). So, v_4 being high implies $||v_4, F|| \geq 2$. If $N(v_4) \cap V(F) \subseteq A$, then there exists $x' \in A$ such that $N(x') \supseteq \{v_1, v_3, v_4\}$ so that $G[\{x', v_1, v_3, v_4\}] \supseteq K_4^-$ contradicting (O1). So, $N(v_4) \cap V(F) \subseteq B$. Then, $G[F - x_1 + v_4] \supseteq K_{3,3}^-$ and $x_1v_1v_2v_{r-1}v_rv_{r-3}v_{r-2}x_1$ is a cycle with chord x_1v_r, producing two disjoint chorded cycles. This completes the case and the proof.

6 Proof of Theorem 6

Recall, we consider the minimum k such that Theorem 6 fails. Now, let G be an edge-maximal counterexample such that $\mathscr{F}$ contains less than k chorded cycles. For any such graph G, let $\mathscr{F}$ be a collection of disjoint chorded cycles chosen by our rules (O1)–(O8). By Sections 3 and 4, we have $k = 2$ and $G[R]$ containing a Hamiltonian path. By the result of Section 5 and Lemma 38, $d_{G[R]}(v_1) = 1$ or $d_{G[R]}(v_r) = 1$. Suppose $d_{G[R]}(v_1) = 1$. By Lemmas 11 and 34, $|R| \geq 4$ and $G[R]$ is not a Hamiltonian cycle and so $v_1v_r \notin E(G)$. Moreover, by Lemma 37, $v_1v_3, v_rv_{r-2} \notin E(G)$.

Case 1. $d_{G[R]}(v_r) = 1$. Then $||\{v_1, v_r\}, F|| \geq 7$. By symmetry, assume $||v_1, F|| \geq 4$. By Lemma 7, $F = K_4$ and so $d_G(v_1) = 5$ and $d_G(v_r) \geq 4$. Since $n \geq 8$, $r \geq 4$. Also, for each $v \in S := \{v_2, v_3, v_{r-2}, v_{r-1}\}$, $d_{G[R]}(v) \leq 3$. Note, $2 \leq |S| \leq 4$ and any vertex $v \in S - v_2$ has $||v, F|| \geq 1$ since $v_1v \notin E(G)$, but similarly $||v_2, F|| \geq 1$ since $v_2v_r \notin E(G)$. If there exists $x \in N(v_2) \cap N(v_{r-1}) \cap V(F)$, then $G[R - v_r + x]$ contains a cycle with chord xv_2 and $G[F - x + v_r] \supseteq K_4^-$. So, we have two disjoint chorded cycles, a contradiction. Hence, let $V(F) = \{x_1, x_2, x_3, x_4\}$ with $N(v_r) \cap V(F) \supseteq \{x_2, x_3, x_4\}$ and $v_{r-1}x, v_2x' \in E(G)$ for distinct $x, x' \in V(F)$. If $x \in N(v_r) \cap V(F)$, then we can choose $x'' \in N(v_r) \cap V(F) - \{x, x'\}$ and get cycles $v_rx''xv_{r-1}v_r$ and $v_1v_2x'x_1v_1$ with chords v_rx and v_1x', resp. Therefore, we are left to handle when $N(v_r) \cap V(F) = \{x_2, x_3, x_4\}$ and $N(v_{r-1}) \cap V(F) = \{x_1\}$, i.e. $x = x_1$. If v_2 or v_{r-1} has a chord on P, then $G[R - v_r + x_1]$ and $G[F - x_1 + v_r]$ each has a chorded cycle. Thus, $d_{G[R]}(v_2) = d_{G[R]}(v_{r-1}) = 2$ which forces $||v_2, F|| \geq 3$ since $v_2v_r \notin E(G)$ and $||v_{r-1}, F|| \geq 2$ by (1), contradicting $N(v_2) \cap N(v_{r-1}) \cap V(F) = \emptyset$.

Case 2. $d_{G[R]}(v_r) = 2$. Say $v_kv_r \in E(G)$ for exactly one $2 \leq k \leq r-3$. By symmetry, $d_{G[R]}(v_{k+1}) = 2$.

First, we claim

$$||v_1, F|| = 3 \tag{65}$$

First, suppose that $||v_1, F|| = 4$. By Lemma 7, $F = K_4$. Note also $d_{G[R]}(v_{r-1}) \leq 3$ and so $||\{v_{k+1}, v_{r-1}, v_r\}, F|| \geq 5$ by (1). Hence, there exists $x \in V(F)$ such that $||\{v_{k+1}, v_{r-1}, v_r\}, x|| \geq 2$. Thus, $G[F - x + v_1] = K_4$ and

$G[P - v_1 + x]$ also has a path of length $|P|$ but with $|E(G[P - v_1 + x])| \geq |E(G[P])| + 1$ due to v_1 essentially being replaced by x, contradicting (O7) or contradicting (O6) if $v_2x \in E(G)$.

Otherwise, $||v_1, F|| \leq 2$. By Lemma 12, $||v_1, F|| = 2$. Then $d_G(v) \geq 6$ for all $v \in \{v_{k+1}, v_{r-1}, v_r\}$. Specifically, $||v_{k+1}, F|| = ||v_r, F|| = 4$ and $||v_{r-1}, F|| \geq 3$. Hence, $F = K_4$, and there exists $x \in V(F) \cap N(v_{r-1}) \cap N(v_r) - N(v_1)$. Then, $G[F - x + v_1] = K_4^-$ and $xv_rv_k \cdots v_{r-1}x$ is a cycle with chord $v_{r-1}v_r$, thus producing two disjoint chorded cycles. This proves (65) so that now we have the four cases of Lemma 7.

Subcase 2.1: $F = K_4$. Then v_{r-1} and v_r have a common neighbor $x \in V(F)$ and so $G[F - x + v_1] \supseteq K_4^-$ but $xv_rv_kv_{k+1} \cdots v_{r-1}x$ is a cycle with chord $v_{r-1}v_r$, so we have two disjoint chorded cycles, a contradiction.

Subcase 2.2: $F = K_4^-$. If $x \in \{x_1, x_3\}$ is such that $x \in N(v_{r-1}) \cap N(v_r)$, then $xv_{r-1} \cdots v_kv_rx$ is a cycle with chord $v_{r-1}v_r$ and $G[F - x + v_1] \supseteq K_4^-$, producing two disjoint chorded cycles. Hence, $N(v_{r-1}) \cap V(F) = \{x_2, x_4\}$ and we may assume $N(v_r) = \{x_1, x_2, x_3\}$. If $v \in \{v_1, v_{k+1}\}$ is such that $N(v) \supseteq \{x_1, x_4, x_3\}$, then $G[\{v_1, \ldots v_{k+1}, x_1, x_4\}]$ contains a chorded cycle since $||\{x_1, x_4\}, \{v_1, v_{k+1}\}|| \geq 3$ and $x_2v_{r-1}v_rx_3x_2$ is a cycle with chord x_2v_r. Hence, for all $v \in \{v_1, v_{k+1}, v_r\}$, $N(v) \cap V(F) = \{x_1, x_2, x_3\}$. Then, $G[F - x_1 + v_1] = K_4^-$ and $G[R - v_1 + x_1]$ contains path $x_1v_r \cdots v_2$ as long as $|P|$. However, $|E(G[R - v_1 + x_1])| \geq |E(G[R])| + 1$, contradicting (O7). This completes the case.

Subcase 2.3: $F = C_5^+$. Since $d_G(v_1) = 4$, $||v_r, F|| \geq 3$. By Lemma 7, $N(v_1) \cap V(F) = N(v_r) \cap V(F) = \{x_1, x_2, x_4\}$. Hence, $v_1x_1v_rx_2v_1$ is a 4-cycle with chord x_1x_2, contradicting (O1).

Subcase 2.4: $F = K_{3,3}$. For definiteness, let $N(v_1) \cap V(F) = A$. Again, $d_G(v_1) = 4$ implies $d_G(v_r) \geq 5$ and so, by the case, $d_G(v_r) = 5$.

Suppose v_2 has a neighbor $z \in F$. If $z \in A$, say $z = x_1$, then G has a 5-cycle $v_1v_2x_1x_2x_3v_1$ with chord v_1x_1, contradicting (O3). Thus we may assume $z = x_2$. Since $v_1x_2 \notin E(G)$, $d_G(x_2) \geq 9 - d_G(v_1) = 5$. Then $G[F - x_2 + v_1] = K_{3,3}$ and for $R' = R - v_1 + x_2$, $G[R']$ has r-vertex path $x_2v_2 \cdots v_r$ where $d_{G[R']}(x_2) \geq 2$, contradicting (O7) at least. Hence, $||v_2, F|| = 0$.

Then $d_G(v_2) \leq 3$. Since $d_G(v_r) = 5 < 9 - d_G(v_2)$, $v_2v_r \in E(G)$, i.e. $k = 2$. Since $G[R]$ has no chorded cycles, path P has no chords aside from v_2v_r. Hence, $v_2v_{r-1} \notin E(G)$ and $d_{G[R]}(v_{r-1}) = 2$. Therefore, $d_G(v_{r-1}) \geq 6$ by (1) and $||v_{r-1}, F|| = 4$. By Lemma 7, $F = K_4$, contradicting the case.

This completes the proof of Theorem 6.

Acknowledgments This research of the author "Alexandr Kostochka" was supported in part by NSF grant DMS-1600592 and grants 18-01-00353A and 19-01-00682 of the Russian Foundation for Basic Research. This research of the author "Derrek Yager" was supported in part by Award RB17164 of the Research Board of the University of Illinois at Urbana-Champaign.

References

1. A. Bialostocki, D. Finkel, A. Gyárfás, Disjoint chorded cycles in graphs. Discrete Mathematics **308**(23), 5886–5890 (2008)
2. S. Chiba, S. Fujita, Y. Gao, G. Li, On a sharp degree sum condition for disjoint chorded cycles in graphs. Graphs Combin. **26**(2), 173–186 (2010)
3. K. Corrádi, A. Hajnal, On the maximal number of independent circuits in a graph. Acta Math. Hungar. **14**(3–4), 423–439 (1963)
4. G.A. Dirac, Some results concerning the structure of graphs. Can. Math. Bull. **6**(183–210), 247 (1963)
5. H. Enomoto, On the existence of disjoint cycles in a graph. Combinatorica **18**(4), 487–492 (1998)
6. D. Finkel, On the number of independent chorded cycles in a graph. Discrete Mathematics **308**(22), 5265–5268 (2008)
7. R.J. Gould, K. Hirohata, P. Horn, Independent cycles and chorded cycles in graphs. J. Combin. **4**(1), 105–122 (2013)
8. R.J. Gould, K. Hirohata, P. Horn, On independent doubly chorded cycles. Discrete Mathematics **338**(11), 2051–2071 (2015)
9. R.J. Gould, P. Horn, C. Magnant, Multiply chorded cycles. SIAM J. Discrete Math. **28**(1), 160–172 (2014)
10. H.A. Kierstead, A.V. Kostochka, T. Molla, D. Yager, An algorithmic answer to the Ore-type version of Dirac's question on disjoint cycles, in *Optimization Problems in Graph Theory*, vol. 139, chapter 8, pp. 149–168, ed. by B. Goldengorin (Springer, 2018)
11. H.A. Kierstead, A.V. Kostochka, T. Molla, E.C. Yeager, Sharpening an ore-type version of the Corrádi-Hajnal theorem. Abh. Math. Semin. Univ. Hamburg **12**, 1–37 (2016)
12. H.A. Kierstead, A.V. Kostochka, E.C. Yeager, The (2k-1)-connected multigraphs with at most k-1 disjoint cycles. Combinatorica **37**(1), 77–86 (2017)
13. H.A. Kierstead, A.V. Kostochka, E.C. Yeager, On the Corrádi–Hajnal theorem and a question of Dirac. J. Combin. Theory B **122**, 121–148 (2017)
14. T. Molla, M. Santana, E.C. Yeager, A refinement of theorems on vertex-disjoint chorded cycles. Graphs Combin. **33**(1), 181–201 (2017)
15. S. Qiao, Neighborhood unions and disjoint chorded cycles in graphs. Discrete Mathematics **312**(5), 891–897 (2012)
16. H. Wang, On the maximum number of independent cycles in a graph. Discrete Mathematics **205**(1–3), 183–190 (1999)

A New Embedding of the $3x + 1$ Dynamical system

John Leventides

Abstract The $3x + 1$ dynamical system T can be studied via the Collatz graph that depicts the trajectories of T in the set of natural numbers $\mathbb{N}^*$. The study of this graph is problematic as there is no evident structure that can be exploited. We embed this graph and its shifted copies in a new fully binary tree and extend T to a new map $\overline{T}$ that all its trajectories converge to a single equilibrium. The new graph resembles that of a shift map yet whole Collatz trajectories exist intact within it. This new structure allows the simultaneous study of all important features of the conjecture, such as Collatz sequences, transition of parity vectors and the double indexed sequence $(-1)^{T^k(n)}$.

1 Introduction

The Collatz conjecture concerns the limiting behaviour of the $3x + 1$ dynamical system in $\mathbb{N}^*$ which is given by

$$x_{i+1} = T(x_i), \tag{1}$$

where

$$T(x) = \begin{cases} \frac{x}{2}, & \text{if } x \text{ is even;} \\ \frac{3x+1}{2}, & \text{if } x \text{ is odd.} \end{cases}$$

The conjecture states that for any initial condition $x_0 = n \in \mathbb{N}^*$ the trajectory of the $3x + 1$ dynamical system given by (1) always reaches the number 1 in a finite number of steps the minimum of which is called stopping time denoted by $s(n)$.

J. Leventides (✉)
Department of Economics, Faculty of Economics and Political Sciences, National and Kapodistrian University of Athens, Athens, Greece
e-mail: ylevent@econ.uoa.gr

A. M. Raigorodskii, M. Th. Rassias (eds.), *Discrete Mathematics and Applications*, Springer Optimization and Its Applications 165,
https://doi.org/10.1007/978-3-030-55857-4_12

Alternatively, it can be stated as $s(n) < +\infty$ for all $n \in \mathbb{N}^*$. Or, equivalently, if we consider the set

$$U_k = \{n \in \mathbb{N}^* \mid s(n) = k\},$$

then the conjecture states that $\cup_{k=0}^{\infty} U_k = \mathbb{N}^*$.

The conjecture obviously rejects the existence of other attractor sets or diverging trajectories. Computational projects ran to verify the conjecture with no counterexample so far. At the same time the efforts for the proof are not successful and the problem is still open. For a review of results we refer to [3] and [4]. Recently, the strongest result concerning almost bounded trajectories was proved by Tao [5] utilizing probabilistic tools.

Related to this conjecture is the Collatz graph that has as vertices the natural numbers $\mathbb{N}^*$ and two vertices n, m are connected with an edge if $m = T(n)$. The graph depicts the trajectories of the $3x+1$ dynamical system in $\mathbb{N}^*$ which provide to the set $\mathbb{N}^*$ the structure of a connected tree ending at the cycle $\{1, 2\}$ if the conjecture is true. The level sets of this graph are the sets U_k which although they grow on average by $\frac{4}{3}$ they have a sparse and irregular structure. Furthermore, numbers that are close in $\mathbb{N}$ in the ordinary norm can have stopping times that are far and vice versa.

On the other hand, if n, m are close in the dyadic norm, i.e. $\|n - m\|_2 \leq \frac{1}{2^k}$, then $T^i(n)$ and $T^i(m)$ behave similarly for $i \leq k$, although if $l > k$, $T^l(n)$ and $T^l(m)$ may diverge. For a fixed i, the sequence $(-1)^{T^i(n)}$, $n \in \mathbb{N}^*$ shows periodicity with the period to double as i increases by 1. The generating function of $(-1)^{T^i(n)}$ for fixed i and variable n is given by

$$g_i(x) = \frac{p_i(x)}{1 + x^{2^i}},$$

where $p_i(x) = \sum_{v=1}^{2^n} (-1)^{T^i(v)} x^v$. The polynomials $p_i(x)$ show irregular and almost random structure with no obvious relationship between $p_i(x)$ and $p_{i+1}(x)$.

The double indexed sequence $(-1)^{T^i(n)}$ is central to the problem as it can be seen that the Collatz conjecture is equivalent to the fact that the sequence $\{(-1)^{T^i(n)}\}_{i=0}^{\infty}$ is eventually periodic with periodic part $\{1, -1\}$ for any $n \in \mathbb{N}^*$. As a result the study of the polynomials $p_i(x)$ stemming from the generating functions of $(-1)^{T^i(n)}$ and the evolution $p_i(x) \to p_{i+1}(x)$ are closely related to the solvability of the conjecture.

The strong irregularity appearing in the Collatz graph, the stopping times $s(n)$ and the polynomials $p_i(x)$ make use of various tools to tackle the Collatz conjecture and to produce significant results extremely difficult.

Opposed to that, if we consider the following dynamical system of shift type on $\mathbb{N}^*$ defined by the map

$$T_1(x) = \begin{cases} \frac{x}{2}, & \text{if } x \text{ is even;} \\ \frac{x+1}{2}, & \text{if } x \text{ is odd,} \end{cases}$$

as this results to decreasing trajectories in $\mathbb{N}^*$ one can easily deduce that for any initial condition the trajectory always descents to 1 and the stopping time is uniformly given by

$$s_1(n) = k, \quad \text{where } 2^{k-1} < n \leq 2^k.$$

Furthermore, the graph for T_1 is a perfect binary graph, the polynomials $p_i(x)$ are given by

$$g_i(x) = \frac{p_i(x)}{1 + x^{2^i}} = \frac{-\sum_{v=1}^{2^i} x^v}{1 + x^{2^i}},$$

and the relationship between $p_i(x)$ and $p_{i+1}(x)$ is given by:

$$p_{i+1}(x) = p_i(x^2) + x^{-1} p_i(x^2).$$

In this paper, we present the following results. Seeking the relationship between $p_i(x)$ and $p_{i+1}(x)$, i.e. $(-1)^{T^i(n)}$ and $(-1)^{T^{i+1}(n)}$, we produce a new map $\overline{T}$ revealing this relationship. This is a two-to-one map between $A_{n+1} \to A_n$, where $A_n = \{1, 2, \ldots, 2^n\}$, and gives rise to an infinite perfect binary tree with level sets the sets A_n.

The new map $\overline{T}$ extends the Collatz map T and defines a new tree $G(\overline{T})$ that extends the Collatz tree. However, the new tree $G(\overline{T})$ contains $n \in \mathbb{N}^*$ in many of its level sets A_i provided $n \leq 2^i$. In this respect, it is not a tree on $\mathbb{N}^*$ but rather on the disjoint union $\sqcup_{i \in \mathbb{N}^*} A_i$. Furthermore, to define uniquely $\overline{T}(n)$ one has to select the leaf A_i that n belongs to. A second characteristic of $\overline{T}$ is that all its trajectories converge to 1. If $x \in A_i$, then the $\overline{T}$-path of depth i, denoted as $\text{path}_{\overline{T}}(x, i) = (x, \overline{T}(x), \ldots \overline{T}^i(x))$, has as terminal value $\overline{T}^i(x)$ the value 1. Under certain circumstances, $\text{path}_{\overline{T}}(x, k) = \text{path}_T(x, k)$ are identical and hence there exist Collatz paths in $G(\overline{T})$. For every $1 \in A_i$ there is a Collatz tree ending to it, denoted by $S_i(G_1)$. In this sense, there are as many shifted Collatz trees as the level sets A_n and hence the new tree extends the Collatz graph. As a result $A_n \supseteq \cup_{k=0}^n U_k$, that is, A_n contains the first n level sets of the ordinary Collatz graph. The n-th level set of the ordinary Collatz tree is U_n which is a relatively small subset of $A_n = \{1, 2, \ldots, 2^n\}$. In the new extended graph $G(\overline{T})$, it is apparent how U_n can be completed to produce the whole A_n which is the n-th level set of $G(\overline{T})$. A result is that A_n can be partitioned as:

$$A_n = E_n^0 \bigcup \left(\cup_{j=1}^{n-1} E_n^j \right) \bigcup \left(\cup_{i=0}^n U_i \right),$$

where E_n^j are those subsets of A_n such that they initiate j-length paths of $\overline{T}$-map which are identical to normal Collatz paths. E_n^0 consists of odd numbers that cannot initiate such paths from the n-th level of $G(\overline{T})$. This decomposition highlights the fact that the new contribution of $\overline{T}$ on top of what we know from T arises from the calculation of $\overline{T}$ on the sublevel sets E_n^0.

Furthermore, the subgraph of $G(\overline{T})$ corresponding to nodes and edges from the original Collatz function T, denoted by $G(T)$, satisfies the following relationship:

$$G(\overline{T}) \supset G(T) \supseteq \overline{G_1^\infty} \supset G_1^\infty,$$

where $G_1^\infty = \cup_i S_i(G_1)$ and $\overline{G_1^\infty}$ denotes the T-hull of G_1^∞. In this setting the Collatz conjecture can be formulated as $G(T) = \overline{G_1^\infty}$. An additional structure of $G(\overline{T})$ is that its level sets A_n can be partitioned to 2^{n-k} equivalence classes according to whether two numbers reaching the same node of the $\overline{T}$-tree after k iterations of the $\overline{T}$-map.

The $\overline{T}$ graph allows also the iterative calculation of the sequence of signs $\{(-1)^{T^i(x)}\}$. In fact the value of $(-1)^{T^i(x)}$ can be expressed in terms of the path $\text{path}_{\overline{T}}(x, i)$, $x \in A_i$. The inverse two-valued map $\overline{T}^{-1} : A_n \to A_{n+1}$ gives rise to an operator $M_i : p_i(x) \to p_{i+1}(x)$. In this respect, $p_i(x)$ can be iteratively calculated by successive applications of M_i. The actions of the operators M_i can be incorporated in the $G(\overline{T})$ tree by the introduction of weights $w_{i,\overline{T}(i)} \in \{1, -1\}$ on every edge of the tree. This way $G(\overline{T})$ becomes a weighted tree. In this tree the concatenated action of the operators M_i may be simplified as multiplication of all the weights of the $\overline{T}$-path connecting two nodes. The evolution of the sequence of signs $(-1)^{T^i(n)}$ as i grows may also be described in terms of cyclotomic roots in a more compact form. This amounts to a change of basis of the operators M_i relating $p_i(x)$, $p_{i+1}(x)$ and all these can be written as a direct system of linear maps giving rise to a direct limit of vector spaces. This direct system allows the study of both the sequence of signs $(-1)^{T^i(n)}$, which is directly related to the Collatz conjecture, as well as the more general dynamics arising from the new graph taking as initial conditions signed sequences other than those arising from the $3x + 1$-map. The limiting properties of the various signed sequences may be viewed by mapping this direct limit to an appropriate Hilbert space.

One of the main contributions of the present paper is that it demonstrates explicitly and in simple terms the interplay between the following three fundamental extensions of the Collatz map:

1. the extension of T to the dyadic numbers $\mathbb{Z}_2$;
2. the linear operator on the group ring $\mathbb{C}(\mathbb{Z}(2^\infty))$ of the Prüfer group $\mathbb{Z}(2^\infty)$ defined by $L_T(f)(g) = f(g^2) + g^{-1/3} f(g^{2/3})$;
3. the extension of T to an appropriate space of analytic functions on the unit disk which is defined as the Riesz projection of a composition operator similar to L_T.

The new structure allows also the successive calculation of indicators related to convergence of Collatz orbits. In this paper, we describe how to assess powers $f(n)$ of the Collatz function in terms of the ability to send an initial value n to $\{1, 2\}$ after $f(n)$ applications of T. In this respect, with the help of the direct system, for an increasing function $f : \mathbb{N}^* \to \mathbb{N}^*$ we define an indicator $|f|_T \geq 0$ measuring the convergence ability of f with respect to T. For the Collatz conjecture to be valid we inquire that there is f such that $|f|_T = 0$. If the Collatz conjecture is true, then such an f is an increasing function enveloping all stopping times $s(n)$.

2 The Extension $\overline{T}$ of the Collatz Map

The Collatz graph that depicts the trajectories of the $3x + 1$ dynamics system in $\mathbb{N}^*$ is not complete. One of the reasons is that the inverse Collatz map $x \mapsto T^{-1}(x)$ is two-valued only if $x = 3k + 2$, in which case $T^{-1}(x) = \left\{2x, \frac{2x-1}{3}\right\}$. In all other cases, one has $T^{-1}(x) = 2x$. This means that only in one out of three numbers (nodes of the graph) the graph is binary. There are ways to correct this by perturbing x and creating value for $T^{-1}(x)$ but there is no unique way to achieve that. Such modification does not produce meaningful results if such an extension is arbitrary. As this stems from the fact that $\frac{2x-1}{3}$ is not always an integer, one can fix it by considering this ratio in $\mathbb{Z}/n\mathbb{Z}$ with $(3, n) = 1$ and in the same time avoiding loosing information of the original T-map.

This requires some type of extension of the original Collatz map which addresses the evolution of Collatz branch selection in the iterative application of T. This selection is depicted in the double indexed sequence $(-1)^{T^i(n)}$ or the equivalent binary sequence (a_{in}), where $T^i(n) = a_{in} \mod 2$. The role of this sequence in the Collatz conjecture is described in the following known lemma.

Lemma 1 *For a given n the trajectory $\{T^i(n)\}_{i=1}^{\infty}$ converges to 1 in a finite number of steps if and only if $\{(-1)^{T^i(n)}\}_{i=1}^{\infty}$ is eventually periodic with periodic part $\{-1, 1\}$.*

Proof If $T^i(n)$ reaches 1 in a finite number of steps $s(n)$, then $T^{2k+s(n)}(n) = 1$ and $T^{2k+1+s(n)}(n) = 2$ for every $k \geq 0$, hence after the $s(n)$-th step the sequence alternates between $\{(-1)^1, (-1)^2\}$.

Conversely, if $T^i(n)$ becomes periodic after $s(n)$ steps with periodic part $\{-1, 1\}$, then from that point and onwards T is applied by its odd branch $\frac{3x+1}{2}$ followed by its even branch $\frac{x}{2}$ a total of $\frac{3x+1}{4}$ which creates a strictly decreasing subsequence in the trajectory unless $x = 1$. Thus the decreasing sequence will eventually reach one in a finite number of r steps after $s(n)$. Furthermore, as the last part of the Collatz graph is

$$(8, 4, 2, 1, 2, 1),$$

this is only possible if $r = 0$ and therefore $T^{s(n)}(n) = 1$.

Additionally, the sequence $\{a_{in}\}_{i=0}^{k-1}$ is central for the calculation of $T^k(n)$. If d is the number of 1's in $\{a_{in}\}_{i=0}^{k-1}$ and $\{a_1, a_2, \ldots, a_d\} \subseteq \{0, 1, \ldots, k-1\}$ is their position in an ascending order in $\{a_{in}\}_{i=0}^{k-1}$, then:

$$T^k(n) = \frac{3^d n + \sum_{i=1}^{d} 3^{d-i} 2^{a_i}}{2^k}. \tag{2}$$

If the Collatz conjecture is true, then $s(n) < \infty$ for all $n \in \mathbb{N}^*$, hence for every n there exist $\{s(n), d(n), (0 \leq a_1(n) < a_2(n) < \cdots < a_{d(n)}(n) < s(n))\}$ such that

$$n = \frac{2^{s(n)}}{3^{d(n)}} - \sum_{i=1}^{d(n)} \frac{2^{a_i(n)}}{3^i}. \tag{3}$$

In this respect the Collatz conjecture amounts to the solvability of a multi-variable exponential diophantine equation of variable structure under inequality constraints, a problem that is not tackled so far. It is worth noting that the right hand expression appearing in (3) does not always take integer values for arbitrary selections of $0 \leq a_1 < a_2 < \cdots < a_d < s$. However, if the Collatz conjecture is true and for those selections satisfying (3) one can define an injection $I : \mathbb{N}^* \to \mathbb{Z}_2$ and a much simpler dynamical system (see also [1, 2]) in $I(\mathbb{N}^*) \subset \mathbb{Z}_2$ conjugate to the Collatz system given by T.

Theorem 1 *If $s(n) < \infty$ for all $n \in \mathbb{N}^*$, there is an injection $I : \mathbb{N}^* \to \mathbb{Z}_2$ given by:*

$$I(n) = \frac{2^{s(n)}}{3} - \sum_{i=1}^{d(n)} 2^{a_i(n)},$$

where $s(n)$, $d(n)$, $a)i(n)$ satisfy (3). *Furthermore, the restriction to $I(\mathbb{N}^*)$ of the dynamical system in $\mathbb{Z}_2$ given by:*

$$x_{n+1} = T_2(x_n)$$

$$T_2(x) = \begin{cases} \frac{x+1}{2}, & \text{if } x \in 2\mathbb{Z}_2 + 1; \\ \frac{x}{2}, & \text{if } x \in 2\mathbb{Z}_2 \end{cases}$$

is conjugate to the Collatz dynamical system by the conjugation

$$T = I^{-1} \circ T_2 \circ I.$$

Proof One can easily see that, for a given $n \in \mathbb{N}^*$, the successive application of T_2 on $I(n)$ reaches $\frac{1}{3}$ in exactly $s(n)$ steps and not less. Hence, if $I(n) = I(n')$, we must have $s(n) = s(n')$. Furthermore, since $I(n') = \frac{2^{s(n')}}{3} - \sum_{i=1}^{d(n')} 2^{a_i(n')}$ and $I(n) = \frac{2^{s(n)}}{3} - \sum_{i=1}^{d(n)} 2^{a_i(n)}$ the relation $I(n) = I(n')$ implies that the two binary expressions $\sum_{i=1}^{d(n')} 2^{a_i(n')}$ and $\sum_{i=1}^{d(n)} 2^{a_i(n)}$ are equal, hence $d(n) = d(n')$ and $a_i(n) = a_i(n')$. This, in connection with the fact that, if $s(n) < \infty$ for all $n \in \mathbb{N}^*$, the tuple $\big(s(n), d(n), a_1(n), a_2(n), \ldots, a_{d(n)}(n)\big)$ uniquely defines n by (3), proves that I is an injection.

Furthermore, if n corresponds to the tuple $t_n = (s(n), d(n), a_1(n), a_2(n), \ldots, a_{d(n)}(n)\big)$ one can see that, if n is even, $a_1(n)$ successive applications of T_2

corresponds t_n to

$$\left(s(n) - a_1(n), d(n), 0, a_2(n) - a_1(n), \ldots, a_{d(n)}(n) - a_1(n)\right).$$

On the other hand, if n is odd, then T_2 corresponds t_n to the tuple

$$\left(s(n) - 1, d(n) - 1, a_2(n) - 1, \ldots, a_{d(n)}(n) - 1\right).$$

The same does T on the expression given by (3). Therefore, one can inductively prove that $I \circ T = T_2 \circ I$.

The 2-adic consideration, although clear, does not allow a description of the evolution of binary sequence on a T-orbit that can be easily utilized. Complementary to the above is the next lemma that describes the evolution of $(-1)^{T^i(n)}$ as i is fixed, which in the sequel will highlight the extension of the Collatz graph and the duality of other approaches using cyclotomic fields.

Lemma 2 *The generating function of the sequence* $\{(-1)^{T^i(n)}\}_{n=1}^{\infty}$ *is given by*

$$g_i(x) = \frac{p_i(x)}{x^{2^i} + 1}, \quad \textit{for } |x| < 1,$$

where $p_i(x) = \sum_{\upsilon=1}^{2^i} (-1)^{T^i(\upsilon)} x^{\upsilon}$.

Proof Write $n = \upsilon + \pi 2^i$, where $0 \le \upsilon < 2^i$. Then

$$T^i(n) = T^i(\upsilon) + 3^{d_\upsilon}\pi,$$

where $d\upsilon$ is the number of times the odd part of T is applied. Hence,

$$(-1)^{T^i(n)} = (-1)^{T^i(\upsilon)} \cdot (-1)^{\pi}.$$

Therefore, we have

$$\sum_{n=1}^{\infty} (-1)^{T^i(n)} x^n = \sum_{\pi=0}^{\infty} \left(\sum_{\upsilon=0}^{2^i-1} (-1)^{T^i(\upsilon)} x^{\upsilon} \right) \left(-x^{2^i}\right)^{\pi}$$

and by excluding the case $n = 0$ (i.e. $\upsilon = 0$, $\pi = 0$), rearranging the indices to $\upsilon = 1, \ldots, 2^i$ and taking into account that $\frac{1}{x^{2^i}+1} = \sum_{\pi=0}^{\infty} (-x^{2^i})^{\pi}$, we get the result.

For a given i it is evident that $(-1)^{T^i(n)}$ depends on $(-1)^{T^i(\upsilon_n)}$, where $\upsilon_n \in A_i = \{1, 2, \ldots, 2^i\}$, as well as to $[n]_i$ which is the digit corresponding to 2^i in the binary expansion of n. To explore the evolution of $(-1)^{T^i(n)}$ as $i \to \infty$ for all n one has to explore the evolution of $g_i(x)$, or $p_i(x)$ as i grows. To this end we will construct a series of operators M_i that map $p_i(x) \mapsto p_{i+1}(x)$. We introduce the following notation:

$$A_k = \{1, 2, \dots, 2^k\}$$

$2A_k \subset A_{k+1}$ the even numbers of A_{k+1}

$2A_k - 1 \subset A_{k+1}$ the odd numbers of A_{k+1}

σ_3^{k+1} is the multiplication by 3 in the set of units $(\mathbb{Z}/2^{k+1}\mathbb{Z})^{\times}$ inducing a permutation map in the set $2A_k - 1 \subset A_{k+1}$

$[n]_i$ is the digit (coefficient) corresponding to 2^i in the binary expansion of n.

Definition 1 Define $\overline{T}: A_{k+1} \to A_k$ by

$$\overline{T}(n) = \begin{cases} \frac{n}{2}, & \text{if } n \in 2A_k; \\ \frac{\sigma_3^{k+1}(n)+1}{2}, & \text{if } n \in 2A_k - 1. \end{cases}$$

We can now prove the next result.

Theorem 2 *The following equation holds:*

$$T^{k+1}(n) = \left(T^k(\overline{T}(n)) + [3n]_{k+1}[n]_0\right) \mod 2.$$

Proof As $T^{k+1}(n) = T^k(T(n))$, we have the following.

Case I If $n \in 2A_k$, then

$$T^{k+1}(n) = T^k\left(\frac{n}{2}\right) = T^k(\overline{T}(n)). \tag{4}$$

Case II If $n \in 2A_k - 1$, then $3n = \sigma_3^{k+1}(n) + \pi 2^{k+1}$, which implies that

$$T(n) = \frac{\sigma_3^{k+1}(n) + 1}{2} + \pi 2^k.$$

Hence,

$$\begin{aligned} T^{k+1}(n) &= T^k\left(\frac{\sigma_3^{k+1}(n) + 1}{2}\right) + \pi 3^d \\ &= \left(T^k(\overline{T}(n)) + \pi\right) \mod 2 \\ &= \left(T^k(\overline{T}(n)) + [3n]_{k+1}\right) \mod 2. \end{aligned} \tag{5}$$

Combining (4) and (5) we get the result.

Corollary 1 *The following equation holds:*

$$T^k(n) = \left(1 + \sum_{i=0}^{k-1}[3\overline{T}^i(n)]_{k-i}[\overline{T}^i(n)]_0\right) \mod 2.$$

Proof Recursively applying Theorem 2 and taking into account that $\overline{T}^k(n) = 1$.

Proposition 1 *The map $\overline{T}\colon A_{k+1} \to A_k$ can be alternatively written as*

$$\overline{T}(x) = \begin{cases} 2^k, & \text{if } x \in \left\{2^{k+1}, \frac{4^{\lfloor k/2 \rfloor+1}-1}{3}\right\}; \\ \alpha_x = T(x) \mod 2^k, & \text{if } x \notin \left\{2^{k+1}, \frac{4^{\lfloor k/2 \rfloor+1}-1}{3}\right\}. \end{cases}$$

Proof We distinguish the following cases:

If $n \in 2A_k$ and $n \neq 2^{k+1}$, then we have $\overline{T}(n) = T(n) = \frac{n}{2} \in A_k \setminus \{2^k\}$. Hence, $T(n) = \overline{T}(n) \mod 2^k$.

If $n = 2^{k+1}$, then $\overline{T}(n) = T(n) = 2^k$ (it cannot be reduced modulo 2 as the value is $0 \notin A_k$).

If n belongs to $2A_k - 1$, then as the function $2A_k - 1 \to A_k$, $\overline{T}(x) = \frac{\sigma_3(x)+1}{2}$ is $1-1$ and onto, there is only one x such that $\overline{T}(x) = 2^k$ which is $\sigma_3^{-1}(2^{k+1}-1) \in 2A_k - 1$. This is given by $x = \frac{4^{\lfloor k/2 \rfloor+1}-1}{3}$. For all the rest $x \in 2A_k - 1$ we have that $\overline{T}(x) < 2^k$, hence $T(x) = \overline{T}(x) \mod 2^k$, which proves the result.

Remark 1 Let $q_k\colon \mathbb{Z} \to \mathbb{Z}/2^k\mathbb{Z}$ be the natural projection from $\mathbb{Z}$ to $\mathbb{Z}/2^k\mathbb{Z}$. Define $\overline{q_k}\colon \mathbb{Z}/2^k\mathbb{Z} \to A_k$ as

$$\{\overline{q_k}(a)\} = q_k^{-1}(a) \cap A_k.$$

Then,

$$\overline{T}(x) = \overline{q_k}\left(T(x)\right).$$

Corollary 2 *For $\overline{T}\colon A_{k+1} \to A_k$, we have*

$$\|T(x) - \overline{T}(x)\|_2 \leq \frac{1}{2^k}.$$

Now the map $\overline{T}$ and Theorem 2 help us define the required relationship between $p_i(x)$ and $p_{i+1}(x)$. As when $x \in 2A_k - 1 \subset A_{k+1}$,

$$\overline{T}(x) = \frac{\sigma_3^{k+1}(x)+1}{2} \in A_k,$$

then $\overline{T}(2\upsilon-1)=\frac{\sigma_3^{k+1}(2\upsilon-1)+1}{2}\in A_k$. Define a map $\overline{\sigma}_3^k(\upsilon)$:

$$\overline{\sigma}_3^k=t^{-1}\circ\sigma_3^{k+1}\circ t\colon A_k\to A_k,$$

where $t(\upsilon)=2\upsilon-1$. This map $\overline{\sigma}_3^k$ is a permutation of A_k induced by its conjugate permutation σ_3^{k+1} of $2A_k-1\subset A_{k+1}$. Additionally, define a map

$$f_k\colon A_k\to\{-1,1\}$$

$$f_k(\lambda)=(-1)^{[3(2\lambda-1)]_{k+1}}.$$

The maps $\overline{\sigma}_3^k$ and f_k induce a map φ_3 on the set of polynomials $p(x)$ of degree 2^k with $p(0)=0$, i.e. $p(x)=\sum_{i\in A_k}a_ix^i=a_1x+a_2x^2+\cdots+a_{2^k}x^{2^k}$, as follows

$$(\varphi_3p)(x)=\sum_{i\in A_k}f_k(i)a_{\overline{\sigma}_3^k(i)}x^i.$$

In this setting we have the following theorem.

Theorem 3 *Let $p_i(x)$ be the numerator polynomial of the generating function $g_i(x)=\frac{p_i(x)}{x^{2^i}+1}$ of the sequence $\{(-1)^{T^i(n)}\}_{n=1}^{\infty}$. Then,*

$$p_{i+1}(x)=p_i(x^2)+x^{-1}(\varphi_3p_i)(x^2). \tag{6}$$

Proof We separate

$$p_{i+1}(x)=p_{i+1}^{\text{even}}(x)+p_{i+1}^{\text{odd}}(x),$$

as a sum of its even and odd parts. Then, as

$$(-1)^{T^{i+1}(2\upsilon)}x^{2\upsilon}=(-1)^{T^i(\upsilon)}x^{2\upsilon}=(-1)^{T^i(\upsilon)}(x^2)^{\upsilon},$$

we have $p_{i+1}^{\text{even}}(x)=p_i(x^2)$.

For the odd part we have, by Theorem 2,

$$(-1)^{T^{i+1}(2\upsilon-1)}x^{2\upsilon-1}=(-1)^{T^i(\overline{T}(2\upsilon-1))}(-1)^{[3(2\upsilon-1)]_{i+1}}x^{2\upsilon-1}.$$

Since, by definition, $\overline{T}(2\upsilon-1)=\overline{\sigma}_3^i(\upsilon)$, we get

$$(-1)^{T^{i+1}(2\upsilon-1)}x^{2\upsilon-1}=(-1)^{T^i(\overline{\sigma}_3^i(\upsilon))}f_i(\upsilon)(x^2)^{\upsilon}x^{-1}.$$

Combining the above equation with the definition of φ_3, we obtain

$$p_{i+1}^{\text{odd}}(x) = x^{-1}(\varphi_3 p_i)(x^2),$$

and hence

$$p_{i+1}(x) = p_i(x^2) + x^{-1}(\varphi_3 p_i)(x^2).$$

Remark 2 The map defined by Equation (6) of Theorem 3 is a linear map from the set of coefficients of p_i to the set of coefficients of p_{i+1} and therefore can be represented by a $2^i \times 2^{i+1}$ matrix.

Definition 2 Define as M_i the $2^i \times 2^{i+1}$ matrix representing the linear relationship (6), i.e.

$$\overline{p}_{i+1}^t = \overline{p}_i^t \cdot M_i,$$

where $\overline{p}_i^t, \overline{p}_{i+1}^t$ are the coefficient vectors of the polynomials $p_i(x)$ and $p_{i+1}(x)$, respectively.

Example 1 Let

$$\begin{aligned} p_1(x) =& a_1 x + a_2 x^2 \\ p_2(x) =& b_1 x + b_2 x^2 + b_3 x^3 + b_4 x^4. \end{aligned}$$

Then σ_3^2 is a permutation of $[\mathbb{Z}/4\mathbb{Z}]^{\times} \approx \{1, 3\}$ represented by the matrix

$$\begin{bmatrix} 0 & 1 \\ 1 & 0 \end{bmatrix}$$

which matrix represents also the same conjugate permutation $\overline{\sigma}_3^1$ of $A_1 = \{1, 2\}$. Furthermore,

$$\begin{aligned} f_1 &\colon A_1 \to \{1, -1\} \\ f_1(1) &= 1 \\ f_1(2) &= 1. \end{aligned}$$

Hence,

$$(\varphi_3 p_1)(x) = a_2 x + a_1 x^2,$$

and therefore

$$p_2(x) = p_1(x^2) + x^{-1}(\varphi_3 p_1)(x^2) = a_1 x^2 + a_2 x^4 + a_2 x + a_1 x^3.$$

Consequently, we have the map

$$[a_1, a_2] \to [a_2, a_1, a_1, a_2]$$

which may be represented by

$$M_1 = \begin{bmatrix} 0\,1\,1\,0 \\ 1\,0\,0\,1 \end{bmatrix}.$$

Similarly,

$$M_2 = \begin{bmatrix} 0\,1 & -1\,0 & 0 & 0\,0\,0 \\ 1\,0 & 0 & 1 & 0 & 0\,0\,0 \\ 0\,0 & 0 & 0 & 0 & 1\,1\,0 \\ 0\,0 & 0 & 0 & -1\,0\,0\,1 \end{bmatrix}.$$

Theorem 4 *The operator* $\frac{1}{\sqrt{2}}M_i$ *is an isometric embedding* $\mathbb{R}^{2^i} \to \mathbb{R}^{2^{i+1}}$, *i.e.* $M_i M_i^t = 2I_{2^i}$.

Proof The operator M_i by construction contains in every row two elements from $\{1, -1\}$. Commuting by Q the columns of M_i so that the odd and the even part of $\overline{T}$ are separated, we have

$$M_i Q = \left[I_{2^i}, \Lambda\right],$$

where Λ is diagonal with the diagonal entries in $\{1, -1\}$. Thus,

$$M_i M_i^t = M_i Q Q^t M_i^t = I_{2^i} + \Lambda^2 = 2I_{2^i}.$$

3 The Binary Graph Arising from the Map $\overline{T}$

The Collatz map T structures $\mathbb{N}^*$ as a graph connecting those $\kappa, \lambda \in \mathbb{N}^*$ such that $T(\kappa) = \lambda$. This is not a full binary graph as $T^{-1}(x) = \left\{2x, \frac{2x-1}{3}\right\}$ if and only if $x = 3k + 2$. For all other x the inverse image of T contains only the number $2x$. The new definition of $\overline{T}\colon A_{n+1} \to A_n$ allows the inversion of $\frac{2x-1}{3}$ in A_{n+1} as $(\sigma_3^{k+1})^{-1}(2x - 1)$ for all $x \in A_n$. This makes $\overline{T}^{-1}$ two-valued for all x and gives rise to a binary tree in $\mathbb{N}^*$ with level sets the sets A_n as it is shown next.

Lemma 3 *For all* $x \in A_n$, *we have*

$$|T^{-1}(x)| = 2.$$

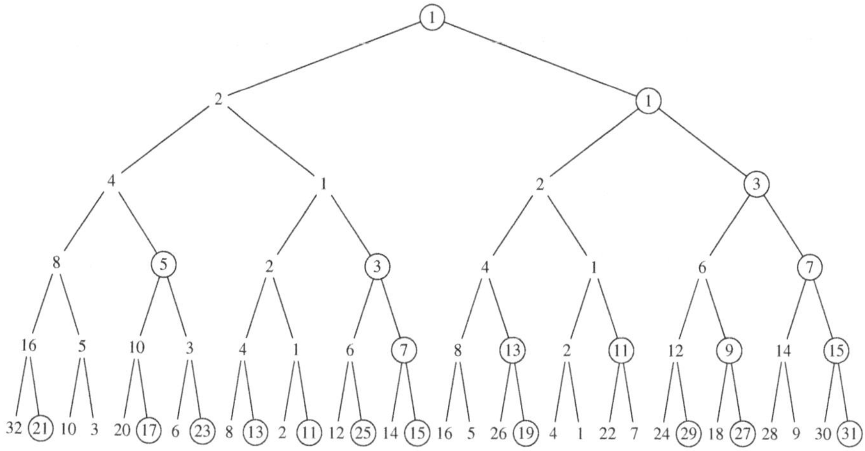

Fig. 1 The tree $G(\overline{T})$ with five layers. The nodes in cycle indicate the set E_n^0

Proof Consider the restrictions of $\overline{T}$:

$$\overline{T}_{\text{even}} = \overline{T}/2A_n \quad \text{and} \quad \overline{T}_{\text{odd}} = \overline{T}/2A_n - 1.$$

Then $\overline{T}_{\text{even}}(x) = \frac{x}{2}$ and therefore $\overline{T}_{\text{even}}: 2A_n \to A_n$ is $1-1$ and onto. The map $\overline{T}_{\text{odd}}$ can be written as the composition of two maps, namely

$$2A_n - 1 \xrightarrow{\sigma_3^{n+1}} 2A_n - 1 \xrightarrow{f} A_n,$$

where $f(x) = \frac{x+1}{2}$. As both σ_3^{n+1} and f are $1-1$ and onto and $|2A_n - 1| = |A_n|$, their composition is $1-1$ and onto and therefore $\overline{T}_{\text{odd}}$ is $1-1$ and onto. Hence, both $\overline{T}_{\text{even}}, \overline{T}_{\text{odd}}$ are invertible. Thus,

$$\overline{T}^{-1}(x) = \left\{\overline{T}_{\text{even}}^{-1}(x), \overline{T}_{\text{odd}}^{-1}(x)\right\}.$$

Remark 3 For every $x \in A_n$ the inverse $\overline{T}^{-1}$ is given by:

$$\overline{T}^{-1}(x) = \left\{2x, (\sigma_3^{n+1})^{-1}(2x-1)\right\}.$$

The map $\overline{T}: A_{n+1} \to A_n$ gives rise to a full binary tree with level sets the sets $\{A_n\}_{n=0}^{\infty}$ which is called the complete Collatz tree or $\overline{T}$-tree. This is depicted in Figure 1.

Lemma 4 *For every $x \in A_n$ we always have*

$$1 \leq \overline{T}^k(x) \leq 2^{n-k}, \quad k = 0, 1, \ldots, n-1$$

and

$$\overline{T}^n(x) = 1.$$

Proof As $\overline{T}$ is a map from A_{r+1} to A_r we have $\overline{T}^k(x) \in A_{n-k}$ and hence $1 \leq \overline{T}^k(x) \leq 2^{n-k}$ and for $k = n$ we get $\overline{T}^n(x) = 1$.

Theorem 5

1. *If $x \in A_{n+1}$, then $T(x) = \overline{T}(x)$ if and only if $T(x) \leq 2^n$.*
2. *$\overline{T}(x) = T(x)$ if and only if either $x \in 2A_n$ or $x \in 2A_n - 1$ and $x \leq \frac{2^{n+1}+4[n]_0-5}{3}$.*

Proof

1. For the even parts of the functions, T and $\overline{T}$ are identical. We need to prove the lemma for the odd parts. As $3x = \sigma_3^{n+1}(x) \mod 2^{n+1}$ for $x \in 2A_n - 1$, we have

$$\begin{aligned} T(x) = \overline{T}(x) &\Leftrightarrow 3x = \sigma_3^{n+1}(x) \\ &\Leftrightarrow 3x \in 2A_n - 1 \\ &\Leftrightarrow T(x) = \frac{3x+1}{2} \in A_n \\ &\Leftrightarrow T(x) \leq 2^n. \end{aligned}$$

2. In the case where $x \in 2A_n$ the result is clear. Assume now that $x \in 2A_n - 1$ and set $x_0 = \max\left\{x \mid \frac{3x+1}{2} \leq 2^n\right\}$. Then $\frac{3x_0+1}{2}$ is one of the numbers $\{2^n, 2^n - 1, 2^n - 2\}$, that is,

$$x_0 \in \left\{\frac{2^{n+1}-1}{3}, \frac{2^{n+1}-3}{3}, \frac{2^{n+1}-5}{3}\right\}.$$

The case $x_0 = \frac{2^{n+1}-3}{3}$ is excluded and we have

$$x_0 = \begin{cases} \frac{2^{n+1}-1}{3}, & \text{if } n \text{ is odd;} \\ \frac{2^{n+1}-5}{3}, & \text{if } n \text{ is even.} \end{cases}$$

Hence, in total,

$$x_0 = \frac{2^{n+1} + 4[n]_0 - 5}{3}.$$

As a result of the previous lemma, we have the following definition.

Definition 3 A path $\text{path}_{\overline{T}}(x, i) = (x, \overline{T}(x), \overline{T}^2(x), \ldots, \overline{T}^i(x))$ with $x \in A_n$ is called a Collatz path if $\overline{T}^k(x) = T^k(x)$ for every $k = 0, 1, \ldots, i$ (i.e. $\text{path}_{\overline{T}}(x, i) = \text{path}_T(x, i)$), or, equivalently, if $T^k(x) \le 2^{n-k}$ for every $k = 0, 1, \ldots, i$.

Example 2 The path $(\ldots, 2^n, 2^{n-1}, \ldots, 1)$ with $2^i \in A_i$ is clearly a Collatz path since $T(2^i) = \overline{T}(2^i) = 2^{i-1}$ for any i. On the other hand, the path $(\ldots, 2^i - 1, 2^{i-1} - 1, 2^{i-2} - 1 \ldots, 3, 1, 1)$, with $2^i - 1 \in A_i$, is not a Collatz path as

$$T(2^i - 1) = 3 \cdot 2^{i-1} - 1 \ne 2^{i-1} - 1 = \overline{T}(2^i - 1).$$

However, the Collatz tree G_1, which is the connected component of the Collatz graph ending to the node $1 \in A_0$ (i.e. all numbers attracted to 1 by T), exists in the $\overline{T}$-tree.

Theorem 6 *For a given n such that $s(n) < \infty$, the path*

$$\text{path}_{\overline{T}}(n, s(n)) = \left(n, \overline{T}(n), \overline{T}^2(n), \ldots, \overline{T}^{s(n)-1}(n), 1\right),$$

where $n \in A_{s(n)}$, is a Collatz path.

Proof Firstly, we prove that $n \le 2^{s(n)}$. Indeed, if we assume that $n > 2^{s(n)}$, then $T^{s(n)}(n - 2^{s(n)}) = T^{s(n)}(n) - 3^d = 1 - 3^d \le 0$ and we have a contradiction.

Now, since $s(T^i(n)) = s(n) - i$, we also have that $T^i(n) \le 2^{s(n)-i}$. Hence, $T^i(n) = \overline{T}^i(n)$ and $\text{path}_{\overline{T}}(n, s(n))$ is a Collatz path.

Definition 4 We define G_1 to be the T-subgraph of the $\overline{T}$-tree with nodes all $U_k \subseteq A_k$ and edges all (x_{k+1}, x_k) such that $x_k \in U_k$, $x_{k+1} \in U_{k+1}$ and $T(x_{k+1}) = x_k$.

Definition 5 We define $S^i(G_1)$ to be the T-subgraph of the $\overline{T}$-tree with nodes $U_k \subseteq A_{k+i}$ and edges all (x_{k+1}, x_k) such that $x_k \in U_k \subseteq A_{k+i}$, $x_{k+1} \in U_{k+1} \subseteq A_{k+i+1}$ and $T(x_{k+1}) = x_k$ and we call it the i-shift of G_1 in the $\overline{T}$-tree.

The tree $S^i(G_1)$ is a copy of G_1 that starts from $1 \in A_i$. We denote by G_1^∞ the union (of edges, nodes) of all $S^i(G_1)$, that is,

$$G_1^\infty = \cup_{i=0}^\infty S^i(G_1).$$

Proposition 2 *The following holds:*

$$\text{nodes}(G_1^\infty) \cap A_n = \cup_{k=0}^n U_k.$$

Proof We have

$$\text{nodes}(G_1^\infty) \cap A_n = \cup_{i=0}^\infty \left(\text{nodes}(S^i(G_1)) \cap A_n\right) = \cup_{i=0}^n \left(\text{nodes}(S^i(G_1)) \cap A_n\right).$$

And as $\text{nodes}(S^i(G_1)) \cap A_n = U_{n-i}$, we get

$$\text{nodes}(G_1^{\infty}) \cap A_n = \cup_{i=0}^{n} U_k.$$

G_1^{∞} is not T or $\overline{T}$-invariant as the last node of every $S^i(G_1)$ is $1 \in A_i$ the link $1 \to 2$, $2 \in A_{i-1}$, does not exist in G_1^{∞} by construction. To this end we give the following definition:

Definition 6 Given a subgraph $SG \subset G(\overline{T})$ define $cl_{\overline{T}}(SG)$ to be a graph with the same nodes as those of SG and all edges from $G(\overline{T})$ that it is possible to use to connect these nodes.

Then it is easy to show the next corollary.

Corollary 3 *The graph $cl_{\overline{T}}(G_1^{\infty})$ is a T-invariant subgraph of $G(\overline{T})$.*

Proof All nodes of G_1^{∞} are connected by the Collatz map T apart from the nodes $1 \in A_i, 2 \in A_{i-1}$ which the $cl_{\overline{T}}$ operator connects. As $\overline{T}(1) = T(1) = 2$, the new edge is also a T-edge, thus all nodes of G_1^{∞} are now connected via T-edges. Hence, $cl_{\overline{T}}(G_1^{\infty})$ is T-invariant.

The graph $cl_{\overline{T}}(G_1^{\infty})$ is not the largest subgraph of $G(\overline{T})$ containing Collatz paths. In fact, every level set A_n may be partitioned into sublevel sets of nodes according to their ability to initiate long Collatz paths.

Definition 7 Define the sets E_n^j as

$$E_n^j = \left\{x \in A_n \mid T^i(x) = \overline{T}^i(x) \text{ for all } i = 1, 2, \ldots, j, \text{ and } T^{j+1}(x) \neq \overline{T}^{j+1}(x)\right\}.$$

As $E_n^n = \cup_{i=0}^{n} U_i$, A_n may be partitioned as

$$A_n = E_n^0 \cup \left(\cup_{j=1}^{n-1} E_n^j\right) \cup \left(\cup_{i=0}^{n} U_i\right).$$

The sublevel set E_n^0 is the only part of A_n that Collatz paths cannot be initiated and also due to Theorem 5 we have that

$$E_n^0 = (2A_{n-1} - 1) \setminus [1, e_n],$$

where $e_n = \frac{2^n + 4[n-1]_0 - 5}{3}$.

Definition 8

1. The T-subtree of $G(\overline{T})$, denoted by $G(T)$, is the subtree of $G(\overline{T})$ with all nodes from $G(\overline{T})$ and those edges $[x, y]$ such that $y = T(x) = \overline{T}(x)$.
2. For every $x \in E_n^0$ denote by G_x the subtree of $G(T)$ ending to x.

Theorem 7 *The following hold:*

1.

$$G(T) = cl_{\overline{T}}(G_1^\infty) \bigcup \left(\bigcup_{x \in E_n^0, x \geq 3} G_x \right).$$

2.

$$cl_{\overline{T}}(G(T)) = G(\overline{T}).$$

Proof

1. As all Collatz paths end to one element of E_n^0, $x \in E_n^0$, we have that

$$G(T) = \bigcup_{x \in E_n^0} G_x.$$

Furthermore, as

$$cl_{\overline{T}}(G_1^\infty) = G_{1 \in E_0^0} \cup G_{1 \in E_1^0},$$

we get the result.

2. As $G(T)$ contains the same nodes with those of $G(\overline{T})$, we obviously have $cl_{\overline{T}}(G(T)) = G(\overline{T})$.

Definition 9 Define the T-hull $\overline{G_1^\infty}$ of G_1^∞ in $G(T)$ as follows

$$\overline{G_1^\infty} = cl_{\overline{T}}(G_1^\infty) \bigcup \left(\bigcup_{x \in E_n^0, x \geq 3, s(x) < \infty} G_x \right).$$

Proposition 3 *The Collatz conjecture is equivalent to*

$$\overline{G_1^\infty} = G(T).$$

Proof As $\cup_n E_n^0 = 2\mathbb{N}^* - 1$, then

$$\overline{G_1^\infty} = G(T) \text{ iff } \forall x \in 2\mathbb{N}^* - 1 \;\; s(x) < \infty$$

and since every $n \in \mathbb{N}^*$ is written as $n = 2^k x$, x odd, we get the result.

With the help of the $\overline{T}$-map we can define an equivalence relation on every A_n such that

$$x \sim_k y \text{ iff } \overline{T}^k(x) = \overline{T}^k(y).$$

For an $x \in A_n$, its equivalence class $[x]_k$ is given by $[x]_k = \overline{T}^{-k}(\overline{T}^k(x))$ and the cardinality of $[x]_k$ is equal to 2^k. In this way, for every $k = 1, 2, \ldots, n$, A_n can be partitioned as $A_n = \cup_{i=1}^{2^{n-k}} D_i^k$, where D_i^k runs through the distinct equivalence classes $[x]_k$. This defines a partial order structure of the subsets D_i^k ordered by inclusion.

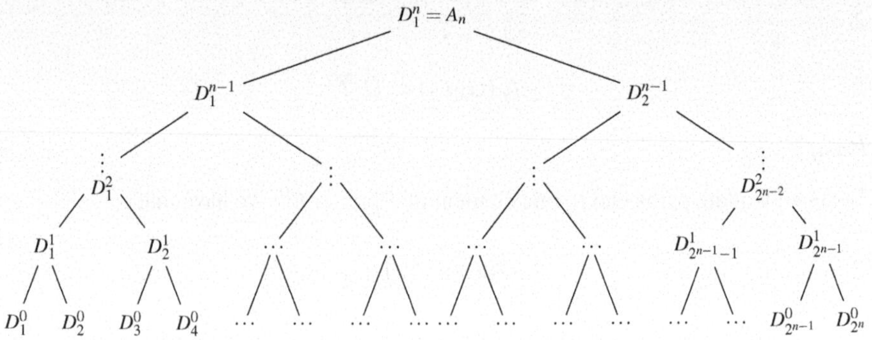

4 The Sequence of Signs $(-1)^{T^i(n)}$ and the $\overline{T}$-Tree $G(\overline{T})$

As it was shown in Section 2, the number $(-1)^{T^k(n)}$ can be calculated from the $\overline{T}$-path

$$\left\{n, \overline{T}(n), \ldots, \overline{T}^{k-1}(n), 1\right\}$$

as follows

$$(-1)^{T^k(n)} = (-1) \cdot \prod_{i=0}^{k-1} (-1)^{[3\overline{T}^i(n)]_{k-i} \cdot [\overline{T}^i(n)]_0}.$$

Hence, to any edge $x \to \overline{T}(x)$, $x \in A_k$, of the tree $G(\overline{T})$ we can assign the weight $w_{x,T(x)}^k (-1)^{[3x]_k \cdot [x]_0}$ and to the bottom node of the tree, i.e. $1 \in A_0$, we assign the number -1. Starting from the sign of $1 \in A_0$ (i.e. -1) and using the weights of the edges, we can calculate a unique sign for every node $x \in A_n$ as described below. Connect $x \in A_n$ with $1 \in A_0$ via a $\overline{T}$-path, i.e. $\left\{x, \overline{T}(x), \ldots, \overline{T}^{k-1}(x), 1\right\}$ and then find the sign of the node x by multiplying all the weights of the path with the sign of the ending node of the path (which is -1). This sign of the node $x \in A_n$ is exactly $(-1)^{T^n(x)}$.

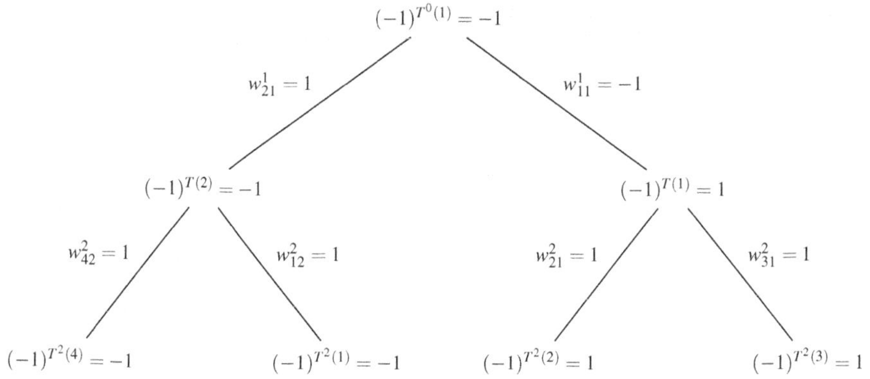

Fig. 2 The tree showing the weights assigned to edges

Example 3 Consider the path:

$$\begin{array}{ccccccccc} A_4 & \xrightarrow{\overline{T}} & A_3 & \xrightarrow{\overline{T}} & A_2 & \xrightarrow{\overline{T}} & A_1 & \xrightarrow{\overline{T}} & A_0 = \{1\} \\ 7 & \xrightarrow{-1} & 3 & \xrightarrow{-1} & 1 & \xrightarrow{1} & 2 & \xrightarrow{1} & 1. \end{array}$$

Then, the sign of the node $7 \in A_4$ is given by the product of the weights

$$w_{73}^4 \cdot w_{31}^3 \cdot w_{12}^2 \cdot w_{21}^1 \cdot w_1 = (-1) \cdot (-1) \cdot 1 \cdot 1 \cdot (-1) = -1.$$

Hence, $(-1)^{T^4(7)} = -1$.

Hence the signs $(-1)^{T^k(x)}$, $x \in A_k$, can be iteratively calculated using the tree showing in Figure 2. That is, the sign $(-1)^{T^k(x)}$, $x \in A_k$ is given by the product of $(-1)^{T^0(1)} = -1$ with the weights corresponding to the edges of the path starting from the bottom node and ending at the node $(-1)^{T^k(x)}$. For instance, we obtain

$$(-1)^{T^2(2)} = w_{21}^2 \cdot w_{11}^1 \cdot (-1) = 1.$$

The above procedure describes the iterative calculation of the vector $\underline{p}_k = ((-1)^{T^k(x)})_{x \in A_k}$ and it may also be depicted in matrix form as

$$\underline{p}_{k+1} = p_k \cdot M_k, \tag{7}$$

where

$$M_0 = [w_{11}^1 \ \ w_{12}^1] = [-1 \ \ 1], \quad M_1 = \begin{bmatrix} 0 & w_{21}^2 & w_{31}^2 & 0 \\ w_{12}^2 & 0 & 0 & w_{42}^2 \end{bmatrix} = \begin{bmatrix} 0\,1\,1\,0 \\ 1\,0\,0\,1 \end{bmatrix}.$$

Proposition 4 *The following holds*

$$\underline{p}_k = (-1)M_0 M_1 \dots M_{k-1}. \tag{8}$$

Proof The proof is straightforward and it follows by recursively applying (7) and then letting $p_0 = -1$.

Theorem 8 *The identity of Corollary 1 for all $n \in A_k$ is equivalent to Equation* (8) *of Proposition 4.*

Proof

The equation of Corollary 1 can be rewritten in the next form

$$T^k(n) = \left(1 + \sum_{i=0, \overline{T}^i(n)\text{odd}}^{k-1} \left[3\overline{T}^i(n)\right]_{k-i}\right) \mod 2.$$

Hence, for all $n \in A_k$,

$$(-1)^{T^k(n)} = -\prod_{i=0, \overline{T}^i(n)\text{odd}}^{k-1} w^{k-i}_{\overline{T}^i(n)\,\overline{T}^{i+1}(n)}.$$

It is apparent that if $\overline{T}^{i+1}(n) = T(\overline{T}^i(n))$, then $w^{k-i}_{\overline{T}^i(n)\overline{T}^{i+1}(n)} = 1$. Therefore, if $x \in E^j_n$, we have that

$$(-1)^{T^n(x)} = (-1)^{T^{n-j}(x)}.$$

Hence, we define an order relation in the disjoint union $\sqcup_n A_n$ as follows.

Definition 10 For $x \in A_n$ and $y \in A_m$, $n \geq m$, we have $x \geq y$ if and only if there is a Collatz path connecting x, y, i.e. $\text{path}_T(x, n-m) = \left(x, T(x), \dots, T^{n-m}(x) = y\right)$.

Then, for every $x \in A_n$, there is a minimal element $m(x) \in E_{n_x}$ with respect to $\geq$.

Theorem 9 *For every $x \in A_n$, we have*

$$(-1)^{T^n(x)} = (-1)^{T^{n_x}(m(x))}.$$

Proof We have $x \in E^{j_x}_n$, therefore $m(x) = T^{j_x} \in E^0_{n-j_x}$. Hence,

$$(-1)^x = (-1)^{T^{n_x}(m(x))}, \quad n_x = n - j_x.$$

Example 4 The signed sequence $\left(T^3(n)\right)_{n\in A_3}$ may be equivalently calculated by

$$\Big((-1)^{T^3(1)}, (-1)^{T^3(2)}, (-1)^{T^3(3)}, (-1)^{T^3(4)},$$
$$(-1)^{T^3(5)}, (-1)^{T^3(6)}, (-1)^{T^3(7)}, (-1)^{T^3(8)}\Big)$$
$$= \Big((-1)^{T(1)}, (-1)^{T^0(1)}, (-1)^{T^3(3)}, (-1)^{T(1)},$$
$$(-1)^{T^3(5)}, (-1)^{T^2(3)}, (-1)^{T^3(7)}, (-1)^{T^0(1)}\Big)$$

A summary of the properties if the $\overline{T}$-tree follows.

1. It is a full binary graph.
2. Its level sets are the sets A_n.
3. For every $x \in A_n$, $\overline{T}^n(x) = 1$. Hence, there is a single equilibrium point.
4. The Collatz graph and its shifted copies are subgraphs of this graph.
5. Let $G_1 \subseteq G$ be the Collatz subgraph of the Collatz graph G containing all $n \in \mathbb{N}^*$ ending to 1 via T. $\overline{G_1^\infty}$ contains all those Collatz corresponding to integers with finite stopping time and, hence, if there are other attractors or divergent trajectories, then $\overline{G_1^\infty}$ is a proper subgraph of $G(T)$.
6. For every level set A_i there is a map $v_i \colon A_i \to \{1, -1\}$ such that $v_i(n) = (-1)^{T^I(n)}$. This map can be iteratively calculated as i grows. This gives rise to an iteratively relation of the polynomials $p_i(x)$, $p_{i+1}(x)$ given by a matrix M_i. This allows to study the periodic properties of the sequence $\{(-1)^{T^i(n)}\}_{i=1}^{\infty}$, which are essential for the conjecture.
7. The number $(-1)^{T(n)}$ can be calculated from the $\overline{T}$-path $\{n, \overline{T}(n), \ldots, \overline{T}^{i-1}(n), 1\}$ starting from $n \in A_i$.

5 Collatz Transition and Cyclotomy

The evolution of the signed sequence $(-1)^{T^i(n)}$ of the $3x+1$ dynamical system was manifested via the relation of the polynomials $p_i(x)$ through the maps M_i. However, the generating function of the signed sequence is $g_i(x) = \frac{p_i(x)}{1+x^{2^i}}$, which suggest that $(-1)^{T^i(n)}$ may be written as a linear combination of elementary exponential functions $\{w_k^n\}_{k=1}^{2^i}$ where w_k are roots of the cyclotomic equation $1 + x^{2^i} = 0$. In such a description the vector $(p_i(w_k))_{k=1}^{2^i}$ plays more important role and may be used in place of the coefficient vector of $p_i(x)$. This also corresponds to a change of bases of the linear operators M_i.

The connection between the Collatz conjecture and cyclotomy stems also form the following fact. The relation of $p_{i+1}(x)$ to $p_i(x)$ via the Collatz $\overline{T}$-tree, by construction of $\overline{T}$, is given by

$$p_{i+1}(x) = \sum_{\upsilon=1}^{2^i} a_\upsilon x^{\overline{T}^{-1}_{\text{even}}(\upsilon)} + \sum_{\upsilon=1}^{2^i} a_\upsilon w^{i+1}_{\overline{T}^{-1}_{\text{odd}}(\upsilon),\upsilon} x^{\overline{T}^{-1}_{\text{odd}}(\upsilon)},$$

where $p_i(x) = \sum_{\upsilon=1}^{2^i} a_\upsilon x^\upsilon$.

This polynomial is somehow awkward to be utilized recursively due to the existence of functions as exponents and due to the existence of weights. The introduction of cyclotomic roots resolves this issue.

The next theorem describes the evolution of $p_i(x)$ in terms of the cyclotomic roots.

Theorem 10 *Let $p_i(x)$ be the i-th level polynomial of the signed sequence $(-1)^{T^i(n)}$, $n \in A_i$. Then:*

$$p_{i+1}(\omega) = p_i(\omega^2) + \omega^{\rho_{i+1}} p_i\left(\omega^{\rho_{i+1}+1}\right), \tag{9}$$

where ω is any root of $\omega^{2^{i+1}} + 1 = 0$ and ρ_{i+1} is the natural projection of $-\frac{1}{3} = \overline{10} \in \mathbb{Z}_2$ to $\mathbb{Z}/2^{i+1}\mathbb{Z}$.

Proof Separate $p_{i+1}(x) = p_{i+1}^{\text{even}}(x) + p_{i+1}^{\text{odd}}(x)$ as previously. For the even part we obviously have $p_{i+1}^{\text{even}}(x) = p_i(x^2)$. Then, as in Theorem 2, for the odd case we have

$$(-1)^{T^{i+1}(2\upsilon-1)} = (-1)^{T^i(\overline{T}(2\upsilon-1))} \cdot (-1)^\pi, \quad \upsilon \in A_i,$$

where

$$3(2\upsilon - 1) = 2\mu - 1 + \pi 2^{i+1}, \quad \mu \in A_i.$$

This is equivalent to

$$3\upsilon = \mu + 1 + \pi 2^i. \tag{10}$$

By definition, we have

$$3\rho_{i+1} = -1 + \pi_1 2^{i+1}. \tag{11}$$

Multiplying (10) with ρ_{i+1} and substituting $3\rho_{i+1}$ from Equation (11), we get

$$\left(-1 + \pi_1 2^{i+1}\right)\upsilon = \rho_{i+1}(\mu + 1) + \pi \rho_{i+1} 2^i,$$

which gives

$$-\upsilon = \rho_{i+1}(\mu + 1) + \pi\rho_{i+1}2^i - \pi_1 2^{i+1}\upsilon. \tag{12}$$

Adding (12) and (10) we get

$$\begin{aligned} 2\upsilon - 1 =& \rho_{i+1}(\mu + 1) + \mu + \pi\rho_{i+1}2^i - \pi_1 2^{i+1}\upsilon + \pi 2^i \\ \Leftrightarrow 2\upsilon - 1 =& (\rho_{i+1} + 1)\mu + \rho_{i+1} + \pi 2^i(\rho_{i+1} + 1) - \pi_1 2^{i+1}\upsilon. \end{aligned}$$

As now $\rho_{i+1} + 1 = 2(2\varphi + 1)$ for some φ and $4 \mid \pi_1$, we get that

$$2\upsilon - 1 = (\rho_{i+1} + 1)\mu + \rho_{i+1} + \pi 2^{i+1}(2\varphi_1 + 1) \quad \text{for some } \varphi_1.$$

Hence,

$$\omega^{2\upsilon-1} = \omega^{(\rho_{i+1}+1)\mu+\rho_{i+1}} \cdot (-1)^{\pi}.$$

Hence,

$$\begin{aligned} (-1)^{T^{i+1}(2\upsilon-1)}\omega^{2\upsilon-1} &= (-1)^{T^i(\mu)}(-1)^{\pi}\omega^{2\upsilon-1} \\ &= (-1)^{T^i(\mu)}(-1)^{2\pi}\omega^{(\rho_{i+1}+1)\mu+\rho_{i+1}} \\ &= (-1)^{T^i(\mu)}\omega^{\rho_{i+1}}\left(\omega^{\mu}\right)^{\rho_{i+1}+1}, \end{aligned}$$

which proves the result.

Consider now the cyclotomic field $\mathbb{Q}(\zeta_{2^{i+1}})$ where $\zeta_{2^{i+1}}$ is the first 2^{i+1}-th primitive root of unity, i.e. $\zeta_{2^{i+1}} = (-1)^{1/2^i}$. Then consider the set of polynomials

$$\mathbb{Q}_i[x] = \left\{p(x) \mid p(x) \in \mathbb{Q}[x],\ \deg p(x) \le 2^i \text{ and } p(0) = 0\right\}$$

as a 2^i-dimensional $\mathbb{Q}$-vector space. Consider now the $\mathbb{Q}$-linear injective map:

$$F\colon \mathbb{Q}_i[x] \to \mathbb{Q}(\zeta_{2^{i+1}})^{2^i}$$

such that

$$F(p(x)) = \frac{1}{2^i}\left(p(w_1), p(w_2), \ldots, p(w_{2^i})\right),$$

where w_i are the distinct roots of $x^{2^i} + 1 = 0$. We call the image of this map X_i which is a $\mathbb{Q}$-vector space and also $X_i \approx \mathbb{Q}_i[x] \approx \mathbb{Q}^{2^i}$.

Theorem 11 *The recursive equation* (9) *relating* $p_i(\omega)$ *and* $p_{i+1}(\omega)$ *of Theorem 10 defines a linear injective map from* X_i *to* X_{i+1}.

Proof Let $w_1, w_2, \ldots, w_{2^{i+1}}$ be the distinct roots of $x^{2^{i+1}} + 1 = 0$. Then Equation (9) implies that:

$$p_{i+1}(w_k) = p_i(w_k^2) + w_k^{\rho_{i+1}} p_i\left((w_k^2)^{\frac{\rho_{i+1}+1}{2}}\right),$$

as $\rho_{i+1} + 1$ is even. Then as w_k runs on the 2^{i+1} cyclotomic roots, w_k^2 runs on all 2^i cyclotomic roots. Furthermore, as $\frac{\rho_{i+1}+1}{2}$ is odd, raising all the roots w_k^2 to this power amounts to a permutation of the 2^i cyclotomic roots. Hence, the above 2^{i+1} relations define a linear map from X_i to X_{i+1} represented by a $2^i \times 2^{i+1}$ matrix L_i acting to the right $F(p_i(x))$ by matrix multiplication as

$$F(p_i(x))L_i = F(p_{i+1}(x)).$$

Furthermore, if we order the 2^{i+1} cyclotomic roots as

$$w_1, w_2, \ldots, w_{2^i}, -w_1, -w_2, \ldots, -w_{2^i},$$

and notice that

$$p_{i+1}(-w_k) = p_i(w_k^2) - w_k^{\rho_{i+1}} p_i\left((w_k^2)^{\frac{\rho_{i+1}+1}{2}}\right),$$

the matrix L_i takes the form

$$L_i = \frac{1}{2}\left[I_{2^i} + R_i, I_{2^i} - R_i\right].$$

For any y in the left kernel of L_i we have

$$y + yR_i = 0 \text{ and } y - yR_i = 0,$$

which implies that $y = 0$. Hence, L_i is injective.

Furthermore, if we normalize the vectors $p_i(x)$ as $\frac{1}{2^{i/2}} p_i(x)$ and the operator M_i by $M_i' = \frac{1}{\sqrt{2}} M_i$, we have the following commutative diagram:

$$\begin{array}{ccc} Q_i[x] & \xrightarrow{M_i'} & Q_{i+1}[x] \\ \downarrow{\scriptstyle F} & & \downarrow{\scriptstyle F} \\ X_i & \xrightarrow{L_i} & X_{i+1} \end{array}$$

and since M_i' is a linear isometry and F are isometric isomorphisms of vector spaces, L_i must also be a linear isometry.

6 The New Structure as a Direct System

So far the $\overline{T}$-tree allows the iterative calculation of the sequence $(-1)^{T^i(n)}$ moving from lower level to the higher applying operators M_i or L_i of successively increasing dimensions. The actions of the operators are applied to a single number in the beginning producing new vectors of the signed sequence which, however, depend only on the initial condition. The proposed structure, however, has wider capabilities as it is possible to use other vectors or starting points which seemingly are not directly related to the problem. As a matter of fact the series of operators M_i (or L_i) may be applied to infinite flags of vector spaces all of them along with the aforementioned operators making a direct system. This is described next.

Consider the infinite series of isometric $\mathbb{Q}$-linear maps:

$$V_0 \xrightarrow{M_0} V_1 \xrightarrow{M_1} V_2 \xrightarrow{M_2} V_3 \xrightarrow{M_3} V_4 \xrightarrow{M_4} \ldots, \tag{13}$$

where $\dim V_i = 2^i$ and M_i isometries. M_i isometrically injects V_i to V_{i+1} and also as it respects inner products preserves any direct sum decomposition from the one level to the next. Therefore, we may start with $V_0 \approx \mathbb{Q}$ and inject it to V_1. This way V_1 decomposes to $\mathrm{Im}M_0 \oplus (\mathrm{Im}M_0)^T$ which we denote as $V_1 = V_{1,1} \oplus V_{1,2}$ each of $V_{1,1}$, $V_{1,2}$ are one dimensional. Subsequently, we repeat this procedure by injecting via M_1, V_1 to V_2 which will inherit the previous decomposition whose isomorphic images in V_2 we call $V_{2,1}$, $V_{2,2}$, whereas we denote by $V_{2,3}$ the residual part of dimension 2. Hence, $V_2 = V_{2,1} \oplus V_{2,2} \oplus V_{2,3}$. For every summand we apply a restriction of M_i which is indexed accordingly. This procedure generates a finer structure of (13) depicted in the following diagram.

$$\begin{array}{cccccccccccc}
V_{01} & \xrightarrow{M_{01}} & V_{11} & \xrightarrow{M_{11}} & V_{21} & \xrightarrow{M_{21}} & V_{31} & \xrightarrow{M_{31}} & V_{41} & \xrightarrow{M_{41}} & \ldots \\
 & & V_{12} & \xrightarrow{M_{12}} & V_{22} & \xrightarrow{M_{22}} & V_{32} & \xrightarrow{M_{32}} & V_{42} & \xrightarrow{M_{42}} & \ldots \\
 & & & & V_{23} & \xrightarrow{M_{23}} & V_{33} & \xrightarrow{M_{33}} & V_{43} & \xrightarrow{M_{43}} & \ldots \\
 & & & & & & V_{34} & \xrightarrow{M_{34}} & V_{44} & \xrightarrow{M_{44}} & \ldots \\
 & & & & & & & & V_{45} & \xrightarrow{M_{45}} & \ldots
\end{array} \tag{14}$$

Characteristic of this structure is that each V_i is decomposed as $V_i = \oplus_{k=1}^{i+1} V_{ik}$ and each M_i as $M_i = \oplus_{k=1}^{i+1} M_{ik}$. For a fixed k, all V_{ik} are isomorphic and M_{ik} are isomorphisms. The dimension of V_{ik} is as follows: $\dim V_{i1} = 1$, $\dim V_{ik} = 2^{k-2}$, $k \neq 1$. Finally, out of the above decomposition, it is only the first row ($\dim = 1$) that is directly related to the Collatz conjecture, in fact

$$V_{01} = \mathrm{span}_{\mathbb{Q}}(-1) \subseteq V_0$$

$$V_{11} = \mathrm{span}_{\mathbb{Q}}((1,-1)) \subseteq V_1$$

$$V_{21} = \mathrm{span}_{\mathbb{Q}}((-1,1,1,-1)) \subseteq V_2$$

$$\vdots$$

$$V_{k1} = \mathrm{span}_{\mathbb{Q}}\left(\left((-1)^{T^k(\upsilon)}\right)_{\upsilon=1}^{2^k}\right) \subseteq V_k.$$

Thus the first row contains the required evolution of the signed sequence. We define

$$\overline{M}_{ij} \colon V_i \to V_i \quad j > i$$

as the composition $\overline{M}_{ij} = M_{j-1} \circ M_{j-2} \circ \ldots \circ M_{i+1} \circ M_i$ and, based on these functions, we also define an equivalence relation:

$$x_i \sim x_k \quad \text{iff} \quad \exists j \geq \max(i,k) \text{ such that } \overline{M}_{ij}(x_i) = \overline{M}_{kj}(x_k).$$

Then the direct limit $\underrightarrow{\lim}\, V_i$ is defined as $\oplus_{i=0}^{\infty} V_i / \sim$.

It is evident from the construction of the above diagram (14) that every element x_i of the starting space $V_{i,i+1}$ of the chain

$$V_{i,i+1} \xrightarrow{M_{i,i+1}} V_{i+1,i+1} \longrightarrow \ldots$$

defines a representative of the equivalence class $[x_i] = \{x_i, M_i(x_i), M_{i+1}M_i(x_i), \ldots\}$ and that the equivalence classes of $\sim$ are exactly of the type $[x_i]$. Finite linear combinations of the classes $[x_{i_1}], [x_{i_2}], \ldots, [x_{i_n}]$ with $x_{i_1} \in V_{i_k}$ and $i_1 < i_2 < \cdots < i_k$ can be defined in the natural way by considering the representatives belonging to the space V_{i_k}, i.e. $M_{i_k} \ldots M_{i_1+1} M_{i_1}(x_{i_1})$ and so on. Hence, the direct limit is endowed with the vector space structure and, more specifically, we have that $\underrightarrow{\lim}\, V_i$ is isomorphic to $\bigoplus_{i=0}^{\infty} V_{i,i+1}$ as $\mathbb{Q}$-vector spaces.

The first line of the direct system is one dimensional and $V_{i1} \subseteq V_i$ is spanned by the vector $p_i^t = \left\{(-1)^{T^i(\upsilon)}\right\}_{\upsilon=1}^{2^i}$ in V_i. Experimental evidence shows that this sequence of signs stabilizes at the very bottom part of p_i^t, i.e. $T^{f(n)}(n) \in \{1,2\}$ when $f(n) = O(n)$. One can see the following result.

Proposition 5 *We have $s(n) < \infty$ for all $n \in \mathbb{N}^*$ if and only if there is an increasing function $f \colon \mathbb{N}^* \to \mathbb{N}^*$ such that $T^{f(n)}(n) \in \{1,2\}$.*

Proof If $T^{f(n)}(n) \in \{1,2\}$ every n has finite stopping time $s(n) \leq f(n) < \infty$.

Conversely, consider the subset Inc of $(\mathbb{N}^*)^{\mathbb{N}^*}$ of all increasing functions $f \colon \mathbb{N}^* \to \mathbb{N}^*$ such that $n \leq 2^{f(n)}$ for all $n \in \mathbb{N}^*$. We also consider an order relation $\leq$ in $(\mathbb{N}^*)^{\mathbb{N}^*}$ such that $g_1 \leq g_2$ if and only if $g_1(n) \leq g_2(n)$ for all $n \in \mathbb{N}^*$

(with equality holding if the two functions are equal). We now define the set

$$Inc_{\geq s} = \{g \in Inc \mid g \geq s\}.$$

This is a closed set in $(\mathbb{N}^*)^{\mathbb{N}^*}$ and bounded below, hence it attains a minimum f. Then f is increasing and $f(n) \geq s(n)$ for any $n \in \mathbb{N}^*$. Therefore, $T^{f(n)}(n) \in \{1, 2\}$.

To study the convergence properties of the Collatz signed sequence, i.e. of the space V_{i1}, we embed V_i in an appropriate inner product space as follows. Consider the weighted sequence space

$$\mathscr{L} = \ell_2^w\left(\bigcup_{i=0}^{\infty} V_i\right) = \left\{(a_i)_{i=1}^{\infty} \mid a_i \in \mathbb{Q},\ \|(a_i)\|_w^2 = \sum_i \frac{1}{2^i} a_i^2 < \infty\right\}.$$

We may also define an inner product in $\mathscr{L}$ as:

$$\langle (a_i), (b_i) \rangle_w = \sum_i \frac{a_i b_i}{2^i}.$$

Then each V_i can be naturally embedded in $\mathscr{L}$ by

$$(a_1, a_2, \ldots, a_{2^i}) \mapsto (a_1, a_2, \ldots, a_{2^i}, 0, 0, \ldots).$$

Consider the infinite projective space of all lines in $\mathscr{L}$ denoted by $P_1(\mathscr{L})$. Then the one dimensional spaces V_{k1} can be viewed as points $l_k \in P_1(\mathscr{L})$ in this projective space. We would like to examine the convergence properties of the sequence $\{l_k\}_{k=1}^{\infty}$ in $P_1(\mathscr{L})$. To this end we define a metric d_w in $P_1(\mathscr{L})$ as

$$d_w\left((a_i), (b_i)\right) = \sqrt{1 - \frac{\langle (a_i), (b_i) \rangle_w^2}{\|(a_i)\|_w^2 \|(b_i)\|_w^2}}.$$

The above formula is independent from the choice of representatives of the lines and therefore it is well defined and provides a metric on the set $P_1(\mathscr{L})$.

If we denote by

$$a^j = \left((-1)^{T^j(1)}, (-1)^{T^j(2)}, \ldots, (-1)^{T^j(2^i)}, 0, 0, \ldots\right)$$
$$a = \left((-1)^{s(1)}, (-1)^{s(2)}, \ldots, (-1)^{s(n)}, \ldots\right)$$

the representatives of the points l_k, l in $P_1(\mathscr{L})$ and also by $1, 1_k$ in $\mathscr{L}$ the vectors

$$1 = (1, 1, 1, \ldots, 1, \ldots)$$
$$1_k = (\underbrace{1, 1, 1, \ldots, 1}_{k}, \ldots, 0, 0, \ldots)$$

then we can state the following theorem.

Theorem 12 *The following are equivalent.*

1. *There exists an increasing function* $f : \mathbb{N}^* \to \mathbb{N}^*$ *with* $n \leq 2^{f(n)}$ *for all* $n \in \mathbb{N}^*$ *such that* $T^{f(x)}(x) \in \{1, 2\}$ *for all* $x \in \mathbb{N}^*$.
2. *The sequence* $\{l_k\}_{k=1}^{\infty}$ *converges in* $P_1(\mathscr{L})$ *(with respect to the norm* d_w*) to a unique limit* $l = \text{span}_{\mathbb{Q}}(a)$, $a \in \{-1, 1\}^{\mathbb{N}^*}$, *which in this case may be written as* $l = \text{span}_{\mathbb{Q}}\left(\left((-1)^{s(n)}\right)_{n=1}^{\infty}\right)$.

Proof (1) ⇒ (2): For every $j \in \mathbb{N}$, select k such that $f(k) \leq j < f(k+1)$. Then $T^j(x) \in \{1, 2\}$ for all x with $1 \leq x \leq k$, and therefore the first k-parts of a^j and a are collinear, denoted by a_k^j and a_k, respectively. Hence, a^j and a can be conformally partitioned as

$$a^j = \left(a_k^j, a_{k+}^j\right)$$

$$a = (a_k, a_{k+})$$

implying that

$$\left\langle a^j, a\right\rangle_w^2 = \left(\pm\|1_k\|_w^2 + \frac{1}{2^k}\left\langle a_{k+}^j, a_{k+}\right\rangle_w\right)^2 \geq \left(\|1_k\|_w^2 - \frac{1}{2^k}\|1\|_w^2\right)^2.$$

Furthermore,

$$\|a^j\|_w^2 = \|1_{2^j}\|_w^2 \leq \|1\|_w^2 \quad \text{and} \quad \|a\|_w^2 = \|1\|_w^2.$$

Hence,

$$d_w(l_j, l) \leq \sqrt{1 - \frac{(\|1_k\|_w^2 - \frac{1}{2^k}\|1\|_w^2)^2}{\|1\|_w^4}}.$$

And as we let $j \to \infty$, we also have $k \to \infty$ and $\|1_k\|_w^2 \to \|1\|_w^2$ proving that $d_w(l_j, l) \to 0$.

(2) ⇒ (1): Let $a \in \{-1, 1\}^{\mathbb{N}}$ such that $l = \text{span}_{\mathbb{Q}}(a)$ is the unique limit that $l_j \to l$ in $P_1(\mathscr{L})$. Then for $k \in \mathbb{N}^*$ select $\epsilon > 0$:

$$\epsilon < \sqrt{1 - \left(1 - \frac{1}{2^{k-1}}\right)^2}.$$

For this ϵ, there exists $j(\epsilon)$ such that for any $j \geq j(\epsilon)$ with $2^l > k$ we have $d(l_j, l) \leq \epsilon$. Then

$$\left\langle a^j, a \right\rangle_w^2 \geq (1 - \epsilon^2) \|a^j\|_w^2,$$

which implies that

$$\left(|\left\langle a_k^j, a_k \right\rangle_w| + \frac{1}{2^k} \|1_{2^j - k}\|_w^2 \right)^2 \geq (1 - \epsilon^2) \|1_{2^j}\|_w^2,$$

and hence

$$\begin{aligned} |\left\langle a_k^j, a_k \right\rangle_w| &\geq \sqrt{(1 - \epsilon^2)} \|1_{2^j}\|_w - \frac{1}{2^k} \|1_{2^j - k}\|_w^2 \\ &\geq \sqrt{(1 - \epsilon^2)} \|1_{2^j}\|_w - \|1_{2^j}\|_w^2 + \|1_k\|_w^2 \\ &\geq \sqrt{(1 - \epsilon^2)} - 1 + \|1_k\|_w^2 \\ &> \|1_k\|_w^2 - \frac{1}{2^{k-1}}. \end{aligned}$$

Consequently, for every $j \geq j(\epsilon)$ we have

$$\|1_k\|_w^2 \geq |\left\langle a_k^j, a_k \right\rangle_w| > \|1_k\|_w^2 - \frac{1}{2^{k-1}}$$

implying that a_k^j, a_k are collinear for any $j \geq j(\epsilon)$. As

$$a_k^j = \left((-1)^{T^j(1)}, (-1)^{T^j(2)}, \ldots, (-1)^{T^j(k)}, 0, 0, \ldots \right)$$

then $(-1)^{T^j(x)}$, $x \leq k$, will follow the change of signs of $(-1)^{T^j(1)}$, for all $j \geq j(\epsilon)$. Hence, $(-1)^{T^j(x)}$ is eventually periodic with period two. Furthermore, this is proven for any $k \in \mathbb{N}^*$ and hence the Collatz conjecture is true.

Finally, in this case every k has finite stopping time, $s(k) < \infty$, and we may define f as:

$$f(n) = \max\{s(1), s(2), \ldots, s(n)\}$$

which is an increasing function such that $T^{f(x)}(x) \in \{1, 2\}$ for any $x \in \mathbb{N}^*$. It is also easy to see that a can be taken as $a = \left((-1)^{s(n)}\right)_{n=1}^{\infty}$.

The second line of the direct system defines a one dimensional linear subspace $\langle V_{12} \rangle$ of $\varinjlim V_i$. As V_{12} is orthogonal to V_{11} (via the Euclidean inner product) and V_{11} is spanned by $(1, -1)$, the spaces $V_{12}, V_{22}, V_{32}, \ldots$ are spanned by $(1, 1), (1, 1)M_1, (1, 1)M_1 M_2 \ldots$, respectively. Hence, the one dimensional flags of subspaces that are collapsed to $\langle V_{12} \rangle$ through the direct limit formation process are given by

$$
\begin{aligned}
V_{12} =& \text{span}_{\mathbb{Q}}(1, 1) \subseteq V_1 \\
V_{22} =& \text{span}_{\mathbb{Q}}((1, 1, 1, 1,)) \subseteq V_2 \\
V_{32} =& \text{span}_{\mathbb{Q}}((1, 1, -1, 1, -1, 1, 1, 1)) \subseteq V_3 \\
&\vdots \\
V_{k2} =& \text{span}_{\mathbb{Q}}\left(\left(\text{sg}_k(\upsilon)(-1)^{T^k(\upsilon)}\right)_{\upsilon=1}^{2^k}\right) \subseteq V_k.
\end{aligned}
$$

The function $\text{sg}_k \colon A_k \to \{1, -1\}$ is defined as:

$$
\text{sg}_k(x) = \begin{cases} -1, & x \in D_2^{k-1}; \\ 1, & x \in D_1^{k-1}. \end{cases}
$$

As in Section 3, $A_k = D_1^{k-1} \cup D_2^{k-1}$ and D_i^{k-1} contains all those $x \in A_k$ such that $\overline{T}^{k-1}(x) = i$. The sign change by $\text{sg}(\upsilon)$ was necessary as the signs in A_1 part of the tree were changed from $(1, -1)$ to $(1, 1)$ and all the operators M_i, $i \geq 1$, transporting the signs to the next level remained the same. This implies that the signed sequence at the k-level $(-1)^{T^k(\upsilon)}$ must change sign for those υ such that $\overline{T}^k(\upsilon) = 2$.

If we denote by

$$
U_{\text{odd}}^n = \bigcup_{k \leq n, k \text{ odd}} U_k \quad \text{and} \quad U_{\text{even}}^n = \bigcup_{k \leq n, k \text{ even}} U_k,
$$

then one can easily see that:

$$
A_n \supset D_1^{n-1} \supset \begin{cases} U_{\text{even}}^n, & \text{if } n \text{ is odd}; \\ U_{\text{odd}}^n, & \text{if } n \text{ is even}. \end{cases}
$$

$$
A_n \supset D_2^{n-1} \supset \begin{cases} U_{\text{odd}}^n, & \text{if } n \text{ is odd}; \\ U_{\text{even}}^n, & \text{if } n \text{ is even}. \end{cases}
$$

In a similar fashion as in Theorem 12, the one dimensional subspaces V_{k2} can be viewed as points l'_k in $P_1(\mathscr{L})$ and we can state the following theorem.

Theorem 13 *The following are equivalent.*

1. *There is an increasing function $f \colon \mathbb{N}^* \to \mathbb{N}^*$, with $n \leq 2^{f(n)}$, such that $T^{f(n)}(n) \in \{1, 2\}$ for all $n \in \mathbb{N}^*$.*
2. *The sequence $\{l'_k\}_{k=1}^{\infty}$ converges in $P_1(\mathscr{L})$ to a point $l' = \text{span}_{\mathbb{Q}}(a)$, where $a \in \{-1, 1\}^{\mathbb{N}^*}$ in which case $a = 1$.*

Proof The proof goes exactly as the one of Theorem 12. We take into account that the corresponding partial sequence

$$a_k^j = \left((-1)^{T^j(n)}\right)_{n=1}^k$$

stabilizes to $\pm a_k$ for $j \geq j(\epsilon)$ where $a_k = \left((-1)^{s(n)}\right)_{n=1}^k$. We also have that the sequence $(\text{sg}_i(n))_{n=1}^{\infty}$ stabilizes to $\left((-1)^{s(n)}\right)_{n=1}^k$ as j increases to ∞. Consequently,

$$b_k^j = \left(\text{sg}_j(n)(-1)^{T^j(n)}\right)_{n=1}^k$$

will stabilize to $\pm\left((-1)^{s(n)}(-1)^{s(n)}\right)_{n=1}^k$, i.e. to $\pm(\underbrace{1, 1, \ldots, 1}_{k})$.

Corollary 4 *The following are equivalent.*

1. *There is a strictly increasing function $f\colon \mathbb{N}^* \to \mathbb{N}^*$ such that $n \leq 2^{f(n)}$ and $T^{f(x)}(x) \in \{1, 2\}$ for all $x \in \mathbb{N}^*$.*
2. *The one dimensional subspace $\langle V_{01}\rangle \subseteq \varinjlim V_i$ has a well-defined correspondence through the limit process of Theorem 12 to a single point in $P_1(\mathcal{L})$.*
3. *The one dimensional subspace $\langle V_{12}\rangle \subseteq \varinjlim V_i$ has a well-defined correspondence through the limit process of Theorem 13 to a single point in $P_1(\mathcal{L})$.*

Alternatively, we may utilize the direct system to study the convergence properties of the Collatz signed sequence by correlating the successive vectors of this sequence and utilizing the operators M_i as i grows.

Definition 11 For every pair (k, r), $k \in A_r$, and every $\underline{x} \in V_r$ we define the quadratic form:

$$Q(k, r, \underline{x}) = \frac{1}{n} \sum_{i \in A_r,\, i \leq n} \underline{x}_i \left(\underline{x} M_r\right)_i,$$

where the index i denotes the i-th coordinate of the vector.

Additionally, we may define a quadratic form in every V_r as follows:

Definition 12 For every pair (n, r) with $n \in A_r$ we define the function

$$g_n(r) = Q(n, r, \underline{p}_r) = \frac{1}{n} \sum_{i \in A_r,\, i \leq n} (-1)^{T^r(i) + T^{r+1}(i)}.$$

In this setting, we have the following theorem.

Theorem 14 *For any increasing function $f\colon \mathbb{N}^* \to \mathbb{N}^*$ such that $x \leq 2^{f(x)}$ the following are equivalent:*

1. $T^{f(x)}(x) \in \{1, 2\}$, *for every* $x \in \mathbb{N}^*$.
2. $g_x(f(x)) = -1, g_x(f(x)+1) = -1, \ldots, g_x(f(x+1)-1) = -1$, *for all* $x \in \mathbb{N}^*$.

Proof (1) ⇒ (2). Since $f(x)$ is increasing and $T^{f(x)}(x) \in \{1, 2\}$, we obtain

$$T^i(j) \in \{1, 2\} \quad \forall j \leq x \text{ and } \forall i \geq f(x).$$

Hence,

$$T^i(j) + T^{i+1}(j) = 3 \quad \forall j \leq x \text{ and } i \geq f(x),$$

which proves that

$$g_x(f(x)+k) = \frac{1}{x}\sum_{i=1}^{x}(-1)^3 = -1 \quad \forall x \text{ and } k = 0, 1, \ldots, f(x+1) - f(x) - 1. \tag{15}$$

Conversely, if (15) is true, then, as -1 is the unique minimum for these types of average functions, we must have

$$(-1)^{T^i(j)} = -(-1)^{T^{i+1}(j)} \quad \text{for all } i \geq f(x) \text{ and } j \leq x.$$

Thus, $\{(-1)^{T^i(x)}\}_{i=0}^{\infty}$ is eventually periodic with period part $\{-1, 1\}$ and from Lemma 1 we obtain that $T^{f(x)}(x) \in \{1, 2\}$.

For every increasing function $f\colon \mathbb{N}^* \to \mathbb{N}^*$ with $n \leq 2^{f(n)}$ for all $n \in \mathbb{N}^*$, we have linear maps $M_{f(n)+i}\colon V_{f(n)+i} \to V_{f(n)+i+1}$ for every $0 \leq i \leq f(n+1) - f(n) - 1$. For every $\underline{x} \in \{1, -1\}^{\mathbb{N}^*} \subseteq \cup_i V_i$ we may define a quadratic form on $V_{f(n)+i}$ as $Q\left(n, f(n)+i, P_{f(n)+i}(\underline{x})\right)$, where $P_{f(n)+i}$ is the natural projection $P_{f(n)+i}\colon \cup_i V_i \to V_{f(n)+i}$. Hence, for every such $\underline{x}$ and f, we define a series of quadratic forms denoted by $\left\{Q_i^f(\underline{x})\right\}_{i=1}^{\infty}$. As we expect that the limiting vector will follow the sign of $(-1)^{T^i(1)}$, we normalize $\underline{x}$ as $(-1)^i\underline{x}$. We are seeking to find $\underline{x} \in \{1, -1\}^{\mathbb{N}^*} \subseteq \cup_i V_i$ such that $Q_i^f((-1)^i\underline{x}) = -1$ which is the actual minimum attained by any quadratic form. We define as *defect*, $D(\underline{f}, \underline{x})$, the sum of all deviations from the required value -1 of all levels, i.e. we have the next definition.

Definition 13 For any increasing function $f\colon \mathbb{N}^* \to \mathbb{N}^*$ with $n \leq 2^{f(n)}$, for all $n \in \mathbb{N}^*$, and for a given $\underline{x} \in \{1, -1\}^{\mathbb{N}^*}$ we set

$$D(f, \underline{x}) = \sum_{i=1}^{\infty}\left(Q_i^f((-1)^i\underline{x}) + 1\right).$$

Definition 14 The absolute defect of f with respect to the Collatz function T, denoted by $|f|_T$, is defined as: $|f|_T = \inf_{\underline{x} \in \{-1,1\}^{\mathbb{N}^*}} D(f, \underline{x})$.

It is apparent that $|f|_T$ makes sense if there is $\underline{x} \in \{1, -1\}^{\mathbb{N}^*}$ such that $D(f, \underline{x}) < \infty$.

Hence, we have a correspondence:

$$Inc \rightarrow \mathbb{R}^+$$
$$f \rightarrow |f|_T.$$

This correspondence defines a total ordering in Inc.

The following theorem is apparent from the definition of $|f|_T$.

Theorem 15 *If the Collatz conjecture is true, then there is $f \in Inc$ such that for every $g \in Inc$, with $g \geq f$, we have $|g|_T = 0$.*

7 Conclusions

We presented a new embedding of the $3x + 1$ dynamical system giving rise to a perfect binary tree extending the Collatz graph. Many algebraic, analytic and structural properties of the Collatz dynamical system may be transferred efficiently in this new tree and may produce new framework for the study of this system.

References

1. D.J. Bernstein, A non-iterative 2-adic statement of the $3x + 1$-conjecture. Proc. Am. Math. Soc. **121**, 405–408 (1992)
2. D.J. Bernstein, J.C. Lagarias, The $3x + 1$-conjugacy map. Can. J. Math. **48**, 1154–1169 (1996)
3. J.C. Lagarias, The $3x+1$ problem: an overview, in *The Ultimate Challenge: The* $3x+1$ *Problem* (American Mathematical Society, Providence, 2010), pp. 3–29. MR2560705
4. J.C. Lagarias, The 3x+1 Problem: An Annotated Bibliography, II (2000–2009), Unpublished. Preprint on arXiv [arxi.org/abs/math/0608208]
5. T. Tao, Almost all orbits of the Collatz map attain almost bounded values (submitted). Preprint on arXiv [arxiv.org/abs/1909.03562]

Diffusion on Dynamical Interbank Loan Networks

John Leventides and Nick Poulios

Abstract In this paper we study the effect of diffusion method to interbank networks in concept of connected, directed and weighted networks. We consider networks of n different banks which they exchange funds (loans) and the main feature is how the leverages of banks can be choosen to improve the financial stability of the network. This is done by considering differential equations of diffusion type. It is well known that banks exchange funds in the form of credit which are supported partly by the banks own capital. The ratio of their assets by the capital constitute the leverage of the bank and for minimization of risk purposes this ratio has to be kept within reasonable limits. The aim of this paper is to show how ideas from diverse domains such as diffusion, differential equations and graph theory can be used to demonstrate how financial risk can be controlled in this type of interbank networks. Diffusion acts as a stabilization process by the flow of funds from banks of higher leverage to those of lower. This process leads to equilibrium and stops either in a state of equal leverages between banks or whenever this is not possible in a final state which is more robust compared to the initial. The relation between the initial and final values of the interbank network may be described by a projection operator.

1 Introduction

Diffusion is the study of motion of atom, molecules, etc. from a region of higher concentration to a region of lower concentration. The idea behind the diffusion has roots primarily in physics (particle diffusion) but we can find it in chemistry, biology, geology, economics/finance and others scientific areas. A few examples are molecular diffusion, Brownian motion and heat diffusion.

J. Leventides · N. Poulios (✉)
Department of Economics, Division of Mathematics and Informatics, National and Kapodistrian University of Athens (NKUA), Athens, Greece
e-mail: ylevent@econ.uoa.gr; npoulios@econ.uoa.gr

A. M. Raigorodskii, M. Th. Rassias (eds.), *Discrete Mathematics and Applications*, Springer Optimization and Its Applications 165,
https://doi.org/10.1007/978-3-030-55857-4_13

The word diffusion derives from the Latin word *diffundere*, which means "to spread out". The first systematic study of diffusion (in gases) as we looking back to the past was given by Thomas Graham from 1831 to 1833. Moreover, a mathematical description of diffusion and Brownian motion was elaborated by Joseph Fourier in 1822, Adolf Fick in 1855 and by Albert Einstein in 1905.

In this study we consider a connected, directed and weighted dynamic network $G(V, E)$ with $V = \{1, 2, \ldots, n\}$ and $E = \{e_{ij}\}, i \neq j$, where V denotes the nodes (banks) with $|V| = n$ and E are the edges $|E| = e$, respectively. The edges represent the fund flow from bank i to bank j, characterized by a positive weight ϕ_{ij} and a direction. Furthermore, to each bank we assign a leverage index (l_i), which is the sum of all outgoing fund flows ϕ_{ij} over the own capital of bank i, (k_i), as described by (1) with $i \neq j$, i.e. self-loops are not allowed (no capital increase).

$$l_i = \frac{\sum_j \phi_{ij}}{k_i} \tag{1}$$

Moreover, we extract from the graph $G(V, E)$ the adjacent operator $\mathbf{A} \in \mathbb{R}^{n\times n}$, where the main diagonal has all the elements equal to zero and off-diagonals are zero or one, zero if there is no connection between two nodes and 1 otherwise, respectively. $\boldsymbol{\Phi}(0)$ is an $n \times n$ graph matrix of the initial loans of the network and $\boldsymbol{\Phi}(t)$ is the evolution of this matrix in time. We assume that $\boldsymbol{\Phi}(t)$ is governed by a diffusion law transferring funds from the banks of higher leverages to those of lower. Thus is described by the following set of matrix differential equations:

$$\dot{\boldsymbol{\Phi}} = [\mathbf{A}, \mathbf{L}] = \mathbf{A}diag(l_i) - diag(l_i)\mathbf{A} \tag{2}$$

where l_i is the leverage of bank i. In equilibrium we have $\dot{\boldsymbol{\Phi}} = 0$ and in a connected network all the leverages of the banks will be equal, i.e. $l_i = l_j \Rightarrow l_i - l_j = 0$, for any $(i, j) \in V$. In addition if $\dot{\boldsymbol{\Phi}} = 0$, then (2) gives a homogeneous system with e the number of unknowns and $n - 1$ equations. Thus the dimension of the solution space of the system (Σ) is $dim(\Sigma) = e - rank(\mathbf{I}_k) = e - (n - 1) = e - n + 1$ where $\mathbf{I}_k$ is a non rectangular operator and in order to have a unique solution should be: $e - n + 1 = 0 \Rightarrow e = n - 1$. Therefore if the dimension of e is $n - 1$, then there exists a unique solution and $\dot{\boldsymbol{\Phi}} = 0$.

Equation (2) may be rewritten as a vector differential equation of all loans $\boldsymbol{\phi}(t) \in \mathbb{R}^e$ as:

$$\dot{\underset{\sim}{\boldsymbol{\phi}}}(t) = \boldsymbol{\Delta}\underset{\sim}{\boldsymbol{\phi}}(t) \tag{3}$$

where $\boldsymbol{\phi}(t)$ is the vector of loans of the network with the solution, described by

$$\underset{\sim}{\boldsymbol{\phi}}(t) = e^{\boldsymbol{\Delta} t}\underset{\sim}{\boldsymbol{\phi}}(0) \tag{4}$$

where $\boldsymbol{\phi}(0)$ is the vector of initial loans $(t = 0)$. Moreover the operator $\underset{\sim}{\boldsymbol{\Delta}}$ contains information from the interconnections and capital of the banks and can be written in the form $\boldsymbol{\Delta} = \mathbf{PLP}^{-1}$ where $\mathbf{P}, \mathbf{L} \in \mathbb{R}^{n\times n}$ are the operators of eigenvectors and eigenvalues of $\boldsymbol{\Delta}$, respectively. In order to have a stable equilibrium on our system the eigenvalues z_i which are derived from $\mathbf{\Delta}$ operator should have negative real parts and at least one equal to zero. The eigenvector of the 0 mode corresponds to an atractor of the system or equilibrium and leads to a minimum dispersion of leverages.

In equilibrium the final solution of differential equation as described by (3) and given by (4) will be a vector $\underset{\sim}{\boldsymbol{\phi}}(+\infty) \in \mathbb{R}^{e\times 1}$ describing all those directed flows $\boldsymbol{\phi}_{ij}$ which all the banks in our network should have at $t \to +\infty$. The leverage of any bank in equilibrium will be given again by (1) but $\boldsymbol{\phi}_{ij}$ is now replaced by its steady state value $\boldsymbol{\phi}_{ij}(+\infty)$. Both the solution of (3) and the equilibrium solution depend on the properties of operator $\mathbf{\Delta}$ and the final equilibrium flows are calculated by applying an appropriate projection operator on the initial values $\underset{\sim}{\boldsymbol{\phi}}(0)$.

2 Diffusion Equations and Equilibrium Points

Let a network of n-banks which exchange funds between them. This can be depicted in terms of a graph (V, E), where V are the nodes of the graph i.e. the banks and E are the edges denoted by $\boldsymbol{\phi}_{ij}$ (the i bank lends to j bank amount $\boldsymbol{\phi}_{ij}$). On every node (bank) the leverage is defined as in (1) and the aim of this paper is to provide means to minimize the dispersion of leverages in the network making it more stable. We propose the diffusion method, i.e. funds are transferred from those banks of high leverage to those of low that is written as:

$$\dot{\boldsymbol{\phi}}_{ij} = l_j - l_i$$

where $\boldsymbol{\phi}_{ij}$ is a function of time t and with initial condition $\boldsymbol{\phi}_{ij}(0)$. These equations can be written in matrix form as (2). We may rewrite them in matrix form by introducing the following notation:

1. Φ is the matrix of loans.

2. $Vec(\Phi)$ is the column vector of all columns of Φ superimposed.

3. $D^{(0)}_{n^2-e}(Vec(\Phi))$ is the vector taken from $Vec(\Phi)$ by deleting the fixed $n^2 - e$ zeros appearing in Φ and $D^{(0)}_{n^2-e}$ is the corresponding selection matrix.

4. $D^{(0)}_{n^2-n}(Vec(diag(l_i)))$ is the $n \times 1$ vector $[l_1, l_2, \ldots, l_n]^T$ and $D^{(0)}_{n^2-n}$ is the corresponding selection matrix.

5. The above two operators will be also applied to matrices as induced by matrix multiplication on $Vec(\Phi)$ and we will loosely apply the same notation to them.

6. $\otimes$ is the matrix tensor product.

Theorem 1 *The diffusion equation (2) may be written in vector form as*

$$\dot{\boldsymbol{\phi}}(t) = \boldsymbol{\Delta}\boldsymbol{\phi}(t) \tag{5}$$

which is a system of homogeneous linear differential equations

Theorem 2 (General Approach) *For any connected, directed and weighted network in the equilibrium all the leverages $l_i, i = \{1, 2, \ldots, n\}$ are equal, i.e. $l_i = l_j, i \neq j$.*

Proof Let the following equation:

$$\dot{\Phi} = \mathbf{A}diag(l_i) - diag(l_i)\mathbf{A} \tag{6}$$

Taking the transpose of (6) we get

$$\dot{\Phi}^T = diag(l_i)\mathbf{A}^T - \mathbf{A}^T diag(l_i) \Rightarrow \dot{\Phi}^T = \mathbf{I}_n diag(l_i)\mathbf{A}^T - \mathbf{A}^T diag(l_i)\mathbf{I}_n \tag{7}$$

From linear algebra and more precisely from Kronecker product we know a property which says that:

$$Vec(\mathbf{AXB}) = (\mathbf{B}^T \otimes \mathbf{A})Vec(\mathbf{X}) \tag{8}$$

Apply the $Vec(\cdot)$ operator to (7) and taking in account (8) we get

$$\begin{aligned}
Vec(\dot{\Phi}^T) &= (\mathbf{A} \otimes \mathbf{I}_n)Vec(diag(l_i)) - (\mathbf{I}_n \otimes \mathbf{A}^T)Vec(diag(l_i)) \\
Vec(\dot{\Phi}^T) &= [(\mathbf{A} \otimes \mathbf{I}_n) - (\mathbf{I}_n \otimes \mathbf{A}^T)]Vec(diag(l_i)) \\
D^{(0)}_{n^2-e}(Vec(\dot{\Phi}^T)) &= D^{(0)}_{n^2-e}\{[(\mathbf{A} \otimes \mathbf{I}_n) - (\mathbf{I}_n \otimes \mathbf{A}^T)]Vec(diag(l_i))\} \\
\dot{\phi}(t) &= D^{(0)}_{n^2-e}[(\mathbf{A} \otimes \mathbf{I}_n) - (\mathbf{I}_n \otimes \mathbf{A}^T)][D^{(0)}_{n^2-n}]^T \\
&\quad \times \underbrace{D^{(0)}_{n^2-n}[Vec(diag(l_i))]}_{A}
\end{aligned} \tag{9}$$

The part (A) of (9) is just the vector of leverages l_i after cancelling out all the zeros. This can be further expressed as:

$$\begin{bmatrix} l_1 \\ l_2 \\ \vdots \\ l_n \end{bmatrix} = diag(1/k_i)(\mathbf{I}_n \otimes \mathbb{1}_n)Vec(\Phi^T) = diag(1/k_i)(\mathbf{I}_n \otimes \mathbb{1}_n)D^{(0)}_{n^2-e}\phi(t)$$

where k_i is the initial capital per bank i. Plug in the last equation to (9) we get

$$\dot{\boldsymbol{\phi}}(t) = \underbrace{D^{(0)}_{f,n^2-e}\{[(\mathbf{A}\otimes\mathbf{I}_n) - (\mathbf{I}_n\otimes\mathbf{A}^T)]\}D^{(0)^T}_{f,n^2-n}}_{\boldsymbol{\Delta}_1\in\mathbb{R}^{e\times n}}\underbrace{diag(1/k_i)}_{\in\mathbb{R}^{n\times n}}$$

$$\times \underbrace{(\boldsymbol{I}_n\otimes\mathbb{1}_n)D^{(0)}_{f,n^2-e}}_{\boldsymbol{\Delta}_2\in\mathbb{R}^{n\times e}}\phi(t)$$

$$\dot{\boldsymbol{\phi}}(t) = \underbrace{\boldsymbol{\Delta}_1 diag(1/k_i)\boldsymbol{\Delta}_2}_{\boldsymbol{\Delta}\in\mathbb{R}^{e\times e}}\boldsymbol{\phi}(t)$$

$$\dot{\boldsymbol{\phi}}(t) = \boldsymbol{\Delta}\boldsymbol{\phi}(t) \tag{10}$$

■

Equation (10) define a set of diffusion differential equations describing the evolution of interbank loans. The state space of this dynamical system is $\mathbb{R}^e$ with dimension equal to the number of loans, and the state matrix $\boldsymbol{\Delta}\in\mathbb{R}^{e\times e}$. The matrix $\boldsymbol{\Delta}$ contains information of the structure of the network and the capitals of the banks. Thus defines an operator semigroup $e^{\boldsymbol{\Delta}t}_{t>0}$ which acts on the initial vector of loans $\underset{\sim}{\boldsymbol{\phi}}(0)$ and the equilibrium is reached at $t=+\infty$ provided $\boldsymbol{\Delta}$ is such that $\lim_{t\to+\infty} e^{\boldsymbol{\Delta}t}$ exists. Similarly we may define evolution equations for the leverages $\underset{\sim}{\boldsymbol{l}} = (l_i)_{i=1}^n$ as follows:

Theorem 3 *The vector $\boldsymbol{l}$ of leverages is governed by a set of linear differential equations $\dot{\underset{\sim}{\boldsymbol{l}}} = \mathscr{L}\underset{\sim}{\boldsymbol{l}}$ where $\mathscr{L}$ is an appropriate Laplacian.*

Proof Consider the set of matrix diffusion equations

$$\dot{\Phi} = \mathbf{A}diag(l_i) - diag(l_i)\mathbf{A} \tag{11}$$

then as $\underset{\sim}{\boldsymbol{l}} = diag(1/k_i)\Phi\mathbb{1}_n$ we get

$$\dot{\underset{\sim}{\boldsymbol{l}}} = diag(1/k_i)\mathbf{A}\underset{\sim}{\boldsymbol{l}} - diag(1/k_i)diag(a_i)\underset{\sim}{\boldsymbol{l}} \tag{12}$$

where

$$\mathbf{A}\mathbb{1}_n = \begin{bmatrix} a_1 \\ a_2 \\ \vdots \\ a_n \end{bmatrix}$$

and $a_i, i\in\{1,2,\ldots,n\}$ are the column degrees of the graph. Thus

$$\dot{\underset{\sim}{\boldsymbol{l}}} = \mathscr{L}\underset{\sim}{\boldsymbol{l}} \tag{13}$$

where $\mathscr{L} = diag(1/k_i)(\mathbf{A} - diag(a_i))$ ■

Thus (13) describes the evolution of leverages, $\mathscr{L} \in \mathbb{R}^{n \times n}$ is the Laplacian for the network and the equilibrium leverages are given by

$$\underset{\sim}{l}(+\infty) = \lim_{t \to +\infty} e^{\mathscr{L}t} \underset{\sim}{l}(0) \tag{14}$$

The equilibria for both (13) and (10) depend on the spectral structure of the operators $\boldsymbol{\Delta}$ and $\mathscr{L}$. We require for both systems to lead to equilibrium so that the steady state values for loans and leverages are obtained.

Theorem 4 *When a network G(V, E) is connected then in the equilibrium the leverages $l_i, i = \{1, 2, \dots, n\}$ are all equal, i.e. $l_1 = l_2 = \cdots = l_n$ or in other words $l_i = l_j$ with $i \neq j$.*

Proof Let the following equation:

$$\dot{\Phi} = \mathbf{A} diag(l_i) - diag(l_i)\mathbf{A} \tag{15}$$

At the point of steady state we have $\dot{\Phi} = 0$ and (15) gives

$$\mathbf{A} diag(l_i) = diag(l_i)\mathbf{A} \tag{16}$$

Therefore at equilibrium we have that $l_i = l_j \forall (i, j) \in E$. Consider now a connected network with adjacent matrix $\mathbf{A}$. Then for every pair $i, j \in V$ there is a path which connects i and j, this means that there are pairs in the graph, i.e. $(i_1, j_1), (i_2, j_2), (i_3, j_3), \dots, (i_k, j_k)$, that connect i and j such that satisfy

$$\{i_k, j_k\} \bigcap \{i_{k+1}, j_{k+1}\} = \{r_k\}.$$

Hence $l_i = l_{r_1} = l_{r_2} = \cdots = l_{r_k} = l_j$ and thus for every pair $i, j \in V$ at equilibrium we have that $l_i = l_j$. Therefore when the network is connected at equilibrium we must have all leverages equal. ■

2.1 Connectivity and Equilibria

It is worth noting that, from an operator we can understand if our network is connected. Suppose that we have an adjacent operator $\mathbf{A}$ with all the elements $a_{ij} = 0$, for all $i = j$.

Definition 1 (https://en.wikipedia.org/wiki/Directed_graph) A directed graph is called weakly connected if replacing all of its directed edges with undirected edges produces a connected (undirected) graph. It is connected if it contains a directed path from u to v or a directed path from v to u for every pair of vertices u, v. It is strongly connected or simply strong if it contains a directed path from u to v and a directed path from v to u for every pair of vertices u, v. The strong components are the maximal strongly connected subgraphs.

2.1.1 Connected and Unconnected Network

Suppose we have a connected, directed and weighted network. In order to characterize it as connected it must satisfy the following four cases-theorems.

Theorem 5 ([1]) *Suppose that a network $G(V, E)$ is directed. Let $\mathbf{A}$ be the adjacent operator of $G(V, E)$. We convert the number of ones from $\mathbf{A}$ as a new node and connecting them in such a way in order to get triangles. Moreover, let $\mathcal{X}_E(G)$ be the Euler characteristic of graph G. If the graph is (weakly) connected, then the Euler characteristic is equal to 1 or $\mathcal{X}_E(G) = 1$, i.e.:*

$$\mathcal{X}_E(G) = |V| - |E| + |F| = 1 \tag{17}$$

where $|V|$, $|E|$ and $|F|$, are the number of vertices, edges and faces, respectively.

Theorem 6 *Let $\mathbf{A}$ be the adjacent operator of a graph $G(V, E)$ and $f_2 : \{0, 1, 2\} \to \{0, 1\}$. We call that the graph G is unconnected if and only if:*

$$\mathbf{S}(\mathbf{A} + \mathbf{A}^T)_{ij} = f_2(a_{ij} + a_{ji}) \ where \ f_2 = \begin{cases} x \ if \ x \in \{0, 1\} \\ 1 \ if \ x = 2 \end{cases}$$

has at least $n - 1$ elements equal to one at symmetric positions based on the main diagonal and more than $n - 2$ secondary diagonal with zeros.

Theorem 7 *Let $\mathbf{A}$ be the adjacent operator of a network. The network is connected if there are $n - 1$ ones in such a way that the pairs of the network compose a connection.*

Theorem 8 *Let $\mathbf{A}$ be an adjacent operator of a connected, directed and weighted network. If there is no way to make $\mathbf{A}$ block diagonal by permutating rows and columns by the same permutation matrix $\mathbf{Q}$, then the graph is connected.*

Theorem 9 *Let G(V, E) be a graph with n vertices and at least $n - 1$ edges. Then we can say that:*

If there are connected components, then the leverages are all equals only in the connected components of the network and not among them.

Proof We define a connected component network G with n nodes and at least $n - 1$ edges. By (2) we have

$$\dot{\Phi} = [\mathbf{A}, \mathbf{L}] = \mathbf{A}diag(l_i) - diag(l_i)\mathbf{A} \tag{18}$$

In the equilibrium $\dot{\Phi} = 0$. Thus (18) gives

$$[\mathbf{A}, \mathbf{L}] = \mathbf{A}diag(l_i) - diag(l_i)\mathbf{A} = 0 \tag{19}$$

We have proved from Theorem 4 that when we have a connected network G the leverages are all equal, i.e. $l_i = l_j$ or $l_i - l_j = 0, \forall (i,j) \in V$ with $i \neq j$. Thus we have the following system (Σ)

$$\underbrace{\left[\begin{array}{c|c|c|c|c|c} \frac{1}{k_1}\mathbb{1}_{a_1} & \frac{-1}{k_2}\mathbb{1}_{a_2} & 0_{a_3} & \dots 0 & 0 & 0 \\ 0 & \frac{1}{k_2}\mathbb{1}_{a_2} & \frac{-1}{k_3}\mathbb{1}_{a_3} & \dots 0 & 0 & 0 \\ \vdots & \vdots & \vdots & \dots \vdots & \vdots & \vdots \\ 0 & 0 & 0 & \dots 0 & \frac{1}{k_{n-1}}\mathbb{1}_{a_{n-1}} & \frac{-1}{k_n}\mathbb{1}_{a_n} \end{array}\right]}_{\mathbf{I}_k} \underset{\sim}{\boldsymbol{\phi}} = 0_{n-1,1}$$

which is equivalent to $\boldsymbol{\Delta\phi} = \underset{\sim}{0}$.

This is a homogeneous system of $n-1$ equations in n unknowns. For n banks to form a connected graph, with no self-loops, we must have $n-1 \leq e \leq n^2 - n$. The solution space of the homogeneous system (Σ) is as linear space of dimension $dim(\Sigma)$ which is equal to the dimension $dim(Ker\mathbf{I}_k)$. Hence:

$$\dim(\Sigma) = e - rank(\mathbf{I}_k) = e - (n-1) = e - n + 1 \tag{20}$$

If we further impose conservation of the total volume of loans in the system, i.e. $\sum_i \boldsymbol{\phi} = c$ and $\mathbf{V}$ is a positive $(e-n+1)e$ bases matrix for (Σ), then a solution for $\boldsymbol{\phi}_{eq}$ for equilibrium is given by

$$\boldsymbol{\phi}_{eq} = \frac{\mathbb{1}^T\Sigma}{\mathbb{1}^T\Sigma\mathbb{1}}c \tag{21}$$

■

Thus, on the one hand if the operator $\mathbf{I}_k$ has at least one element per line not equal to zero, then all the leverages in equilibrium are equal, i.e. $l_1 = l_2 = \cdots = l_n = l$ but on the other hand if there exists one line with no outgoing flows, then it is true that the leverages are all equal and more precisely equals to zero, i.e. $l_1 = l_2 = \cdots = l_n = 0$.

Suppose that we have a connected, directed and weighted network of banks where they exchange loans at some unit of time. By (10) we have

$$\dot{\underset{\sim}{\boldsymbol{\phi}}}(t) = \boldsymbol{\Delta}\underset{\sim}{\boldsymbol{\phi}}(t) \tag{22}$$

with solution:

$$\underset{\sim}{\boldsymbol{\phi}}(t) = \boldsymbol{\Delta}\underset{\sim}{\phi}(0) \tag{23}$$

where $\underset{\sim}{\boldsymbol{\phi}}(0)$ is the $n \times 1$ vector of initial values of loans ($t = 0$), and at the same time the leverages (l_i) of banks are not all equal or in other words the banking system is not in balance. Then through the proposed diffusion process the network will approach balance, as $t \to +\infty$. Furthermore, at equilibrium all the loan flows, $\boldsymbol{\phi}$, are describing a vector space, as the next formula says

$$\mathbf{V}_b = \{\boldsymbol{\phi} : \boldsymbol{\Delta\phi} = 0\} \tag{24}$$

In addition this imbalance in the network can be described by the norm of $||\boldsymbol{\Delta}\phi(t)||$ and if this measure increases, then we can say that the leverages among the banks diverge and on the other hand if this measure is equal to zero, then we have an equilibrium in the network. Our main target is to make a projection of the initial flows $\boldsymbol{\phi}(0)$ on $\mathbf{V}_b$ produces a row required vector of flow $\boldsymbol{\phi}_b$. This is done by taking an orthonormal basis of $\mathbf{V}_b$ and $\boldsymbol{\phi}_b$ is given by

$$\boldsymbol{\phi}_b = \mathbf{B}_b\mathbf{B}_b^T\boldsymbol{\phi}(0) \tag{25}$$

where $\mathbf{B}_b\mathbf{B}_b^T = \mathbf{P}_{\mathbf{V}b} \in \mathbb{R}^{e \times e}$ is a projection operator to that vector space $\mathbf{V}_b$ so that $||\boldsymbol{\Delta}\phi_b||$ is minimal and $\mathbf{B}_b$ is an orthonormal basis of $\mathbf{V}_b$. $\mathbf{P}_b$ also can be alternatively calculated as:

1. by diffusion:

$$\lim_{t \to +\infty} e^{\boldsymbol{\Delta}t} = \mathbf{P}_b$$

2. by operational calculus:

$$\lim_{\epsilon \to 0} \epsilon(\epsilon\mathbf{I}_e - \boldsymbol{\Delta})^{-1} = \mathbf{P}_b$$

3 The Basic Two Structural Cases by Toy-Examples: Calculations of Equilibrium Points

3.1 *Case 1: At Least One Non-zero Element per Line to the Adjacent Operator*

Next we present two simple examples of graphs demonstrating how diffusion can be utilised to reach network equilibrium. The two graphs have different topologies as the first exhibits cycles and the second is linear.

Let the following graph G(V, E), with the set of vertices as $V = \{1, 2, 3, 4, 5\}$, $|V| = 5$ and edges, $E = \{e_{12}, e_{25}, e_{31}, e_{35}, e_{41}, e_{54}\}$, $|E| = 6$.

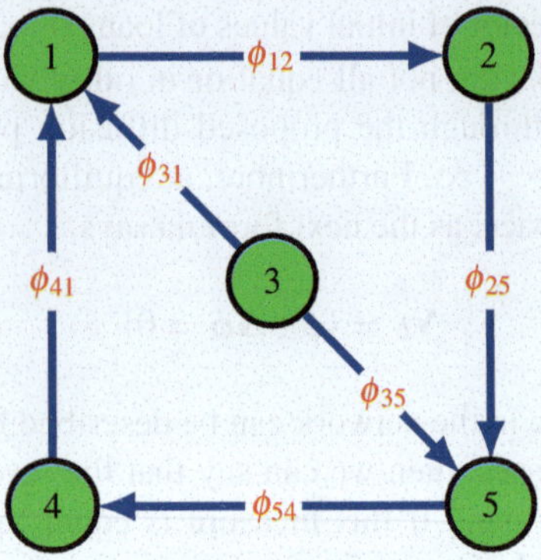

Graph 1, G(5, 6)

Every vertex on Graph 1 represented by a number is a bank. The arrows show us the direction of the loans which means which bank gives a loan to another and we denote by ϕ_{ij}. Furthermore, every bank is associated with a capital k_i and we calculate the initial leverages for any bank-vertex i. Thus, let us say that the initial per bank capitals are: $k_1 = 81, k_2 = 9, k_3 = 27, k_4 = 4, k_5 = 3$. Thus, the leverages are

$$l_1 = \frac{\phi_{12}}{k_1}, l_2 = \frac{\phi_{25}}{k_2}, l_3 = \frac{\phi_{31} + \phi_{35}}{k_3}, l_4 = \frac{\phi_{41}}{k_4}, l_5 = \frac{\phi_{54}}{k_5} \tag{26}$$

The adjacent operator which corresponds to Graph 1 is

$$\mathbf{A}_{adj} = \begin{bmatrix} 0\,1\,0\,0\,0 \\ 0\,0\,0\,0\,1 \\ 1\,0\,0\,0\,1 \\ 1\,0\,0\,0\,0 \\ 0\,0\,0\,1\,0 \end{bmatrix}$$

According to the Theorem 2.2, in the equilibrium we have that: $l_1 = l_2 = l_3 = l_4 = l_5$ and from the last expression we get

$$l_1 - l_2 = l_2 - l_3 = l_3 - l_4 = l_4 - l_5 = 0$$

Due to (26) if we replace l_i to the above equations and then write them in matrix form we get an $m \times n, n > m$ operator, $\mathbf{A_1}$ which is a special type of incident operator. Thus we get

$$\underbrace{\begin{bmatrix} 1/k_1 & -1/k_2 & 0 & 0 & 0 & 0 \\ 0 & 1/k_2 & -1/k_3 & -1/k_3 & 0 & 0 \\ 0 & 0 & 1/k_3 & 1/k_3 & -1/k_4 & 0 \\ 0 & 0 & 0 & 0 & 1/k_4 & -1/k_5 \end{bmatrix}}_{\mathbf{A_1}} \begin{bmatrix} \phi_{12} \\ \phi_{25} \\ \phi_{31} \\ \phi_{35} \\ \phi_{41} \\ \phi_{54} \end{bmatrix} = \begin{bmatrix} 0 \\ 0 \\ 0 \\ 0 \end{bmatrix} \quad (\mathbf{\Sigma})$$

If we replace k_i to the operator $\mathbf{A_1}$ and do some algebraic operations, finally we get

$$\mathbf{A_1} = \begin{bmatrix} 1/81 & -1/9 & 0 & 0 & 0 & 0 \\ 0 & 1/9 & 0 & 0 & -1/4 & 0 \\ 0 & 0 & 1/27 & 1/27 & -1/4 & 0 \\ 0 & 0 & 0 & 0 & 1/4 & -1/3 \end{bmatrix}$$

For those values of operator $\mathbf{A_1}$ the system $(\mathbf{\Sigma})$ gives $\phi_{12} = (81/4)\phi_{41}, \phi_{25} = (9/4)\phi_{41}, \phi_{31} = (27/4)\phi_{41} - \phi_{35}, \phi_{54} = (3/4)\phi_{41}$ and the right kernel corresponding to equilibrium is

$$\begin{bmatrix} \phi_{12} \\ \phi_{25} \\ \phi_{31} \\ \phi_{35} \\ \phi_{41} \\ \phi_{54} \end{bmatrix} = \phi_{41} \underbrace{\begin{bmatrix} 81/4 \\ 9/4 \\ 27/4 \\ 0 \\ 1 \\ 3/4 \end{bmatrix}}_{\underset{\sim}{\mathbf{a}}_1} + \phi_{35} \underbrace{\begin{bmatrix} 0 \\ 0 \\ -1 \\ 1 \\ 0 \\ 0 \end{bmatrix}}_{\underset{\sim}{\mathbf{a}}_2}$$

Consequently we have found a set of two elements—vectors, $\underset{\sim}{\mathbf{a}}_1, \underset{\sim}{\mathbf{a}}_2$ which constitute the basis of $\mathbf{V}$ and $\phi_{45}, \phi_{35}, \in \mathbb{R}$. Suppose that $\lambda \equiv \phi_{45}$, $\mu \equiv \phi_{35}$ and if we want the total, current number of loans into the interbank system to be, for instance, 93 monetary units, then we have $(81/4)\lambda + (9/4)\lambda + (27/4)\lambda + \lambda + (3/4)\lambda - \mu + \mu = 93 \Rightarrow \lambda = 3$. This gives rise to equilibrium flows as:

$$\begin{aligned} \phi_{12} &= (81/4) \times 3 = 243/4 \\ \phi_{25} &= (9/4) \times 3 = 27/4 \\ \phi_{31} &= (27/4) \times 3 = 81/4 \\ \phi_{35} &= 0 \\ \phi_{41} &= 1 \times 3 = 3 \\ \phi_{54} &= (3/4) \times 3 = 9/4 \end{aligned} \tag{27}$$

By (26) the initial leverages were: $l_1 = 1/4, l_2 = 27/16, l_3 = 27/16, l_4 = 1/4, l_5 = 3/16$. But now, if we replace in (26) the new flows which are given by (27) we get the final leverages which are: $l_1 = l_2 = l_3 = l_4 = l_5 = 3/4$.

3.1.1 Conclusion

Equilibrium is achieved by transferring funds from banks of higher leverage to those of lower.

3.2 Case 2: Adjacent Operator with Two Lines Equals to Zero

Let the following network $G(V, E)$ with a set of vertices $V = \{1, 2, 3, 4, 5\}$, $|V| = 5$ and edges $E = \{e_{12}, e_{32}, e_{34}, e_{45}\}$, $|E| = 4$.

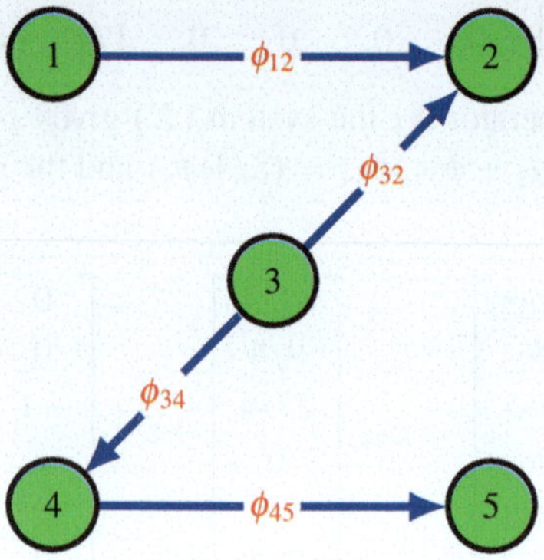

Graph 2, G(5, 4)

Suppose that the initial per bank capitals are $k_1 = 81, k_2 = 9, k_3 = 27, k_4 = 3, k_5 = 3$. Moreover, the initial leverages for those banks which are given from Graph 2 are

$$l_1 = \frac{\phi_{12}}{k_1}, l_2 = 0, l_3 = \frac{\phi_{32} + \phi_{34}}{k_3}, l_4 = \frac{\phi_{45}}{k_4}, l_5 = 0 \tag{28}$$

The adjacent operator which can be extracted from the Graph 2 is the following:

$$\mathbf{A}_{adj} = \begin{bmatrix} 0\,1\,0\,0\,0 \\ 0\,0\,0\,0\,0 \\ 0\,1\,0\,1\,0 \\ 0\,0\,0\,0\,1 \\ 0\,0\,0\,0\,0 \end{bmatrix}$$

In the equilibrium as we already mentioned at Theorem 2.2, that all the leverages are equal, thus we have that: $l_1 = l_2 = l_3 = l_4 = l_5$ or equivalently:

$$l_1 = 0, l_3 = 0, l_3 - l_4 = 0, l_4 = 0$$

By (28) if we replace l_i to the above equations and write them in matrix form we get an $m \times n, n > m$ operator $\mathbf{A_2}$, which is a special type of incident operator:

$$\underbrace{\begin{bmatrix} 1/81 & 0 & 0 & 0 \\ 0 & 1/27 & 1/27 & 0 \\ 0 & 1/27 & 1/27 & -1/3 \\ 0 & 0 & 0 & 1/3 \end{bmatrix}}_{\mathbf{A_2}} \begin{bmatrix} \phi_{12} \\ \phi_{32} \\ \phi_{34} \\ \phi_{45} \end{bmatrix} = \begin{bmatrix} 0 \\ 0 \\ 0 \\ 0 \end{bmatrix} (\boldsymbol{\Sigma})$$

We can observe that $\mathbf{A_2}$ is an upper-triangular operator with $rank = 3$, i.e. as the number of non-zero rows. Thus, from the system (Σ) above we get $\phi_{12} = 0, \phi_{32} = -\phi_{34}, \phi_{45} = 0$ or

$$\begin{bmatrix} \phi_{12} \\ \phi_{32} \\ \phi_{34} \\ \phi_{45} \end{bmatrix} = \begin{bmatrix} 0 \\ -\phi_{34} \\ \phi_{34} \\ 0 \end{bmatrix} = \phi_{34} \underbrace{\begin{bmatrix} 0 \\ -1 \\ 1 \\ 0 \end{bmatrix}}_{\underset{\sim}{\mathbf{v}}} (\boldsymbol{\Sigma})$$

Any vector from $\Lambda(\boldsymbol{\Sigma})$ always gives rise to zero equilibrium leverages for all the banks.

3.2.1 Conclusion

If we have a connected, directed and weighted Graph G(V, E) and the adjacent operator of the graph has at least one row with all the elements equals to zero, then the vertices (banks) of the graph at equilibrium will have all the leverages equal to zero.

4 Differential Equation and Its Solution

In what it follows we do an approach of a more general solution in matrix form of the equation:

$$\dot{\boldsymbol{\Phi}} = [\mathbf{A}, \mathbf{L}] \Leftrightarrow \dot{\phi_{ij}} = l_j - l_i \tag{29}$$

The incident operator should have at least one non-zero element in every row. Furthermore, we are studying connected, directed and weight networks. Thus we have

$$\dot{\boldsymbol{\phi}} = \boldsymbol{\Delta\phi}$$

with solution is given by

$$\boldsymbol{\phi}(t) = e^{\boldsymbol{\Delta} t}\boldsymbol{\phi}(0).$$

Next we will calculate the steady state values of $\boldsymbol{\phi}(t)$ by exploiting the spectral structure of the operator $\mathbf{\Delta}$. In our analysis we are interested to those cases where on the one hand the operator $\mathbf{A}$ has all the eigenvalues λ_i negative and at least one equal to zero, i.e. $\{\lambda_i \in \mathbb{R}_- : \lambda_i \leq 0\}$ and on the other hand, if it has complex eigenvalues the real part should be negative, i.e. $\{\lambda_i \in \mathbb{C} : \mathfrak{Re}(\lambda_i) \leq 0\}$. Thus let, $\lambda_1, \lambda_2, \ldots, \lambda_n$ and $e_1, e_2, \ldots, e_n$ are the eigenvalues and eigenvectors, respectively. Furthermore, the solution of the system will be of the form:

$$\dot{\underset{\sim}{\boldsymbol{\phi}}}(t) = e^{\mathbf{\Delta}t}\underset{\sim}{\boldsymbol{\phi}}(0) \tag{30}$$

where $\underset{\sim}{\boldsymbol{\phi}}_{ij}(0)$ are the initial conditions. As we know the rectangular operator can be written in the form as:

$$\mathbf{\Delta} = \mathbf{P\Lambda P}^{-1} \tag{31}$$

where $\mathbf{P}$, $\mathbf{\Lambda}$ are the eigenvectors and eigenvalues matrices, respectively. Then we have

$$\dot{\underset{\sim}{\boldsymbol{\phi}}}(t) = \mathbf{P}e^{\mathbf{\Lambda}t}\mathbf{P^{-1}}\underset{\sim}{\boldsymbol{\phi}}(0) \tag{32}$$

and provided $\mathbf{\Lambda}$ has simple structure with only zero eigenvalues or eigenvalues with negative real part. Thus $\lim\limits_{t\to\infty} e^{\mathbf{\Lambda t}}$ can be calculated as:

$$\lim_{t\to+\infty} e^{\mathbf{\Lambda}t} = \begin{bmatrix} \lim\limits_{t\to+\infty} e^{\lambda_1 t} & 0 & \ldots & 0 \\ 0 & \lim\limits_{t\to+\infty} e^{\lambda_2 t} & \ddots & 0 \\ \vdots & \vdots & \ddots & \vdots \\ 0 & 0 & \ldots & \lim\limits_{t\to+\infty} e^{\lambda_e t} \end{bmatrix}$$

$$\underbrace{\left[\begin{array}{cccc|cccc} 0 & & & & & & & \\ & 0 & & & & & \mathbf{0} & \\ & & \ddots & & & & & \\ & & & 0 & & & & \\ \hline & & & & 1 & & & \\ & \mathbf{0} & & & & 1 & & \\ & & & & & & \ddots & \\ & & & & & & & 1 \end{array}\right]}_{\mathbf{diag}(\lambda_i)}$$

and hence $\boldsymbol{\phi}(+\infty) = \mathbf{P}_b\boldsymbol{\phi}(0)$ where $\mathbf{P}_b$ is the required projection operator.

5 Solution by Diffusion Three Structural Examples

5.1 *Case 1: Two Real Negatives Eigenvalues and One Zero*

The operator $\mathbf{\Delta}$ of differential equation as well as the Laplacian of this problem is not symmetric hence their eigenvalues may be complex and also it is possible that $\mathbf{\Delta}$ has no zero eigenvalues. This is demonstrated in the following examples.

Suppose the next graph, which describes an interbanking loan system among three banks as V = $\{1, 2, 3\}$ and the exchange transfer loans among them as E = $\{e_{12}, e_{23}, e_{32}\}$, respectively. Furthermore, each bank has an initial capital $k_i, i = \{1, 2, 3\}$ and more precisely $k_1 = 2, k_2 = 4$ and $k_3 = 3$. Moreover, suppose the initial leverage, i.e. $\phi_{ij}(0) = 10$ for all those three banks.

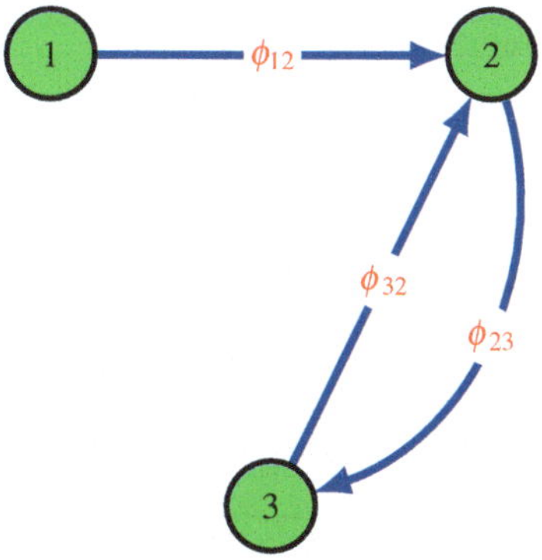

Graph 1, G(3, 3)

The adjacent operator from Graph 1 is given by

$$\mathbf{A}_{adj} = \begin{bmatrix} 0 & 1 & 0 \\ 0 & 0 & 1 \\ 0 & 1 & 0 \end{bmatrix}$$

According to (29) we can write the differential equations for Graph 1 as follows:

$$\dot{\phi}_{12} = \frac{\phi_{23}}{k_2} - \frac{\phi_{12}}{k_1}$$

$$\dot{\phi}_{23} = \frac{\phi_{32}}{k_3} - \frac{\phi_{23}}{k_2}$$

$$\dot{\phi}_{32} = \frac{\phi_{23}}{k_2} - \frac{\phi_{32}}{k_3}$$

or in matrix form:

$$\begin{bmatrix} \dot{\phi}_{12} \\ \dot{\phi}_{23} \\ \dot{\phi}_{32} \end{bmatrix} = \underbrace{\begin{bmatrix} (-1)/k_1 & 1/k_2 & 0 \\ 0 & (-1)/k_2 & 1/k_3 \\ 0 & 1/k_2 & (-1)/k_3 \end{bmatrix}}_{\mathbf{\Delta}} \begin{bmatrix} \phi_{12} \\ \phi_{23} \\ \phi_{32} \end{bmatrix} \tag{33}$$

and the solution as given by (30). Replacing all k_i as given into the above operator $\mathbf{\Delta}$ we get

$$\mathbf{\Delta} = \begin{bmatrix} -1/2 & 1/4 & 0 \\ 0 & -1/4 & 1/3 \\ 0 & 1/4 & -1/3 \end{bmatrix}$$

The eigenvalues of operator $\mathbf{\Delta}$ are

$$\lambda_1 = \frac{-7}{12}, \lambda_2 = \frac{-1}{12}, \lambda_3 = 0$$

we rewrite the operator $\mathbf{\Delta}$ in the form of (31) as:

$$\mathbf{\Delta} = \underbrace{\begin{bmatrix} 3 & 1 & 2/3 \\ -1 & 0 & 4/3 \\ 1 & 0 & 1 \end{bmatrix}}_{\mathbf{P}} \underbrace{\begin{bmatrix} -7/12 & 0 & 0 \\ 0 & -1/12 & 0 \\ 0 & 0 & 0 \end{bmatrix}}_{\mathbf{\Lambda}} \underbrace{\begin{bmatrix} 0 & -3/7 & 4/7 \\ 1 & 1 & -2 \\ 0 & 3/7 & 3/7 \end{bmatrix}}_{\mathbf{P}^{-1}} \tag{34}$$

and we are studying the system of differential equations given by (32) in the equilibrium, i.e. for time $t \to +\infty$. Thus we calculate the $\lim_{t\to+\infty} e^{\mathbf{\Delta}t} = \lim_{t\to+\infty} \mathbf{P}e^{\mathbf{\Lambda}t}\mathbf{P}^{-1}$, where $\mathbf{\Lambda}$ is the eigenvalues operator of the system and $\mathbf{P}, \mathbf{P}^{-1}$ are the operator and inverse operator of eigenvectors, respectively.

Taking the limit to infinity of $e^{\mathbf{\Lambda}t}$ we get

$$\lim_{t\to+\infty} e^{\mathbf{\Lambda}t} = \begin{bmatrix} 0 & 0 & 0 \\ 0 & 0 & 0 \\ 0 & 0 & 1 \end{bmatrix} \tag{35}$$

Finally to find the solution we replace (35) to $\dot{\boldsymbol{\phi}}(t) = e^{\mathbf{\Delta}t}\boldsymbol{\phi}(0)$ which is the solution to the system as was mentioned in (30) to get

$$\boldsymbol{\phi}(+\infty) = \lim_{t\to+\infty} e^{\mathbf{\Delta}t}\boldsymbol{\phi}(0) = \begin{bmatrix} 0 & 2/7 & 2/7 \\ 0 & 4/7 & 4/7 \\ 0 & 3/7 & 3/7 \end{bmatrix} \begin{bmatrix} 10 \\ 10 \\ 10 \end{bmatrix} = \begin{bmatrix} 40/7 \\ 80/7 \\ 60/7 \end{bmatrix} \tag{36}$$

Table 1 Initial and final leverages (l_i) per bank

l_i	$\sum_j \phi_{ij}/k_i$	Initial values	Equilibrium values
l_1	$\frac{\phi_{12}}{k_1}$	$\frac{10}{2}$	$\frac{20}{7}$
l_2	$\frac{\phi_{23}}{k_2}$	$\frac{10}{4}$	$\frac{20}{7}$
l_3	$\frac{\phi_{32}}{k_3}$	$\frac{10}{3}$	$\frac{20}{7}$

The initial leverages $l_i, i = \{1, 2, 3\}$ for any of those three vertices-banks V_i and the final leverages after replacing the final $\dot{\underset{\sim}{\phi}}(t)$ by (36) are presented in the following matrix:

To sum up we can say that as diffusion governs the network the leverages tend in the long run to obtain a single value. In other words as $t \to +\infty$ all banks tend to have equal leverages (Table 1).

5.2 *Case 2: Two Complex Eigenvalues with Negative Real Part and One Zero*

Suppose the next graph which depicts 3 vertices-banks and three edges, i.e. loans flow from one bank to another. Furthermore, all those banks in the graph have an initial capital per bank $k_i, i = \{1, 2, 3\}$, which is given by $k_1 = 2, k_2 = 4, k_3 = 3$, respectively.

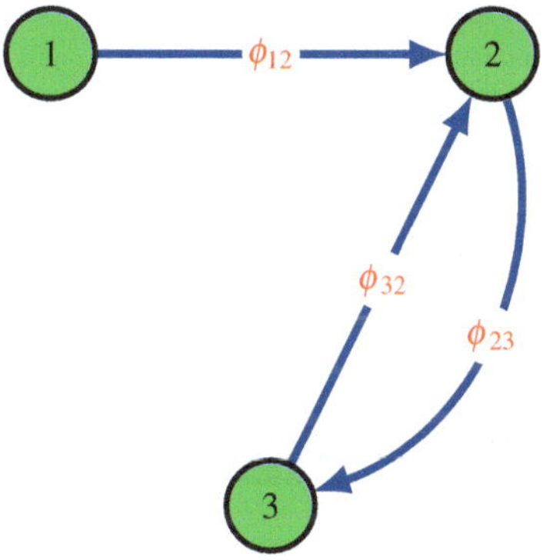

Graph 1, G(3, 3)

The adjacent operator of the Graph 1 above is

$$\mathbf{A}_{adj} = \begin{bmatrix} 0 & 1 & 0 \\ 0 & 0 & 1 \\ 1 & 0 & 0 \end{bmatrix}$$

We apply (29) to write down the differential equations of the Graph 1 which are

$$\dot{\phi}_{12} = \frac{\phi_{23}}{k_2} - \frac{\phi_{12}}{k_1}$$

$$\dot{\phi}_{23} = \frac{\phi_{31}}{k_3} - \frac{\phi_{23}}{k_2}$$

$$\dot{\phi}_{31} = \frac{\phi_{12}}{k_1} - \frac{\phi_{31}}{k_3}$$

or we can rewrite in matrix form as $\dot{\underset{\sim}{\phi}} = \mathbf{\Delta}\underset{\sim}{\phi}$ where $\mathbf{\Delta} \in \mathbb{R}^{3\times3}$:

$$\begin{bmatrix} \dot{\phi}_{12} \\ \dot{\phi}_{23} \\ \dot{\phi}_{31} \end{bmatrix} = \underbrace{\begin{bmatrix} -1/k_1 & 1/k_2 & 0 \\ 0 & -1/k_2 & 1/k_3 \\ 1/k_1 & 0 & -1/k_3 \end{bmatrix}}_{\mathbf{\Delta}} \begin{bmatrix} \phi_{12} \\ \phi_{23} \\ \phi_{31} \end{bmatrix} \tag{37}$$

If we replace the k_i into the operator $\mathbf{\Delta}$ we get

$$\mathbf{\Delta} = \begin{bmatrix} -1/2 & 1/4 & 0 \\ 0 & -1/4 & 1/3 \\ 1/2 & 0 & -1/3 \end{bmatrix}$$

The eigenvalues of operator $\mathbf{\Delta}$ are

$$\lambda_1 = \frac{1}{24}\left(-13 + i\sqrt{47}\right), \lambda_2 = \frac{1}{24}\left(-13 - i\sqrt{47}\right), \lambda_3 = 0 \tag{38}$$

We write down the operator $\mathbf{P}, \in \mathbb{R}^{3\times3}$ with elements the eigenvectors of above operator $\mathbf{\Delta}$.

$$\mathbf{P} = \begin{bmatrix} 1/12(-5 + i\sqrt{47}) & 1/12(-5 - i\sqrt{47}) & 2/3 \\ 1/12(-7 - i\sqrt{47}) & 1/12(-7 + i\sqrt{47}) & 4/3 \\ 1 & 1 & 1 \end{bmatrix}$$

Thus as we already know by (31), we can rewrite the operator $\mathbf{\Delta}$ in the form of:

$$\mathbf{\Delta} = \mathbf{P} \underbrace{\begin{bmatrix} 1/24(-13 + i\sqrt{47}) & 0 & 0 \\ 0 & 1/24(-13 - i\sqrt{47}) & 0 \\ 0 & 0 & 0 \end{bmatrix}}_{\mathbf{\Lambda}=diag(\lambda_i)} \mathbf{P}^{-1}$$

Table 2 Initial and final leverages (l_i) per bank

l_i	$\sum_j \phi_{ij}/k_i$	Initial values	Equilibrium values
l_1	$\frac{\phi_{12}}{k_1}$	$\frac{10}{2}$	$\frac{10}{3}$
l_2	$\frac{\phi_{23}}{k_2}$	$\frac{10}{4}$	$\frac{10}{3}$
l_3	$\frac{\phi_{31}}{k_3}$	$\frac{10}{3}$	$\frac{10}{3}$

As in the previous case 1, we calculate $e^{\Lambda t}$ as $t \to +\infty$:

$$e^{\Lambda t} = \begin{bmatrix} \lim_{t\to+\infty} e^{1/24(-13+i\sqrt{47})t} & 0 & 0 \\ 0 & \lim_{t\to+\infty} e^{1/24(-13-i\sqrt{47})t} & 0 \\ 0 & 0 & 1 \end{bmatrix} \tag{39}$$

Thus, we have

$$\lim_{t?+\infty} e^{\Delta t} = \mathbf{P} \lim_{t?+\infty} e^{\Lambda t}\mathbf{P}^{-1} = \begin{bmatrix} 188/3 & 188/3 & 188/3 \\ 376/3 & 376/3 & 376/3 \\ 94 & 94 & 94 \end{bmatrix} \tag{40}$$

The solution to the system is given by (30) and in our case we have

$$\dot{\underline{\phi}}(+\infty) = \begin{bmatrix} 94/423 & 94/423 & 94/423 \\ 188/423 & 188/423 & 188/423 \\ 1/3 & 1/3 & 1/3 \end{bmatrix} \begin{bmatrix} 10 \\ 10 \\ 10 \end{bmatrix} = \begin{bmatrix} 20/3 \\ 40/3 \\ 10 \end{bmatrix} \tag{41}$$

To sum up the initial leverages $l_i, i = \{1, 2, 3\}$ for any of those vertices-banks V_i and the final leverages after replacing the final $\dot{\underline{\phi}}(t)$ by (41) are presented in Table 2.

Again as in the previous case, the diffusion works in a sense of stabilizer. In other words comparing the results in Table 2, we can observe that at $t = 0$ the leverage was different for every $V_i, i = \{1, 2, 3\}$ bank and as $t \to +\infty$, i.e. in the equilibrium, all the banks will have equal leverages. The main idea is that diffusion directs funds from high leveraged to low leveraged banks in order the banking network-system to converge in an equilibrium and all of them to have the same leverage ratio.

5.3 *Case 3: Three Real Negative Eigenvalues*

Suppose the next network where we have four nodes (banks) $V_i, i = \{1, 2, 3, 4\}$ and three edges e_{ij} with $i \neq j$ as the loans flows, E = $\{e_{12}, e_{32}, e_{24}\}$ as shown in the following Graph 1. Furthermore, all those four banks have an initial capital, i.e. $k_1 = 2, k_2 = 4, k_3 = 3, k_4 = 5$, respectively.

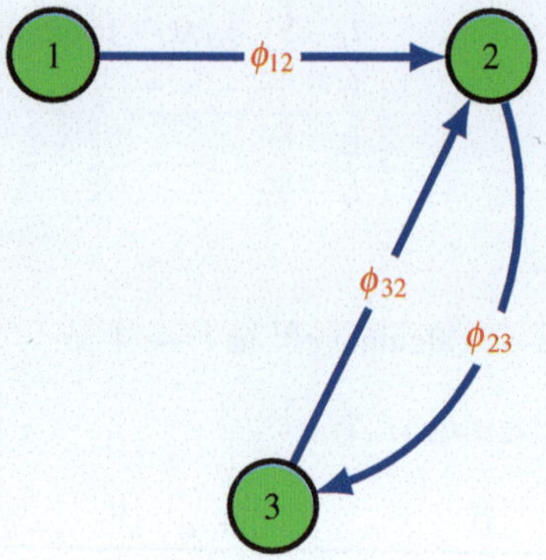

Graph 1, G(3, 3)

The adjacent operator of the Graph 1 above is

$$\mathbf{A}_{adj} = \begin{bmatrix} 0\,1\,0\,0 \\ 0\,0\,0\,1 \\ 0\,1\,0\,0 \\ 0\,0\,0\,0 \end{bmatrix}$$

By (29) we can get the system of differential equations which is

$$\begin{aligned} \dot{\phi_{12}} &= \frac{\phi_{24}}{k_2} - \frac{\phi_{12}}{k_1} \\ \dot{\phi_{32}} &= \frac{\phi_{24}}{k_2} - \frac{\phi_{32}}{k_3} \\ \dot{\phi_{24}} &= \frac{-\phi_{24}}{k_2} \end{aligned} \tag{42}$$

or equivalence the former system of differential equations (42) we can rewrite in matrix form as:

$$\dot{\underset{\sim}{\mathbf{z}}} = \mathbf{\Delta}\underset{\sim}{\mathbf{z}} \tag{43}$$

$$\begin{bmatrix} \dot{\phi}_{12} \\ \dot{\phi}_{32} \\ \dot{\phi}_{24} \end{bmatrix} = \underbrace{\begin{bmatrix} -1/k_1 & 0 & 1/k_2 \\ 0 & -1/k_3 & 1/k_2 \\ 0 & 0 & -1/k_2 \end{bmatrix}}_{\mathbf{\Delta}} \begin{bmatrix} \phi_{12} \\ \phi_{32} \\ \phi_{24} \end{bmatrix} \tag{44}$$

Table 3 Initial and final leverages (l_i) per bank

l_i	$\sum_j \phi_{ij}/k_i$	Initial values	Equilibrium values
l_1	$\frac{\phi_{12}}{k_1}$	$\frac{10}{2}$	0
l_2	$\frac{\phi_{23}}{k_2}$	$\frac{10}{4}$	0
l_3	$\frac{\phi_{32}}{k_3}$	$\frac{10}{3}$	0

If we replace the initial capital to the operator $\mathbf{\Delta}$ we get

$$\mathbf{\Delta} = \begin{bmatrix} -1/2 & 0 & 1/4 \\ 0 & -1/3 & 1/4 \\ 0 & 0 & -1/4 \end{bmatrix}$$

The operator $\mathbf{\Delta}$ has the following eigenvalues:

$$\lambda_1 = \frac{-1}{2}, \lambda_2 = \frac{-1}{3}, \lambda_3 = \frac{-1}{4} \tag{45}$$

Taking the $\lim_{t \to +\infty}(e^{\mathbf{\Delta} t})$ then we get the zero operator, i.e.:

$$\lim_{t \to +\infty}(e^{\mathbf{\Delta} t}) = \begin{bmatrix} 0\,0\,0 \\ 0\,0\,0 \\ 0\,0\,0 \end{bmatrix} \tag{46}$$

Thus taking in account (30) we get

$$\dot{\phi}(+\infty) = \begin{bmatrix} 0\,0\,0 \\ 0\,0\,0 \\ 0\,0\,0 \end{bmatrix} \begin{bmatrix} 10 \\ 10 \\ 10 \end{bmatrix} = \begin{bmatrix} 0 \\ 0 \\ 0 \end{bmatrix} \tag{47}$$

To sum up, by comparing the results in Table 3, we observe that the new leverages are all equal to zero due to the fact that the eigenvalues of the related operator are all negative (none of them is zero) and the final projection operator at infinity is identically zero i.e. $\mathbf{\Lambda} = diag(\lambda_i)$ goes to zero.

5.3.1 Conclusion

In all these three case studies presented at Section 5 we can point out a general observation. First of all in this framework we consider a connected, directed and weighted network which is not necessarily symmetric and hence general theorems on operators or Laplacians on symmetric graphs do not apply. To achieve the goal for equilibrium in the banking leverages of the network we must have the real parts of all the operator eigenvalues to be negative and at least one equal to zero. Furthermore, this observation can be true only within connected components in the network, i.e. the steady state values may be different in every connected component.

5.4 Solutions to the Former Three Cases Without Solving Differential Equations

5.4.1 Review Case 5.1

We are trying to find the leverages in the equilibrium without solving any differential equation as we did in 5.1 up to 5.3 sections. Suppose we have the same graph (Graph 1) as was described in Section 5.1. According to the Theorem 4 in equilibrium all the leverages are equal or in other words: $l_1 = l_2 = l_3$ or we get $l_1 - l_2 = 0, l_2 - l_3 = 0$ and we have the following system:

$$\begin{aligned} \phi_{23} &= \ 2\phi_{12} \\ \phi_{23} &= 4/3\phi_{32} \\ \phi_{12} &= 2/3\phi_{32} \end{aligned}$$

or in matrix form we get

$$\begin{bmatrix} \phi_{12} \\ \phi_{23} \\ \phi_{32} \end{bmatrix} = \phi_{32} \underbrace{\begin{bmatrix} 2/3 \\ 4/3 \\ 1 \end{bmatrix}}_{\underset{\sim}{\mathbf{v}}}$$

Thus, the solution to the system is: $\Lambda(\Sigma) = \{(2/3, 4/3, 1)^T | \phi_{32} \in \mathbb{R}\}$. Next, denote by $\lambda \equiv \phi_{32}$ and the total amount of transactions into the banking network suppose that is equal to 90 monetary units. Then we have $\lambda \begin{bmatrix} 2/3 \\ 4/3 \\ 1 \end{bmatrix} = \begin{bmatrix} \phi_{12} \\ \phi_{23} \\ \phi_{32} \end{bmatrix} \Rightarrow \lambda = 30.$

The new loan flows now are: $\phi_{32} = 30, \phi_{23} = 40, \phi_{12} = 20$. We wrap them up all the results into the next table. In Table 4 we conclude that in equilibrium all the banks will have the same leverage and equals to 10.

Table 4 Final leverages (l_i) in equilibrium, taking account the *new* $\boldsymbol{\phi}_{ij}$ per bank

l_i	$\sum_j \phi_{ij}/k_i$	Initial values	Equilibrium values
l_1	ϕ_{12}/k_1	5	20/2 = 10
l_2	ϕ_{23}/k_2	10/4	40/4 = 10
l_3	ϕ_{32}/k_3	10/3	30/3 = 10

Table 5 Final leverages (l_i) in equilibrium, taking account the *new* $\boldsymbol{\phi}_{ij}$ per bank

l_i	$\sum_j \phi_{ij}/k_i$	Initial values	Equilibrium values
l_1	ϕ_{12}/k_1	1/2	0
l_2	ϕ_{23}/k_2	10/4	0
l_3	ϕ_{32}/k_3	10/3	0
l_4	0	0	0

5.4.2 Review Case 5.2

Here we are trying to resolve the case 5.3 without any use of the tools of differential equations. As was described in the last subsection again here we have a particularity about the diffusion process. If we look closely on the Graph 1 in case 5.3 we can see that the nodes 1, 3 and 4 are nodes that only give and take from and to the node 2, respectively. This is somehow paradox because if we look more closely on the graph we observe that the nodes 1, 3 provide with loans to the node 2 but on the other hand they cannot absorb any amount of loans from this node. More analytically and taking account the Theorem 2.4 suppose that: $l_1 = l_2 = l_3 = l_4 \Rightarrow l_1 - l_2 = 0, l_2 - l_3 = 0, l_3 - l_4 = 0$.

$$\begin{aligned} \phi_{12}/k_1 - \phi_{24}/k_2 &= 0 \Rightarrow \phi_{12} = 0 \\ \phi_{24}/k_2 - \phi_{32}/k_3 &= 0 \Rightarrow \phi_{24} = 0 \\ \phi_{32}/k_3 \qquad\quad &= 0 \Rightarrow \phi_{32} = 0 \end{aligned}$$

According to Table 5 we conclude that in equilibrium all the banks will have the same leverage and equals to 0.

6 Case Study in a Banking Network

In this last section we study $G(10, 13)$, which has the characteristics of directed, connected and weighted network. Any node, v_i from the set $V = \{1, 2, \ldots, 10\}$ represents a bank. Furthermore we have a set E of directed edges, with elements e_{ij}, i.e. $E = \{e_{12}, e_{23}, \ldots, e_{109}\}$, and weights $\boldsymbol{\phi}_{ij}$ which are showing to us the loan(s) transfer of bank i to bank j, (with $i \neq j$) as depicted in the following network. Finally, the initial capitals per bank are: $k_1 = 20, k_2 = 30, k_3 = 20, k_4 = 20, k_5 = 50, k_6 = 100, k_7 = 30, k_8 = 40, k_9 = 30, k_{10} = 100$.

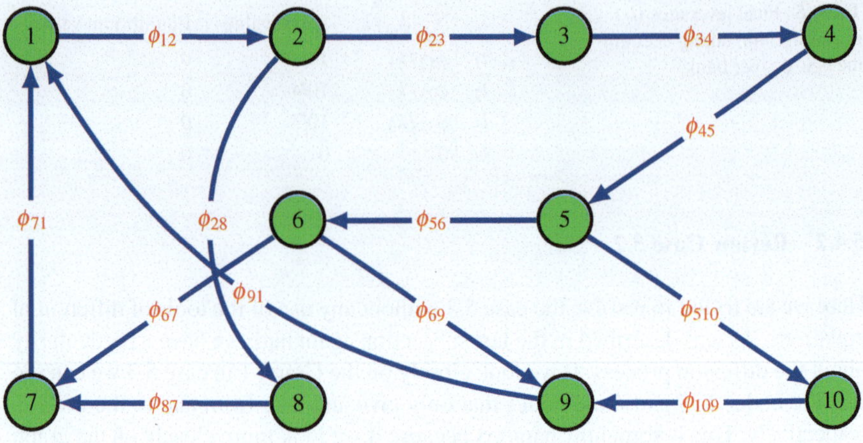

Graph 1, G(10, 13)

Table 6 Initial leverages (l_i) per bank

l_i	$\sum_j \phi_{ij}/k_i$	Results
l_1	ϕ_{12}/k_1	1 / 2
l_2	$(\phi_{28} + \phi_{23})/k_2$	4 / 3
l_3	ϕ_{34}/k_3	3 / 2
l_4	ϕ_{45}/k_4	1 / 2
l_5	$(\phi_{510} + \phi_{56})/k_5$	3 / 5
l_6	$(\phi_{69} + \phi_{67})/k_7$	3 / 5
l_7	ϕ_{71}/k_7	1 / 3
l_8	ϕ_{87}/k_8	1
l_9	ϕ_{91}/k_9	1
l_{10}	ϕ_{109}/k_{10}	1 / 2

So from the Graph 1, we can find the initial leverages, $l_i, i = \{1, 2, \ldots, 10\}$ per bank (Table 6) and sort them in a descending order as Figure 1 shows.

The next step is to write down the differential equations as the differences amongst leverages. In other words, we take the difference between two nodes which are connected by a directed edge as the final point minus the initial point as given by (48).

$$\dot{\phi}_{12} = l_2 - l_1 = (\phi_{28} + \phi_{23})/k_2 - \phi_{12}/k_1$$

$$\dot{\phi}_{23} = l_3 - l_2 = \phi_{34}/k_3 - (\phi_{28} + \phi_{23})/k_2$$

$$\dot{\phi}_{28} = l_8 - l_2 = \phi_{87}/k_8 - (\phi_{28} + \phi_{23})/k_2$$

$$\dot{\phi}_{34} = l_4 - l_3 = \phi_{45}/k_4 - \phi_{34}/k_3$$

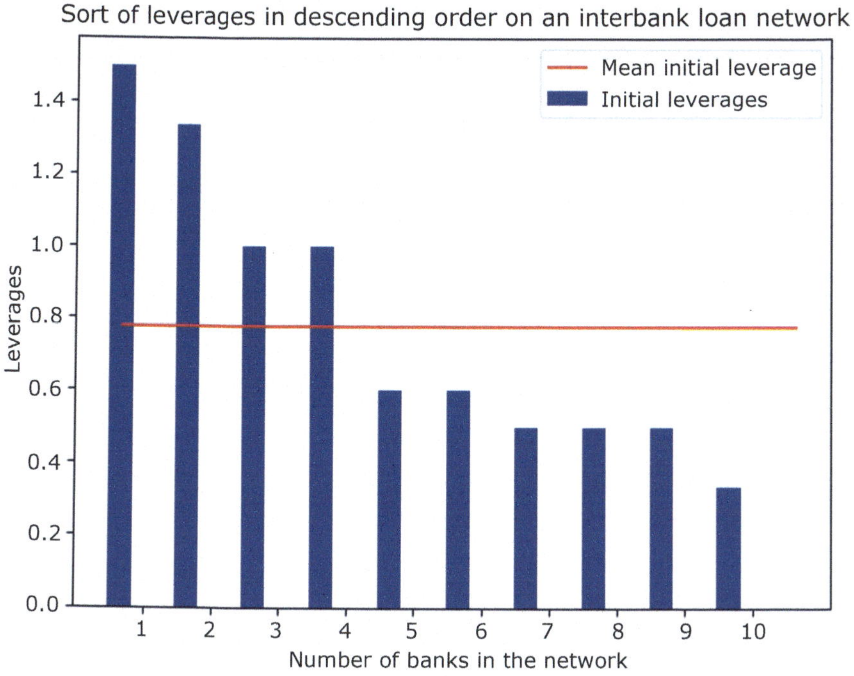

Fig. 1 Sort list of leverages in descending order

$$
\begin{aligned}
\dot{\phi}_{45} &= l_5 - l_4 = (\phi_{510} + \phi_{56})/k_5 - \phi_{45}/k_4 \\
\dot{\phi}_{510} &= l_{10} - l_5 = \phi_{109}/k_{10} - (\phi_{510} + \phi_{56})/k_5 \\
\dot{\phi}_{56} &= l_6 - l_5 = (\phi_{69} + \phi_{67})/k_6 - (\phi_{510} + \phi_{56})/k_5 \\
\dot{\phi}_{69} &= l_9 - l_6 = \phi_{91}/k_9 - (\phi_{69} + \phi_{67})/k_6 \\
\dot{\phi}_{67} &= l_7 - l_6 = \phi_{71}/k_7 - (\phi_{69} + \phi_{67})/k_6 \\
\dot{\phi}_{71} &= l_1 - l_7 = \phi_{12}/k_1 - \phi_{71}/k_7 \\
\dot{\phi}_{87} &= l_7 - l_8 = \phi_{71}/k_7 - \phi_{87}/k_8 \\
\dot{\phi}_{91} &= l_1 - l_9 = \phi_{12}/k_1 - \phi_{91}/k_9 \\
\dot{\phi}_{109} &= l_9 - l_{10} = \phi_{91}/k_9 - \phi_{109}/k_{10}
\end{aligned} \tag{48}
$$

The adjacent operator of Graph 1 is given by $\mathbf{A}_{adj}$:

$$
\mathbf{A}_{adj} = \begin{bmatrix}
0&1&0&0&0&0&0&0&0&0\\
0&0&1&0&0&0&0&1&0&0\\
0&0&0&1&0&0&0&0&0&0\\
0&0&0&0&1&0&0&0&0&0\\
0&0&0&0&0&1&0&0&0&1\\
0&0&0&0&0&0&1&0&1&0\\
1&0&0&0&0&0&0&0&0&0\\
0&0&0&0&0&0&1&0&0&0\\
1&0&0&0&0&0&0&0&0&0\\
0&0&0&0&0&0&0&0&1&0
\end{bmatrix}
$$

In the equilibrium, as Theorem 4 states, we have all the leverages l_i equals, i.e. $l_1 = l_2 = \cdots = l_{10}$ and we can write

$$
\begin{aligned}
l_1 - l_2 = l_2 - l_3 = l_3 - l_4 = l_4 - l_5 = l_5 - l_6 = l_6 - l_7 \\
= l_7 - l_8 = l_8 - l_9 = l_9 - l_{10} = 0
\end{aligned} \tag{49}
$$

If we replace every leverage into (49), as described by (1) and rewrite them as operators we have

$$
\underbrace{\begin{bmatrix}
1/k_1 & 0 & 0 & -1/k_3 & 0 & \cdots & 0 & 0\\
0 & 1/k_2 & 1/k_2 & -1/k_3 & 0 & \cdots & 0 & 0\\
0 & 0 & 0 & 1/k_3 & -1/k_4 & \cdots & 0 & 0\\
0 & 0 & 0 & 0 & 1/k_4 & \cdots & 0 & 0\\
0 & 0 & 0 & 0 & 0 & \ddots & 0 & 0\\
0 & 0 & 0 & 0 & 0 & \cdots & 0 & 0\\
0 & 0 & 0 & 0 & 0 & \cdots & 0 & 0\\
0 & 0 & 0 & 0 & 0 & \cdots & -1/k_9 & 0\\
0 & 0 & 0 & 0 & 0 & \cdots & 1/k_9 & -1/k_{10}
\end{bmatrix}}_{\mathbf{A_1} \in \mathbb{R}^{9\times 13}}
\underbrace{\begin{bmatrix}
\phi_{12}\\ \phi_{23}\\ \phi_{28}\\ \phi_{34}\\ \phi_{45}\\ \phi_{56}\\ \phi_{510}\\ \vdots\\ \phi_{109}
\end{bmatrix}}_{\Phi \in \mathbb{R}^{1\times 13}}
= \mathbf{0}
$$

The above system has the following parametric solution.

$$
\begin{aligned}
\phi_{12} &= \phi_{34}\\
\phi_{23} &= \phi_{91} - \phi_{28}\\
\phi_{34} &= \phi_{45}\\
\phi_{45} &= (2/3)\phi_{91}\\
\phi_{67} &= (10/3)\phi_{91} - \phi_{69}\\
\phi_{71} &= \phi_{91}
\end{aligned}
$$

$$\phi_{109} = (10/3)\phi_{91} \tag{50}$$

Moreover the basis matrix $\mathbf{V}$ with columns the vectors which are taken from the solution of the system (50) is:

$$\mathbf{V}_{(13\times 4)} = \begin{bmatrix} 2/3 & 0 & 0 & 0 \\ 1 & -1 & 0 & 0 \\ 0 & 1 & 0 & 0 \\ 2/3 & 0 & 0 & 0 \\ 2/3 & 0 & 0 & 0 \\ 5/3 & 0 & -1 & 0 \\ 0 & 0 & 1 & 0 \\ 10/3 & 0 & 0 & -1 \\ 0 & 0 & 0 & 1 \\ 1 & 0 & 0 & 0 \\ 4/3 & 0 & 0 & 0 \\ 1 & 0 & 0 & 0 \\ 1 & 0 & 0 & 0 \end{bmatrix} \tag{51}$$

Suppose that we have $\phi_{91} \equiv k$, $\phi_{28} \equiv \lambda$, $\phi_{510} \equiv \mu$ and $\phi_{69} \equiv \nu$. Furthermore, we make a hypothesis that the total amount of current loans in the interbank system must not exceed the value of 296 monetary units. Thus we found that $k = 24$. Therefore, now we are ready to calculate the *new* ϕ_{ij}, i.e. in equilibrium.

$$\begin{bmatrix} \phi_{12} \\ \phi_{23} \\ \phi_{28} \\ \phi_{34} \\ \phi_{45} \\ \phi_{56} \\ \phi_{510} \\ \phi_{67} \\ \phi_{69} \\ \phi_{71} \\ \phi_{87} \\ \phi_{91} \\ \phi_{109} \end{bmatrix} = 24 \begin{bmatrix} 2/3 \\ 1 \\ 0 \\ 2/3 \\ 2/3 \\ 5/3 \\ 0 \\ 10/3 \\ 0 \\ 1 \\ 4/3 \\ 1 \\ 10/3 \end{bmatrix} = \begin{bmatrix} 16 \\ 24 \\ 0 \\ 16 \\ 16 \\ 40 \\ 0 \\ 80 \\ 0 \\ 24 \\ 32 \\ 24 \\ 80 \end{bmatrix} \tag{52}$$

Therefore, the *new*, (final), leverages in the equilibrium to our interbank loan network are given by the next Table 7.

We can present all the previous analysis of leverages in the next Figure 2, in order to compare the initial, mean and final value of leverage of our interbank network.

Table 7 Final leverages (l_i) in equilibrium, taking account the *new* $\boldsymbol{\phi}_{ij}(t)$ per bank

l_i	$\sum_j \phi_{ij}/k_i$	Equilibrium results
l_1	ϕ_{12}/k_1	4 / 5
l_2	$(\phi_{28}+\phi_{23})/k_2$	4 / 5
l_3	ϕ_{34}/k_3	4 / 5
l_4	ϕ_{45}/k_4	4 / 5
l_5	$(\phi_{510}+\phi_{56})/k_5$	4 / 5
l_6	$(\phi_{69}+\phi_{67})/k_7$	4 / 5
l_7	ϕ_{71}/k_7	4 / 5
l_8	ϕ_{87}/k_8	4 / 5
l_9	ϕ_{91}/k_9	4 / 5
l_{10}	ϕ_{109}/k_{10}	4 / 5

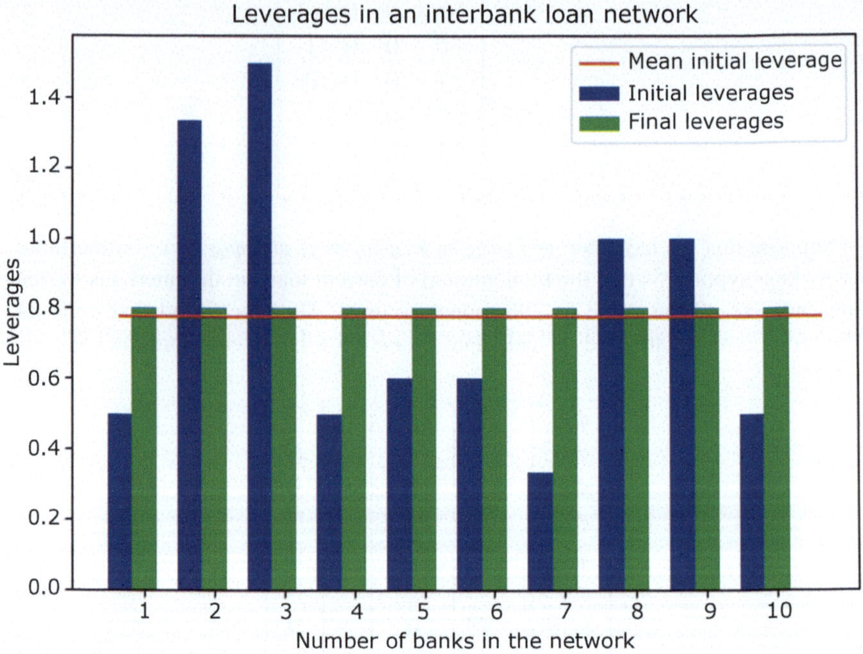

Fig. 2 The mean initial leverage (l_M) is equal to 7/9. At the point of equilibrium $l_i = l_j = l \in$ R, $i \neq j$ and $l \geq l_M$

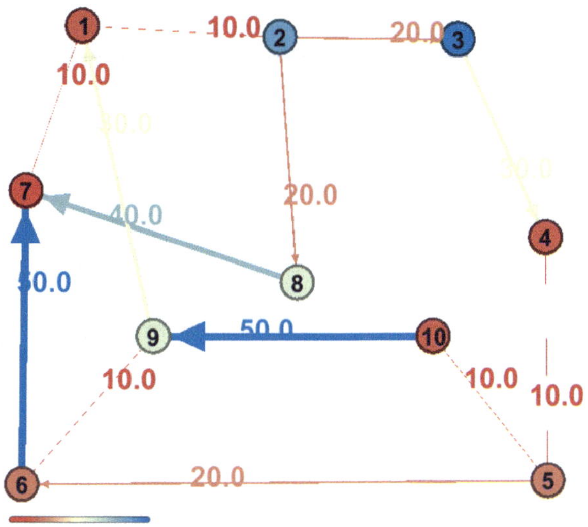

Fig. 3 From dark blue to red we are moving from the highest level of leverage per bank to the lowest and furthermore from the highest level of weights to the lowest, for the nodes and edges, respectively

In Figure 3 we achieve a double target. The first one is to order the nodes taking account the initial leverage per bank and the second one is to order the edges of the interbank network taking account the weights of any edge.

Reference

1. R.J. Trudeau, *Introduction to Graph Theory* (Dover, Illinois, 1993)

The Dynamics of Interbank Networks

John Leventides, Maria Livada, and Costas Poulios

Abstract Given an interbank network (X, C, E_0), where X is the set of banks, C is the vector of capitals, and E_0 is the bilateral exposures matrix, the bankruptcy set U_x for each $x \in X$ is defined. The set U_x contains all possible combinations of institutions whose failure would result in the default of the bank x. It turns out that the minimal elements of the sets $U_x, x \in X$ play a prominent role in the study of the problem of default contagion in an interbank market. Several aspects of this problem can be characterized in terms of the minimal elements of U_x, $x \in X$, for example, the propagation of an initial shock through the network as well as the equilibrium points of the network.

1 Introduction

Systemic risk in financial markets and the problem of default contagion have emerged as major research topics in economic studies. It has become clear that an event at the company level or the inability of some institution to service a scheduled debt payment may lead to a cascade of defaults and create severe instability to the global economic network. As a result, the regulators and policy makers of each interbank market are under pressure, since they are in charge of eliminating the systemic risk and fortifying the financial system against the default contagion.

Although it is generally accepted that inadequate capital levels in connection with large degree of interconnectedness entail the propagation of a crisis through the interbank market, finding the most suitable proactive measures for strengthening the resilience of the system is not always clear. The difficulty stems, in part, from the fact that the problem of default contagion depends on a plethora of factors, which may also behave quite differently under seemingly similar situations.

J. Leventides · M. Livada · C. Poulios (✉)
Department of Economics, Faculty of Economics and Political Sciences,
National and Kapodistrian University of Athens, Athens, Greece
e-mail: ylevent@econ.uoa.gr; maria.livada@city.ac.uk; konpou@econ.uoa.gr

A. M. Raigorodskii, M. Th. Rassias (eds.), *Discrete Mathematics and Applications*,
Springer Optimization and Its Applications 165,
https://doi.org/10.1007/978-3-030-55857-4_14

For instance, interconnectedness among institutions, which has been raised over the last decades and it is expected to be developed even further, has a two-pronged aim: from the one hand it contributes to the stability of the market by distributing funds in an effective way: banks facing liquidity problems can borrow from banks with liquidity surpluses. On the other hand, it can make the system prone to financial contagion through the interbank linkages. Inevitably, interconnectedness, and more generally the topology of the interbank system, has been studied in many papers (indicatively, we refer to [1, 4, 6, 7, 12, 29, 34]).

Another significant challenge in the problem of financial contagion is the quantitative analysis of this problem. This general statement can be specified as follows: (1) to relate, in a precise manner the propagation of default with measurable features of the network (e.g. capitals, bilateral exposures, balance sheet size, etc.); (2) to introduce adequate indicators which objectively assess the whole network and each one of the banks and to classify them from the most vulnerable to the most robust; (3) to provide the supervisors and regulators of the interbank market with new tools for the suitable policy formulation. (For instance, see [5, 16, 19, 21, 27, 28, 32].)

Network theory offers a suitable framework for the study of interbank markets and has been utilized by many researchers so far (see, for example, [1, 5, 10, 15, 16, 19, 20, 32, 33]). In such network structure every node represents a bank and connections between banks are represented by edges. Hence, the interbank network can be depicted by a directed weighted graph.

Once the interbank network has been identified, one is interested in the propagation of an initial shock through the network and evaluating every node for its robustness and resilience to systemic risk. In order to study these problems, in this work we adopt the following approach. Given a bank (node) x of the network, we examine the least combinations of banks whose failure would lead the bank x to collapse (see Section 4 for the formal definition). This approach, however basic it may seem, it turns out to be of great significance. Using the aforementioned "least combinations," we are able to describe the development of a default in the network as well as other characteristics such as the equilibrium points. Furthermore, these least combinations constitute a measure for evaluating each node. Indeed, in the case of a robust institution, these combinations are expected to be only a few, involving perhaps a great deal of other banks. On the contrary, if a bank is vulnerable, then its failure is expected to be the result of the default of many combinations of other banks and each combination may contain a small number of elements.

The rest of the paper is organized as follows. In Section 3 we present the basic elements of the interbank networks and the mechanism of default contagion. In Section 4 we define the bankruptcy set U_x for every node x of the network and we study its basic properties. In Section 5 we consider the contagion map (introduced in [32]) which describes the propagation of a financial shock through the network for all possible combinations of initial default. We show that this map can be described in terms of the bankruptcy sets U_x. This result is improved in Section 6, where we use Boolean dynamical systems in order to describe the default contagion in the network and it is shown that the contagion depends actually on the minimal elements of $U_x, x \in X$. In Section 7, we focus our attention to the equilibrium points of the interbank system, which play a prominent role since they correspond to the points where the contagion terminates. Closely related to the equilibrium points is the *global function* (introduced in Section 8) whose image contains all these points. In Section 9, we present some basic tools for the assessment of a network and of the institutions contained in it. Finally, in Section 10 we illustrate the ideas and methods of the paper using some example.

2 Literature Review

A network approach for the analysis and study of systematic risk and financial stability of banks could be rather useful, as it can be quite helpful in capturing several externalities that an institution may generate for the whole system as a result of risk. A better and in more depth understanding of these network externalities will facilitate the implementation of a more sufficient financial supervisory system. These emerging externalities appear when the actions of institutions (i.e. banks) in risk affect other institutions and the economy as a whole system. Measuring the systemic importance of individual banks, i.e. their capacity to produce contagion in the rest of the system, is quite important in this context, hence, analyzing, modelling, and evaluating the interbank network can provide helpful guidance. Financial networks literature is quite broad, reviewing linkages between financial institutions: banks, but also hedge funds, insurance companies, etc. Interbank networks are part of these financial networks and the associated literature provides a detailed view of the current understanding of interbank networks, how network features affect the interbank contagion and how banks form links when faced with the possibility of contagion and systemic risk. A network approach is quite useful in the representation of structural features of the banking system. A network not only can reflect the fact that the banks have heterogeneous properties, but it can also capture the course of ties between banks, since it is economically important whether a bank borrows from another bank or lends it to another. In addition, a network approach of these financial systems may also include the various types of connections between banks. As Langfield and Soramaki [31] note, banks engage with each other in hundreds of different types of transactions, such as interbank lending, repurchase agreements or derivatives. Multilayer financial networks, where different layers represent different types of transactions, will catch this multitude of potential link forms (Aldasoro and Alves, [2]; Bargigli et al., [8]; Langfield et al., [30]; Montagna and Kok, [35]).

The effect that the structure of the network might have in interbank contagion has been addressed by many scientists. Some argue that the number of defaults and their severity depend primarily on the topology of the network. Gai and Kapadia [25] state that system's fragility rests on where the initially affected node is located within the network. Intuitively, central bank defaults are more damaging than a peripheral bank failure. Finally, another key point is that due to too many linkages, there is a trade-off between risk sharing by transactions with other banks and danger of contagion. While there is no dispute about the nature of the trade-off, there is no consensus as to whether a full network dampens (Allen and Gale, [4]; Freixas et al.,[23]) or contagion from fuels (Battiston et al., [9]; Vivier-Lirimont, [40]). Other researchers suggest that intermediate levels of connectivity are higher, because of the non-monotonic impact of connectivity on contagion, for example (Gai et al., [26]; Nier et al., [36]). Furthermore, Acemoglu et al. [1] shows that a large network is more resilient to minor shocks while smaller connections are more able to prevent contagion in the face of major shocks. Moreover, empirical studies have also brought new insights into the interbank network structure. A significant result is that interbank networks exhibit a core-peripheral structure with a highly connected core and few connections to and between peripheral banks (Craig and von Peter, [17]; Fricke and Lux, [24]; Langfield et al., [30]; Veld and van Lelyveld, [39]).

Considering the growing body of literature on financial networks, notable few surveys have been carried out on the subject as well. An early survey by Allen and Babus [3] discusses the emerging financial network literature as well as the methodologies used and explains why the complexity of financial systems can be "naturally captured" using a presentence network. In their survey, Chinazzi and Fagiolo [13] focus on the connection between connectivity and stability in financial networks. Other recent surveys focus primarily on contagion models of financial networks and their implementations. Upper [38] reviews methods for simulation testing for contagion in empirical interbank networks as well as methods for constructing interbank networks from various data sources. Elsinger et al. [22] also study empirical financial networks simulated in systemic risk analysis and their applications. Langfield and Soramaki [31] review from a systemic risk perspective the empirical literature relating to interbank markets. They also study the literature on the topology of empirical interbank networks, as well as the literature for identifying systemically important banks, in addition to discussing network simulations. Summer's survey [37] also focuses on the study of empirical simulation. Ultimately, the De Bandt et al. [18] and Benoit et al. [11] studies on systemic risk also include a review of simulations of financial networks alongside other modelling methods used in that literature.

3 Interbank Networks and Default Contagion

Financial systems can be represented as directed weighed graphs, or networks. An interbank network is a triple (X, E_0, C), where

1. the vertex set is $X = \{1, 2, \ldots, n\}$ and its elements represent financial institutions;
2. $C = (c_1, c_2, \ldots, c_n)$ denotes the vector of capitals, i.e. c_i is the capital of the institution i representing its ability to absorb losses;
3. $E_0 = (E_{ij})_{i,j=1}^{n}$ is the matrix of bilateral exposures. That is, E_{ij} stands for the exposure of institution i to institution j defined as the value of all liabilities of j

to i at the date of computation. Clearly, E_0 is an $n \times n$ matrix whose diagonal line contains only 0, since banks do not lend to themselves.

Hence, this network can be depicted as a directed weighted graph, whose nodes represent institutions and (weighted) links represent exposures. If we sum up the i-th column of the matrix E_0, then we obtain the total interbank liabilities of the bank i. Hence, the **interbank liabilities vector** can be introduced:

$$L = (L_i)_{i=1}^n = \left(\sum_{j=1}^n E_{ji} \right)_{i=1}^n .$$

On the other hand, if we take horizontal summation, then we obtain the total interbank assets of each institution. Hence,

$$A = (A_i)_{i=1}^n = \left(\sum_{j=1}^n E_{ij} \right)_{i=1}^n ,$$

is the **interbank assets vector**.

The mechanism of the financial contagion is as follows. Some institutions, which form a subset $B \subseteq X$, default when they fail to fulfill a legal obligation, for instance some scheduled debt payment, or when they are unable to service a loan. The defaulted institutions cannot meet their liabilities and create loan losses to their creditors. These losses are imputed directly to the capital of the creditors, leading to a total loss of $\sum_{i \in B} E_{ji}$ for the creditor j. In the case where the loss exceeds the capital, i.e. $\sum_{i \in B} E_{ji} > c_j$, then the bank j becomes insolvent. Insolvency may not necessarily imply default, since the bank j may be able to find some liquidity to serve its obligations. However, financial institutions are funded through short-term debt, which must be constantly renewed and depends to the solvency and creditworthiness of the institutions. Consequently, insolvent banks are extremely difficult to raise up their levels of liquidity and it is generally accepted that inadequate capital levels entail the instability of the financial network. Hence, in this work we consider insolvent banks as defaulted. Furthermore, we recognize that other scenarios to default may exist which entail the phenomenon of contagion. From this point of view, this work serves as a lower bound of the contagion.

Finally, in line with various studies, we adopt the zero recovery assumption for shock transmission, i.e. the creditor banks loss all their interbank assets held against a defaulting bank. This assumption is a realistic one in the short run and it has been used in the literature to analyze worst case scenarios (see [14, 25, 33]).

Summarizing, the failure of one or more nodes of the graph may lead other institutions to default and the shock propagation continues creating a cascade of defaults.

4 The Bankruptcy Set of the Institution *x*

Assume that $x \in X$ is any member of the interbank network. By the above description it becomes clear that some families of banks, i.e. some subsets of X, are "poisonous" for x and, consequently, this bank should "keep an eye" on them. These are the subsets $A \in 2^X$ with the following property: if the banks of the set A declare bankruptcy, then the bank x will bankrupt as well. The collection of all sets A with this property forms a subset of the powerset 2^X. This collection is denoted by U_x and it is called the **bankruptcy set of** x. Hence, we have $U_x \subseteq 2^X$ and, by the definition,

$$A \in U_x \Longleftrightarrow \text{if the banks of the set } A \text{ fail, then the bank } x \text{ fails as well.}$$

Remark 1 We have to clarify the case where the set A contains the bank x itself. In this case, when the banks of the set A fail, then x also defaults. Therefore, according to the above definition, the set A belongs to the bankruptcy set U_x. Hence, we adopt the convention that U_x contains the singleton $\{x\}$ and all its supersets, although this may seem meaningless. However, this convention allows us to treat all the sets U_x, $x \in X$, as subsets of 2^X. Otherwise, we would have to consider U_x in the set $2^{X\setminus\{x\}}$, i.e. every U_x would belong to a different powerset. What is more, this convention adds to every U_x the same number of sets (all the supersets of $\{x\}$), and therefore it does not disturb the analogy between the bankruptcy sets U_x's.

The sets U_x, $x \in X$, play a central role in the development of a crisis through the interbank network and in the assessment of the institutions belonging to the network. Indeed, if U_x contains plenty of elements, then the bank x is prone to financial contagion, since there are a lot of combinations of banks whose failure results in the bankruptcy of x. On the contrary, when the cardinality of U_x is small, then the bank is resistant to the failure of other banks. Hence, the smaller the cardinality of U_x the more robust the bank is and vice versa, i.e. the cardinality of the bankruptcy set constitutes a measure for the robustness of the corresponding bank. Our next purpose is to study these sets.

First of all, we give a description of the bankruptcy sets U_x, $x \in X$, in terms of the vector of capitals C and the exposures matrix E_0. Such description facilitates the detection of the sets belonging to U_x. In order to achieve this goal, we need a slight modification of E_0. Namely, we denote by E the matrix which is formed by replacing the zero diagonal of E_0 with the vector C of capitals, i.e.

$$E = \begin{bmatrix} E_{11} & E_{12} & \dots & E_{1n} \\ E_{21} & E_{22} & \dots & E_{2n} \\ \vdots & \vdots & \vdots & \vdots \\ E_{n1} & E_{n2} & \dots & E_{nn} \end{bmatrix},$$

where $E_{ii} = c_i$ for any $i = 1, 2, \dots, n$.

Proposition 1 *The subset* $A \in 2^X$ *belongs to the bankruptcy set* U_x *of the bank* x *if and only if*

$$\sum_{j \in A} E_{xj} \geq c_x.$$

Proof We distinguish the following two cases.

Case 1: If $x \in A$, then A belongs to the set U_x. In this case, we observe that $\sum_{j \in A} E_{xj} \geq E_{xx} = c_x$.

Case 2: Assume that x is not in A. Then, according to the definition, A belongs to U_x if and only if the failure of the banks A lead to the default of x. By the description of Section 3 this happens if and only if $\sum_{j \in A} E_{xj} \geq c_x$.

Hence, in every case we have the desired result.

4.1 Structure of the Bankruptcy Sets U_x, $x \in \{1, 2, \ldots, n\}$

Recall that a subset U of a partially ordered set $(\mathfrak{Y}, \leq)$ is said to be an upper set if for any $y \in U$ and any $y \leq z$ we have that z belongs to U. It is easy to see that the bankruptcy sets U_x, $x \in X$, share this property. Indeed, if $A \in U_x$, then the failure of the banks of the set A results in the bankruptcy of x. Clearly, this is also true for any B with $A \subseteq B$, and hence B belongs to U_x. Therefore, the set U_x, for any $x \in X$, is an upper set in the partially ordered set $(2^X, \subseteq)$ and we will also say that U_x is the upper set corresponding to x.

4.2 Maximal and Minimal Elements of the Bankruptcy set U_x

Our next concern is to describe the maximal and minimal elements of the upper set U_x, for any $x \in X$. Let us start with the maximal sets, where the situation is trivial and uninteresting. Each upper set U_x contains of course the set X itself and this is the only maximal element of U_x.

On the contrary, the minimal elements of the set U_x seem to play a prominent role in the study of the problem of financial contagion. Indeed, the minimal sets of U_x represent the least combinations which lead the bank x to failure. Of course, there is always one minimal set, the singleton $\{x\}$. However, there may be other minimal sets as well. We will denote the minimal sets of U_x by $\omega_{x1} = \{x\}, \omega_{x2}, \ldots, \omega_{xm_x}$, where the number m_x depends on the bank x. (Observe that we always reserve the symbol ω_{x1} for the singleton $\{x\}$.) These sets determine the whole upper set U_x as the next proposition shows.

Proposition 2 *For any subset* A *of* X *the following equivalence holds*

$$A \in U_x \iff A \supseteq \omega_{xi} \quad \textit{for some } i = 1, 2, \ldots, m_x.$$

Proof Assume that A belongs to U_x. If A is not a minimal element, then we can find A_1 in U_x such that $A \supsetneqq A_1$. If A_1 is not minimal, then there is A_2 in U_x such that $A_1 \supsetneqq A_2$. Since U_x contains finitely many elements, the previous procedure cannot continue for infinitely many steps. Hence, at some stage we will find a minimal set $A_k = \omega_{xi}$ (for some $i = 1, 2, \ldots, m_x$) such that $A_k \subseteq A$.

Conversely, if $\omega_{xi} \subseteq A$, for some i, then A belongs to U_x, since U_x is an upper set.

For any subset A of X we denote by W_A the upper set defined by A, i.e. the collection of all supersets of A, $W_A = \{B \subseteq X \mid A \subseteq B\}$. Then, the previous proposition can be written in the following form.

Corollary 1 *Let $\omega_{x1}, \omega_{x2}, \ldots, \omega_{xm_x}$ be the minimal elements of the upper set U_x. Then,*

$$U_x = \bigcup_{i=1}^{m_x} W_{\omega_{xi}}.$$

The importance of the minimal sets will be revealed gradually in the paper. At this point, let us highlight two essential features of these sets, which are: their total number and the length (cardinality) of each minimal set. When the total number is small, then there are only a few combinations of defaulted banks that lead the bank x to bankruptcy. For example, in the extreme case where there is only one minimal set, $\{x\}$, then the bank is robust since it is not affected by the failure of any other institutions. Therefore, the less the number the more stable the bank and vice versa.

On the other hand, when the length of a minimal set ω_{xi} is small, this implies that the failure of a small number of institution causes the default of x. For instance, in the case where $\sharp\omega_{xi} = 1$ and $\omega_{xi} \neq \{x\}$, then there is one bank whose failure is sufficient for the bankruptcy of x. Hence, the smaller the cardinality of ω_{xi}, the more prone the bank is to financial contagion.

Finally, we show that the cardinality of the upper set U_x, which gives us a measure for the robustness of the bank x, can be calculated directly from the minimal elements of U_x. For this result, we need some notation. If $\omega_{xi_1}, \omega_{xi_2}, \ldots, \omega_{xi_k}$ are minimal sets of U_x, then we denote by $\omega_{xi_1 i_2 \ldots i_k}$ their union, i.e. $\omega_{xi_1 i_2 \ldots i_k} = \bigcup_{j=1}^{k} \omega_{xi_j}$.

Theorem 1 *Let $\omega_{x1}, \omega_{x2}, \ldots, \omega_{xm_x}$ be the minimal elements of the upper set U_x. Then, the cardinality of U_x is given by*

$$\sharp U_x = 2^n \left(\sum_{i=1}^{m_x} \frac{1}{2^{\sharp\omega_{xi}}} - \sum_{1 \le i < j \le m_x} \frac{1}{2^{\sharp\omega_{xij}}} + \sum_{1 \le i < j < k \le m_x} \frac{1}{2^{\sharp\omega_{xijk}}} - \cdots + (-1)^{m_x - 1} \frac{1}{2^{\sharp\omega_{x12\ldots m_x}}} \right).$$

Proof In order to simplify the notation, throughout this proof we fix a node x and we denote the minimal elements of U_x by $\omega_1, \omega_2, \ldots, \omega_m$, where $m = m_x$. By Corollary 1, we obtain

$$U_x = \bigcup_{i=1}^{m} W_{\omega_i}.$$

The well known formula giving the cardinality of the union of sets implies that:

$$\sharp U_x = \sum_{i=1}^{m} \sharp W_{\omega_i} - \sum_{1 \le i < j \le m} \sharp(W_{\omega_i} \cap W_{\omega_j}) + \sum_{1 \le i < j < k \le m} \sharp(W_{\omega_i} \cap W_{\omega_j} \cap W_{\omega_k})$$
$$- \cdots + (-1)^{m-1} \sharp \bigcap_{i=1}^{m} W_{\omega_i}.$$

However, $W_{\omega_i} \cap W_{\omega_j}$ is the collection of all subsets of X containing both ω_i and ω_j, that is the subsets of X containing the union $\omega_i \cup \omega_j = \omega_{ij}$. In other words,

$$W_{\omega_i} \cap W_{\omega_j} = W_{\omega_{ij}},$$

and this is also true for more than two sets. Hence,

$$\sharp U_x = \sum_{i=1}^{m} \sharp W_{\omega_i} - \sum_{1 \le i < j \le m} \sharp W_{\omega_{ij}} + \sum_{1 \le i < j < k \le m} \sharp W_{\omega_{ijk}} - \cdots + (-1)^{m-1} \sharp W_{\omega_{12\ldots m}}.$$

Now it is easy to see that, for any $A \subseteq X$, the cardinality of the set V_A is equal to the cardinality of the powerset of $X \setminus A$, that is,

$$\sharp W_A = 2^{n - \sharp A} = \frac{2^n}{2^{\sharp A}}.$$

Thus, we obtain

$$\sharp U_x = \sum_{i=1}^{m} \frac{2^n}{2^{\sharp \omega_i}} - \sum_{1 \le i < j \le m} \frac{2^n}{2^{\sharp \omega_{ij}}} + \sum_{1 \le i < j < k \le m} \frac{2^n}{2^{\sharp \omega_{ijk}}} - \cdots + (-1)^{m-1} \frac{2^n}{2^{\sharp \omega_{12\ldots m}}}$$

and the formula of the theorem can be deduced easily.

In the case where the minimal sets $\omega_{x1}, \omega_{x2}, \ldots, \omega_{xm_x}$ are pairwise disjoint, the formula of the previous theorem can be simplified as the next corollary shows.

Corollary 2 *Assume that the minimal elements $\omega_{x1}, \omega_{x2}, \ldots, \omega_{xm_x}$ of the upper set U_x are pairwise disjoint. Then, the cardinality of U_x is given by*

$$\sharp U_x = 2^n \left[1 - \prod_{i=1}^{m_x} \left(1 - \frac{1}{2^{\sharp \omega_{xi}}} \right) \right].$$

Proof We fix a node x of the interbank network and we denote by $\{\omega_i\}_{i=1}^m$ the minimal elements of the bankruptcy set U_x. Since $\{\omega_i\}_{i=1}^m$ are pairwise disjoint, it follows that

$$\sharp \omega_{i_1 i_2 \dots i_k} = \sharp \omega_{i_1} + \sharp \omega_{i_2} + \dots + \sharp \omega_{i_k},$$

for any $k \leq m$ and any $i_1 < i_2 < \dots < i_k$. Hence, the formula of the previous theorem is written in the form:

$$\sharp U_x = 2^n \left(\sum_{i=1}^m \frac{1}{2^{\sharp \omega_i}} - \sum_{1 \leq i < j \leq m} \frac{1}{2^{\sharp \omega_i}} \cdot \frac{1}{2^{\sharp \omega_j}} + \sum_{1 \leq i < j < k \leq m} \frac{1}{2^{\sharp \omega_i}} \cdot \frac{1}{2^{\sharp \omega_j}} \cdot \frac{1}{2^{\sharp \omega_k}} \right.$$
$$\left. - \dots + (-1)^{m-1} \frac{1}{2^{\sharp \omega_1}} \cdot \frac{1}{2^{\sharp \omega_2}} \cdot \frac{1}{2^{\sharp \omega_m}} \right).$$

Recall that for any $a_1, a_2, \dots, a_m \in \mathbb{R}$ the following equation holds

$$\prod_{i=1}^m (1 - a_i) = 1 - \sum_{i=1}^m a_i + \sum_{1 \leq i < j \leq m} a_i \cdot a_j$$
$$- \sum_{1 \leq i < j < k \leq m} a_i \cdot a_j \cdot a_k + \dots + (-1)^m a_1 \cdot a_2 \dots a_m.$$

Setting $a_i = \frac{1}{2^{\sharp \omega_i}}$, we obtain

$$\sum_{i=1}^m \frac{1}{2^{\sharp \omega_i}} - \sum_{1 \leq i < j \leq m} \frac{1}{2^{\sharp \omega_i}} \cdot \frac{1}{2^{\sharp \omega_j}} + \sum_{1 \leq i < j < k \leq m} \frac{1}{2^{\sharp \omega_i}} \cdot \frac{1}{2^{\sharp \omega_j}} \cdot \frac{1}{2^{\sharp \omega_k}}$$
$$- \dots + (-1)^{m-1} \frac{1}{2^{\sharp \omega_1}} \cdot \frac{1}{2^{\sharp \omega_2}} \cdot \frac{1}{2^{\sharp \omega_m}} = 1 - \prod_{i=1}^m \left(1 - \frac{1}{2^{\sharp \omega_i}} \right),$$

and the proof of the corollary follows immediately.

5 The Contagion Map

The contagion map was introduced in [32] and its purpose is to describe the transmission of financial contagion through an interbank network for all possible combinations of initial defaults. It is a map $T : 2^X \to 2^X$ on the powerset of X and it is defined as follows: for any $A \in 2^X$, we have

$$T(A) = \{x \in X \mid \text{the bank } x \text{ defaults after the failure of the banks belonging to } A\}$$
$$= \left\{ x \in X \mid \sum_{j \in A} E_{xj} \geq c_x \right\}.$$

It should be pointed out that $T(A)$ describes the first stage of the transmission of financial contagion after the collapse of the banks of the set A. The set $T(A)$ may contain new defaulted institutions, i.e. $T(A) \setminus A$ may be non-empty. These institutions create loan losses to their counterparties and a new round of defaults may occur. The second stage of the transmission is described by the set $T(T(A))$. The contagion continues up to the point where we meet an equilibrium set of T, that is a set $B \subseteq X$ such that $T(B) = B$. Consequently, these equilibrium sets are of great importance for the study of the dynamics of an interbank network.

By Proposition 1 and the definition of the contagion map, it is easy to observe that the map T depends on the bankruptcy sets U_x, $x \in X$. Therefore, these sets determine the contagion of a financial shock through the network.

Proposition 3 *For any $A \subseteq X$ the following equivalence holds*

$$x \in T(A) \Leftrightarrow A \in U_x.$$

In other words, $T(A) = \{x \in X \mid A \in U_x\}$.

Proof Clearly, by the definition of T and Proposition 1, we have the equivalences

$$x \in T(A) \Longleftrightarrow \sum_{j \in A} E_{xj} \geq c_x \Longleftrightarrow A \in U_x.$$

As it was commented above, the equilibrium sets of the contagion map T, that is the sets $A \in 2^X$ such that $T(A) = A$, are fundamental for the study of economic networks, since they show the point up to which the contagion continues. These sets can be characterized in terms of the bankruptcy sets U_x, $x \in X$, and such a characterization is provided with the next proposition.

Proposition 4 *The set $A \in 2^X$ is an equilibrium set of the contagion map T if and only if A has the following property:*

$$\text{if } A \in U_x \text{ for some } x \in X, \text{ then } x \text{ belongs to } A.$$

Proof Assume that $T(A) = A$ and that $A \in U_x$ for some $x \in X$. By Proposition 3, it follows that $x \in T(A)$, i.e. $x \in A$.

Conversely, let A be any subset of X with the desired property. For any $x \in T(A)$, we have (by Proposition 3) that $A \in U_x$, and hence $x \in A$. Therefore, $T(A) \subseteq A$, which implies that $T(A) = A$.

5.1 *The Contagion Graph*

The contagion map T introduces a graph structure in the powerset 2^X (see [32]). Hence, the latter becomes a directed graph $(2^X, ED)$, where the set ED of edges is given by

$$ED = \left\{(A, T(A)) \mid A \in 2^X\right\}.$$

The directed graph $(2^X, ED)$ is called the **contagion graph**. It consists of several connected components. Every equilibrium set A of the map T determines a component containing all sets B with the property: there is $m \in \mathbb{N}$ such that $T^m(B) = A$. Furthermore, for any set $B \subseteq X$, the path

$$\{B, T(B), \ldots, T^m(B)\},$$

where $m \geq 0$ is the least integer such that $T^{m+1}(B) = T^m(B)$, describes the propagation of an initial shock through the network. Thus, the contagion graph concentrates all the information for the contagion of a financial crisis in the network for all possible combinations of initial defaults.

5.2 *The Contagion Vector*

The contagion vector was introduced in [32] and provides a method for the assessment of the robustness of the banks which consist an interbank network. This vector is defined as follows. For any subset $A \subseteq X$, we consider the vector $\mathbf{1_A} \in \mathbb{Z}_2^n$ whose entries are zero, except the entries corresponding to the numbers of A which are equal to 1. Formally, we have

$$\mathbf{1_A}(i) = \begin{cases} 1, & i \in A; \\ 0, & i \notin A. \end{cases}$$

Then, the contagion vector is given by the following sum:

$$V = \sum_{A \in 2^X} (\mathbf{1_{T(A)}} - \mathbf{1_A}).$$

If the bank i is prone to financial contagion, then i appears in the difference $T(A)\setminus A$ for many sets $A \in 2^X$. This implies that the i-th coordinate of the vector V is large. Consequently, the bigger the coordinate of V the more vulnerable the corresponding institution is and vice versa.

The contagion vector is closely related to our basic tool for the assessment of banks, which is the cardinality of the upper sets U_x, $x \in X$. This connection becomes accurate with the next theorem.

Theorem 2 *For any $x \in \{1, 2, \dots, n\}$, the x-th coordinate of the contagion vector is given by*

$$V_x = \sharp U_x^o,$$

where U_x^o is the subset of the upper set U_x consisting of all the sets $A \in U_x$ that do not contain x, that is:

$$U_x^o = \{A \in U_x \mid x \notin A\} = U_x \setminus W_{\{x\}}.$$

Proof Fix $x \in \{1, 2, \dots, n\}$. We firstly observe that for any subset $A \subseteq X$, the x-th coordinate of the vector $\mathbf{1_{T(A)}} - \mathbf{1_A}$ is either zero or one. (Since $A \subseteq T(A)$ the coordinates of this vector cannot take the value -1.) This implies that each set $A \subseteq X$ contributes either 1 or nothing to the x-th coordinate of the contagion vector V and hence the coordinate V_x is equal to the number of the sets A whose contribution is 1.

Secondly, we have that the set A contributes 1 to the coordinate V_x if and only if

$$\begin{aligned}
\mathbf{1_{T(A)}}(x) - \mathbf{1_A}(x) = 1 &\Leftrightarrow \mathbf{1_{T(A)}}(x) = 1 \text{ and } \mathbf{1_A}(x) = 0 \\
&\Leftrightarrow x \in T(A) \quad \text{and} \quad x \notin A \\
&\Leftrightarrow A \in U_x \quad \text{and} \quad x \notin A \quad \text{(by Proposition 3)} \\
&\Leftrightarrow A \in U_x^o.
\end{aligned}$$

By the above remarks, it follows immediately that $V_x = \sharp U_x^o$ and the proof is complete.

6 Boolean Dynamical Systems

Our next purpose is to show that the contagion map T can be described in terms of the minimal sets of the bankruptcy sets U_x, $x \in X$. This description will also reveal the significance of the minimal sets, since the propagation of a financial crisis through the network actually depends on these sets.

In order to achieve our goal we appeal to Boolean dynamical systems. It is well-known that the powerset 2^X of the set $X = \{1, 2, \dots, n\}$ can be identified with the set $\mathbb{Z}_2^n$ of all vectors of length n whose coordinates are 0 or 1. More precisely, the function

$$\mathfrak{h}\colon 2^X \to \mathbb{Z}_2^n$$

$$A \mapsto \mathbf{1_A}$$

defines a one-to-one and onto correspondence. Therefore, the contagion map $T\colon 2^X \to 2^X$ corresponds (via $\mathfrak{h}$) to a map $F: \mathbb{Z}_2^n \to \mathbb{Z}_2^n$, such that the following diagram commutes

$$\begin{array}{ccc} 2^X & \xrightarrow{T} & 2^X \\ \downarrow \mathfrak{h} & & \downarrow \mathfrak{h} \\ \mathbb{Z}_2^n & \xrightarrow{F} & \mathbb{Z}_2^n. \end{array}$$

In order to describe the induced map F, we define, for any subset A of X, the Boolean function $\chi_A\colon \mathbb{Z}_2^n \to \mathbb{Z}_2$ with

$$\chi_A(\mathbf{x}) = \begin{cases} 1, \text{ if } A \subseteq \{j \mid \mathbf{x}_j = 1\}; \\ 0, \text{ otherwise.} \end{cases}$$

When A is a singleton, $A = \{i\}$, then we denote χ_A by χ_i and we have

$$\chi_i(\mathbf{x}) = \begin{cases} 1, \text{ if } \mathbf{x}_i = 1; \\ 0, \text{ otherwise.} \end{cases}$$

It is not hard to verify that for any subset A of X the following relation holds

$$\chi_A = \prod_{i \in A} \chi_i.$$

Let $x \in X$ be a bank and let $\omega_{x1}, \omega_{x2}, \ldots, \omega_{xm_x}$ be the minimal elements of the upper set U_x. Then, we define the *membership function of* x as follows:

$$F_x\colon \mathbb{Z}_2^n \to \mathbb{Z}_2$$

$$F_x = 1 + (1 + \chi_{\omega_{x1}})(1 + \chi_{\omega_{x2}}) \ldots (1 + \chi_{\omega_{xm_x}}).$$

The term membership function is justified by the next lemma.

Lemma 1 *For any $A \in 2^X$, we have $F_x(\mathbf{1_A}) = 1$ if and only if A belongs to the upper set U_x.*

Proof By Proposition 2, we obtain that the set A belongs to U_x if and only if $\omega_{xi} \subseteq A$ for some $i \in \{1, 2, \ldots, m_x\}$. This is equivalent to say that $\omega_{xi} \subseteq \{j \in \{1, 2, \ldots, n\} \mid \mathbf{1_A}(j) = 1\}$, which in turn implies that $\chi_{\omega_{xi}}(\mathbf{1_A}) = 1$. Therefore, $1 + \chi_{\omega_i}(\mathbf{1_A}) = 0$, and thus $F_x(\mathbf{1_A}) = 1$. The inverse implications are also true, and the proof is complete.

Given the membership functions F_x, $x \in X$, we consider the Boolean function $F : \mathbb{Z}_2^n \to \mathbb{Z}_2^n$ as follows:

$$F = (F_1, F_2, \ldots, F_x, \ldots, F_n).$$

The function F coincides with the contagion map T when the latter is seen as a map from $\mathbb{Z}_2^n$ to $\mathbb{Z}_2^n$. This statement becomes more accurate with the next theorem.

Theorem 3 *The function F is related to the contagion map T via the following equation:*

$$F = \mathfrak{h} \circ T \circ \mathfrak{h}^{-1}.$$

Proof Assume that A is a subset of X. We have to prove that $F(\mathbf{1_A}) = \mathfrak{h} \circ T \circ \mathfrak{h}^{-1}(\mathbf{1_A})$, or equivalently

$$F(\mathbf{1_A}) = \mathbf{1_{T(A)}}.$$

To this end, it suffices to show that, for any $x \in X = \{1, 2, \ldots, n\}$, the following equivalence holds

$$F_x(\mathbf{1_A}) = 1 \Longleftrightarrow \mathbf{1_{T(A)}}(x) = 1$$

that is,

$$F_x(\mathbf{1_A}) = 1 \Longleftrightarrow x \in T(A).$$

However, the above equivalence follows immediately, since by Lemma 1 and Proposition 3 the expressions "$F_x(\mathbf{1_A}) = 1$" and "$x \in T(A)$" are both equivalent to the fact that A belongs to the upper set U_x.

Corollary 3 *The contagion map T depends only on the minimal elements of the upper sets U_x, $x \in X$.*

Remark 2 With a slight abuse of terminology, we will also refer to the function $F : \mathbb{Z}_2^n \to \mathbb{Z}_2^n$ as the contagion map of the interbank network. We may also make no distinction between the set $A \subseteq X$ and the corresponding vector $\mathbf{1_A}$.

7 Fixed Points of the Function F

As it has been observed in Section 5, the equilibrium sets of the contagion map T, that is the sets $B \in 2^X$ such that $T(B) = B$, play a major role in the study of economic networks, since they show up to which point the propagation of an economic crisis continues. Our aim now is to take a closer look at these sets. We start with the next proposition giving the relation between the equilibrium sets and the contagion map F.

Proposition 5 *Let A be a subset of X. Then the following holds*

$$T(A) = A \Longleftrightarrow F(\mathbf{1_A}) = \mathbf{1_A}.$$

The proof is straightforward, since $F(\mathbf{1_A}) = \mathfrak{h} \circ T \circ \mathfrak{h}^{-1}(\mathbf{1_A}) = \mathfrak{h} \circ T(A) = \mathfrak{h}(A) = \mathbf{1_A}$. Consequently, the equilibrium sets of the contagion map T correspond to vectors $\mathbf{x} \in \mathbb{Z}_2^n$ such that $F(\mathbf{x}) = \mathbf{x}$, that is fixed points of the Boolean function F.

The above result also facilitates the calculation of the equilibrium sets. In order to find the equilibrium points, it suffices to find the fixed points of the function F, that is to solve in $\mathbb{Z}_2^n$ the equation:

$$F(\mathbf{x}) = \mathbf{x}.$$

This equation is equivalent to the following system of n equations in $\mathbb{Z}_2$ with n variables:

$$F_k(x_1, x_2, \ldots x_n) = x_k \quad \text{for any } k = 1, 2, \ldots, n,$$

that is,

$$\begin{aligned}
1 + (1 + x_1)\left(1 + \prod_{j \in \omega_{12}} x_j\right) \cdots \left(1 + \prod_{j \in \omega_{1m_1}} x_j\right) &= x_1 \\
1 + (1 + x_2)\left(1 + \prod_{j \in \omega_{22}} x_j\right) \cdots \left(1 + \prod_{j \in \omega_{2m_2}} x_j\right) &= x_2 \\
&\vdots \\
1 + (1 + x_n)\left(1 + \prod_{j \in \omega_{n2}} x_j\right) \cdots \left(1 + \prod_{j \in \omega_{nm_n}} x_j\right) &= x_n.
\end{aligned}$$

It is not hard to observe that the above system of equations can also be written in the next form which facilitates the procedure of solution:

$$(1 + x_1)\left[1 + \left(1 + \prod_{j \in \omega_{12}} x_j\right) \cdots \left(1 + \prod_{j \in \omega_{1m_1}} x_j\right)\right] = 0$$

$$(1+x_2)\left[1+\left(1+\prod_{j\in\omega_{22}} x_j\right)\cdots\left(1+\prod_{j\in\omega_{2m_2}} x_j\right)\right]=0$$

$$\vdots$$

$$(1+x_n)\left[1+\left(1+\prod_{j\in\omega_{n2}} x_j\right)\cdots\left(1+\prod_{j\in\omega_{nm_n}} x_j\right)\right]=0.$$

The above equations can be considered as polynomial equations of n variables, under the additional assumptions $x_i^2 = x_i$ for any $i \in \{1, 2, \ldots, n\}$ and $1 + 1 = 0$. The system of these n equations can be solved with numerical methods, e.g. Groebner bases.

The previous procedure of calculating the fixed points of the interbank system can also be formulated in terms of subsets of the powerset 2^X. Assume that $A \subseteq \{1, 2, \ldots, n\} = X$ is an equilibrium set and $\mathbf{x} = \mathbf{1}_A$ is the corresponding vector in $\mathbb{Z}_2^n$. Then $\mathbf{x}$ is one of the solutions of the aforementioned system. That is for any $i = 1, 2, \ldots, n$, the coordinates of $\mathbf{x}$ satisfy the equation:

$$(1+x_i)\left[1+\left(1+\prod_{j\in\omega_{i2}} x_j\right)\cdots\left(1+\prod_{j\in\omega_{im_i}} x_j\right)\right]=0.$$

This implies that

$$\text{either } 1 + x_i = 0 \text{ or } 1 + \prod_{j\in\omega_{ik}} x_j = 1 \ \forall k = 1, 2, \ldots, m_i,$$

that is,

$$\text{either } x_i = 1 \text{ or } \prod_{j\in\omega_{ik}} x_j = 0 \ \forall k = 1, 2, \ldots, m_i.$$

Returning to the set A, the above relations means that

$$\text{either } i \in A \text{ or } \omega_{ik} \cap \left(2^{X\setminus\{i\}} \setminus A\right) \neq \emptyset \ \forall k = 1, 2, \ldots, m_i.$$

We now set W_i to be the upper set of $\{i\}$ in X, i.e. W_i consists of all subsets of X containing i. Let also $L_i = 2^{X\setminus\{i\}} \setminus U_i'$, where $U_i' = U_i \setminus W_i$. Since, U_i' is an upper set in $2^{X\setminus\{i\}}$, it follows that L_i is a lower set (i.e. if $A \in L_i$ and $B \subseteq A$, then B also belongs to L_i). Then, we observe that:

- If $i \in A$, then A belongs to the upper set W_i.
- If $i \notin A$, then $\omega_{ik} \cap \left(2^{X \setminus \{i\}} \setminus A\right) \neq \emptyset$ for any $k = 1, 2, \ldots, m_i$. Therefore, A does not belong to the upper set U_i and hence it belongs to the set L_i.

Consequently, a set A is an equilibrium set if and only if $A \in \bigcap_i (W_i \cup L_i)$. Thus the set

$$\bigcap_{i=1}^{n} (W_i \cup L_i)$$

contains all the fixed points of the interbank system.

8 The Global Function

We denote by Fix the set of fixed points of the map F, i.e.

$$\text{Fix} = \left\{\mathbf{x} \in \mathbb{Z}_2^n \mid F(\mathbf{x}) = \mathbf{x}\right\} \subseteq \mathbb{Z}_2^n.$$

Every $\mathbf{x} \in \mathbb{Z}_2^n$ is associated via F with a unique fixed point $y_{\mathbf{x}}$ as the next result shows.

Theorem 4 *For every* $\mathbf{x} \in \mathbb{Z}_2^n$ *there is a unique* $\mathbf{y_x} \in Fix$ *such that* $F^n(\mathbf{x}) = \mathbf{y_x}$ *eventually for all* $n \in \mathbb{N}^*$.

Proof By its construction, the function F has the property:

$$\mathfrak{h}^{-1}(\mathbf{x}) \subseteq \mathfrak{h}^{-1}(F(\mathbf{x})) \quad \text{for all } \mathbf{x} \in \mathbb{Z}_2^n.$$

That is, the number of 1's in the vector $F(\mathbf{x})$ is greater than or equal to the number of 1's appearing in $\mathbf{x}$. Since, 2^X (or equivalently $\mathbb{Z}_2^n$) is finite, we cannot find a strictly increasing sequence in this set. Consequently, the sequence

$$\left(\mathbf{x}, F(\mathbf{x}), F^2(\mathbf{x}), \ldots\right)$$

stabilizes, that is there is n such that $F^{n+1}(\mathbf{x}) = F^n(\mathbf{x})$. If $n_{\mathbf{x}}$ is the least positive integer with this property and $\mathbf{y_x} = F^{n_{\mathbf{x}}}(\mathbf{x})$, then $\mathbf{y_x}$ is a fixed point of F. The uniqueness of $\mathbf{y_x}$ follows immediately.

Motivated by the above theorem, we can now define the *global function* G as follows.

Definition 1 The **global function** G maps each $\mathbf{x} \in \mathbb{Z}_2^n$ to the fixed point $\mathbf{y_x}$ which $\mathbf{x}$ reaches after finitely many iterated applications of the contagion map F, that is

$$G: \mathbb{Z}_2^n \to \text{Fix}$$

$$G(\mathbf{x}) = \mathbf{y}_\mathbf{x}.$$

Hence, in terms of interbank contagion, the map G describes the final situation of the interbank network after the collapse of the banks $A = \mathfrak{h}^{-1}(\mathbf{x})$ and omits the intermediate steps (which are given by the contagion map F).

Corollary 4 *Let* $n_0 = \max\{n_\mathbf{x} \mid \mathbf{x} \in \mathbb{Z}_2^n\}$. *Then, we have* $G = F^{n_0}$.

Proof By the definition of G we have that $G(\mathbf{x}) = F^{n_\mathbf{x}}(\mathbf{x})$ for all $\mathbf{x} \in \mathbb{Z}_2^n$. Furthermore, $F^n(\mathbf{x}) = F^{n_\mathbf{x}}(\mathbf{x})$ for all $n \geq n_\mathbf{x}$. It follows that $G(\mathbf{x}) = F^{n_0}(\mathbf{x})$ for all $\mathbf{x} \in \mathbb{Z}_2^n$ as we desire.

The above corollary also provides us with a method for calculating the fixed points of F. The method is as follows. One has to calculate the iterated powers $F, F^2, F^3, \ldots$, where we recall that F^n is the composition of F with itself. In this calculation, we use the fact that $x_i^2 = x_i \cdot x_i = x_i$ for any $x_i \in \mathbb{Z}_2^n$. According to Corollary 4, after a finite number of steps (which coincides with the length of the maximal chain in the contagion graph), we find an integer n such that $F^n = F^{n+1}$. Let us consider the minimum n with this property. Then F^n is the global function G and, hence, its image is the set of fixed points for F.

9 Assessment of Banks and Networks

One of the most important aspects of the study of economic networks is the assessment of a network and the banks of which it consists. This assessment is crucial for the formulation of suitable policies and the taking of proactive measures for the strengthening of the network and the reduction of systemic risk. Towards this direction, the analysis of the previous sections indicates the following methods.

9.1 Assessment of Banks

The assessment of an individual bank $x \in X = \{1, 2, \ldots, n\}$ can be achieved through the cardinality of the upper set U_x or equivalently through the coordinates of the contagion vector. As it was analyzed in Section 4, the more elements the set U_x contains, the more vulnerable the bank x is and vice versa. This remark can be quantified with the following metric. For each bank $x \in X = \{1, 2, \ldots, n\}$, we define

$$I_x = \frac{\sharp U_x}{2^n}, \tag{1}$$

where 2^n is the cardinality of the powerset 2^X. Since U_x contains the supersets of $\{x\}$, which implies that $\sharp U_x \geq 2^{n-1}$, it follows that the metric I_x takes values between $\frac{1}{2}$ and 1. When $I_x = \frac{1}{2}$, then U_x coincides with the set of all supersets of $\{x\}$ and the only minimal element of U_x is the singleton $\{x\}$. Thus, the failure of any other bank does not imply the bankruptcy of x. In the other extreme case, i.e. $I_x = 1$, we have that $U_x = 2^X$, that is any combination of defaults, even the empty set, results in the default of x. Hence, the bank x has already collapsed. If this is not the situation, then U_x cannot contain the empty set, and therefore the cardinality of U_x is less or equal to $2^n - 1$ and the maximum value of I_x is $1 - \frac{1}{2^n}$. Consequently, the closer to $\frac{1}{2}$ the metric I_x is, the more robust the bank is and vice versa.

By Theorem 1 and Corollary 2, the metric I_x can be calculated from the minimal sets of U_x. Namely, we have

Corollary 5 *Let $\omega_{x1}, \omega_{x2}, \ldots, \omega_{xm_x}$ be the minimal elements of U_x. Then, the metric I_x is given by*

$$I_x = \sum_{i=1}^{m_x} \frac{1}{2^{\sharp\omega_{xi}}} - \sum_{1 \leq i < j \leq m_x} \frac{1}{2^{\sharp\omega_{xij}}} + \sum_{1 \leq i < j < k \leq m_x} \frac{1}{2^{\sharp\omega_{xijk}}} - \cdots + (-1)^{m_x - 1} \frac{1}{2^{\sharp\omega_{x12\ldots m_x}}}.$$

In the case where $\omega_{x1}, \omega_{x2}, \ldots, \omega_{xm_x}$ are pairwise disjoint, we have the formula:

$$I_x = 1 - \prod_{i=1}^{m_x} \left(1 - \frac{1}{2^{\sharp\omega_{xi}}}\right).$$

9.2 Assessment of Interbank Networks

In order to assess the whole network, we utilize the number of equilibrium sets of the contagion map T, equivalently the number of fixed points of the Boolean function F. There are at least two fixed points: the point $\mathbf{x} = (0, 0, \ldots, 0)$ corresponding to the empty set, and the point $\mathbf{x} = (1, 1, \ldots, 1)$ corresponding to the whole set X. Hence, if f stands for the total number of fixed points in the interbank system, then we have $2 \leq f \leq 2^n$. Now two extreme cases appear:

The worst-case scenario is when $f = 2$, i.e. we have only two fixed points. In this case we have total contagion and the interbank network is completely vulnerable. The failure of any combination of banks, even of a single bank, results in the collapse of the whole network.

The best case scenario is when $f = 2^n$, i.e. every point $\mathbf{x} \in \mathbb{Z}_2^n$ is a fixed point. In this case, the default of any combination of banks does not affect any other bank. Consequently, there is no contagion at all and the network is completely robust.

The above remarks can be quantified by considering the following metric:

$$m = \frac{f}{2^n}.$$

The quantity m takes values between $\frac{1}{2^{n-1}}$ and 1 and assess the robustness of the whole network. The bigger the metric, the more robust the network is and vice versa.

Additionally, by combining the indexes I_x, $x \in X$, which assess the robustness of each individual institution, we can provide indicators for the whole network. More specifically, taking also into account the capital and liabilities of each institution, we define

$$m_c = \frac{\sum_{x=1}^{n} c_x \cdot I_x}{\sum_{x=1}^{n} c_x},$$

$$m_L = \frac{\sum_{x=1}^{n} L_x \cdot I_x}{\sum_{x=1}^{n} L_x}.$$

Since, $\frac{1}{2} \leq I_x \leq 1$ for every $x \in X$, it follows that the indicators m_c and m_L take values between $\frac{1}{2}$ and 1. Furthermore, the closer to $\frac{1}{2}$ these indicators are, the more robust the network and vice versa.

10 Example

In this section, we illustrate the ideas and techniques of the present work with some example. We consider an interbank network consisting of three banks, i.e. $X = \{1, 2, 3\}$. The other two characteristic features of the network, that is the vector of capitals and the exposures matrix are given as follows:

$$C = (10, 8, 5) \quad \text{and} \quad E = \begin{bmatrix} 10 & 7 & 4 \\ 9 & 8 & 0 \\ 3 & 2 & 5 \end{bmatrix},$$

where in the exposures matrix we have replaced the zero diagonal with the vector of capitals. The interbank network is depicted in Figure 1.

Using Proposition 1, we can easily detect the bankruptcy sets, which for each node of the network are the following:

$$U_1 = \Big\{\{1\}, \{2, 3\}, \{1, 2\}, \{1, 3\}, \{1, 2, 3\}\Big\}$$

$$U_2 = \Big\{\{1\}, \{2\}, \{1, 2\}, \{1, 3\}, \{2, 3\}\{1, 2, 3\}\Big\}$$

$$U_3 = \Big\{\{1, 2\}, \{3\}, \{1, 3\}, \{2, 3\}, \{1, 2, 3\}\Big\}.$$

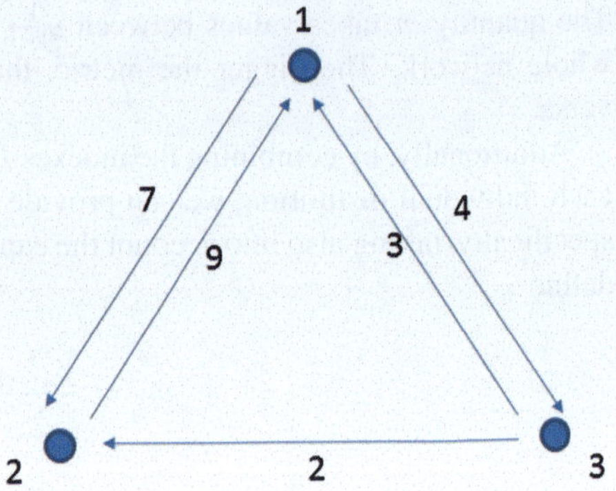

Fig. 1 The directed weighed graph representing the interbank network that is used in the example. The weights of the edges corresponds to the exposures between the banks of the market

The minimal elements of the upper sets $U_x, x = 1, 2, 3$, are, respectively

$$\omega_{11} = \{1\},\ \omega_{12} = \{2, 3\},$$
$$\omega_{21} = \{2\},\ \omega_{22} = \{1\},$$
$$\omega_{31} = \{3\},\ \omega_{32} = \{1, 2\}.$$

Consequently, the membership function of each node can be deduced directly from the above:

$$F_1 = 1 + (1 + \chi_{\omega_{11}})(1 + \chi_{\omega_{12}}) = 1 + (1 + \chi_1)(1 + \chi_2\chi_3)$$
$$F_2 = 1 + (1 + \chi_{\omega_{21}})(1 + \chi_{\omega_{22}}) = 1 + (1 + \chi_2)(1 + \chi_1)$$
$$F_3 = 1 + (1 + \chi_{\omega_{31}})(1 + \chi_{\omega_{32}}) = 1 + (1 + \chi_3)(1 + \chi_1\chi_2)$$

that is

$$F_1(x_1, x_2, x_3) = 1 + (1 + x_1)(1 + x_2x_3)$$
$$F_2(x_1, x_2, x_3) = 1 + (1 + x_2)(1 + x_1)$$
$$F_3(x_1, x_2, x_3) = 1 + (1 + x_3)(1 + x_1x_2).$$

Therefore, the contagion map $F : \mathbb{Z}_2^n \to \mathbb{Z}_2^n$ is given by

$$F(x_1, x_2, x_3) = (1+(1+x_1)(1+x_2x_3), 1+(1+x_2)(1+x_1), 1+(1+x_3)(1+x_1x_2)).$$

Easy calculations shows that

$$F(0, 0, 0) = (0, 0, 0)\ F(0, 1, 0) = (0, 1, 0)\ F(1, 1, 0) = (1, 1, 1)\ F(0, 1, 1) = (1, 1, 1)$$
$$F(1, 0, 0) = (1, 1, 0)\ F(0, 0, 1) = (0, 0, 1)\ F(1, 0, 1) = (1, 1, 1)\ F(1, 1, 1) = (1, 1, 1).$$

Hence, the contagion graph corresponding to the interbank network can be constructed and it is depicted in Figure 2.

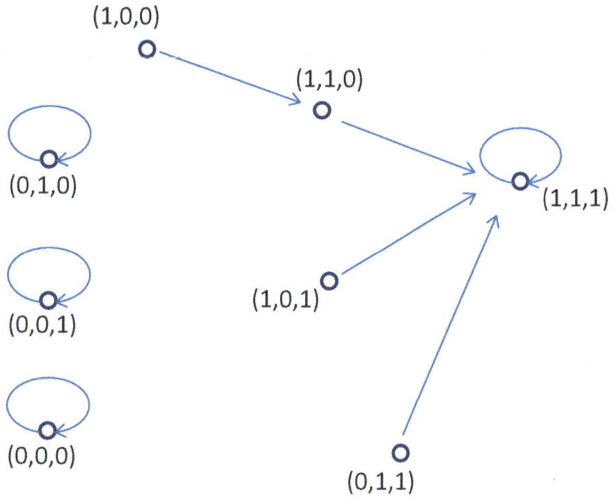

Fig. 2 The contagion graph corresponding to the interbank system of the example

At first glance, it may not be clear for the regulatory system in which bank it should intervene. However, if we calculate the contagion vector, we find that $V = (1, 2, 1)$. Hence, the bank 2 is the most vulnerable. Indeed, the failure of the bank 1 leads the second bank to default and then the entire system is collapsed. Consequently, the second bank should take proactive measures, for instance, to raise its capitals or to reduce its exposure to the bank 1.

10.1 Fixed Points of the Network

We next focus our attention on the equilibrium points of the interbank network of the example. From the contagion graph (Figure 2) we can immediately observe that there are four equilibrium sets, namely the trivial ones, i.e. $(0, 0, 0)$ and $(1, 1, 1)$, and also the points $(0, 1, 0)$, $(0, 0, 1)$. However, we have developed a number of techniques for calculating the fixed points and it is not necessary to construct the contagion graph.

10.1.1 First Method

According to the results of Section 7, the calculation of the fixed points is achieved through the solution of the following system of three equation with three variables:

$$\left\{ \begin{array}{l} F_1(x_1, x_2, x_3) = x_1 \\ F_2(x_1, x_2, x_3) = x_2 \\ F_3(x_1, x_2, x_3) = x_3 \end{array} \right\} \Leftrightarrow \left\{ \begin{array}{r} 1 + (1 + x_1)(1 + x_2x_3) = x_1 \\ 1 + (1 + x_1)(1 + x_2) = x_2 \\ 1 + (1 + x_1x_2)(1 + x_3) = x_3 \end{array} \right\}.$$

As we have commented, the above system can be written in the form:

$$\left\{\begin{array}{c}(1+x_1)(1+1+x_2x_3)=0\\(1+x_2)(1+1+x_1)=0\\(1+x_3)(1+1+x_1x_2)=0\end{array}\right\}$$

and in this specific example it takes the form:

$$\left\{\begin{array}{r}(1+x_1)x_2x_3=0\\(1+x_2)x_1=0\\(1+x_3)x_1x_2=0\end{array}\right\}.$$

The solution is now straightforward:

$$\left\{\begin{array}{ccccc}x_1=1 & \text{or} & x_2=0 & \text{or} & x_3=0\\ & & x_2=1 & \text{or} & x_1=0\\ x_3=1 & \text{or} & x_1=0 & \text{or} & x_2=0\end{array}\right\}.$$

It follows immediately that the equilibrium points of the interbank network are

$$(0,0,0),\quad (1,1,1),\quad (0,0,1),\quad (0,1,0).$$

10.1.2 Second Method

The nature of the second method, which is also described in Section 7, is set-theoretic. One has to calculate, for each bank i of the network, the set $W_i \cup L_i$, where W_i is the upper set of $\{i\}$ and L_i is the lower set $L_i = 2^{X\setminus\{i\}} \setminus U_i'$ with $U_i' = U_i \setminus W_i$. Then the fixed points are provided by the intersection $\cap_i(W_i \cup L_i)$.

In this particular example, for the first bank we obtain

$$W_1 = \Big\{\{1\},\{1,2\},\{1,3\},\{1,2,3\}\Big\} \quad \text{and} \quad U_1' = \Big\{\{2,3\}\Big\}.$$

Hence, $L_1 = \Big\{\emptyset,\{2\},\{3\}\Big\}$, and

$$W_1 \cup L_1 = \Big\{\emptyset,\{1\},\{2\},\{3\},\{1,2\},\{1,3\},\{1,2,3\}\Big\}.$$

In a similar fashion, we find the following:

$$W_2 \cup L_2 = \Big\{\emptyset, \{2\}, \{3\}, \{1,2\}, \{2,3\}, \{1,2,3\}\Big\}$$
$$W_3 \cup L_3 = \Big\{\emptyset, \{1\}, \{2\}, \{3\}, \{1,3\}, \{2,3\}, \{1,2,3\}\Big\}.$$

Finally, it suffices to find the elements of $W_1 \cup L_1$ that belong to every set $W_i \cup L_i$ for all i. In this example, these are the sets

$$\bigcap_{i=1}^{3} (W_i \cup L_i) = \Big\{\emptyset, \{2\}, \{3\}, \{1,2,3\}\Big\}.$$

These are the equilibrium points of the contagion map T. The corresponding fixed points of F are the vectors $(0,0,0)$, $(0,1,0)$, $(0,0,1)$, and $(1,1,1)$, respectively.

10.1.3 Third Method

Finally, our last technique for finding the fixed points of the interbank network is provided in Section 8. According to this method, one has to calculate (taking into account that $x_i^2 = x_i$ for any i) the iterated applications of the contagion map F, that is $F^n = F \circ F \circ \ldots \circ F$ up to the point where this function stabilizes, i.e. we have $F^{n+1} = F^n$. Then F^n is the global function G and its image gives the fixed points.

In this particular example, we have that

$$F_1(F_1(\mathbf{x}), F_2(\mathbf{x}), F_3(\mathbf{x})) = x_1 + x_2x_3 + x_1x_2x_3$$
$$F_2(F_1(\mathbf{x}), F_2(\mathbf{x}), F_3(\mathbf{x})) = x_1 + x_2 + x_1x_2$$
$$F_3(F_1(\mathbf{x}), F_2(\mathbf{x}), F_3(\mathbf{x})) = x_1 + x_3 + x_1x_3$$

that is

$$F^2(\mathbf{x}) = (x_1 + x_2x_3 + x_1x_2x_3, x_1 + x_2 + x_1x_2, x_1 + x_3 + x_1x_3).$$

Another application of the contagion map F reveals that $F^3 = F^2$. Hence, the global function G is equal to F^2. The image of G can now be calculated and we obtain that the set of fixed points is

$$\text{Fix} = \{(0,0,0),\ (1,1,1),\ (0,0,1),\ (0,1,0)\}.$$

References

1. D. Acemoglu, A. Ozdaglar, A. Tahbaz-Salehi, Systemic risk and stability in financial networks. Am. Econ. Rev. **105**(2), 564–608 (2015)
2. I. Aldasoro, I. Alves, Multiplex interbank networks and systemic importance: An application to European data. SAFE Working Paper, No. 102 (2015)
3. F. Allen, A. Babus, Networks in finance, in *The Network Challenge: Strategy, Profit, and Risk in An Interlinked World*, ed. by P.R. Kleindorfer, Y. Wind (Prentice Hall Professional, Upper Saddle River, 2009), pp. 367–382
4. F. Allen, D. Gale, Financial contagion. J. Polit. Econ. **108**, 1–33 (2000)
5. H. Amini, R. Cont, A. Minca, Resilience to contagion in financial networks. Math. Financ. **26**, 329–365 (2016)
6. K. Anand, B. Craig, G. Von Peter, Filling in the blanks: network structure and interbank contagion. Quant. Finance **15**(4), 625–636 (2015)
7. M. Bardoscia, S. Battiston, F. Caccioli, G. Caldarelli, Pathways towards instability in financial networks. Nat. Commun. **8**, 14416 (2017). https://doi.org/10.1038/ncomms14416
8. L. Bargigli, G. di Iasio, L. Infante, F. Lillo, F. Pierobon, The multiplex structure of interbank networks. Quant. Finance **15**(4), 673–691 (2014)
9. S. Battiston, D.D. Gatti, M. Gallegati, B. Greenwald, J.E. Stiglitz, Liaisons dangereuses: increasing connectivity, risk sharing, and systemic risk. J. Econ. Dyn. Control. **36**(8), 1121–1141 (2012)
10. S. Battiston, M. Puliga, R. Kaushik, P. Tasca, G. Caldarelli, DebtRank: too central to fail? Financial networks, the fed and systemic risk. Sci. Rep. **2**, 541 (2012)
11. S. Benoit, J.-E. Colliard, C. Hurlin, C. Pérignon, Where the risks lie: A survey on systemic risk. HEC Paris Research Paper, No. FIN-2015-1088 (2015)
12. F. Caccioli, T.A. Catanach, J.D. Farmer, Heterogeneity, correlations and financial contagion. Adv. Complex Syst. **15**(Suppl. 02), 1250058 (2012)
13. M. Chinazzi, G. Fagiolo, Systemic risk, contagion, and financial networks: A survey. LEM Working Paper Series, No. 8 (2013)
14. M. Chinazzi, S. Pegoraro, G. Fagiolo, Defuse the bomb: Rewiring interbank networks. LEM Working Paper Series, Institute of Economics, Scuola Superiore Sant' Anna (2015)
15. G. Cimini, T. Squartini, D. Garlaschelli, A. Gabrielli, Systemic risk analysis on reconstructed economic and financial networks. Sci. Rep. **5**, 15758 (2015)
16. R. Cont, A. Moussa, E.B. Santos, *Network Structure and Systemic Risk in Banking Systems* (2010). Available at SSRN: https://ssrn.com/abstract=1733528 or http://dx.doi.org/10.2139/ssrn.1733528
17. B. Craig, G. von Peter, Interbank tiering and money center banks. J. Financial Int. **23**(3), 322–347 (2014)
18. O. De Bandt, P. Hartmann, J. Peydro, in *Systemic Risk in Banking: An Update*, ed. by A. Berger, P. Molyneux, J. Wilson. The Oxford Handbook of Banking (Oxford University Press, Oxford, 2012), pp. 633–672
19. S.R.S. De Souza, T.C. Silva, B.M. Tabak, S.M. Guerra, Evaluating systemic risk using bank default probabilities in financial networks. J. Econ. Dyn. Control **66**, 54–75 (2016)
20. M. Elliott, B. Golub, M.O. Jackson, Financial networks and contagion. Am. Econ. Rev. **104**(10), 3115–3153 (2014)
21. H. Elsinger, A. Lehar, M. Summer, Risk assessment for banking systems. Manage. Sci. **52**(9), 1301–1314 (2006)
22. H. Elsinger, A. Lehar, M. Summer, in *Network Models and Systemic Risk Assessment*, ed. by J.-P. Fouque, J.A. Langsam. Handbook on Systemic Risk (Cambridge University Press, Cambridge, 2013), pp. 287–305
23. X. Freixas, B.M. Parigi, J.-C. Rochet, Systemic risk, interbank relations, and liquidity provision by the central bank. J. Money Credit Bank. **32**(3), 611–638 (2000)

24. D. Fricke, T. Lux, Core-periphery structure in the overnight money market: evidence from the e-MID trading platform. Comput. Econ. **45**, 359–395 (2015)
25. P. Gai, S. Kapadia, Contagion in financial networks. P. Roy. Soc. A Math. Phy. **466**, 2401–2423 (2010)
26. P. Gai, A. Haldane, S. Kapadia, Complexity, concentration and contagion. J. Monetary Econ. **58**(5), 453–470 (2011)
27. P. Glasserman, H.P. Young, Contagion in financial networks. J. Econ. Lit. **54**(3), 779–831 (2016)
28. X. Huang, H. Zhou, H. Zhu, Assessing the systemic risk of a heterogeneous portfolio of banks during the recent financial crisis. J. Finan. Stab. **8**(3), 193–205 (2012)
29. G. Iori, S. Jafarey, F. Padilla, Systemic risk on the interbank market. J. Econ. Behav. Organ. **61**, 525–542 (2006)
30. S. Langfield, Z. Liu, T. Ota, Mapping the UK interbank system. J. Banking Finance **45**, 288–303 (2014)
31. S. Langfield, K. Soramaki, Interbank exposure networks. Comput. Econ. **47**, 3–17 (2014)
32. J. Leventides, C. Poulios, E. Camouzis, A mathematical approach to interbank financial contagion dynamics, Submitted
33. J. Leventides, K. Loukaki, V.G. Papavassiliou, Simulating financial contagion dynamics in random interbank networks. J. Econ. Behav. Organ. **158**, 500–525 (2019). https://doi.org/10.1016/j.jebo.2018.12.017
34. P.E. Mistrulli, Assessing financial contagion in the interbank market: maximum entropy versus observed interbank lending patterns. J. Bank. Financ. **35**(5), 1114–1127 (2011)
35. M. Montagna, C. Kok, Multi-layered interbank model for assessing systemic risk. Kiel Working Paper 1873 (2013)
36. E. Nier, J. Yang, T. Yorulmazer, A. Alentorn, Network models and financial stability. J. Econ. Dyn. Control **31**(6), 2033–2060 (2007)
37. M. Summer, Financial contagion and network analysis. Ann. Rev. Financial Econ. **5**(1), 277–297 (2013)
38. C. Upper, Simulation methods to assess the danger of contagion in interbank markets. J. Finan. Stab. **7**(3), 111–125 (2011)
39. D.T. Veld, I. van Lelyveld, Finding the core: network structure in interbank markets. J. Banking Finance **49**, 27–40 (2014)
40. S. Vivier-Lirimont, Contagion in interbank debt networks, Working Paper (2006)

Prime Avoidance Property of k-th Powers of Prime Numbers with Beatty Sequence

Helmut Maier and Michael Th. Rassias

Abstract In a previous paper, the authors establish the prime avoidance property of k-th powers of prime numbers. In this paper the authors extend this result by considering k-th powers of prime numbers with Beatty sequences.

1 Introduction

In their paper [3], K. Ford, D. R. Heath-Brown and S. Konyagin, prove the existence of infinitely many "prime-avoiding" perfect k-th powers for any positive integer k.

Their result is closely related to questions on large gaps between prime numbers. Results on this problem since the paper [2] of P. Erdős and [14] of R. A. Rankin have been based on modifications of the Erdős–Rankin method, named after these authors.

Let

$$g_1(m) := \frac{\log m \log_2 m \log_4 m}{(\log_3 m)^2} .$$

Here $\log_k x := \log(\log_{k-1} x)$.

H. Maier
Department of Mathematics, University of Ulm, Ulm, Germany
e-mail: helmut.maier@uni-ulm.de

M.Th. Rassias (✉)
Institute of Mathematics, University of Zurich, Zurich, Switzerland

Moscow Institute of Physics and Technology, Dolgoprudny, Russia

Institute for Advanced Study, Program in Interdisciplinary Studies, Princeton, NJ, USA
e-mail: michail.rassias@math.uzh.ch

A. M. Raigorodskii, M. Th. Rassias (eds.), *Discrete Mathematics and Applications*, Springer Optimization and Its Applications 165,
https://doi.org/10.1007/978-3-030-55857-4_15

P. Erdős [2] and R. A. Rankin [14] obtained the following result:
Infinitely often

$$p_{n+1} - p_n \geq c\, g_1(n)\,,$$

where $c > 0$ is a fixed constant.

They construct a long interval of integers which all have greatest common divisor larger than 1 with

$$P(x) := \prod_{p<x} p\,.$$

A modified version of this method was applied by the authors of [3] on the problem of prime-avoiding k-th powers. They give the following definition of prime avoidance: an integer m is called *prime-avoiding* with constant c, if $m+u$ is composite for all integers u satisfying

$$|u| \leq c\, g_1(m) \tag{1.1}$$

with c being a positive constant.

In [11] the authors of the present paper extend this result by proving the existence of infinitely many prime-avoiding k-th powers of prime numbers. For the sake of simplicity they only treat one-sided prime avoidance. They prove the following:

There are infinitely many n, such that

$$p_{n+1} - p_n \geq c\, g_1(n)$$

and the interval $[p_n,\ p_{n+1}]$ contains the k-th power of a prime number.

Their method of proof consists in a combination of the method of [3] with the matrix method of the first author [9]. The matrix $\mathcal{M}$ employed in this technique is of the form:

$$\mathcal{M} := (a_{r,u})_{\substack{1\leq r\leq P(x)^{D-1}\\ u\in\mathcal{B}}}\,,$$

where D is a fixed positive integer, where the *rows*

$$R(r) := \{a_{r,u}\ :\ u \in \mathcal{B}\}$$

of the matrix are *translates*—in closer or wider sense—of the *base-row* $\mathcal{B}$.

The base-row $\mathcal{B}$ is constructed by another modification of the Erdős–Rankin method. Whereas in the papers of Erdős and Rankin [2, 14] and several other papers [10, 11, 15] the base-row is completely free of numbers not coprime to $P(x)$, here $\mathcal{B}$ contains some coprime numbers. The columns

$$\mathcal{C}(u) = (a_{r,u})_{1 \le r \le P(x)^{D-1}}, \quad \text{with } (a, P(x)) = 1$$

are called *admissible columns.*

The appearance of prime numbers in these columns, which are arithmetic progressions (or—in the case of [11]—shifted powers of elements of arithmetic progressions), can be studied using results on primes in arithmetic progressions. Until recently, all improvements of the result (1.1) only concerned the constant c. It was a famous prize problem of Erdős to replace the function g_1 by a function of higher order of magnitude. This problem could finally be solved in two independent papers by K. Ford, B. J. Green, S. Konyagin and T. Tao in [4] and by J. Maynard in [13]. Later all five authors improved on this result in a joint paper. In fact K. Ford, B. J. Green, S. Konyagin, J. Maynard and T. Tao in [5] have proved the following result:

$$p_{n+1} - p_n \ge c\, g_2(n)$$

infinitely often, where

$$g_2(m) := \frac{\log m \ \log_2 m \ \log_4 m}{\log_3 m} .$$

In the paper [12] the authors of the present paper combined the methods of the papers [3, 5] and [9] to obtain the following theorem.

Theorem 1.1 of [12] *There is a constant $c > 0$ and infinitely many n, such that*

$$p_{n+1} - p_n \ge c\, g_2(n)$$

and the interval $[p_n, p_{n+1}]$ contains the k-th power of a prime.

In the present paper, we modify the above by replacing the sequence of k-th powers of primes by k-th powers of primes of a special type: prime numbers from non-homogeneous Beatty sequences.

We adopt the following definitions from [1]:

Definition 1.1 For two fixed real numbers α and β, the corresponding non-homogeneous Beatty sequence is the sequence of integers defined by

$$\mathscr{B}_{\alpha,\beta} := (\lfloor \alpha n + \beta \rfloor)_{n=1}^{\infty}$$

($\lfloor \cdot \rfloor$ denotes the floor function: $\lfloor u \rfloor$ the largest integer $\le u$).

Definition 1.2 For an irrational number γ we define its type τ by the relation

$$\tau := \sup\{\rho \in \mathbb{R} \,:\, \liminf n^{\rho} \|\gamma n\| = 0\} .$$

Remark It easily follows from Dirichlet's approximation theorem that $\tau \geq 1$ for every real irrational number. By the theorems of Khinchin [8] resp. Roth [16] we have $\tau = 1$ for almost all real (in the sense of the Lebesgue measure) resp. all irrational algebraic numbers γ.

In this paper we shall prove the following result:

Theorem 1.3 *Let $k \geq 2$ be an integer. Let α, β be fixed real numbers with α being a positive irrational and of finite type. Then there is a constant $C_0 > 0$, depending only on α and β, such that for infinitely many n we have*

$$p_{n+1} - p_n \geq C_0\, g_2(n)$$

and the interval $[p_n, p_{n+1}]$ contains the k-th power of a prime $\tilde{p} \in \mathscr{B}_{\alpha,\beta}$.

2 Construction of the Matrix $\mathcal{M}$

The construction of the matrix $\mathcal{M}$ closely follows the paper [12]. We recall definitions and facts from that paper.

Definition 2.1 (Definition 3.6 of [12]) Let us call an integer $q > 1$ a "good" modulus, if $L(s, \chi) \neq 0$ for all characters $\chi \bmod q$ and all $s = \sigma + it$ with

$$\sigma > 1 - \frac{C_0}{\log\left[q(|t| + 1)\right]}.$$

This definition depends on the size of $C_0 > 0$.

Lemma 2.2 *There is a constant $C_0 > 0$, such that in terms of C_0, there exist arbitrarily large values of x, for which the modulus*

$$P(x) = \prod_{p<x} p$$

is good.

Proof This is Lemma 3.7 of [12]. □

Lemma 2.3 *Let q be a good modulus. Then*

$$\pi(x; q, a) \gg \frac{x}{\phi(q)\log x}, \tag{2.1}$$

uniformly for $(a, q) = 1$ and $x \geq q^{D_0}$.

Here the constant D_0 depends only on the value of C_0 in Lemma 2.2.

Proof This result, which is due to Gallagher [6], is Lemma 3.8 from [12]. □

Lemma 2.4 *Let x be sufficiently large.*

$$y = C_1\, x\, \frac{\log x \log_3 x}{\log_2 x}\,, \tag{2.2}$$

$C_1 > 0$ being a sufficiently small constant. Then there is an integer m_0 satisfying

$$1 \le m_0 < P(x) \tag{2.3}$$

and an exceptional set V satisfying

$$\# V \ll x^{1/2+\epsilon}\,, \tag{2.4}$$

such that

$$(m_0 + 1)^k + u - 1 \tag{2.5}$$

is composite for $2 \le u \le y$, unless $u \in V$.

Proof This is Lemma 3.10 of [12]. □

The construction of V, which is due to Ford, Heath-Brown and Konyagin [3], is described in Lemma 3.5 of [12]. We now construct the matrix $\mathcal{M}$.

Definition 2.5 Choose x, such that $P(x)$ is a good modulus. Let y satisfy (2.2), V satisfy (2.4) and m_0 satisfy (2.3) and (2.5). We let

$$\mathcal{M} := (a_{r,u})_{\substack{1 \le r \le P(x)^{D-1} \\ 1 \le u \le y}}\,,$$

where

$$a_{r,u} := (m_0 + 1 + r P(x))^k + u - 1$$

and the constant D will be specified later.

For $1 \le r \le P(x)^{D-1}$ we denote by

$$R(r) := (a_{r,u})_{0 \le u \le y}$$

the r-th row of $\mathcal{M}$ and for $1 \le u \le y$ we denote by

$$C(u) := (a_{r,u})_{1 \le r \le P(x)^{D-1}}$$

the u-th column of $\mathcal{M}$.

3 Prime Numbers with Beatty Sequences

Lemma 3.1 *Let α and β be fixed real numbers with α a positive irrational and of finite type. Then there is a constant $\kappa > 0$, such that for all integers*

$$0 \leq a < q \leq N^{\kappa} \tag{3.1}$$

with $(a, q) = 1$, we have

$$\sum_{\substack{n \leq N \\ \lfloor \alpha n+\beta \rfloor \equiv a \ (\bmod\ q)}} \Lambda(\lfloor \alpha n + \beta \rfloor) = \alpha^{-1} \sum_{\substack{m \leq \lfloor \alpha N+\beta \rfloor \\ m \equiv a \ (\bmod\ q)}} \Lambda(m) + O(N^{1-\kappa}),$$

where the implied constant depends only on α and β.

Proof This is Theorem 5.4 of [1]. □

4 Conclusion of the Proof

Definition 4.1 We let $l(r) := m_0 + 1 + rP(x)$,

$$\mathcal{R}_1(\mathcal{M}) := \{r : 1 \leq r \leq P(x)^{D-1},\ l(r) \text{ prime},\ l(r) \in \mathscr{B}_{\alpha,\beta}\},$$

$$\mathcal{R}_2(\mathcal{M}) := \{r : 1 \leq r \leq P(x)^{D-1},\ r \in \mathcal{R}_1(\mathcal{M}),\ R(r) \text{ contains a prime number}\}.$$

Remark We observe that each row $R(r)$ with $r \in \mathcal{R}_1(\mathcal{M})$ has as its first element $a_{r,1} = l(r)^k$, the k-th power of the prime $l(r) \in \mathscr{B}_{\alpha,\beta}$. If $r \in \mathcal{R}_1(\mathcal{M}) \setminus \mathcal{R}_2(\mathcal{M})$, $a_{r,1}$ is the k-th power of a prime of the desired kind.

Lemma 4.2 *We have*

$$\#\mathcal{R}_1(\mathcal{M}) \gg_{\alpha} \frac{P(x)^D}{\Phi(P(x)) \log(P(x)^D)}.$$

Proof We apply Lemma 2.3 with $P(x)^D$ instead of x, $q = P(x)$ and Lemma 3.1 with $q = P(x)$. We choose D, such that the conditions (2.1) resp. (3.1) from Lemma 2.3 resp. Lemma 3.1 are satisfied.

Let

$$\mathcal{R}_0(\mathcal{M}) := \left\{r : 1 \leq r \leq P(x)^{D-1},\ l(r) \text{ prime}\right\}.$$

Then by Lemma 2.3

$$\mathcal{R}_0(\mathcal{M}) \gg \frac{P(x)^D}{\Phi(P(x))\log(P(x)^D)}.$$

By Lemma 3.1

$$\frac{\#\mathcal{R}_1(\mathcal{M})}{\#\mathcal{R}_0(\mathcal{M})} = \alpha^{-1}(1 + o(1)).$$

This proves Lemma 4.2. □

We obtain an upper estimate for $\mathcal{R}_2(\mathcal{M})$ by the observation that, if $R(r)$ contains a prime number, then $l(r)$ and $l(r)^k + v$ are primes for some $v \in V$. The condition $l(r) \in \mathscr{B}_{\alpha,\beta}$ can be discarded.

The number

$$t(v) := \#\{r \;:\; 1 \le r \le P(x)^{D-1},\ l(r) \text{ and } l(r)^k + v - 1 \text{ are primes}\}$$

is estimated by standard sieves as in Lemma 6.1 in [7].

This concludes the proof of our Theorem.

Open Problem Are there analogous results for k-th powers of other types of prime numbers, for example, Piatetski-Shapiro prime numbers or prime numbers with missing digits.

Acknowledgments M. Th. Rassias: I would like to express my gratitude to the John S. Latsis Public Benefit Foundation for their financial support provided under the auspices of my current "Latsis Foundation Senior Fellowship" position.

References

1. W.D. Banks, I.E. Shparlinski, Prime numbers with Beatty sequences. Colloq. Math. **15**(2), 147–157 (2009)
2. P. Erdős, On the difference of consecutive primes. Quart. J. Math. Oxford Ser. **6**, 124–128 (1935)
3. K. Ford, D.R. Heath-Brown, S. Konyagin, Large gaps between consecutive prime numbers containing perfect powers, in *Analytic Number Theory*. In honor of Helmut Maier's 60th birthday (Springer, New York, 2015)
4. K. Ford, B.J. Green, S. Konyagin, T. Tao, Large gaps between consecutive prime numbers. Ann. Math. **183**, 935–974 (2016)
5. K. Ford, B.J. Green, S. Konyagin, J. Maynard, T. Tao, Long gaps between primes. J. Amer. Math. Soc. **31**(1), 65–105 (2018)
6. P.X. Gallagher, A large sieve density estimate near $\sigma = 1$. Invent. Math. **11**, 329–339 (1970)
7. H. Halberstam, H.-E. Richert, *Sieve Methods* (Academic Press, London, 1974)
8. A.Y. Khinchin, Zur metrischen Theorie der diophantischen approximationen. Math. Z. **24**(4), 706–714 (1926)
9. H. Maier, Chains of large gaps between consecutive primes. Adv. Math. **39**, 257–269 (1981)

10. H. Maier, C. Pomerance, Unusually large gaps between consecutive primes. Trans. Amer. Math. Soc. **322**(1), 201–237 (1990)
11. H. Maier, M.Th. Rassias, Large gaps between consecutive prime numbers containing square-free numbers and perfect powers of prime numbers. Proc. Amer. Math. Soc. **144**, 3347–3354 (2016)
12. H. Maier, M.Th. Rassias, Large gaps between consecutive prime numbers containing perfect k-th powers of prime numbers. J. Funct. Anal. **272**, 2659–2696 (2017)
13. J. Maynard, Large gaps between primes. Ann. Math. **183**, 915–933 (2016)
14. R.A. Rankin, The difference between consecutive prime numbers. J. London Math. Soc. **13**, 242–247 (1938)
15. R.A. Rankin, The difference between consecutive prime numbers, V. Proc. Edinburgh Math. Soc. **13**(2), 331–332 (1962/1963)
16. K.F. Roth, Rational approximations to algebraic numbers. Mathematatika **2**, 1–20 (1955)

A Survey of Hypergraph Ramsey Problems

Dhruv Mubayi and Andrew Suk

Abstract The classical *hypergraph Ramsey number* $r_k(s, n)$ is the minimum N such that for every red-blue coloring of the k-tuples of $\{1, \ldots, N\}$, there are s integers such that every k-tuple among them is red, or n integers such that every k-tuple among them is blue. We survey a variety of problems and results in hypergraph Ramsey theory that have grown out of understanding the quantitative aspects of $r_k(s, n)$. Our focus is on recent developments and open problems.

1 Introduction

A k-uniform hypergraph H (k-graph for short) with vertex set V is a collection of k-element subsets of V. We write $K_n^{(k)}$ for the complete k-graph on an n-element vertex set. The *Ramsey number* $r_k(s, n)$ is the minimum N such that every red-blue coloring of the edges of $K_N^{(k)}$ contains a monochromatic red copy of $K_s^{(k)}$ or a monochromatic blue copy of $K_n^{(k)}$. The existence of $r_k(s, n)$ follows from the celebrated theorem of Frank Ramsey from 1930 [88]. However, the asymptotic behavior of $r_k(s, n)$ is still not well understood.

In this survey we focus on open problems and results related to generalizations and extensions of $r_k(s, n)$ in the *hypergraph* case, i.e., when $k \geq 3$ (we refer the reader to [27] for a survey of Graph Ramsey theory). Our emphasis is on recent results and although we believe we have touched on most important developments in this area, this survey is not an exhaustive compendium of all work in hypergraph Ramsey theory.

D. Mubayi
Department of Mathematics, Statistics, and Computer Science, University of Illinois, Chicago, IL, USA
e-mail: mubayi@uic.edu

A. Suk (✉)
Department of Mathematics, University of California at San Diego, La Jolla, CA, USA
e-mail: asuk@ucsd.edu

A. M. Raigorodskii, M. Th. Rassias (eds.), *Discrete Mathematics and Applications*, Springer Optimization and Its Applications 165,
https://doi.org/10.1007/978-3-030-55857-4_16

2 General Notation

The full statement of Ramsey's theorem extends to multiple colors and to general hypergraphs as follows. Given an integer $q \geq 2$ and k-uniform hyergraphs $H_1, \ldots, H_q$, there is a minimum $r_k(H_1, \ldots, H_q) = N$, such that every q-coloring of the edges of $K_N^{(k)}$ contains H_i as a subgraph in the ith color for some i. In the special case that $H = H_1 = \cdots = H_q$, we simply write

$$r_k(H; q) = r_k(\underbrace{H, \ldots, H}_{q\text{times}}).$$

If $H_i = K_{n_i}^{(k)}$, we use the simpler notation $r_k(n_1, \ldots, n_q)$ and $r_k(n; q) = r_k(\underbrace{n, \ldots, n}_{q\text{times}})$.

3 Diagonal Ramsey Numbers

Diagonal Ramsey numbers refer to the special case when $s = n$, i.e. $r_k(n, n)$, and have been studied extensively over the past 80 years. Classic results of Erdős and Szekeres [48] and Erdős [39] imply that $2^{n/2} < r_2(n, n) \leq 2^{2n}$ for every integer $n > 2$. While small improvements have been made in both the upper and lower bounds for $r_2(n, n)$ (see [15, 91]), the constant factors in the exponents have not changed over the last 70 years.

Unfortunately for 3-graphs, our understanding of $r_3(n, n)$ is much less than in the graph case. A result of Erdős, Hajnal, and Rado [50] gives the best known lower and upper bounds for $r_3(n, n)$,

$$2^{c_1 n^2} < r_3(n, n) < 2^{2^{c_2 n}},$$

where c_1 and c_2 are absolute constants. Another proof of the lower bound above was given by Conlon, Fox, and Sudakov in [16], which will be discussed in more detail in Section 4. For $k \geq 4$, there is also a difference of one exponential between the known lower and upper bounds for $r_k(n, n)$, that is,

$$\mathrm{twr}_{k-1}(c_1 n^2) \leq r_k(n, n) \leq \mathrm{twr}_k(c_2 n), \tag{1}$$

where the *tower function* $\mathrm{twr}_k(x)$ is defined by $\mathrm{twr}_1(x) = x$ and $\mathrm{twr}_{i+1}(x) = 2^{\mathrm{twr}_i(x)}$ (see [45, 46, 48]). A notoriously difficult conjecture of Erdős, Hajnal, and Rado states that the upper bound in (1) is essentially the truth, that is, there are constructions which demonstrate that $r_k(n, n) > \mathrm{twr}_k(cn)$, where $c = c(k)$. The crucial case is when $k = 3$, since a double exponential lower bound for $r_3(n, n)$

would verify the conjecture for all $k \geq 4$ by using the well-known stepping-up lemma of Erdős and Hajnal (see [54]).

Conjecture 3.1 (Erdős) *For $n \geq 4$ we have $r_3(n, n) > 2^{2^{cn}}$, where c is an absolute constant.*

It is worth mentioning that Erdős offered a \$500 reward for a proof of this conjecture (see [12]), and his conjecture is supported by the fact that a double exponential lower bound is known if one allows four colors. More precisely, Erdős and Hajnal (see [54]) showed that $r_3(n; 4) > 2^{2^{cn}}$, and for three colors, the best known lower bound for $r_3(n; 3)$ is due to Conlon, Fox, and Sudakov [16] who showed that $r_3(n; 3) > 2^{n^{c \log n}}$. There is some evidence that perhaps Conjecture 3.1 is false, and we refer the interested reader to [17, 23] for two results in this direction.

4 Off-Diagonal Ramsey Numbers

Off-diagonal Ramsey numbers, $r_k(s, n)$, refer to the special case when k, s are fixed and n tends to infinity. It is known [1, 5, 6, 60] that $r_2(3, n) = \Theta(n^2/\log n)$, and more generally for fixed $s > 3$, $r_2(s, n) = n^{\Theta(1)}$. For 3-graphs, Conlon et al. [16] proved that there are absolute constants $c, c' > 0$ such that for all $4 \leq s \leq n$,

$$2^{csn \log(\frac{n}{s}+1)} < r_3(s, n) < 2^{(c'n/s)^{s-2} \log(n/s)}.$$

For $s = n$, this gives another proof that $r_3(n, n) > 2^{cn^2}$.

For k-graphs, where $s > k \geq 4$, it is known that $r_k(s, n) \leq \text{twr}_{k-1}(n^c)$, where $c = c(s)$ [46]. Erdős and Hajnal proved that

$$r_k(s, n) > \text{twr}_{k-1}(c'n), \tag{2}$$

for $k \geq 4$ and $s \geq 2^{k-1} - k + 3$, where $c' = c'(s)$. They conjectured that a similar bound should hold for smaller s as follows.

Conjecture 4.1 (Erdős–Hajnal [45]) *Fix $4 \leq k < s$. There are constants c and c' such that*

$$\text{twr}_{k-1}(cn) < r_k(s, n) < \text{twr}_{k-1}(c'n).$$

Actually, this was part of a more general conjecture that they posed in that paper which will be discussed in Section 4. Erdős and Hajnal (see [54]) showed that $r_4(7, n) > 2^{2^{cn}}$, and the authors [83] and Conlon et al. [24] independently verified the conjecture for $k \geq 4$ and $s \geq k + 3$ (using different constructions). However, showing that $r_4(5, n)$ and $r_4(6, n)$ grows double exponentially in a power of n seemed to be much more difficult.

Just as for diagonal Ramsey numbers, a double exponential in n^c lower bound for $r_4(5, n)$ and $r_4(6, n)$ would imply $r_k(k+1, n) > \mathrm{twr}_{k-1}(n^{c'})$ and $r_k(k+2, n) > \mathrm{twr}_{k-1}(n^{c'})$, respectively, for all fixed $k \geq 5$, by a variant of the Erdős–Hajnal stepping-up lemma. In [81], the authors established the following lower bounds for $r_4(5, n)$ and $r_4(6, n)$, which represents the current best bounds: for all $n \geq 6$,

$$r_4(5, n) > 2^{n^{c \log n}} \quad \text{and} \quad r_4(6, n) > 2^{2^{cn^{1/5}}},$$

where $c > 0$ is an absolute constant. More generally, for $n > k \geq 5$, there is a $c = c(k) > 0$ such that

$$r_k(k+1, n) > \mathrm{twr}_{k-2}(n^{c \log n}) \quad \text{and} \quad r_k(k+2, n) > \mathrm{twr}_{k-1}(cn^{1/5}).$$

A standard argument in Ramsey theory together with results in [16] for 3-graphs yields

$$r_k(k+2, n) < \mathrm{twr}_{k-1}(c'n^3 \log n),$$

so we now know the tower growth rate of $r_k(k+2, n)$. It remains an open problem to prove that $r_4(5, n)$ is double exponential in a power of n.

Conjecture 4.2 *For $n \geq 5$, there is an absolute constant $c > 0$ such that $r_4(5, n) > 2^{2^{n^c}}$.*

In [83], the authors established a connection between diagonal and off-diagonal Ramsey numbers, by showing that a solution to Conjecture 3.1 implies a solution to Conjecture 4.2 (see Section 10 for more details).

5 The Erdős–Hajnal Problem

As mentioned in previous sections, it is a major open problem to determine if $r_3(n, n)$ and $r_4(5, n)$ grow double exponentially in a power of n. In order to shed more light on these questions, Erdős and Hajnal [45] in 1972 considered the following more general parameter.

Definition 5.1 For integers $2 \leq k < s < n$ and $2 \leq t \leq \binom{s}{k}$, let $r_k(s, t; n)$ be the minimum N such that every red/blue coloring of the edges of $K_N^{(k)}$ results in a monochromatic blue copy of $K_n^{(k)}$ or has a set of s vertices which contains at least t red edges.

The function $r_k(s, t; n)$ encompasses several fundamental problems which have been studied for a while. Clearly $r_k(s, n) = r_k(s, \binom{s}{k}; n)$ so $r_k(s, t; n)$ includes classical Ramsey numbers. In addition to off-diagonal and diagonal Ramsey numbers already mentioned, the function $r_k(k+1, k+1; k+1)$ has been studied in the context of the Erdős–Szekeres theorem and Ramsey numbers of ordered tight-paths by several researchers [36, 38, 53, 71, 72], the more general function $r_k(k+1, k+1; n)$ is related to high dimensional tournaments [67], and even the very special case $r_3(4, 3; n)$ has tight connections to quasi-random hypergraph constructions [4, 61, 65, 66].

The main conjecture of Erdős and Hajnal [45] for $r_k(s, t; n)$ is that, as t grows from 1 to $\binom{s}{k}$, there is a well-defined value $t_1 = h_1^{(k)}(s)$ at which $r_k(s, t_1 - 1; n)$ is polynomial in n, while $r_k(s, t_1; n)$ is exponential in a power of n, another well-defined value $t_2 = h_2^{(k)}(s)$ at which it changes from exponential to double exponential in a power of n and so on, and finally a well-defined value $t_{k-2} = h_{k-2}^{(k)}(s) < \binom{s}{k}$ at which it changes from twr_{k-2} to twr_{k-1} in a power of n. They were not able to offer a conjecture as to what $h_i^{(k)}(s)$ is in general, except when $i = 1$ and when $s = k + 1$.

- When $i = 1$, Erdős and Hajnal made the following conjecture. Let $g_k(s)$ be the number of edges in the k-graph obtained by taking a complete k-partite k-graph on s vertices with almost equal part sizes, and repeating this construction recursively within each part. They proved that $h_1^{(k)}(s) \geq g_k(s) + 1$ and conjectured this to be tight. Erdős even offered a $500 reward for a proof or disproof of this conjecture (see [12]). For $k = 3$, this was settled for many values of s, including powers of three, by Conlon, Fox, and Sudakov [16]. Very recently, Mubayi and Razborov proved it for all $s \geq k \geq 4$.

Theorem 5.2 (Mubayi-Razborov [79]) *Fix $s \geq k \geq 4$. Then $h_1^{(k)}(s) = g_k(s) + 1$.*

- When $s = k + 1$, Erdős and Hajnal conjectured that $h_i^{(k)}(k+1) = i + 2$, that is, $r_k(k+1, 2; n)$ is polynomial in n, $r_k(k+1, 3; n)$ is exponential in a power of n, $r_k(k+1, 4; n)$ is double exponential in a power of n, etc. such that at the end, both $r_k(k+1, k; n)$ and $r_k(k+1, k+1; n)$ are twr_{k-1} in a power of n. They proved this for $i = 1$ via the following:

Theorem 5.3 (Erdos–Hajnal [45]) *For $k \geq 3$, there are positive $c = c(k)$ and $c' = c'(k)$ such that*

$$r_k(k+1, 2; n) < cn^{k-1} \qquad \text{and} \qquad r_k(k+1, 3; n) > 2^{c'n}.$$

Results of Rödl–Šinajová [89] on partial Steiner systems, and of Kostochka–Mubayi–Verstraëte [64] on independent sets in hypergraphs, determine the order of magnitude of the function $r_k(k+1, 2; n)$ as follows. For each $k \geq 3$ there exist positive $c = c_k$ and $c' = c'(k)$ such that

$$c'n^{k-1}/\log n < r_k(k+1, 2; n) < c\,n^{k-1}/\log n.$$

For the $t = 3$ case, the authors in [84] showed that for $k \geq 3$, there are positive $c = c(k)$ and $c' = c'(k)$ such that

$$2^{cn^{k-2}} \leq r_k(k+1, 3; n) \leq 2^{c'n^{k-2}\log n}. \quad (3)$$

For general t, the methods of Erdős and Rado [46] show that there exists $c = c(k, t) > 0$ such that

$$r_k(k+1, t; n) \leq \mathrm{twr}_{t-1}(n^c), \quad (4)$$

for $3 \leq t \leq k$. Erdős and Hajnal conjectured that this upper bound is the correct tower growth rate for $r_k(k+1, t; n)$.

Conjecture 5.4 (Erdos–Hajnal [45]) *For $k \geq 3$ and $2 \leq t \leq k$, there exists $c = c(k, t) > 0$ such that*

$$r_k(k+1, t; n) \geq \mathrm{twr}_{t-1}(cn).$$

Note that when $t = k+1$, the results from the previous section yield $r_k(k+1, k+1; n) = r_k(k+1, n) \leq \mathrm{twr}_{k-1}(n^{c'})$ where $c' = c'(k, t)$.

Hence for 3-graphs, $r_3(4, t; n)$ is fairly well understood. We know that $r_3(4, 2; n)$ is polynomial in n, and both $r_3(4, 3; n)$ and $r_3(4, 4; n)$ are exponential in $n^{1+o(1)}$. See [16] for more results on $h_1^{(3)}(s)$ for $s > 4$. Unfortunately for 4-graphs, we do not have a good understanding of $r_4(5, t; n)$ when $4 \leq t \leq 5$. The best known upper and lower bounds for $r_4(5, 4; n)$ are obtained by (3) and (4), which give $2^{cn^2} < r_4(5, 4; n) < 2^{2^{n^c}}$. Notice that Conjecture 5.4 states that $r_4(5, 4; n)$ grows double exponential in a power of n, but we do not even know if $r_4(5, 5; n) = r_4(5, n)$ is double exponential in a power of n. Likewise, for 5-graphs, not much is known about $r_5(6, 4; n)$ and $r_5(6, 5; n)$. Combining (3) and (4) gives

$$2^{c'n^3} < r_5(6, 4; n) < 2^{2^{n^c}} \quad \text{and} \quad 2^{c'n^3} < r_5(6, 5; n) < 2^{2^{2^{n^c}}}. \quad (5)$$

Problem 5.5 Determine the tower growth rate of $r_4(5, 4; n)$, $r_5(6, 4; n)$, and $r_5(6, 5; n)$.

However for k-graphs, when $k \geq 6$, the authors in [84] settled Conjecture 5.4 in almost all cases in a strong form, by determining the correct tower growth rate, and in half of the cases also determining the correct power of n within the tower.

Theorem 5.6 (Mubayi–Suk [84]) *For $k \geq 6$ and $4 \leq t \leq k-2$, there are positive $c = c(k, t)$ and $c' = c'(k, t)$ such that*

$$\mathrm{twr}_{t-1}(c'n^{k-t+1}\log n) \geq r_k(k+1, t; n) \geq \begin{cases} \mathrm{twr}_{t-1}(cn^{k-t+1}) & \text{if } k-t \text{ is even} \\ \mathrm{twr}_{t-1}(cn^{(k-t+1)/2}) & \text{if } k-t \text{ is odd.} \end{cases}$$

When $k \geq 6$ and $t \in \{k-1, k\}$, Conjecture 5.4 remains open. We note that the upper bound in Theorem 5.6 also holds when $k-1 \leq t \leq k$. The best known upper and lower bounds for $r_k(k+1, k-1; n)$ and $r_k(k+1, k; n)$, also due to the authors [84], are

$$\mathrm{twr}_{k-3}(c\,n^3) \le r_k(k+1, k-1;\ n) \le \mathrm{twr}_{k-2}(c'\,n^2),$$

and

$$\mathrm{twr}_{k-3}(c\,n^3) \le r_k(k+1, k;\ n) \le \mathrm{twr}_{k-1}(c'\,n).$$

In fact, by using the stepping-up lemma established in [84], any improvement in the lower bound for $r_5(6, 4; n)$ and $r_5(6, 5; n)$ in (5) would imply a better lower bound for $r_k(k+1, k-1;\ n)$ and $r_k(k+1, k;\ n)$, respectively.

6 The Erdős–Rogers Problem

An s-independent set in a k-graph H is a vertex subset that contains no copy of $K_s^{(k)}$. So if $s = k$, then it is just an independent set. Let $\alpha_s(H)$ denote the size of the largest s-independent set in H.

Definition 6.1 For $k \le s < t < N$, the Erdős–Rogers function $f_{s,t}^k(N)$ is the minimum of $\alpha_s(H)$ taken over all $K_t^{(k)}$-free k-graphs H of order N.

To prove the lower bound $f_{s,t}^k(N) \ge n$, one must show that every $K_t^{(k)}$-free k-graph on N vertices contains an s-independent set with n vertices. On the other hand, to prove the upper bound $f_{s,t}^{(k)}(N) < n$, one must construct a $K_t^{(k)}$-free k-graph H of order N with $\alpha_s(H) < n$.

The problem of determining $f_{s,t}^k(n)$ extends that of finding Ramsey numbers. Formally,

$$r_k(s, n) = \min\{N : f_{k,s}^k(N) \ge n\}.$$

For $k = 2$, the above function was first considered by Erdős and Rogers [47] only for $t = s+1$, which is perhaps the most interesting case. So in this case we wish to construct a K_{s+1}-free graph on N vertices such that the s-independence number is as small as possible. Since then the function has been studied by several researchers culminating in the work of Wolfowitz [94] and Dudek et al. [35] who proved the upper bound that follows (the lower bound is due to Dudek and the first author [34]): for every $s \ge 3$ there are positive constants c_1 and $c_2 = c_2(s)$ such that

$$c_1 \left(\frac{N \log N}{\log\log N}\right)^{1/2} < f_{s,s+1}^2(N) < c_2(\log N)^{4s^2} N^{1/2}. \tag{6}$$

The problem of estimating the Erdős–Rogers function for $k > 2$ appears to be much harder. Let us denote

$$g(k, N) = f^k_{k+1,k+2}(N).$$

In other words, $g(k, N)$ is the minimum n such that every $K^{(k)}_{k+2}$-free k-graph on N vertices has the property that every n-set of vertices has a copy of $K^{(k)}_{k+1}$. With this notation, the bounds in (6) for $s = 3$ imply that $g(2, N) = N^{1/2+o(1)}$.

Dudek and the first author [34] proved that $(\log N)^{1/4+o(1)} < g(3, N) < O(\log N)$, and more generally, there are positive $c_1 = c_1(k)$ and $c_2 = c_2(k)$ with

$$c_1(\log_{(k-2)} N)^{1/4} < g(k, N) < c_2(\log N)^{1/(k-2)}, \tag{7}$$

where $\log_{(i)}$ is the log function iterated i times. The exponent 1/4 in (7) was improved to 1/3 by Conlon et al. [26]. Both sets of authors asked whether the upper bound could be improved (presumably to an iterated log function). This was achieved by the current authors [82] who proved that for $k \geq 14$,

$$g(k, N) = O(\log_{(k-13)} N).$$

It remains an open problem to determine the correct number of iterations (which may well be $k - 2$). We pose this as a conjecture.

Conjecture 6.2 *For all $k \geq 3$, there are $c_1, c_2 > 0$ such that*

$$c_1 \log_{(k-2)} N < g(k, N) < c_2 \log_{(k-2)} N.$$

7 The Erdős–Gyárfás–Shelah Problem

A (p, q)-coloring of $K^{(k)}_N$ is an edge-coloring of $K^{(k)}_N$ that gives every copy of $K^{(k)}_p$ at least q colors. Let $f_k(N, p, q)$ be the minimum number of colors in a (p, q)-coloring of K^k_N.

The problem of determining $f_k(N, p, q)$ for fixed k, p, q has a long history, beginning with its introduction by Erdős and Shelah [40, 41], and subsequent investigation (for graphs) by Erdős and Gyárfás [44]. Since

$$f_k(N, p, 2) = t \quad \Longleftrightarrow \quad r_k(p; t) \geq N + 1 \quad \text{and} \quad r_k(p; t - 1) \leq N,$$

most of the effort on determining $f_k(N, p, q)$ has been for $q > 2$. As mentioned above, Erdős and Gyárfás [44] initiated a systematic study of this parameter for graphs and posed many open problems. One main question was to determine the minimum q such that $f_2(N, p, q) = N^{o(1)}$ and $f_2(N, p, q + 1) > N^{c_p}$ for some

$c_p > 0$. For $p = 3$ this value is clearly $q = 2$ as $f_2(N, 3, 2) = O(\log N)$ due to the easy bound $r_2(3; t) > 2^t$, while $f_2(N, 3, 3) = \chi'(K_N) \geq N - 1$. Erdős and Gyárfás proved that $f_2(N, p, p) > N^{c_p}$ and asked whether $f_2(N, p, p-1) = N^{o(1)}$. The first open case was $f_2(N, 4, 3)$, which was shown to be $N^{o(1)}$ by the first author [74] and later $\Omega(\log N)$ (see [52, 62]). The same upper bound was shown for $f(N, 5, 4)$ in [37]. Conlon et al. [21] recently extended this construction considerably by proving that $f_2(N, p, p-1) = N^{o(1)}$ for all fixed $p \geq 4$. Their result is sharp in the sense that $f_2(N, p, p) = \Omega(N^{1/(p-2)})$. The exponent $1/(p-2)$ was shown to be sharp for $p = 4$ by the first author [75] and recently also for $p = 5$ by Cameron and Heath [10] via explicit constructions.

The first nontrivial hypergraph case is $f_3(N, 4, 3)$ and this function has tight connections to Shelah's breakthrough proof [90] of primitive recursive bounds for the Hales–Jewett numbers. Answering a question of Graham, Rothschild, and Spencer [54], Conlon, Fox, Lee, and Sudakov showed that

$$f_3(N, 4, 3) = N^{o(1)}.$$

They also posed a variety of basic questions about $f_k(N, p, q)$ and related parameters including the following generalization of the Erdős–Gyárfás problem for hypergraphs. Using a variant of the pigeonhole argument for hypergraph Ramsey numbers due to Erdős and Rado, [20] proved that

$$f_k\left(N, p, \binom{p-i}{k-i} + 1\right) = \Omega(\log_{(i-1)} N^{c_{p,k,i}}),$$

where $\log_{(0)}(x) = x$ and, as usual, $\log_{(i+1)} x = \log \log_{(i)} x$ for $i \geq 0$.

Problem 7.1 (Conlon–Fox–Lee–Sudakov [20]) For $p > k \geq 3$ and $0 < i < k$ prove that $f_k(N, p, \binom{p-i}{k-i})$ is substantially smaller than $f_k(N, p, \binom{p-i}{k-i} + 1)$, in particular, prove that $f_k(N, p, \binom{p-i}{k-i})$ is much smaller than $\log_{(i-1)} N$.

One natural way to interpret this problem is that it asks whether

$$f_k\left(N, p, \binom{p-i}{k-i}\right) = (\log_{(i-1)} N)^{o(1)}?$$

The case $k = 2$ is precisely the Erdős–Gyárfás problem and the case $k = 3, p = 4, i = 1$ is to prove that $f_3(N, 4, 3) = N^{o(1)}$ which was established in [20]. The next open case is $k = 3, p = 5, i = 2$, which asks whether $f_3(N, 5, 3) = (\log N)^{o(1)}$. This was solved with a better bound by the first author [76], who showed that

$$f_3(N, 5, 3) = e^{O(\sqrt{\log \log N})} = (\log N)^{O(1/\sqrt{\log \log N})}.$$

No other nontrivial cases of Problem 7.1 have been solved. We refer the reader to [20] for related problems and results.

8 More Off-diagonal Problems

In this section we consider k-graph Ramsey numbers of the form $r_k(H, n) := r_k(H, K_n^{(k)})$ where H is a (fixed) k-graph and n grows.

8.1 $K_4^{(3)}$ Minus an Edge and a Generalization

Let $K_4^{(3)} \backslash e$ denote the 3-graph on four vertices, obtained by removing one edge from $K_4^{(3)}$. A simple argument of Erdős and Hajnal [45] implies $r(K_4^{(3)} \backslash e, K_n^{(3)}) < (n!)^2$. This was generalized in [82] as follows. A *k-half-graph*, denote by $B = B^{(k)}$, is a k-graph on $2k - 2$ vertices, whose vertex set is of the form $S \cup T$, where $|S| = |T| = k - 1$, and whose edges are all k-subsets that contain S, and one k-subset that contains T. So $B^{(3)} = K_4^{(3)} \setminus e$. Write $r_k(B, n) = r(B^{(k)}, K_n^{(k)})$. It was shown in [82] that for each $k \geq 4$ there exists $c = c_k$ such that

$$2^{cn} < r_k(B, n) < (n!)^{k-1}.$$

A problem that goes back to the 1972 paper of Erdős and Hajnal (for $k = 3$) is to improve the lower bound above. Indeed, $r_3(B, n) = r_3(4, 3; n)$ and this is therefore a very special case of the Erdős–Hajnal problem discussed earlier.

Problem 8.1 Show that for each $k \geq 3$ there exists $c = c_k$ such that $r_k(B, n) > 2^{cn \log n}$.

8.2 Independent Neighborhoods

Definition 8.2 A k-uniform triangle $T^{(k)}$ is a set of $k + 1$ edges $b_1, \ldots, b_k, a$ with $b_i \cap b_j = R$ for all $i < j$ where $|R| = k - 1$ and $a = \cup_i (b_i - R)$. In other words, k of the edges share a common $(k - 1)$-set of vertices, and the last edge contains the remaining point in all these previous edges.

When $k = 2$, then $T^{(2)} = K_3$, so in this sense $T^{(k)}$ is a generalization of a graph triangle. We may view a $T^{(k)}$-free k-graph as one in which all neighborhoods are independent sets, where the neighborhood of an $R \in \binom{V(H)}{k-1}$ is $\{x : R \cup \{x\} \in H\}$. As usual, write $r_k(T, n)$ for $r(T^{(k)}, K_n^{(k)})$.

Bohman et al. [8] proved that for fixed $k \geq 2$, there are positive constants c_1 and c_2 with

$$c_1 \frac{n^k}{(\log n)^{k/(k-1)}} < r_k(T, n) < c_2 n^k.$$

They conjectured that the upper bound could be improved to $o(n^k)$ and believed that the log factor in the lower bound could also be improved. Results of Kostochka–Mubayi–Verstraëte [64] proved this and then Bohman–Mubayi–Picollelli [9] achieved a matching lower bound by analyzing the hypergraph independent neighborhood process. This may be viewed as a hypergraph generalization of the results of Ajtai–Komlós–Szemerédi [1] for graphs.

Theorem 8.3 (Kostochka–Mubayi–Verstraëte [64], Bohman–Mubayi–Picollelli [9]) *For fixed $k \geq 3$ there are positive constants c_1 and c_2 with*

$$c_1 \frac{n^k}{\log n} < r_k(T, n) < c_2 \frac{n^k}{\log n}.$$

8.2.1 Unordered Tight-Paths Versus Cliques

An (unordered) 3-uniform tight-path $TP_s^{(3)} = TP_s$ is the 3-graph with vertex set $\{v_1, \ldots, v_s\}$ and edge set $\{\{v_i, v_{i+1}, v_{i+2}\} : i \in \{1, \ldots, s-2\}\}$. Note that the vertex set $\{v_1, \ldots, v_s\}$ is not ordered. Results of Phelps and Rödl [86] imply that there are c_1 and c_2 such that

$$c_1 n^2/\log n < r_3(TP_4, n) < c_2 n^2/\log n.$$

It is easy to prove that for all $s \geq 5$, there is $c = c_s$ such that $r_3(TP_s, n) < cn^2$. A matching lower bound for $s \geq 6$ was provided by Cooper and Mubayi [31] with the following construction. Let H be a 3-graph where $V(H) = [n] \times [n]$, and $E(H) = \{\{ab, ac, db\} \in [n] \times [n] : c > b, d > a\}$. It is easy to see that H is TP_6-free and $\alpha(H) < 2n$. Thus, For $s \geq 6$ there exists $c = c_s$ such that $r_3(TP_s, n) > cn^2$. The construction above has many copies of TP_5 so this leaves open the case $s = 5$. Using the trivial lower bound $r_3(TP_4, n)$ we thus have $c_1 n^2/\log n < r_3(TP_5, n) < c_2 n^2$.

Problem 8.4 ([31]) Determine the order of magnitude of $r_3(TP_5, n)$.

The corresponding problems for k-graphs when $k > 3$ are wide open.

8.3 Cycles Versus Cliques

For fixed $s \geq 3$ the graph Ramsey number $r(C_s, n) = r(C_s, K_n)$ has been extensively studied. The case $s = 3$ is one of the oldest questions in Ramsey theory and it is known that $r(C_3, K_n) = \Theta(n^2/\log n)$ (see [1, 60] and [7, 51] for recent improvements). The next case $r(C_4, K_n)$ seems substantially more difficult. An old open problem of Erdős [42] asks whether there is a positive ϵ for which $r(C_4, K_n) = O(n^{2-\epsilon})$. The current best upper bound $r(C_4, K_n) = O(n^2/\log^2 n)$ is an unpublished result of Szemerédi which was reproved in [11] and the current best lower bound is $\Omega(n^{3/2}/\log n)$ from [6]. For longer cycles, the best known bounds can be found in [6, 92], and the order of magnitude of $r(C_s, K_n)$ is not known for any fixed $s \geq 4$.

There are several natural ways to define a cycle in hypergraphs. The two that have been investigated the most are tight cycles and loose cycles.

8.3.1 Loose Cycles Versus Cliques

For $s \geq 3$, the loose cycle $LC_s^{(k)}$ is the k-graph with vertex set $\mathbb{Z}_{(k-1)s}$ and edge set $\{e_1, e_2, \ldots, e_s\}$ where $e_i = \{i(k-1) - k + 2, \ldots, i(k-1) + 1\}$. In other words, consecutive edges intersect in exactly one vertex and non-consecutive edges are pairwise disjoint. As usual, write $r_k(LC_s, n)$ for $r(LC_s^{(k)}, K_n^{(k)})$. Since loose cycles are k-partite, it is easy to see that $r_k(LC_s, n)$ has polynomial growth rate for fixed k, s so the question here is to determine the correct power of n.

Theorem 8.5 (Kostochka–Mubayi–Verstraëte [63]) *There exists $c_1, c_2 > 0$ such that for all $n \geq 1$,*

$$\frac{c_1 n^{3/2}}{(\log n)^{3/4}} \leq r_3(LC_3, n) \leq c_2 n^{3/2}.$$

For $k \geq 3$, we also have $r_k(LC_3, n) = n^{3/2+o(1)}$.

Analogous to the basic result $r(3, n) = O(n^2/\log n)$ due to Ajtai et al. [1], the authors conjectured something similar for hypergraphs.

Conjecture 8.6 ([63]) *For all fixed $k \geq 3$, we have $r_k(LC_3, n) = o(n^{3/2})$.*

Define the 3-graph $F = \{abc, abd, cde\}$. Cooper and Mubayi [30] proved the following weaker version of Conjecture 8.6 in the case $k = 3$:

$$r_3(\{LC_3, F, K_4^{(3)} - e\}, n) = O\left(\frac{n^{3/2}}{(\log n)^{1/2}}\right). \tag{8}$$

Notice that the three forbidden 3-graphs in (8) are all types of triangles, comprising three edges that cyclically share a vertex. Conjecture 8.6 asks that we forbid only one of these three triangles, the loose triangle.

For longer cycles, the following general lower bounds were proved in [63] which improve the bounds given by the standard probabilistic deletion method:

$$r_k(LC_s, n) > n^{1+1/(3s-1)+o(1)}.$$

Furthermore, there exists $c = c_k$ such that

$$r_k(LC_5, n) > c\left(\frac{n}{\log n}\right)^{5/4}.$$

Conjecture 8.7 ([63]) *For each $k \geq 3$, there exists $c = c_k$ such that $r_k(LC_5, n) < cn^{5/4}$.*

Méroueh [70] recently proved that $r_3(LC_5, n) < cn^{4/3}$ and more generally that

$$r_3(LC_s, n) < c_s n^{1+\frac{1}{\lfloor (s+1)/2 \rfloor}}.$$

He also proved that for odd $s \geq 5$ and $k \geq 4$, $r_k(LC_s, n) < c_{k,s} n^{1+1/\lfloor s/2 \rfloor}$ which slightly improved the exponent $1 + 1/(\lfloor s/2 \rfloor - 1)$ proved by Collier-Cartaino et al. [14] for all $k \geq 3$ and $s \geq 4$.

8.3.2 Tight Cycles Versus Cliques

For $k \geq 2$ and $s > 3$, the *tight cycle* $TC_s^{(k)}$ is the k-graph with vertex set $\mathbb{Z}_s$ (integers modulo s) and edge set

$$\{\{i, i+1, \ldots, i+k-1\} : i \in \mathbb{Z}_s\}.$$

We can view the vertex set of $TC_s^{(k)}$ as s points on a circle and the edge set as the s subintervals each containing k consecutive vertices.

When $s \equiv 0 \pmod 3$ the tight cycle $TC_s^{(3)}$ is 3-partite, and in this case it is trivial to observe that $r_3(TC_s, n) := r(TC_s^{(3)}, K_n^{(3)})$ grows polynomially in n. The growth rate of this polynomial is not known for any $s > 3$. When $s \not\equiv 0 \pmod 3$ the Ramsey number is exponential in n.

Theorem 8.8 (Mubayi–Rödl [77, 80]) *Fix $s \geq 5$ and $s \not\equiv 0 \pmod 3$. There are positive constants c_1 and c_2 such that*

$$2^{c_1 n} < r_3(TC_s, n) < 2^{c_2 n^2 \log n}.$$

If $s \not\equiv 0 \pmod 3$ and $s \geq 16$ or $s \in \{8, 11, 14\}$, there is a positive constant c_s such that

$$r_3(TC_s, n) < 2^{c_s n \log n}.$$

Note that when $s = 4$, the cycle $TC_4^{(3)}$ is $K_4^{(3)}$ and in this case the lower bound was proved much earlier by Erdős and Hajnal [45], and in fact has been improved to $2^{c_1 n \log n}$ more recently by Conlon et al. [16].

Problem 8.9 Prove similar bounds for $s \in \{4, 5, 7, 10, 13\}$ and determine whether the log factor in the exponent is necessary.

The problem of determining $r_k(TC_s, n) := r_k(TC_s^{(k)}, K_t^{(k)})$ for fixed $s > k > 3$ seems harder as k grows. It was shown in [80] that we have a lower bound

$$r_k(TC_s, n) > 2^{c_{k,s}n^{k-2}}.$$

The best upper bound that is known (for fixed $s > k$ and all n) is the trivial one $r_k(s, n)$. Consequently, we have

$$2^{c_{k,s}n^{k-2}} < r_k(TC_s, n) < r_k(s, n) < \mathrm{twr}_{k-1}(n^{d_{s,k}}).$$

Closing the gap above seems to be a very interesting open problem. For the case $s = k + 1$, one has a substantially better lower bound as

$$r_k(TC_{k+1}, n) = r_k(k + 1, n) > \mathrm{twr}_{k-2}(bn^{\log n})$$

where $b = b_k$.

Problem 8.10 For fixed $s > k + 1 > 4$ (in particular for $s = k + 2$), determine whether $r_k(TC_s, n)$ is at least a tower function in a power of n where the tower height grows with k.

9 Bounded Degree Hypergraphs

In the special case that G is a sparse graph, much better upper bounds are known for the Ramsey number $r_2(G, G)$. A famous result due to Chvátal et al. [13] says that if G is a graph on n vertices with maximum degree Δ, then $r_2(G, G) \le c_\Delta n$ where c_Δ depends only on Δ. This was later extended to 3-graphs by Cooly et al. [28], Nagle et al. [85], and Ishigami [58] independently, where the degree of a vertex v in a hypergraph H is the number of edges which contain v. Using different methods, Cooly et al. [29] and Conlon et al. [25] extended this to general k-graphs.

Theorem 9.1 (Cooly et al. [29], Conlon–Fox–Sudakov [25]) *For all $\Delta, k \ge 1$, there is a $c(\Delta, k)$ such that for any k-graph H on n vertices with maximum degree Δ, we have*

$$r_k(H, H) \le c(\Delta, k)n.$$

In the special case that $H = LC_n^{(3)}$ or $TC_n^{(3)}$, the asymptotics for both $r_3(LC_n, LC_n)$ and $r_3(TC_n, TC_n)$ were determined in [56] and [57], respectively.

Theorem 9.2 (Haxell et al. [56]) $r_3(LC_n, LC_n) = \frac{5n}{2}(1 + o(1))$.

Theorem 9.3 (Haxell et al. [57]) *For n divisible by 3, we have $r_3(TC_n, TC_n) = \frac{4n}{3}(1+o(1))$. Otherwise, if n is not divisible by 3, we have $r_3(TC_n, TC_n) = 2n(1+o(1))$.*

For q-colors, Gyárfás and Raeisi [55] proved that

$$q + 5 \leq r_3(LC_3; q) \leq 3q + 1.$$

It seems to be an interesting open problem to determine which bound above is closer to the truth. The lower bound appears more likely to be the answer. There has been some work on determining $r_3(P; q)$ where $P = \{123, 345, 567\}$ is the 3-uniform loose path of length three. In particular, the lower bound $r_3(P; q) \geq q + 6$ for all $q \geq 3$ is sharp for all $q \leq 9$ (see [55, 87]). The general upper bound which comes from the Turán number of P is again $3q + 1$. This upper bound was improved by Luczak and Polcyn [68] first to $(2+o(1))q$ and more recently to $\lambda q + O(\sqrt{q})$ where the constant $\lambda = 1.97466..$ is the solution to a particular cubic equation.

10 Ordered Hypergraph Ramsey Problems

In this section, we discuss several Ramsey-type results for ordered hypergraphs. An *ordered N-vertex k-graph H* is a hypergraph whose vertex set is $[N] = \{1, \ldots, N\}$. Given two ordered k-graphs G and H with vertex set $[n]$ and $[N]$, respectively, we say that *H contains G* if there is a function $\phi : [n] \to [N]$ such that $\phi(i) < \phi(j)$ for all $1 \leq i < j \leq n$, and $(v_1, \ldots, v_k) \in E(G)$ implies that $(\phi(v_1), \ldots, \phi(v_k)) \in E(H)$. Given q ordered k-graphs $H_1, \ldots, H_q$, the *ordered Ramsey number $\overline{r}_k(H_1, \ldots, H_n)$* is the minimum integer N, such that every q-coloring of the edges of the complete k-graph with vertex set $[N]$ contains a copy of H_i in the ith color.

10.1 Tight-Paths and Cliques in Hypergraphs

An *ordered tight-path $P_s^{(k)}$* is an ordered k-graph with vertex set $[s]$, whose edges are of the form $(i, i+1, \ldots, i+k-1)$, for $1 \leq i \leq s-k+1$. The *length* of an ordered tight-path $P_s^{(k)}$ is the number of edges it contains, that is, $s - k + 1$. In order to avoid the excessive use of superscripts, we write $P_s = P_s^{(k)}$ when the uniformity is already implied. Two famous theorems of Erdős and Szekeres in [48], known as the monotone subsequence theorem and the cups-caps theorem, imply that $\overline{r}_2(P_s, P_n) = (n-1)(s-1) + 1$ and $\overline{r}_3(P_s, P_n) = \binom{n+s-4}{s-2} + 1$. In [53], Fox, Pach, Sudakov, and Suk extended their results to k-graphs and determined the correct tower growth rate for $\overline{r}_k(P_s, P_s)$. Their results gave a geometric application related to the Happy Ending Theorem.[1] A few years later, Moshkovitz and Shapira [72] sharpened the bounds for $\overline{r}_k(P_s, P_s)$ by determining an exact formula for $\overline{r}_3(P_s, \ldots, P_s) = \overline{r}_3(P_s; q)$ with q colors.

Theorem 10.1 (Moshkovitz–Shapira [72]) *Let $P_{q-1}(s)$ denote the number of $s \times \cdots \times s$ $(q-1)$-dimensional partitions with entries $\{0, 1, \ldots, s\}$. Then*

$$\overline{r}_3(P_s; q) = P_{q-1}(s) + 1.$$

Soon after, Milans–Stolee–West [71] obtained an exact formula for $\overline{r}_k(P_{s_1}, \ldots, P_{s_q})$ for all $k, q \geq 2$, and $s_i \geq k$ (see [32, 36] for some related results).

Theorem 10.2 (Milans–Stolee–West [71]) *Let $k, q \geq 2$, and $s_i > k$ for all $i \in [q]$. Let J_1 be the poset comprising disjoint chains $C_1, \ldots, C_q$, with $|C_i| = s_i - k$ for $i \in [q]$ and for $i \geq 1$, let J_{i+1} be the poset whose elements are the ideals (down sets) of J_i with order defined by containment. Then*

$$\overline{r}_k(P_{s_1}, \ldots, P_{s_q}) = |J_k| + 1.$$

Just as before, we will use the simpler notation $\overline{r}_k(P_s, n) = \overline{r}_k(P_s^{(k)}, K_n^{(k)})$, and note that there is only one ordered complete hypergraph $K_n^{(k)}$ up to isomorphism. Interestingly, the proof of the Erdős–Szekeres monotone subsequence theorem [48] (see also Dilworth's Theorem [33]) actually implies that $\overline{r}_2(P_s, n) = (n-1)(s-1)+1$. For $k \geq 3$, estimating $\overline{r}_k(P_s, n)$ appears to be more difficult. Clearly we have

$$\overline{r}_k(P_s, n) \leq r_k(s, n) \leq \mathrm{twr}_{k-1}(O(n^{s-2} \log n)). \tag{9}$$

In [83], the authors established the following connection between the ordered Ramsey number $\overline{r}_k(P_s, n)$ and the classical multi-color Ramsey number $r_k(n; q)$

Theorem 10.3 (Mubayi-Suk [83]) *Let $k \geq 2$ and $s \geq k+1$. Then for $q = s-k+1$, we have*

$$r_{k-1}(\lfloor n/q \rfloor; q) \leq \overline{r}_k(P_s, n) \leq r_{k-1}(n; q).$$

The upper bound in Theorem 10.3 follows from the following argument. Let $q = s - k + 1$, $N = r_{k-1}(n; q)$, and suppose χ is a red/blue coloring on the k-tuples of $[N]$. We can assume χ does not produce a red tight-path of length q, since otherwise we would have a red P_s and be done. We define the coloring $\phi : \binom{[N]}{k-1} \rightarrow \{0, 1, \ldots, q-1\}$ on the $(k-1)$-tuples of $[N]$, where $\phi(i_1, \ldots, i_{k-1}) = j$ if the longest red tight-path ending in vertices $(i_1, \ldots, i_{k-1})$ has length j. Since

[1] The main result in [48], known as the Happy Ending Theorem, states that for any positive integer n, any sufficiently large set of points in the plane in general position has a subset of n members that form the vertices of a convex polygon.

$N = r_{k-1}(n; q)$, by Ramsey's theorem, we have a monochromatic clique of size n in color j for some $j \in \{0, 1, \dots, q-1\}$. However, this clique would correspond to a blue clique with respect to χ. For the lower bound, set $N = r_{k-1}(\lfloor n/q \rfloor; q) - 1$, and let χ be a q coloring on the $(k-1)$-tuples of $[N]$ with colors $\{1, 2, \dots, q\}$, such that χ does not produce a monochromatic clique of size $\lfloor n/q \rfloor$. Then let $\phi : \binom{[N]}{k} \to \{\text{red, blue}\}$ such that for $i_1 < \cdots < i_k$, $\phi(i_1, \dots, i_k)$ is red if and only if $\chi(i_1, \dots, i_{k-1}) < \chi(i_2, \dots, i_k)$. It is easy to see that ϕ does not produce a red tight-path P_s. With a slightly more complicated argument, one can show by contradiction that ϕ also does not produce a monochromatic blue clique of size n.

The arguments above can be easily extended to obtain the following result for multiple colors [83]. Let $k \geq 2$ and $s_1, \dots, s_t \geq k+1$. Then for $q = (s_1 - k + 1) \cdots (s_t - k + 1)$, we have

$$r_{k-1}(\lfloor n/q \rfloor; q) \leq \overline{r}_k(P_{s_1}, \dots, P_{s_t}, n) \leq r_{k-1}(n; q).$$

Together with known bounds for $r_{k-1}(n; q)$, Theorem 10.3 has several consequences. First, we can considerably improve the upper bound for $\overline{r}_k(P_s, n)$ in (9) to $\overline{r}_k(P_s, n) \leq \mathrm{twr}_{k-1}(O(sn \log s))$. In the other direction, the authors in [83] showed that $\overline{r}_3(P_4, n) > 2^{cn}$, and Theorem 10.3 implies that for $k \geq 4$ and $n > 3k$,

1. $\overline{r}_k(P_{k+3}, n) \geq \mathrm{twr}_{k-1}(cn)$,
2. $\overline{r}_k(P_{k+2}, n) \geq \mathrm{twr}_{k-1}(c \log^2 n)$,
3. $\overline{r}_k(P_{k+1}, n) \geq \mathrm{twr}_{k-2}(cn^2)$.

We conjecture the following strengthening of the Erdős–Hajnal conjecture.

Conjecture 10.4 *For $k \geq 4$ fixed, $\overline{r}_k(P_{k+1}, n) \geq \mathrm{twr}_{k-1}(\Omega(n))$.*

For $s = k+1$ in Theorem 10.3, we have $r_{k-1}(\lfloor n/2 \rfloor, \lfloor n/2 \rfloor) \leq \overline{r}_k(P_{k+1}, n) \leq r_{k-1}(n, n)$. Hence, we obtain the following corollary which relates $\overline{r}_4(P_5, n)$ to the diagonal Ramsey number $r_3(n, n)$.

Corollary 10.5 ([83]) *Conjecture 3.1 holds if and only if there is a constant $c > 0$ such that*

$$\overline{r}_4(P_5, n) \geq 2^{2^{cn}}.$$

For the case when the size of P_s tends to infinity and the size of K_n is fixed, the first author in [78] showed that $\overline{r}_3(P_s, 4) < s^{21}$, and more generally for each $k \geq 3$, there exists $c > 0$ such that for s large,

$$\mathrm{twr}_{k-2}(s^c) < \overline{r}_k(P_s, k+1) < \mathrm{twr}_{k-2}(s^{62}).$$

Unfortunately much less is known about $\overline{r}_k(P_s, k+2)$. The main open problem here is to prove that $\overline{r}_3(P_s, 5)$ has polynomial growth rate, and more generally, that $\overline{r}_3(P_s, n)$ has polynomial growth rate for all fixed $n > 4$. The corresponding results for higher uniformity follow easily from the case $k = 3$.

We next consider a version of the Erdős–Hajnal hypergraph Ramsey problem with respect to tight-paths.

Definition 10.6 For integers $2 \leq k < s < n$ and $2 \leq t \leq \binom{s}{k}$, let $\overline{r}_k(s, t; P_n)$ be the minimum N such that every red/blue coloring of the k-sets of $[N]$ results in a monochromatic blue copy of P_n or has a set of s vertices which induces at least t red edges.

Of course, $\overline{r}_k(s, \binom{s}{k}; P_n) = \overline{r}_k(s, P_n)$. We will focus our attention on the smallest case $s = k + 1$. The following conjecture which parallels the Erdős–Hajnal conjecture for cliques was posed in [78].

Conjecture 10.7 ([78]) *For $3 \leq t \leq k$, there are positive $c = c(k, t)$ and $c' = c'(k, t)$ such that*

$$\mathrm{twr}_{t-2}(n^c) < \overline{r}_k(k + 1, t; P_n) < \mathrm{twr}_{t-2}(n^{c'}).$$

This conjecture seems more difficult than the original problem of Erdős and Hajnal. The current best lower bound is only an exponential function; unfortunately the constructions used for Theorem 5.6 fail. Standard arguments yield an upper bound of the form $\mathrm{twr}_{t-1}(n^c)$ for Conjecture 10.7. This upper bound was improved in [78] to $\mathrm{twr}_{t-2}(n^c)$. Some further minor progress towards Conjecture 10.7 was made in [78] for the cases $t = 3$ and $(k, t) = (4, 4)$.

10.2 Ordered ℓ-Power Paths in Graphs

As mentioned above, the proof of Dilworth's theorem shows that $\overline{r}_2(P_s, n) = \overline{r}(P_s, P_n) = (s - 1)(n - 1) + 1$. On the other hand, we know that the classical Ramsey number $r_2(n, n)$ grows exponentially in $\Theta(n)$. One can consider the case of ordered graphs that are denser than paths but sparser than cliques.

Definition 10.8 Given $\ell \geq 1$, the ℓth power P_s^ℓ of a path P_s has ordered vertex set $v_1 < \cdots < v_s$ and edge set $\{v_i v_j : |i - j| \leq \ell\}$. In particular, $P_s^1 = P_s$. The ordered Ramsey number $r(P_s^\ell, P_n^\ell)$ is the minimum N such that every red/blue coloring of $\binom{[N]}{2}$ results in a red copy of P_s^ℓ or a blue copy of P_n^ℓ.

In [78] it was shown that the problem of determining $\overline{r}(P_n^\ell, P_n^\ell)$ is closely related to the hypergraph ordered Ramsey function $\overline{r}_3(s, P_n)$. Conlon–Fox–Lee–Sudakov [22] asked whether $\overline{r}(P_n^\ell, P_n^\ell)$ is polynomial in n for every fixed $\ell \geq 1$. Actually, the problem in [22] is about the Ramsey number of ordered graphs with bandwidth at most ℓ but P_n^ℓ contains all such graphs so an upper bound for P_n^ℓ provides an upper bound for the bandwidth problem. This question was answered by Balko–Cibulka-Král–Kynčl [3]. Later a better bound was proved in [78] for $\ell = 2$.

Theorem 10.9 (Balko–Cibulka–Král–Kynčl [3] ($\ell \geq 3$), Mubayi [78] ($\ell = 2$)) *There is an absolute constant $c > 0$ and for every $\ell > 0$ there exists $c = c_\ell$ such that*

$$\overline{r}(P_n^\ell, P_n^\ell) < \begin{cases} c\,n^{19.487} & \textit{for } \ell = 2 \\ c_\ell\, n^{128\ell} & \textit{for } \ell \geq 3. \end{cases} \tag{10}$$

The main open problem here is to improve the exponents above. To our knowledge, there are no nontrivial lower bounds published for this problem.

Problem 10.10 (Balko–Cibulka–Král-Kynčl [3]) Determine the growth rate of $\overline{r}(P_n^\ell, P_n^\ell)$ for every fixed $\ell \geq 2$.

11 A Bipartite Hypergraph Ramsey Problem of Erdős

We end with an old problem of Erdős that was perhaps posed to gain a better understanding of the growth rate of the diagonal Ramsey numbers.

Definition 11.1 Let $S_{a,b} = (U, V, E)$ be the 3-graph with vertex set $U \cup V$, where $|U| = a$ and $|V| = b$, such that $E(S_{a,b}) = \{(x, y, z) : x \in U \text{and} y, z \in V\}$. Write $S_n := S_{n,n}$.

An old result due to Erdős (see [43]) says that $r_3(S_n, S_n) = 2^{O(n^2)}$, which is tight up to a constant factor in the exponent by the standard probabilistic method. We were not able to find a published proof of this result and we therefore present a proof below (of a stronger result).

Theorem 11.2 *For every $c > 0$ and sufficiently large n,*

$$r_3(S_n, S_n) < r_3(S_{2^{cn},n}, S_{2^{cn},n}) < 2^{3n^2}.$$

Proof We begin with the simple observation that $r_3(S_{1,n}, S_{1,n}) < 1 + r_2(n, n) < 4^n$. Indeed, if $r = 1 + r_2(n, n)$ and $\binom{[r]}{3}$ is 2-colored by χ, then we have an induced 2-coloring χ' of $\binom{[r-1]}{2}$ where $\chi'(ij) = \chi(ijr)$. Because $r - 1 = r_2(n, n)$ we have a monochromatic n-set under χ' and this yields a monochromatic $S_{1,n}$ under χ with $U = \{r\}$.

Now we use a simple supersaturation trick to prove the result. Suppose that $N = 2^{c'n^2}$ and χ is a 2-coloring of $\binom{[N]}{3}$. For every r-set of $[N]$, where $r = 4^n$, there is a monochromatic copy of $S_{1,n}$ in χ. Hence the number of monochromatic copies of $S_{1,n}$ in χ is at least

$$\frac{\binom{N}{r}}{\binom{N-n-1}{r-n-1}} = \frac{(N)_{n+1}}{(r)_{n+1}}.$$

At least half of these monochromatic copies of $S_{1,n}$ have the same color, say blue. Now, to each of these blue copies of $S_{1,n}$ with parts $|U| = 1$ and $|V| = n$, we associate the n-set V. A short calculation and the fact that n is large show that

$$\frac{(N)_{n+1}}{(r)_{n+1}} > \left(\frac{2^c Ne}{n}\right)^n > 2^{cn}\binom{N}{n}.$$

Consequently, by the pigeonhole principle, there are at least 2^{cn} blue copies of $S_{1,n}$ associated with the same n-set V. These blue copies together form a blue copy of S_n as desired. □

Erdős stated that an important and difficult problem is to decide if his result can be strengthened to imply all triples that meet both U and V.

Definition 11.3 Let $B_n = (U, V, E)$ be the 3-graph with vertex set $U \cup V$, where $|U| = |V| = n$, such that $E(B_n) = \{(x, y, z) : x, y \in U, z \in V \text{or} x, y \in V, z \in U\}$.

Clearly we have

$$2^{cn^2} < r_3(B_n, B_n) \leq r_3(n, n) \leq 2^{2^{c'n}},$$

where the lower bound follows from the probabilistic method.

Problem 11.4 (Erdős [43]) Improve the upper or lower bounds for $r_3(B_n, B_n)$.

Acknowledgments D. Mubayi's research partially supported by NSF grant DMS-1300138. A. Suk was supported by NSF grant DMS-1800736, an NSF CAREER award, and an Alfred Sloan Fellowship.

References

1. M. Ajtai, J. Komlós, E. Szemerédi, A note on Ramsey numbers. J. Combin. Theory Ser. A **29**, 354–360 (1980)
2. M. Axenovich, A. Gyárfás, H. Liu, D. Mubayi, Multicolor Ramsey numbers for triple systems Discrete Math. **322**, 69–77 (2014)
3. M. Balko, J. Cibulka, K. Král, J. Kynčl, Ramsey numbers of ordered graphs (submitted). arXiv:1310.7208. Extended abstract in *Proceedings of The Eight European Conference on Combinatorics, Graph Theory and Applications* (EuroComb 2015). Electronic Notes in Discrete Mathematics, vol. 49 (2015), pp. 419–424
4. V. Bhat, V. Rödl, Note on upper density of quasi-random hypergraphs. Electron. J. Combin. **20**(2), 8 pp. (2013). Paper 59
5. T. Bohman, The triangle-free process. Adv. Math. **221**, 1653–1677 (2009)
6. T. Bohman, P. Keevash, The early evolution of the H-free process. Invent. Math. **181**, 291–336 (2010)
7. T. Bohman, P. Keevash, Dynamic concentration of the triangle-free process (submitted). http://arxiv.org/abs/1302.5963`arXiv.1302.5963`
8. T. Bohman, A. Frieze, D. Mubayi, Coloring H-free hypergraphs. Random Struct. Algoritm. **36**, 11–25 (2010)

9. T. Bohman, D. Mubayi, M. Picollelli, The independent neighborhoods process. Israel J. Math. **214**, 333–357 (2016)
10. A. Cameron, E. Heath, A (5, 5)-coloring of the complete graph with few colors. https://arxiv.org/abs/1702.06227
11. Y. Caro, Y. Li, C. Rousseau, Y. Zhang, Asymptotic bounds for some bipartite graph: complete graph Ramsey numbers. Discrete Math. **220**, 51–56 (2000)
12. F.R.K. Chung, Open problems of Paul Erdős in graph theory. J. Graph Theory **25**, 3–36 (1997)
13. V. Chvátal, V. Rödl, E. Szemerédi, W.T. Trotter Jr., The Ramsey number of a graph with bounded maximum degree. J. Combin. Theory Ser. B **34**, 239–243 (1983)
14. C. Collier-Cartaino, N. Graber, T. Jiang, Linear Turán numbers of r-uniform linear cycles and related Ramsey numbers (2014). arXiv:1404.5015v2
15. D. Conlon, A new upper bound for diagonal Ramsey numbers. Ann. Math. **170**, 941–960 (2009)
16. D. Conlon, J. Fox, B. Sudakov, Hypergraph Ramsey numbers. J. Am. Math. Soc. **23**, 247–266 (2010)
17. D. Conlon, J. Fox, B. Sudakov, Large almost monochromatic subsets in hypergraphs. Israel J. Math. **181**, 423–432 (2011)
18. D. Conlon, J. Fox, B. Sudakov, On two problems in graph Ramsey theory. Combinatorica **32**, 513–535 (2012)
19. D. Conlon, J. Fox, B. Sudakov, An improved bound for the stepping-up lemma. Discrete Appl. Math. **161**, 1191–1196 (2013)
20. D. Conlon, J. Fox, C. Lee, B. Sudakov, On the grid Ramsey problem and related questions. Int. Math. Res. Not. **2015**, 8052–8084 (2015)
21. D. Conlon, J. Fox, C. Lee, B. Sudakov, The Erdős-Gyárfás problem on generalized Ramsey numbers. Proc. Lond. Math. Soc. **110**, 1–18 (2015)
22. D. Conlon, J. Fox, C. Lee, B. Sudakov, Ordered Ramsey numbers. J. Combin. Theory Ser. B. **122**, 353–383 (2017)
23. D. Conlon, J. Fox, V. Rödl, Hedgehogs are not colour blind. J. Combin. **8**, 475–485 (2017)
24. D. Conlon, J. Fox, B. Sudakov, personal communication
25. D. Conlon, J. Fox, B. Sudakov, Ramsey numbers of sparse hypergraphs. Random Struct. Algoritm. **35**, 1–14 (2009)
26. D. Conlon, J. Fox, B. Sudakov, Short proofs of some extremal results. Combin. Probab. Comput. **23**, 8–28 (2014)
27. D. Conlon, J. Fox, B. Sudakov, Recent developments in graph Ramsey theory. In A. Czumaj, A. Georgakopoulos, D. Král, V. Lozin, & O. Pikhurko (Eds.), Surveys in Combinatorics 2015 (London Mathematical Society Lecture Note Series, pp. 49–118). Cambridge: Cambridge University Press
28. O. Cooley, N. Fountoulakis, D. Kühn, D. Osthus, 3-uniform hypergraphs of bounded degree have linear Ramsey numbers. J. Combin. Theory Ser. B **98**, 484–505 (2008)
29. O. Cooley, N. Fountoulakis, D. Kühn, D. Osthus, Embeddings and Ramsey numbers of sparse k-uniform hypergraphs. Combinatorica **29**, 263–297 (2009)
30. J. Cooper, D. Mubayi, List coloring triangle-free hypergraphs. Rand. Struct. Algorithm. **47**, 487–519 (2015)
31. J. Cooper, D. Mubayi, Sparse hypergraphs with low independence number. Combinatorica **37**, 31–40 (2017)
32. C. Cox, D. Stolee, Ordered Ramsey numbers of loose paths and matchings. Discrete Math. **339**(2), 499–505 (2016)
33. R.P. Dilworth, A decomposition theorem for partially ordered sets. Ann. Math. **51**, 161–166 (1950)
34. A. Dudek, D. Mubayi, On generalized Ramsey numbers for 3-uniform hypergraphs. J. Graph Theory **76**, 217–223 (2014)
35. A. Dudek, T. Retter, V. Rödl, On generalized Ramsey numbers of Erdős and Rogers. J. Combin. Theory Ser. B **109**, 213–227 (2014)

36. D. Duffus, H. Lefmann, V. Rödl, Shift graphs and lower bounds on Ramsey numbers $r_k(l; r)$. Discrete Math. **137**, 177–187 (1995)
37. D. Eichhorn, D. Mubayi, Edge-coloring cliques with many colors on subcliques. Combinatorica **20**, 441–444 (2000)
38. M. Eliáš, J. Matoušek, Higher-order Erdős-Szekeres theorems. Adv. Math. **244**, 1–15 (2013)
39. P. Erdős, Some remarks on the theory of graphs. Bull. Am. Math. Soc. **53**, 292–294 (1947)
40. P. Erdős, Problems and results on finite and infinite graphs, in *Recent Advances in Graph Theory. Proceedings of the Second Czechoslovak Symposium*, Prague, 1974 (Academia, Prague, 1975), pp. 183–192
41. P. Erdős, Solved and unsolved problems in combinatorics and combinatorial number theory, in *Proceedings of the Twelfth Southeastern Conference on Combinatorics, Graph Theory and Computing, Vol. I*, Baton Rouge, La. Congressus Numerantium, vol. 32 (1981), pp. 49–62
42. P. Erdős, Extremal problems in number theory, combinatorics and geometry, in *Proceedings of the International Congress of Mathematicians*, vols. 1, 2, Warsaw, 1983 (PWN, Warsaw, 1984), pp. 51–70
43. P. Erdős, Problems and results on graphs and hypergraphs: similarities and differences, in *Mathematics of Ramsey Theory*, ed. by J. Nešetřil, V. Rödl. Algorithms Combinatorics, vol. 5, pp. 12–28 (Springer, Berlin, 1990)
44. P. Erdős, A. Gyárfás, A variant of the classical Ramsey problem. Combinatorica **17**, 459–467 (1997)
45. P. Erdős, A. Hajnal, On Ramsey like theorems, problems and results, in *Combinatorics (Proc. Conf. Combinatorial Math., Math. Inst., Oxford, 1972)* (The Institute of Mathematics, Southend-on-Sea, 1972), pp. 123–140
46. P. Erdős, R. Rado, Combinatorial theorems on classifications of subsets of a given set. Proc. Lond. Math. Soc. **3**, 417–439 (1952)
47. P. Erdős, C.A. Rogers, The construction of certain graphs. Can. J. Math. **14**, 702–707 (1962)
48. P. Erdős, G. Szekeres, A combinatorial problem in geometry. Compos. Math. **2**, 463–470 (1935)
49. P. Erdős, G. Szekeres, On some extremum problems in elementary geometry. Ann. Univ. Sci. Budapest. Eötvös Sect. Math. **3–4**, 53–62 (1960–1961)
50. P. Erdős, A. Hajnal, R. Rado, Partition relations for cardinal numbers. Acta Math. Acad. Sci. Hungar. **16**, 93–196 (1965)
51. G. Fiz Pontiveros, S. Griffiths, R. Morris, The triangle-free process and $R(3, k)$. (submitted). http://arxiv.org/abs/1302.6279 arXiv.1302.6279
52. J. Fox, B. Sudakov, Ramsey-type problem for an almost monochromatic K_4. SIAM J. Discrete Math. **23**, 155–162 (2008)
53. J. Fox, J. Pach, B. Sudakov, A. Suk, Erdős-Szekeres-type theorems for monotone paths and convex bodies. Proc. Lond. Math. Soc. **105**, 953–982 (2012)
54. R.L. Graham, B.L. Rothschild, J.H. Spencer, *Ramsey Theory*, 2nd edn. Wiley Interscience Series in Discrete Mathematics and Optimization (Wiley, New York, 1990)
55. A. Gyárfás, G. Raeisi, The Ramsey number of loose triangles and quadrangles in hypergraphs, Electron. J. Combin. **19**(2), #R30 (2012)
56. P. Haxell, T. Luczak, Y. Peng, V. Rödl, A. Ruciński, M. Simonovits, J. Skokan, The Ramsey number for hypergraph cycles I. J. Combin. Theory Ser. A **113**, 67–83 (2006)
57. P. Haxell, T. Luczak, Y. Peng, V. Rödl, A. Ruciński, J. Skokan, The Ramsey number for hypergraph cycles II. Combin. Probab. Comput. **18**, 165–203 (2009)
58. Y. Ishigami, Linear Ramsey numbers for bounded-degree hypergraphs. Electronic Notes Discrete Math. **29**, 47–51 (2007)
59. G. Károlyi, P. Valtr, Point configurations in d-space without large subsets in convex position. Disc. Comp. Geom. **30**, 277–286 (2003)
60. J.H. Kim, The Ramsey number $R(3, t)$ has order of magnitude $t^2/\log t$. Random Struct. Algoritm. **7**, 173–207 (1995)
61. Y. Kohayakawa, B. Nagle, V. Rödl, M. Schacht, Weak hypergraph regularity and linear hypergraphs. J. Combin. Theory Ser. B **100**, 151–160 (2010)

62. A. Kostochka, D. Mubayi, When is an almost monochromatic K_4 guaranteed? Combin. Probab. Comput. **17**, 823–830 (2008)
63. A. Kostochka, D. Mubayi, J. Verstraëte, Hypergraph Ramsey numbers: triangles versus cliques. J. Combin. Theory Ser. A **120**, 1491–1507 (2013)
64. A. Kostochka, D. Mubayi, J. Verstraëte, On independent sets in hypergraphs. Random Struct. Algoritm. **44**, 224–239 (2014)
65. J. Lenz, D. Mubayi, The poset of hypergraph quasirandomness. Random Struct. Algoritm. **46**, 762–800 (2015)
66. J. Lenz, D. Mubayi, Eigenvalues and linear quasirandom hypergraphs. Forum Math. Sigma **3**, e2, 26 pp. (2015)
67. N. Linial, A. Morgenstern, On high-dimensional acyclic tournaments. Discrete Comput. Geom. **50**, 1085–1100 (2013)
68. T. Luczak, J. Polcyn, The multipartite Ramsey number for the 3-path of length three. https://arxiv.org/abs/1706.08937
69. J. Matoušek, *Lectures in Discrete Geometry* (Springer, Berlin, Heidelberg, 2002)
70. A. Méroueh, The Ramsey number of loose cycles versus cliques. https://arxiv.org/abs/1504.03668
71. K.G. Milans, D. Stolee, D. West, Ordered Ramsey theory and track representations of graphs. J. Combin **6**, 445–456 (2015)
72. G. Moshkovitz, A. Shapira, Ramsey-theory, integer partitions and a new proof of the Erdős-Szekeres theorem. Adv. Math. **262**, 1107–1129 (2014)
73. T. Motzkin, Cooperative classes of finite sets in one and more dimensions. J. Combin. Theory **3**, 244–251 (1967)
74. D. Mubayi, Edge-coloring cliques with three colors on all 4-cliques. Combinatorica **18**, 293–296 (1998)
75. D. Mubayi, An explicit construction for a Ramsey problem. Combinatorica **24**, 313–324 (2004)
76. D. Mubayi, Coloring triple systems with local conditions. J. Graph Theory **81**, 307–311 (2016)
77. D. Mubayi, Improved bounds for the Ramsey number of tight cycles versus cliques. Combin. Probab. Comput. **25**, 791–796 (2016)
78. D. Mubayi, Variants of the Erdős-Szekeres and Erdős-Hajnal Ramsey problems. Eur. J. Combin. **62**, 197–205 (2017)
79. D. Mubayi, A. Razborov, Polynomial to exponential transition in Ramsey theory. To appear in Proc. Lond. Math. Soc.
80. D. Mubayi, V. Rödl, Hypergraph Ramsey numbers: tight cycles versus cliques. Bull. Lond. Math. Soc. **48**, 127–134 (2016)
81. D. Mubayi, A. Suk, New lower bounds for hypergraph Ramsey numbers, Bulletin of the London Mathematical Society **50**, 189–201 (2018)
82. D. Mubayi, A. Suk, Constructions in Ramsey theory, Journal of the London Mathematical Society **97**, 247–257 (2018)
83. D. Mubayi, A. Suk, Off-diagonal hypergraph Ramsey numbers, Journal of Combinatorial Theory, Series B **125**, 168–177 (2017)
84. D. Mubayi, A. Suk, The Erdos-Hajnal hypergraph Ramsey problem, Journal of the European Mathematical Society **22**, 1247–1259 (2020)
85. B. Nagle, S. Olsen, V. Rödl, M. Schacht, On the Ramsey number of sparse 3-graphs. Graphs Combin. **24**, 205–228 (2008)
86. K.T. Phelps, V. Rödl, Steiner triple systems with minimum independence number. ARS Combin. **21**, 167–172 (1986)
87. J. Polcyn, A. Ruciński, Refined Turan numbers and Ramsey numbers for the loose 3-uniform path of length three. Discrete Math. **340**, 107–118 (2017)
88. F. Ramsey, On a problem of formal logic. Proc. Lond. Math. Soc. **30**, 264–286 (1930)
89. V. Rödl, E. Šinajová, Note on independent sets in Steiner systems. Random Struct. Algoritm. **5**, 183–190 (1994)

90. S. Shelah, Primitive recursive bounds for van der Waerden numbers. J. Am. Math. Soc. **1**, 683–697 (1989)
91. J. Spencer, Ramsey's theorem - a new lower bound. J. Combin. Theory Ser. A **18**, 108–115
92. B. Sudakov, A new lower bound for a Ramsey-type problem. Combinatorica **25**, 487–498 (2005)
93. A. Suk, On the Erdős-Szekeres convex polygon problem. J. Am. Math. Soc. **30**, 1047–1053 (2017)
94. G. Wolfowitz, K_4-free graphs without large induced triangle-free subgraphs. Combinatorica **33**, 623–631 (2013)

Factorization Method for Solving Multipoint Problems for Second Order Difference Equations with Polynomial Coefficients

I. N. Parasidis and P. Hahamis

Abstract This paper is devoted to the study of second order linear difference equations with polynomial coefficients subject to multipoint boundary conditions. We provide necessary and sufficient conditions for the existence and uniqueness of solutions and find the unique solution in closed form by using factorization techniques.

1 Introduction

Modeling of processes in economics, computer science, sciences and engineering in many cases leads to solution of difference equations subject to multipoint boundary conditions. Over the last three decades there has been a growing interest in investigating this kind of problems; see, for example [1, 3, 6, 7, 12, 18]. This paper is concerned with the existence of solutions and the construction of the unique solution in closed form of the following discrete boundary value problem.

Let $\mathbb{N} = \{1, 2, 3, \ldots\}$ denote the set of positive non-zero integers, $u(k) : \mathbb{N} \to \mathbb{R}$ be a discrete function (sequence), and S be the space of all bounded functions $u(k)$ defined on $\mathbb{N}$. We consider the second order linear difference equation

$$u(k+2) + a(k)u(k+1) + b(k)u(k) = f(k), \quad k = 1, 2, 3, \ldots, \tag{1}$$

submitted to multipoint boundary conditions

$$\mu_{i1}u(1) + \mu_{i2}u(2) + \cdots + \mu_{il}u(l) = \beta_i, \quad i = 1, 2, \tag{2}$$

I. N. Parasidis (✉)
University of Thessaly, Larissa, Greece
e-mail: paras@teilar.gr

P. Hahamis
Department of Business Information Management and Operations, Westminster Business School, University of Westminster, London, UK
e-mail: hahamip@westminster.ac.uk

A. M. Raigorodskii, M. Th. Rassias (eds.), *Discrete Mathematics and Applications*, Springer Optimization and Its Applications 165,
https://doi.org/10.1007/978-3-030-55857-4_17

where the coefficients $a(k)$, $b(k) \in S$ are polynomials in k and $b(k) \neq 0$, the nonhomogeneous term $f(k) \in S$, $\mu_{ij}, \beta_i \in \mathbb{R}$, $i = 1, 2$, $j = 1, \ldots, l$, $l \geq 2$, and $u(k) \in S$ is an unknown function.

We propose a method involving the decomposition of the boundary value problem (1) and (2) into two first order initial value problems, and then the construction of the solution by solving exactly these two initial value problems. This is an advancement of the work by the first author [13–15], in conjunction with the factorization of difference operators; see, for example [4, 8] and the references therein. It is noticed that, contrary to what is the case for the difference equations with constant coefficients, obtaining closed-form solutions of difference equations with variable coefficients of order $n \geq 2$ is generally a daunting task [5]. In recent years, some individual explicit representations of the solution of nth order difference equations with variable coefficients and initial conditions have been reported in [2, 9–11, 17]. Algorithms for obtaining explicit solutions of certain classes of recurrence equations with polynomial coefficients are surveyed in [16].

In Section 2, we investigate the factorability of the difference operator in (1). In Section 3, we first state a theorem for factorizing the corresponding initial value problem and obtaining its exact solution, and then present the main result for decomposing and solving the multipoint problem (1) and (2) in closed form. In Section 4, the method proposed is applied to solve two illustrative examples. Finally, some conclusions are quoted in Section 5.

2 Preliminaries

Let the second order linear difference operator $A : S \to S$ defined by

$$Au(k) = u(k+2) + a(k)u(k+1) + b(k)u(k), \quad D(A) = S, \tag{3}$$

where $a(k)$, $b(k) \in S$ are given polynomials in k and $b(k) \neq 0$, and $u(k) \in S$. Equation (3) can be written also as

$$Au(k) = \left(E^2 + a(k)E + b(k)\right) u(k), \quad D(A) = S, \tag{4}$$

by employing the shift operator $Eu(k) = u(k+1)$. The idea is to factorize the operator A into a product of two first order linear difference operators as

$$Au(k) = (E + r(k))\,(E + s(k))\,u(k), \tag{5}$$

where the polynomials $r(x)$, $s(x) \in S$ are to be determined. By taking into account the commutation rule $E \cdot u(k) = u(k+1)E$ and expanding the composition of operators in (5), we get

$$Au(k) = \left(E^2 + (r(k) + s(k+1))\,E + r(k)s(k)\right)u(k), \quad D(A) = S. \tag{6}$$

Comparison of (6) term by term with (4) yields that the polynomials $r(k),\ s(k)$ must satisfy the following system of nonlinear difference equations:

$$r(k) + s(k+1) = a(k), \tag{7}$$

$$r(k)s(k) = b(k). \tag{8}$$

Hence, we can state the following proposition.

Proposition 1 *Let the second order difference operator A in (3) and the coefficients $a(k),\ b(k) \in S$ be polynomials of degree* $\deg a(k)$ *and* $\deg b(k)$, *respectively, with*

$$\deg a(k) \le \deg b(k). \tag{9}$$

Let there exist two polynomials $r(k),\ s(k) \in S$ of degree $\deg r(k)$ *and* $\deg s(k)$, *respectively, which satisfy the constrain equations (7) and (8). Then:*

(i) The $\deg s(k)$ *may be* $0, 1, \ldots,$ *or* $\deg b(k)$ *and* $\deg r(k) = \deg b(k) - \deg s(k)$ *subject to the constrain*

$$\deg a(k) \le \max\left(\deg r(k),\ \deg s(k)\right) \le \deg b(k). \tag{10}$$

(ii) The operator A can be decomposed into two first order difference operators $A_1, A_2 : S \to S$ such that

$$Au(k) = A_1A_2u(k), \tag{11}$$

where

$$A_1y(k) = (E + r(k))\,y(k), \quad D(A_1) = \{y(k) : y(k) \in S\}, \tag{12}$$

$$A_2u(k) = (E + s(k))\,u(k), \quad D(A_2) = \{u(k) : u(k) \in S\}. \tag{13}$$

3 Main Results

Let $\widehat{A} : S \to S$ be a restriction of the operator A defined by

$$\widehat{A}u(k) = Au(k),$$

$$D(\widehat{A}) = \{u(k) : u(k) \in D(A),\ \ u(1) = \beta_1,\ \ u(2) = \beta_2\}, \tag{14}$$

where $\beta_1,\ \beta_2 \in \mathbb{R}$. Then the standard discrete problem for second order difference equations with two initial conditions can be expressed by

$$\widehat{A}u(k) = f(k), \tag{15}$$

where $f(k) \in S$ and $u(k) \in S$ is an unknown function. We prove the next theorem.

Theorem 1 *Let the second order difference operator A in (3) and $\widehat{A}$ be a restriction of A as in (14). Suppose that the assumptions in Proposition 1 are satisfied. Then:*

(i) The operator $\widehat{A}$ can be decomposed into two first order operators $\widehat{A}_1, \widehat{A}_2 : S \to S$ as follows:

$$\widehat{A}u(k) = \widehat{A}_1\widehat{A}_2u(k), \tag{16}$$

where

$$\begin{aligned} &\widehat{A}_1y(k) = (E + r(k))\, y(k),\\ &D(\widehat{A}_1) = \{y(k) : y(k) \in S,\ \ y(1) = \beta_2 + \beta_1 s(1)\}, \end{aligned} \tag{17}$$

and

$$\begin{aligned} &\widehat{A}_2u(k) = (E + s(k))\, u(k),\\ &D(\widehat{A}_2) = \{u(k) : u(k) \in S,\ \ u(1) = \beta_1\}. \end{aligned} \tag{18}$$

(ii) The unique solution of the initial value problem (15) is obtained in closed form by

$$u(k) = \widehat{A}^{-1} f(k) = \widehat{A}_2^{-1}\widehat{A}_1^{-1} f(k). \tag{19}$$

Proof

(i) From Proposition 1 follows that the second order difference operator A can be factorized as in (11)–(13). Note that the operator $\widehat{A}$ is a restriction of A, while the operators $\widehat{A}_1, \widehat{A}_2$ are restrictions of the operators A_1, A_2, respectively. Therefore it suffices to prove that $D(\widehat{A}) = D(\widehat{A}_1\widehat{A}_2)$. From (17) and (18), we have

$$\begin{aligned} D(\widehat{A}_1\widehat{A}_2) &= \big\{u(k) : u(k) \in S,\ \ u(k) \in D(\widehat{A}_2),\ \ y(k) = \widehat{A}_2u(k) \in D(\widehat{A}_1)\big\}\\ &= \{u(k) : u(k) \in S,\ \ u(1) = \beta_1,\\ &\qquad y(1) = u(2) + s(1)u(1) = \beta_2 + \beta_1 s(1)\}\\ &= \{u(k) : u(k) \in S,\ u(1) = \beta_1,\ u(2) = \beta_2\}\\ &= D(\widehat{A}). \end{aligned} \tag{20}$$

(ii) It is known that the first order initial value problem $\widehat{A}_1 y(k) = f(k)$ possesses exactly one solution, namely $y(k) = \widehat{A}_1^{-1} f(k)$. Also, the problem $\widehat{A}_2 u(k) = y(k)$ has the unique solution $u(k) = \widehat{A}_2^{-1} y(k)$. Thus, we obtain formula (19). □

Let us now define the operator $B : S \to S$ as follows:

$$\begin{aligned} &Bu(k) = Au(k), \\ &D(B) = \{u(k) : u(k) \in D(A), \\ &\qquad \mu_{i1}u(1) + \mu_{i2}u(2) + \cdots + \mu_{il}u_l = \beta_i,\ i = 1, 2,\ l \geq 2\}, \end{aligned} \tag{21}$$

which is a restriction of the operator A in (3), and μ_{ij}, $\beta_i \in \mathbb{R}$, $i = 1, 2$, $j = 1, \ldots, l$. Then the multipoint boundary value problem (1) and (2) may be written in the form

$$Bu(k) = f(k), \quad k = 1, 2, 3, \ldots, \tag{22}$$

where $f(k) \in S$ and $u(k) \in S$ is an unknown function.

We now state the following theorem which provides sufficient conditions for the factorability of the operator B into two first order operators B_1, B_2, and the unique solvability of the boundary value problem (22).

Theorem 2 *Let the second order difference operator A as in (3) and B be a restriction of A defined in (21). Assume that all requirements of Proposition 1 are fulfilled. Additionally, let*

$$\det \mathbf{D} = \det \begin{bmatrix} b(1) & a(1) & 1 & 0 & \cdots & 0 \\ 0 & b(2) & a(2) & 1 & \ddots & \vdots \\ \vdots & \ddots & \ddots & \ddots & \ddots & 0 \\ 0 & \cdots & 0 & b(l-2) & a(l-2) & 1 \\ \mu_{11} & \mu_{12} & \cdots & \mu_{1,l-2} & \mu_{1,l-1} & \mu_{1l} \\ \mu_{21} & \mu_{22} & \cdots & \mu_{2,l-2} & \mu_{2,l-1} & \mu_{2l} \end{bmatrix} \neq 0. \tag{23}$$

Then:

(i) The operator B can be decomposed into two first order operators, viz.

$$Bu(k) = B_1 B_2 u(k), \tag{24}$$

where

$$\begin{aligned} &B_1 y(k) = (E + r(k))\, y(k) = f(k), \\ &D(B_1) = \{y(k) : y(k) \in S,\ \ y(1) = u^*(2) + s(1)u^*(1)\}, \end{aligned} \tag{25}$$

and

$$B_2u(k) = (E + s(k))\,u(k) = y(k),$$
$$D(B_2) = \{u(k) : u(k) \in S,\ \ u(1) = u^*(1)\}. \tag{26}$$

In (25) and (26),

$$u(i) = u^*(i) = \sum_{j=1}^{l-2} \tilde{d}_{ij} f(j) + \sum_{j=l-1}^{l} \tilde{d}_{ij}\beta_{j-l+2}, \quad i = 1, 2, \tag{27}$$

where $\tilde{d}_{ij}$, $i = 1, 2,\ j = 1, \ldots, l$ denote the elements of the first two rows of the matrix $\mathbf{D}^{-1} = [\tilde{d}_{ij}]$.

(ii) *The unique solution of the multipoint boundary value problem (22) is given in closed form by*

$$u(k) = B^{-1} f(k) = B_2^{-1} B_1^{-1} f(k). \tag{28}$$

Proof We form the following augmented algebraic system of a total l equations in l unknowns,

$$\begin{bmatrix} b(1) & a(1) & 1 & 0 & \cdots & 0 \\ 0 & b(2) & a(2) & 1 & \ddots & \vdots \\ \vdots & \ddots & \ddots & \ddots & \ddots & 0 \\ 0 & \cdots & 0 & b(l-2) & a(l-2) & 1 \\ \mu_{11} & \mu_{12} & \cdots & \mu_{1,l-2} & \mu_{1,l-1} & \mu_{1l} \\ \mu_{21} & \mu_{22} & \cdots & \mu_{2,l-2} & \mu_{2,l-1} & \mu_{2l} \end{bmatrix} \begin{pmatrix} u(1) \\ u(2) \\ \vdots \\ u(l-2) \\ u(l-1) \\ u(l) \end{pmatrix} = \begin{pmatrix} f(1) \\ f(2) \\ \vdots \\ f(l-2) \\ \beta_1 \\ \beta_2 \end{pmatrix}, \tag{29}$$

where the first $l - 2$ equations come from the difference equation (1) for the values of $k = 1, 2, \ldots, l - 2$, while the last two rows are the two multipoint boundary conditions (2). Denoting by $\mathbf{D}$ the $l \times l$ matrix and by $\mathbf{u}$ the $l \times 1$ vector on the left-hand side of (29), and by $\mathbf{f}$ the $l \times 1$ vector on the right-hand side of (29), we have

$$\mathbf{Du} = \mathbf{f}. \tag{30}$$

By assumption $\det \mathbf{D} \neq 0$ and hence the system (30) can be inverted to obtain

$$\mathbf{u} = \mathbf{D}^{-1}\mathbf{f} = \tilde{\mathbf{D}}\mathbf{f} = \begin{bmatrix} \tilde{d}_{11} & \cdots & \tilde{d}_{1l} \\ \vdots & \ddots & \vdots \\ \tilde{d}_{l1} & \cdots & \tilde{d}_{ll} \end{bmatrix} \mathbf{f}, \tag{31}$$

from where we get (27). By making use of (27), the boundary value problem (22) degenerates to initial value problem

$$\begin{aligned} Bu(k) &= Au(k) = u(k+2) + a(k)u(k+1) + b(k)u(k) = f(k), \\ D(B) &= \{u(k) : u(k) \in S,\ \ u(1) = u^*(1),\ \ u(2) = u^*(2)\}. \end{aligned} \tag{32}$$

Application now of Theorem 1 yields the solution (28). □

4 Example Problems

Example 1 (Amer. Math. Monthly Problem No. 10375) Find the solution of the second order difference equation

$$u(k+2) - 2(2k+3)^2 u(k+1) + 4(k+1)^2(2k+1)(2k+3)u(k) = 0, \tag{33}$$

subject to initial conditions $u(1) = 2$ and $u(2) = 24$.

We take

$$\begin{aligned} \widehat{A}u(k) &= u(k+2) - 2(2k+3)^2 u(k+1) + 4(k+1)^2(2k+1)(2k+3)u(k) = 0, \\ D(\widehat{A}) &= \{u(k) \in S,\ \ u(1) = 2,\ \ u(2) = 24\}, \end{aligned}$$

where the polynomials

$$a(k) = -2(2k+3)^2, \quad b(k) = 4(k+1)^2(2k+1)(2k+3),$$

and $f(k) = 0$.

Observe that $\deg a(k) = 2 \leq \deg b(k) = 4$ and therefore the first factorability criterion is fulfilled. Also, from all possible polynomial factorizations of $b(k)$, the polynomials

$$r(k) = -2(k+1)(2k+3), \quad s(k) = -2(k+1)(2k+1),$$

meet the second factorability condition

$$\begin{gathered} \deg a(k) = 2 \leq \max\,(\deg r(k) = 2,\ \deg s(k) = 2) \leq \deg b(k) = 4, \\ r(k) + s(k+1) = a(k), \quad r(k)s(k) = b(k). \end{gathered}$$

Then, according to Theorem 1, $\widehat{A}u(k) = \widehat{A}_1\widehat{A}_2 u(k)$ where

$$\begin{aligned} \widehat{A}_1 y(k) &= y(k+1) - 2(k+1)(2k+3)y(k), \\ D(\widehat{A}_1) &= \{y(k) \in S,\ \ y(1) = 24 + 2s(1) = 0\}, \end{aligned}$$

and

$$\widehat{A}_2 u(k) = u(k+1) - 2(k+1)(2k+1)u(k),$$
$$D(\widehat{A}_2) = \{u(k) \in S,\ \ u(1) = 2\}.$$

The first order initial value problem $\widehat{A}_1 y(k) = f(k) = 0$ admits only the trivial solution

$$y(k) = 0.$$

Using this result, we solve the second first order initial value problem $\widehat{A}_2 u(k) = y(k) = 0$ whose unique solution is given by the formula

$$u(k) = \left[\prod_{i=1}^{k-1} -s(i)\right] u(1) = 2\prod_{i=1}^{k-1} 2(i+1)(2i+1) = (2k)!,$$

which is the unique solution of (33) subject to $u(1) = 2,\ u(2) = 24$.

Example 2 Consider the second order multipoint boundary value problem

$$u(k+2) - u(k+1) - k^2 u(k) = (-1)^k k!\,,$$
$$2u(1) - u(3) = 1,$$
$$3u(2) + 4u(4) + u(5) = -10. \tag{34}$$

We define

$$Bu(k) = u(k+2) - u(k+1) - k^2 u(k) = (-1)^k k!\,,$$
$$D(B) = \{u(k) \in S,\ \ 2u(1) - u(3) = 1,\ \ 3u(2) + 4u(4) + u(5) = -10\}\,,$$

where the polynomials

$$a(k) = -1, \quad b(k) = -k^2,$$

and $f(k) = (-1)^k k!$.

First, we check the solvability of the problem. By using the difference equation and the multipoint conditions, we form the system

$$\mathbf{Du} = \begin{bmatrix} -1 & -1 & 1 & 0 & 0 \\ 0 & -4 & -1 & 1 & 0 \\ 0 & 0 & -9 & -1 & 1 \\ 2 & 0 & -1 & 0 & 0 \\ 0 & 3 & 0 & 4 & 1 \end{bmatrix} \begin{pmatrix} u(1) \\ u(2) \\ u(3) \\ u(4) \\ u(5) \end{pmatrix} = \begin{pmatrix} -1 \\ 2 \\ -6 \\ 1 \\ -10 \end{pmatrix} = \mathbf{f}.$$

The det $\mathbf{D} = 51 \neq 0$ and hence, the problem (34) has only one solution.

Next, we examine the factorability of the difference operator B. Since $\deg a(k) = 0 \leq \deg b(k) = 2$, the first requirement of Proposition 1 is fulfilled. From the possible polynomial factorizations of $b(k)$, the polynomials

$$r(k) = k, \quad s(k) = -k,$$

satisfy the conditions

$$\deg a(k) = 0 \leq \max(\deg r(k) = 1,\ \deg s(k) = 1) \leq \deg b(k) = 2,$$
$$r(k) + s(k+1) = a(k), \quad r(k)s(k) = b(k).$$

Thus, from Theorem (2) follows that $Bu(k) = B_1 B_2 u(k)$ where

$$\begin{aligned} B_1 y(k) &= y(k+1) + ky(k), \\ D(B_1) &= \{y(k) \in S,\ y(1) = u^*(2) + u^*(1)s(1) = 0\}, \end{aligned}$$

and

$$\begin{aligned} B_2 u(k) &= u(k+1) - ku(k), \\ D(B_2) &= \{u(k) \in S,\ u(1) = u^*(1) = 0\}, \end{aligned}$$

where $u^*(1),\ u^*(2)$ are the first two elements of the vector $\mathbf{u} = \mathbf{D}^{-1}\mathbf{f}$. The first order initial value problem $B_1 y(k) = f(k)$ admits a unique solution which is given by the formula

$$\begin{aligned} y(k) &= \left[\prod_{i=1}^{k-1}(-i)\right] y(1) + \sum_{j=1}^{k-1}\left[\prod_{i=j+1}^{k-1}(-i)\right] f(j) \\ &= (-1)^{k-1}(k-1)(k-1)!\,. \end{aligned}$$

Using this result, we solve the second first order initial value problem $B_2 u(k) = y(k)$ whose unique solution is given by the formula

$$\begin{aligned} u(k) &= \left[\prod_{i=1}^{k-1} i\right] u(1) + \sum_{j=1}^{k-1}\left[\prod_{i=j+1}^{k-1} i\right] y(j) \\ &= (k-1)!\sum_{j=1}^{k-1}\frac{(-1)^{j-1}(j-1)}{j}, \end{aligned}$$

which is the unique solution to problem (34).

5 Concluding Remarks

We have presented an effective method for obtaining closed-form solutions to a category of discrete boundary value problems for second order difference equations with polynomial coefficients and multipoint conditions. The method is based on the factorization of the polynomial coefficients and consequently the decomposition of the second order multipoint boundary value problem into first order initial value problems whose solutions can be computed explicitly. No fundamental solutions of the corresponding homogeneous equation and a particular solution of the nonhomogeneous equation are required. The implementation of the method has been elucidated by solving two representative examples and has been proved to work efficiently. The procedure can be easily incorporated into a computer algebra system.

References

1. R.P. Agarwal, D. O'Regan, Boundary value problems for discrete equations. Appl. Math. Lett. **10**, 83–89 (1997)
2. F.G. Boese, On ordinary difference equations with variable coefficients. J. Math. Anal. Appl. **273**, 378–408 (2002). https://doi.org/10.1016/S0022-247X(02)00244-5
3. K.A. Chrysafis, B.K. Papadopoulos, G. Papaschinopoulos, On the fuzzy difference equations of finance. Fuzzy Sets Syst. **159**, 3259–3270 (2008). https://doi.org/10.1016/j.fss.2008.06.007
4. A. Dobrogowska, M.N. Hounkonnou, Factorization method and general second order linear difference equation, in *Differential and Difference Equations with Applications. ICDDEA 2017*, ed. by S. Pinelas, T. Caraballo, P. Kloeden, J. Graef. Springer Proceedings in Mathematics and Statistics, vol. 230, pp. 67–77. Springer, Cham (2018). https://doi.org/10.1007/978-3-319-75647-9_6
5. S. Elaydi, An Introduction to Difference Equations. Springer, Berlin (2005)
6. P.W. Eloe, Difference equations and multipoint boundary value problems. Proc. Am. Math. Soc. **86**, 253–259 (1982)
7. J. Henderson, R. Luca, On a multi-point discrete boundary value problem. J. Differ. Equ. Appl. **19**, 690–699 (2013). https://doi.org/10.1080/10236198.2012.678839
8. K. Janglajew, K. Valeev, The factorization of the difference operator. Comput. Math. Appl. **42**, 729–733 (2001). https://doi.org/10.1016/S0898-1221(01)00192-4
9. R.K. Kittappa, A representation of the solution of the nth order linear difference equation with variable coefficients. Linear Algebra Appl. **193**, 211–222 (1993). https://doi.org/10.1016/0024-3795(93)90278-V
10. R.K. Mallik, On the solution of a second order linear homogeneous difference equation with variable coefficients. J. Math. Anal. Appl. **215**, 32–47 (1997). https://doi.org/10.1006/jmaa.1997.5601
11. R.K. Mallik, Solutions of linear difference equations with variable coefficients. J. Math. Anal. Appl. **222**, 79–91 (1998). https://doi.org/10.1006/jmaa.1997.5903
12. G. Papaschinopoulos, C.J. Schinas, On the dynamics of two exponential type systems of difference equations. Comput. Math. Appl. **64**, 2326–2334, (2012). https://doi.org/10.1016/j.camwa.2012.04.002
13. I.N. Parasidis, E. Providas, Factorization method for the second order linear nonlocal difference equations, in *International Conference Polynomial Computer Algebra'2018*, ed. by N.N. Vassiliev. Euler International Mathematical Institute, St. Petersburg (2018), pp. 85–89

14. I.N. Parasidis, E. Providas, Closed-form solutions for some classes of loaded difference equations with initial and nonlocal multipoint conditions, in *Modern Discrete Mathematics and Analysis*, ed. by N. Daras, T. Rassias. Springer Optimization and Its Applications, vol. 131 (Springer, Berlin, 2018), pp. 363–387. https://doi.org/10.1007/978-3-319-74325-7_19
15. I.N. Parasidis, E. Providas, An exact solution method for a class of nonlinear loaded difference equations with multipoint boundary conditions. J. Differ. Equ. Appl. **24**, 1649–1663 (2018). https://doi.org/10.1080/10236198.2018.1515928
16. M. Petkovšek, H. Zakrajšek, Solving linear recurrence equations with polynomial coefficients. In: *Computer Algebra in Quantum Field Theory. Texts and Monographs in Symbolic Computation*. ed. by C. Schneider, J. Blümlein. A Series of the Research Institute for Symbolic Computation, Johannes Kepler University, Linz (Springer, Vienna, 2013), pp. 259–284. https://doi.org/10.1007/978-3-7091-1616-6_11
17. J. Popenda, One expression for the solutions of second order difference equations. Proc. Am. Math. Soc. **100**, 87–93 (1987). https://doi.org/10.1090/S0002-9939-1987-0883406-X
18. J. Rodríguez, K.K. Abernathy, Nonlocal boundary value problems for discrete systems. J. Math. Anal. Appl. **385**, 49–59 (2012). https://doi.org/10.1016/j.jmaa.2011.06.028

New Construction Machines of Generating Fuzzy Implications

Maria N. Rapti and Basil K. Papadopoulos

Abstract In this paper we present five construction machines of generating fuzzy implications. In these formulas we present some examples by using known fuzzy implications and fuzzy negation as well and increasing function. Through these new formulas, we study the natural propagation of basic properties of given fuzzy implications from the initial ones to the constructed implications. Desirable properties as the left boundary condition (LB), right boundary condition (RB), neutrality property (NP), identity property (IP), and ordering property (OP) are preserved under some minimal conditions in the new fuzzy implications that are created.

1 Introduction

It has been about five decades since Lotfi Zadeh introduced the theory of fuzzy logic. Fuzzy logic in an extension of Boolean logic, allowed us to use more values to express reality. Important part of research in fuzzy logic focuses on extending the classical binary operators, negations ($\neg$), conjunctions ($\bigwedge$), disjunction ($\bigvee$), and implications ($\rightarrow$) to fuzzy logic operators. The extensions of the classical binary implications to fuzzy logic are called fuzzy implications. Also, t-norms and t-conorms are generalizations of the classical conjunction and disjunction, respectively. There exist many families of fuzzy implications which have been reported in detail in the literature.

In the present work we present five constructed machines of fuzzy implications. If the negations and the given implications are selected with certain properties, basic properties as the left boundary condition (LB), right boundary condition (RB), neutrality property (NP), identity property (IP), and ordering property (OP)

M. N. Rapti · B. K. Papadopoulos (✉)
Department of Civil Engineering, Section of Mathematics, Informatics and General Courses, Democritus University of Thrace, Xanthi, Greece
e-mail: marapti@civil.duth.gr; papadob@civil.duth.gr

A. M. Raigorodskii, M. Th. Rassias (eds.), *Discrete Mathematics and Applications*, Springer Optimization and Its Applications 165,
https://doi.org/10.1007/978-3-030-55857-4_18

are preserved in the new fuzzy implications that are created. In four of the five methods we use two fuzzy implications in order to produce new fuzzy implications. One method is generalized by using an increasing function g instead of a variable. Moreover there is a method that produces fuzzy implications by using two negations and an increasing function g. Finally, the last method produces fuzzy implications by using two fuzzy implications, an aggregation function (max), a decreasing function f and an increasing function g. The produced fuzzy implications fulfill the necessary properties required to be fuzzy implications. (see [1] *Definition 1.1.1.*). So, these new constructed machines can be added to the total of all methods used in production of new fuzzy implications [1–10] which will be useful in applications, especially in robotics and artificial intelligent [11–13]. We have already developed various ways of selecting fuzzy implications for certain implications and therefore we will be able to choose from these fuzzy implications.

The paper is organized as follows. In Section 2, we recall the basic concepts and definitions used in the paper. In Section 3, we study the new constructed methods of fuzzy implications. Firstly, we present a constructed method using one negation N and two fuzzy implications I, J then a second method which is a generalization of the first one, using an increasing function g instead of the variable y. Third method uses two fuzzy negations N_1, N_2 and two fuzzy implications. Also we present a method using aggregation function (max), one increasing g and one decreasing f function and two fuzzy implications. Finally, we complete the presentation of the methods with a formula that uses one increasing function and two fuzzy negations. It is worth noting that g, f are denoted $[0, 1] \to [0, 1]$.

2 Preliminaries

In order to help the reader to be familiar with the theory, we recall here some of the concepts and results employed in the rest of the paper.

Definition 2.1 (see [1] Definition 1.1.1)

A function $I : [0, 1]^2 \to [0, 1]$ is called a fuzzy implication if it satisfies, for all x, $x_1, x_2, y, y_1, y_2 \in [0, 1]$,the following conditions:

$$x_1 \le x_2 \text{ then } I\left(x_1, y\right) \ge I\left(x_2, y\right), \text{ i.e., } I\left(\cdot, y\right) \text{ is decreasing.} \tag{I1}$$

$$y_1 \le y_2 \text{ then } I\left(x, y_1\right) \le I\left(x, y_2\right), \text{ i.e., } I\left(x, \cdot\right) \text{ is increasing.} \tag{I2}$$

$$I(0, 0) = 1 \tag{I3}$$

$$I(1, 1) = 1 \tag{I4}$$

$$I(1, 0) = 0 \tag{I5}$$

The set of all fuzzy implications will be denoted by FI.

Fuzzy negation N is a generalization of the classical complement or negation $\neg$, whose truth table consists of the two conditions: $\neg 0 = 1$ and $\neg 1 = 0$.

Definition 2.2 (see [1] Definition 1.4.1)

A function $N : [0, 1] \rightarrow [0, 1]$ is called a fuzzy negation if it satisfies the following conditions:

$$N(0) = 1, N(1) = 0 \tag{N1}$$

$$\text{N is decreasing} \tag{N2}$$

Definition 2.3 (see [1] Definition 1.4.2 (i))

A fuzzy negation N is called strict if in addition:

$$\text{N is strictly decreasing} \tag{N3}$$

$$\text{N is continuous} \tag{N4}$$

Definition 2.4 (see [1] Definition 1.4.2 (ii))

A fuzzy negation N is called strong if it is an involution, i.e.,

$$N\left(N(x)\right) = x, x \in [0, 1] \tag{N5}$$

Definition 2.5 (see [1] Definition 1.4.2 (ii))

A fuzzy negation N is said to be non-vanish if $N(x) = 0 \Longleftrightarrow x = 1$.

Definition 2.6 (see [1] Definition 1.4.2 (ii))

A fuzzy negation N is said to be non-filling if $N(x) = 1 \Longleftrightarrow x = 0$.

Definition 2.7 (see [1] Definition 1.4.15 (ii))

Let $I \in$ FI. The function $N_I : [0, 1] \longrightarrow [0, 1]$ defined by $N_I(x) :- I(x, 0)$, $x \in [0, 1]$ is called the natural negation of I.

Example 2.8 (see [1] example 1.4.4, [2] Section 2.1 Example 1)

Important negations that will be used throughout this paper are the standard negation $N_C = 1 - x$, the least or Godel, and the greatest or dual Godel fuzzy negations given respectively by

$$N_{D_1}(x) = \begin{cases} 1 \text{ if } x = 0 \\ 0 \text{ if } x \in (0, 1] \end{cases} \qquad N_{D_2}(x) = \begin{cases} 1 \text{ if } x \in [0, 1) \\ 0 \text{ if } x = 1 \end{cases}$$

Definition 2.9 (see [1] Definitions 1.3.1, and 1.5.1)

A fuzzy Implication I is said to satisfy

(i) The left neutrality property if:

$$I(1, y) = y, y \in [0, 1] \tag{NP}$$

(ii) The exchange principle if:

$$I(x, I(y, z)) = I(y, I(x, z)), x, y, z \in [0, 1] \tag{EP}$$

(iii) The identity principle if:

$$I(x, x) = 1, x \in [0, 1] \tag{IP}$$

(iv) The ordering property if:

$$I(x, y) = 1 \iff x \leq y, x, y \in [0, 1] \tag{OP}$$

(v) The Law of Contraposition with respect to N if:

$$I(x, y) = I(N(y), N(x)), x, y \in [0, 1] \tag{CP}$$

(vi) The Law of Left Contraposition with respect to N if:

$$I(N(x), y) = I(N(y), x), x, y \in [0, 1] \tag{L-CP}$$

(vii) The Law of right Contraposition with respect to N if:

$$I(x, N(y)) = I(y, N(x)), x, y \in [0, 1] \tag{R-CP}$$

3 New Results

3.1 A Method of Generating Fuzzy Implications from Two Fuzzy Implications and a Fuzzy Negation

Theorem 3.1.1 Let two fuzzy implications $I, J : [0, 1]^2 \longrightarrow [0, 1]$ and a fuzzy negation N:$[0, 1] \longrightarrow [0, 1]$. Then the function $H : [0, 1]^2 \longrightarrow [0, 1]$ is a fuzzy implication, where:

$$H(x, y) = J(N(y), I(x, y))$$

Proof Let $y, x_1, x_2 \in [0, 1]$

Table 1 Examples of basic fuzzy implications

1.	Lukasiewicz $I_{\mathrm{LK}}(x,y) = \min(1,1\text{-}x + y)$ or $\begin{cases} 1, x \le y \\ 1 - x + y, x > y \end{cases}$
2.	Godel $I_{\mathrm{GD}}(x,y) = \begin{cases} 1, x \le y \\ y, x > y \end{cases}$
3.	Reichenbach $I_{\mathrm{RC}}(x,y) = 1 - x + xy$
4.	Goguen $\mathrm{I}_{\mathrm{GG}}(x,y) = \begin{cases} 1, x \le y \\ \frac{x}{y}, x > y \end{cases}$ or $\begin{cases} 1, x = 0 \\ \min\left(1, \frac{x}{y}\right), x > 0 \end{cases}$
5.	Yager $I_{\mathrm{YG}}(x,y) = \begin{cases} 1, x = 0 \text{ and } y = 0 \\ y^x, x > 0 \text{ or } y > 0 \end{cases}$
6.	Weber $I_{\mathrm{WB}}(x,y) = \begin{cases} 1, x < 1 \\ y, x = 1 \end{cases}$
7.	Fodor $I_{\mathrm{FD}}(x,y) = \begin{cases} 1, x \le y \\ \max(1 - x, y), x > y \end{cases}$
8.	Kleene-Dienes $I_{\mathrm{KD}}(x,y) = \max(1 - x,y)$
9.	Rescher $I_{\mathrm{RS}}(x, y) = \begin{cases} 1, \text{if } x \le y \\ \max(1 - x, y), \text{if } x > y \end{cases}$

$$\text{for } x_1 \le x_2 \Rightarrow I(x_1, y) \ge I(x_2, y) \Rightarrow J(N(y), I(x_1, y)) \ge J(N(y), I(x_2, y)) \tag{I1}$$

Let $x, y_1, y_2 \in [0, 1]$
for $y_1 \le y_2 \Rightarrow I(x, y_1) \le I(x, y_2)(1)$ and $N(y_1) \ge N(y_2)$ (2)
Then we have

$$J(N(y_1), I(x, y_1)) \overset{(2)}{\le} J\Big(N(y_2, I(x, y_1)) \overset{(1)}{\le} J(N(y_2), I(x, y_2)) \tag{I2}$$

$$H(0, 0) = J(N(0), I(0, 0)) = J(1, 1) = 1 \tag{I3}$$

$$H(1, 1) = J(N(1), I(1, 1)) = J(0, 1) = 1 \tag{I4}$$

$$H(1, 0) = J(N(0), I(1, 0)) = J(1, 0) = 0 \tag{I5}$$

Proposition 3.1.2 *The fuzzy implication $H(x, y) = J(N(y), I(x, y))$ satisfies the left boundary condition (LB), if the fuzzy implication I satisfies the left boundary condition (LB) as well and the fuzzy implication J satisfies the right boundary condition (RB).*

Proof I satisfies left boundary condition (LB): $I(0, y) = 1$,(1)
J satisfies right boundary condition (RB): so $J(x, 1) = 1$, (2)

$$H(0, y) = J(N(y), I(0, y)) \stackrel{(1)}{=} J(N(y), 1) \stackrel{(2)}{=} 1$$

Proposition 3.1.3 *The fuzzy implication $H(x, y) = J(N(y), I(x, y))$ satisfies the right boundary condition (RB), if the fuzzy implication I satisfies the right boundary condition (RB) as well.*

Proof I satisfies right boundary condition (RB), so $I(x, 1) = 1$, (1)

$$H(x, 1) = J(N(1), I(x, 1)) = J(0, I(x, 1)) \stackrel{(1)}{=} J(0, 1) = 1$$

Proposition 3.1.4 *The fuzzy implication $H(x, y) = J(N(y), I(x, y))$ satisfies neutrality property (NP) if I, J satisfy neutrality property (NP) and N is the greatest fuzzy negation*:

$$N(x) = N_{D_2}(x) = \begin{cases} 1 \text{ if } x \in [0, 1) \\ 0 \text{ if } x = 1 \end{cases}$$

Proof If $y \in [0, 1)$, then $H(1, y) = J(N(y), I(1, y)) = J(1, y) = y$
If $y = 1$, then $H(1, 1) = J(N(1), I(1, 1)) = J(0, 1) = 1 = y$

Proposition 3.1.5 *The fuzzy implication $H(x, y) = J(N(y), I(x, y))$ satisfies identity property (IP) if I satisfies it as well and J satisfies right boundary condition (RB).*

Proof I satisfies identity property (IP): $I(x, x) = 1$,(1)
J satisfies right boundary condition (RB): $J(x, 1) = 1$,(2)

$$H(x, x) = J(N(x), I(x, x)) \stackrel{(1)}{=} J(N(x), 1) \stackrel{(2)}{=} 1$$

Proposition 3.1.6 *The fuzzy implication $H(x, y) = J(N(y), I(x, y))$ satisfies the ordering property (OP) if it satisfies identity property (IP) as well, J satisfies neutrality property (NP), N is the greatest fuzzy negation N_{D_2}, and I satisfies ordering property (OP).*

Proof

$$N(x) = N_{D_2}(x) = \begin{cases} 1 \text{ if } x \in [0, 1) \\ 0 \text{ if } x = 1 \end{cases}$$

J satisfies neutrality property (NP): $J(1, y) = y$ (1)

I satisfies identity property (OP): $I(x, y) = 1 \Longleftrightarrow x \leq y$ (2)

If $H(x, y) = 1 \Rightarrow J(N(y), I(x, y)) = 1$. If $y \in [0, 1)$ then $J(N(y), I(x, y)) = J(1, I(x, y)) \overset{(1)}{=} I(x, y) = 1 \overset{(2)}{\Rightarrow} x \leq y$

If $x \leq y \Rightarrow H(x, x) \leq H(x, y) \Rightarrow 1 \leq H(x, y) \Rightarrow H(x, y) = 1$

Proposition 3.1.7 *Let the fuzzy implication $H(x, y) = J(N(y), I(x, y))$. If the fuzzy implication J satisfies the neutrality property (NP), then $N_H = N_I$, where N_H and N_I are the natural negations of the fuzzy implications H, I, respectively.*

Proof J satisfies neutrality property (NP): $J(1, y) = y$ (1)

$$N_H(x) = H(x, 0) = J(N(0), I(x, 0)) = J(1, I(x, 0)) \overset{(1)}{=} I(x, 0) = N_I(x)$$

Example 3.1.8 Let $N(x) = N_{D_1}(x) = \begin{cases} 1 \text{ if } x = 0 \\ 0 \text{ if } x \in (0, 1] \end{cases}$ be the least fuzzy negation.

$$I(x, y) = J(x, y) = 1 - x + xy = I_{RC}(x)$$

Then, the new fuzzy implication of the Theorem 3.1.1 is:

$$H(x, y) = \begin{cases} 1 - x, y = 0 \\ 1, y \in (0, 1] \end{cases}$$

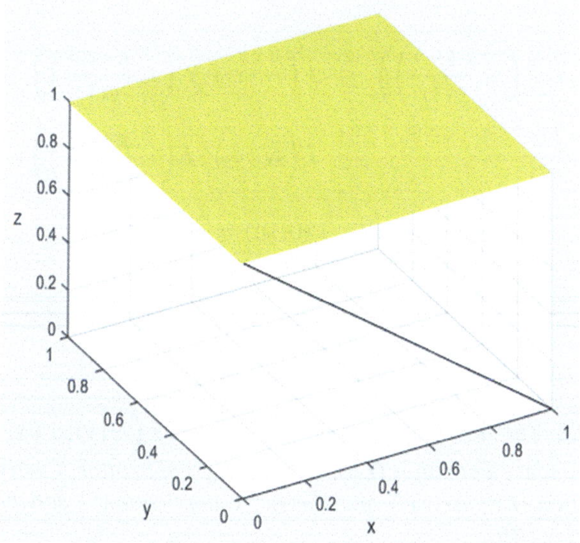

3.2 A Method of Generating Fuzzy Implications from Two Fuzzy Implications, a Fuzzy Negation, and an Increasing Function

Theorem 3.2.1 *Let two fuzzy implications* $I, J : [0, 1]^2 \longrightarrow [0, 1]$ *a fuzzy negation* N*:* $[0, 1] \longrightarrow [0, 1]$ *and an increasing function* $g : [0, 1] \longrightarrow [0, \infty)$ *with* $g(0) = 0$*. Then the function* $H : [0, 1]^2 \longrightarrow [0, 1]$ *is a fuzzy implication, where:*

$$H(x, y) = J\left(N(y), I\left(x, \frac{g(y)}{g(1)}\right)\right)$$

Proof Let $x_1, x_2 \in [0, 1]$

$$\text{for } x_1 \leq x_2 \Rightarrow I\left(x_1, \frac{g(y)}{g(1)}\right) \geq I\left(x_2, \frac{g(y)}{g(1)}\right) \Rightarrow J\left(N(y), I\left(x_1, \frac{g(y)}{g(1)}\right)\right) \geq J\left(N(y), I\left(x_2 \frac{g(y)}{g(1)},\right)\right) \tag{I1}$$

Let $y_1, y_2 \in [0, 1]$
for $y_1 \leq y_2 \Rightarrow \frac{g(y_1)}{g(1)} \leq \frac{g(y_2)}{g(1)} \Rightarrow I\left(x, \frac{g(y_1)}{g(1)}\right) \leq I\left(x, \frac{g(y_2)}{g(1)}\right)$ (1)
and for $y_1 \leq y_2 \Rightarrow N(y_1) \geq N(y_2)$ (2)
Then we have

$$J\left(N(y_1), I\left(x, \frac{g(y_1)}{g(1)}\right)\right) \overset{(2)}{\leq} J\left(N(y_2), I\left(x, \frac{g(y_1)}{g(1)}\right)\right) \overset{(1)}{\leq} J\left(N(y_2), I\left(x, \frac{g(y_2)}{g(1)}\right)\right) \tag{I2}$$

$$H(0, 0) = J(N(0), I(0, 0)) = J(1, 1) = 1 \tag{I3}$$

$$H(1, 1) = J(N(1), I(1, 1)) = J(0, 1) = 1 \tag{I4}$$

$$H(1, 0) = J(N(0), I(1, 0)) = J(1, 0) = 0 \tag{I5}$$

Proposition 3.2.2 *The fuzzy implication* $H(x, y) = J\left(N(y), I\left(x, \frac{g(y)}{g(1)}\right)\right)$ *satisfies the left boundary condition (LB), if the fuzzy implication I satisfies the left boundary condition (LB) as well and the fuzzy implication J satisfies the right boundary condition (RB).*

Proof I satisfies left boundary condition (LB): $I(0, y) = 1$,(1)
J satisfies right boundary condition (RB): $J(x, 1) = 1$, (2)

$$H(0, y) = J\left(N(y), I\left(0, \frac{g(y)}{g(1)}\right)\right) \stackrel{(1)}{=} J(N(y), 1) \stackrel{(2)}{=} 1$$

Proposition 3.2.3 *The fuzzy implication* $H(x, y) = J\left(N(y), I\left(x, \frac{g(y)}{g(1)}\right)\right)$ *satisfies the right boundary condition (RB), if the fuzzy implication I satisfies the right boundary condition (RB) as well.*

Proof I satisfies right boundary condition (RB), so $J(x, 1) = 1$, (1)

$$H(x, 1) = J\left(N(1), I\left(x, \frac{g(1)}{g(1)}\right)\right) = J(0, I(x, 1)) \stackrel{(1)}{=} J(0, 1) = 1$$

Proposition 3.2.4 *The fuzzy implication* $H(x, y) = J\left(N(y), I\left(x, \frac{g(y)}{g(1)}\right)\right)$ *satisfies neutrality property (NP) if I, J satisfy neutrality property (NP), N is the greatest fuzzy negation, and* $g(x) = x$.

Proof If $g(x) = x$, then $H(x, y) = J(N(y), I(x, y))$ is the fuzzy implication of the Theorem 3.1.1. So the proof is the same as the Proposition 3.1.4

Proposition 3.2.5 *The fuzzy implication* $H(x, y) = J\left(N(y), I\left(x, \frac{g(y)}{g(1)}\right)\right)$ *satisfies identity property (IP) if I satisfies it as well, J satisfies right boundary condition (RB), and* $g(x) = x$.

Proof If $g(x) = x$, then $H(x, y) = J(N(y), I(x, y))$ is the fuzzy implication of the Theorem 3.1. So the proof is the same as the Proposition 3.1.5

Proposition 3.2.6 *If the fuzzy implication J satisfies the neutrality property (NP), then* $N_H = N_I$, *where* N_H *and* N_I *are the natural negations of the fuzzy implications H, I respectively.*

Proof J satisfies neutrality property (NP), so $J(1, y) = y$, (1)

$$N_H(x) = H(x, 0) = J\left(N(0), I\left(x, \frac{g(0)}{g(1)}\right)\right) = J(1, I(x, 0)) \stackrel{(1)}{=} I(x, 0) = N_I(x)$$

Example 3.2.7 Let $N(x) = 1 - x$, $g(x) = x^2$

$$I(x, y) = I_{LK}(x, y) = \begin{cases} 1, x \leq y \\ 1 - x + y, x > y \end{cases}$$

Then, the new fuzzy implication of the Theorem 3.2.1 is: $H(x, y) = \begin{cases} 1, x \leq y^2 + y \\ y^2 + y + 1 - x, x > y^2 + y \end{cases}$

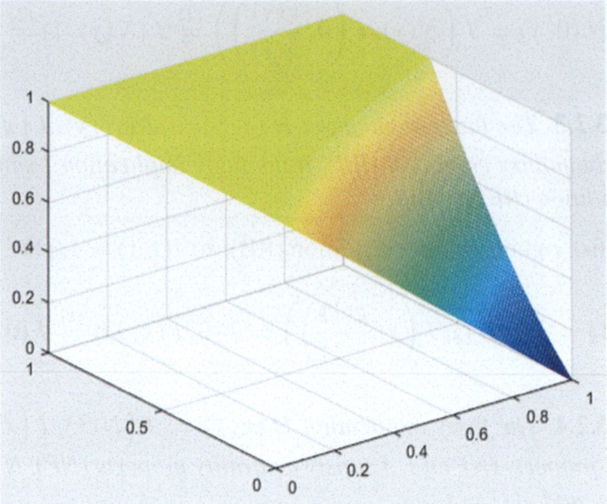

3.3 A Method of Generating Fuzzy Implications from Two Fuzzy Implications and Two Fuzzy Negations

Theorem 3.3.1 *Let two fuzzy implications* I, $J : [0,1]^2 \longrightarrow [0,1]$ *and two fuzzy negations* N_1, $N_2 : [0,1] \longrightarrow [0,1]$. *Then the function* $H : [0,1]^2 \longrightarrow [0,1]$ *is a fuzzy implication, where:*

$$H(x, y) = J\left(x \cdot N_1(y), I\left(x, y \cdot N_2(x)\right)\right)$$

Proof Let $y, x_1, x_2 \in [0,1]$

for $x_1 \le x_2 \Rightarrow N_2(x_1) \ge N_2(x_2) \Rightarrow yN_2(x_1) \ge \mathrm{y}\mathrm{N}_2(x_2)$ (1)$\Rightarrow$

$$I\left(x_1, yN_2(x_1)\right) \ge I\left(x_2, yN_2(x_1)\right) \overset{(1)}{\ge} I\left(x_2, yN_2(x_2)\right) \quad (2)$$

Also for $x_1 \le x_2 \Rightarrow$

$$x_1N_1(y) \le x_2N_1(y) \Rightarrow J\left(x_1N_1(y), I\left(x_1, yN_2(x_1)\right)\right)$$

$$\ge J\left(x_2N_1(y), I\left(x_1, yN_2(x_1)\right)\right) \overset{(1)}{=} J\left(x_2N_1(y), I\left(x_2, \mathrm{y}\mathrm{N}_2(x_2)\right)\right)$$

i.e., $I(\cdot, y)$ is decreasing (I1)

Let $x, y_1, y_2 \in [0,1]$

for $y_1 \le y_2 \Rightarrow N_1(y_1) \ge N_1(y_2) \Rightarrow xN_1(y_1) \ge xN_1(y_2)$ (3)

Also for $y_1 \le y_2 \Rightarrow y_1N_2(x) \le y_2N_2(x) \Rightarrow I(x, y_1N_2(x)) \le I(x, y_2N_2(x))$ (4)

So we have,

$$J\left(xN_1\left(y_1\right), I\left(x, y_1N_2(x)\right)\right) \overset{(3)}{\leq} J\left(xN_1\left(y_2\right), I\left(x, y_1N_2(x)\right)\right)$$
$$\overset{(4)}{\leq} J\left(xN\left(y_2\right), I\left(x, y_2N(x)\right)\right)$$

i.e., $I\left(x,\cdot\right)$ is increasing (I2)

$$H\left(0,0\right) = J\left(0, I\left(0,0\right)\right) = J\left(0,1\right) = 1 \tag{I3}$$

$$H\left(1,1\right) = J\left(0, I\left(1,0\right)\right) = J\left(0,0\right) = 1 \tag{I4}$$

$$H\left(1,0\right) = J\left(1, I\left(1,0\right)\right) = J\left(1,0\right) = 0 \tag{I5}$$

Proposition 3.3.2 *The fuzzy implication of the Theorem 3.3.1 satisfies the left boundary condition (LB) if the fuzzy implication I satisfies (LB) as well.*

Proof I satisfies the left boundary condition (LB): $I\left(0, y\right) = 1$, (1)

$$H\left(0, y\right) = J\left(0, I\left(0, y\right)\right) \overset{(1)}{=} J\left(0,1\right) = 1$$

Proposition 3.3.3 *The fuzzy implication of the Theorem 3.3.1 satisfies the right boundary condition (RB), if I satisfies the right boundary condition (RB) as well and* $N_2(x) = N_{D_2}(x) = \begin{cases} 1\ if\ x \in [0,1) \\ 0\ if\ x = 1 \end{cases}$ *is the greatest fuzzy negation.*

Proof I satisfies the right boundary condition (RB): $I\left(x, 1\right) = 1$, (1)

If $x \in [0,1)$, then $N_2(x) = 1 \Rightarrow H\left(x, 1\right) = J\left(0, I\left(x, 1\right)\right) \overset{(1)}{=} J\left(0, 1\right) = 1$
If $x = 1$, then $N_2(1) = 0 \Rightarrow H(1,1) = J(0, I(1,0)) = J(0,0) = 1$
As we see, the right boundary condition is satisfied.

Proposition 3.3.4 *The fuzzy implication of the Theorem 3.3.1 satisfies the identity principle (IP), if I satisfies the identity principle (IP) as well, the fuzzy implication J satisfies the right boundary condition (RB) and* $N_1(x) = N_2(x) = N_{D_2}(x)$*the greatest fuzzy negation.*

Proof

$$N_1(x) = N_2(x) = N_{D_2}(x) = \begin{cases} 1 \text{ if } x \in [0,1) \\ 0 \text{ if } x = 1 \end{cases}$$

I satisfies the identity principle (IP): $I(x, x) = 1$ (1)
J satisfies the right boundary condition (RB): $J(x, 1) = 1$ (2)
If $x \in [0,1)$, then $N_1(x) = N_2(x) = 1 \Rightarrow H\left(x, x\right) = J\left(x \cdot N_1(x), I\left(x, y \cdot N_2(x)\right)\right) =$ $J\left(x, I\left(x, x\right)\right) \overset{(1)}{=} J\left(x, 1\right) \overset{(2)}{=} 1$
If $x = 1$, then $N_1(x) = N_2(x) = 0 \Rightarrow H(1,1) = J(0, I(1,0)) = J(0,0) = 1$
So, the identity principle (IP) is satisfied.

Proposition 3.3.5 *The natural negation N_H of the implication of the Theorem 3.3.1 is $N_H(x) = J(x, N_I(x))$, where N_I is the natural negation of the fuzzy implication I.*

Proof

$$N_H(x) = H(x, 0) = J(x, I(x, 0)) = J(x, N_I(x))$$

Example 3.3.6 Let,

$$I(x, y) = J(x, y) = I_{RC}(x, y) = 1 - x + xy$$

and $N_1(x) = N_2(x) = N_C(x) = 1 - x$

Then, the new fuzzy implication of the Theorem 3.3.1 is:

$$H(x, y) = 1 - x^2 + 2x^2y - x^2y^2 - x^3 + x^3y^2$$

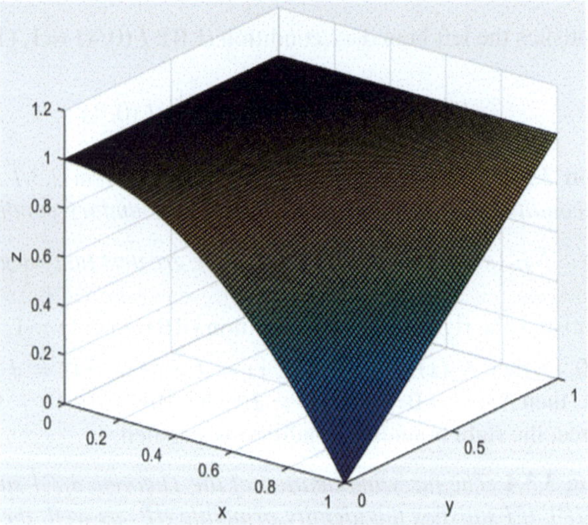

3.4 A Method of Generating Fuzzy Implications from Two Fuzzy Negations and an Increasing Function

Theorem 3.4.1 *Let two fuzzy negations $N_1(x)$, $N_2(x)$: $[0, 1] \longrightarrow [0, 1]$ and an increasing function $g : [0, 1] \rightarrow [0, \infty)$ with $g(0) = 0$. Then the function $H : [0, 1]^2 \longrightarrow [0, 1]$ is a fuzzy implication, where:*

$$H(x, y) = \frac{g(y)}{g(1)} + \frac{N_1(x) \cdot (g(1) - g(y))}{g(1)(2 - N_2(x))}$$

Proof g is an increasing function, so if $1 \geq y \Rightarrow g(1) \geq g(y) \Rightarrow g(1) \geq g(y) \geq 0$
Let $x_1, x_2, y \in [0, 1]$
for $x_1 \leq x_2 \Rightarrow N_1(x_1) \geq N_1(x_2) \Rightarrow N_1(x_1)(g(1) - g(y)) \geq N_1(x_2)(g(1) - g(y))$ (1)
Also for $x_1 \leq x_2 \Rightarrow$

$$N_2(x_1) \geq N_2(x_2) \Rightarrow 2 - N_2(x_1) \leq 2 - N_2(x_2) \Rightarrow \frac{1}{g(1)(2 - N_2(x_1))} \geq \frac{1}{g(1)(2 - N_2(x_2))}$$

$$\overset{(1)}{\Rightarrow} \frac{N_1(x_1)(g(1) - g(y))}{g(1)(2 - N_2(x_1))} \geq \frac{N_1(x_2)(g(1) - g(y))}{g(1)(2 - N_2(x_2))} \Rightarrow \frac{g(y)}{g(1)} + \frac{N_1(x_1)(1 - g(y))}{2 - N_2(x_1)} \geq \frac{g(y)}{g(1)} + \frac{N_1(x_2)(1 - g(y))}{2 - N_2(x_2)}$$

$$\Rightarrow H(x_1, y) \geq H(x_2, y) \text{ i.e., } I(\cdot, y) \text{ is decreasing} \tag{I1}$$

Also, $N_1(x) + N_2(x) \leq 2 \Rightarrow \frac{2 - N_1(x) - N_2(x)}{2 - N_2(x)} \geq 0$ (2)
A different form of H is $H(x, y) = \frac{g(y)}{g(1)} \cdot \left(\frac{2 - N_1(x) - N_2(x)}{2 - N_2(x)}\right) + \frac{N_1(x)}{2 - N_2(x)}$
Let $y_1, y_2 \in [0, 1]$ for $y_1 \leq y_2 \Rightarrow g(y_1) \leq g(y_2) \Rightarrow$

$$\frac{g(y_1)}{g(1)}\left(\frac{2 - N_1(x) - N_2(x)}{2 - N_2(x)}\right) + \frac{N_1(x)}{2 - N_2(x)} \overset{(2)}{\leq} \frac{g(y_2)}{g(1)}\left(\frac{2 - N_1(x) - N_2(x)}{2 - N_2(x)}\right) + \frac{N_1(x)}{2 - N_2(x)} \Rightarrow$$

$$\Rightarrow H(x, y_1) \leq H(x, y_2) \text{ i.e., } H(x) \text{ is increasing} \tag{I2}$$

$$H(0, 0) = 1 \tag{I3}$$

$$H(1, 1) = 1 \tag{I4}$$

$$H(1, 0) = 0 \tag{I5}$$

Proposition 3.4.2 *The fuzzy implication of the Theorem 3.4.1 satisfies left boundary (LB) condition.*

Proof

$$H(0, y) = \frac{g(y)}{g(1)} + \frac{N_1(0)\cdot(g(1) - g(y))}{g(1)(2 - N_2(0))} = \frac{g(y) + g(1) - g(y)}{g(1)} = \frac{g(1)}{g(1)} = 1$$

Proposition 3.4.3 *The fuzzy implication of the Theorem 3.4.1 satisfies right boundary (RB) condition.*

Proof

$$H(x, 1) = \frac{g(1)}{g(1)} + \frac{N_1(x)\cdot(g(1) - g(1))}{g(1)(2 - N_2(x))} = 1$$

Proposition 3.4.4 *If $g(x) = x$, then the implication of the Theorem 3.4.1 satisfies the neutrality property (NP).*

Proof

$$H(1, y) = \frac{g(y)}{g(1)} + \frac{N_1(1)\cdot(g(1) - g(y))}{g(1)(2 - N_2(1))} = \frac{g(y)}{g(1)} = \frac{y}{1} = y$$

Proposition 3.4.5 *If $N_1(x) = N_2(x) = N_{D_2}(x)$ (greatest fuzzy negation), then the fuzzy implication of the Theorem 3.4.1 satisfies the identity property (IP).*

Proof

$$N_1(x) = N_2(x) = N_{D_2}(x) = \begin{cases} 1, \text{ if } x \in [0, 1) \\ 0, \text{ if } x = 1 \end{cases}$$

If $x \in [0, 1)$, then $N_1(x) = N_2(x) = 1 \Rightarrow$

$$H(x, x) = \frac{g(x)}{g(1)} + \frac{N_1(x)\cdot(g(1) - g(x))}{g(1)(2 - N_2(x))} = \frac{g(x)}{g(1)} + \frac{g(1) - g(x)}{g(1)} = 1$$

If $x = 1$, then

$$N_1(x) = N_2(x) = 0 \Rightarrow H(x, x) = H(1, 1) = \frac{g(1)}{g(1)} + \frac{N_1(1)\cdot(g(1) - g(x))}{g(1)(2 - N_2(1))} = \frac{g(1)}{g(1)} = 1$$

Proposition 3.4.6 *The natural negation N_H of the fuzzy implication of the Theorem 3.4.1 is $N_H(x) = \frac{N_1(x)}{2 - N_2(x)}$.*

Proof

$$N_H(x) = H(x, 0) = \frac{g(0)}{g(1)} + \frac{N_1(x)\cdot(g(1) - g(0))}{g(1)(2 - N_2(x))} = \frac{N_1(x)}{2 - N_2(x)}$$

Example 3.4.7 Let $g(x) = x^2$, $N_1(x) = 1 - x$, $N_2(x) = \frac{1-x}{1+2x}$
Then the new fuzzy implication of the Theorem 3.4.1 is:

$$H(x) = y^2 + \frac{(1-x)\left(1-y^2\right)}{2 - \frac{1-x}{1+2x}}$$

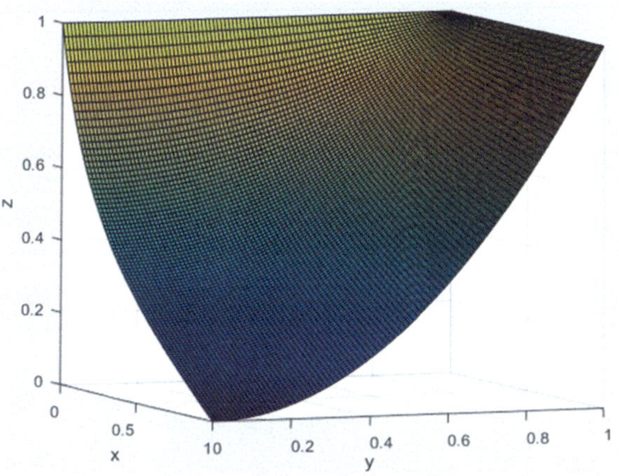

3.5 A Method of Generating Fuzzy Implications from a t-Conorm, an Increasing Function, a Decreasing Function, and Two Fuzzy Implications

Theorem 3.5.1 *Let two fuzzy implications $I, J : [0,1] \to [0,1]$, an increasing function $g : [0,1] \longrightarrow [0,\infty)$ and a decreasing function $f : [0,1] \to [0,\infty)$, then the function $H : [0,1] \to [0,1]^2$ is a fuzzy implication, where*

$$H(x,y) = \max\left\{\frac{g(y)}{g(1)} I(x,y), \frac{f(x)}{f(0)} J(x,y)\right\}$$

Proof Let $x_1, x_2 \in [0,1]$

for $x_1 \leq x_2 \Rightarrow f(x_1) \geq f(x_2) \Rightarrow \frac{f(x_1)}{f(0)} \geq \frac{f(x_2)}{f(0)}$ (1) For $x_1 \leq x_2 \Rightarrow J(x_1,y) \geq J(x_2,y)$ (2)$\Rightarrow$

$$\overset{(1),(2)}{\Rightarrow} \frac{f(x_1)}{f(0)} J(x_1,y) \geq \frac{f(x_2)}{f(0)} J(x_2,y) \quad (3)$$

Also, $x_1 \leq x_2 \Rightarrow I(x_1,y) \geq I(x_2,y) \Rightarrow \frac{g(y)}{g(1)} I(x_1,y) \geq \frac{g(y)}{g(1)} I(x_2,y)$ (4)

$$H(x_1, y) = \max\left\{\frac{g(y)}{g(1)} I(x_1, y), \frac{f(x_1)}{f(0)} J(x_1, y)\right\} \overset{(4)}{\geq}$$

$$\overset{(4)}{\geq} \max\left\{\frac{g(y)}{g(1)} I(x_2, y), \frac{f(x_1)}{f(0)} J(x_1, y)\right\} \overset{(3)}{\geq} \max\left\{\frac{g(y)}{g(1)} I(x_2, y), \frac{f(x_2)}{f(0)} J(x_2, y)\right\}$$

$$= H(x_2, y)$$

$$\Rightarrow H(x_1, y) \geq H(x_2, y) \text{ .So i.e., I } (\cdot, y) \text{ is decreasing} \quad \text{(I1)}$$

Let $y_1, y_2 \in [0, 1]$ for $y_1 \leq y_2 \Rightarrow g(y_1) \leq g(y_2) \Rightarrow \frac{g(y_1)}{g(1)} \leq \frac{g(y_2)}{g(1)}$ (5)

for $y_1 \leq y_2 \Rightarrow I(x, y_1) \leq I(x, y_2)$ (6) $\overset{(5),(6)}{\Rightarrow} \frac{g(y_1)}{g(1)} I(x, y_1) \leq \frac{g(y_2)}{g(1)} I(x, y_2)$ (7)

Also, for $y_1 \leq y_2 \Rightarrow J(x, y_1) \leq J(x, y_2) \Rightarrow \frac{f(x)}{f(0)} I(x, y_1) \leq \frac{f(x)}{f(0)} I(x, y_2)$ (8)

$H(x, y_1) = \max\left\{\frac{g(y_1)}{g(1)} I(x, y_1), \frac{f(x)}{f(0)} J(x, y_1)\right\} \overset{(7)}{\leq} \max\left\{\frac{g(y_2)}{g(1)} I(x, y_2), \frac{f(x)}{f(0)} J(x, y_1)\right\} \overset{(8)}{\leq} \max\left\{\frac{g(y_2)}{g(1)} I(x, y_2), \frac{f(x)}{f(0)} J(x, y_2)\right\} = H(x, y_1) \Rightarrow H(x, y_1) \leq H(x, y_2)$. i.e., $I(x, \cdot)$ is increasing (I2)

$$H(0, 0) = \max\{0, 1\} = 1, \quad \text{(I3)}$$

$$H(1, 1) = \max\{1, 0\} = 1, \quad \text{(I4)}$$

$$H(1, 0) = \max\{0, 0\} = 0, \quad \text{(I5)}$$

Proposition 3.5.2 *The fuzzy implication of the Theorem 3.5.1 satisfies the left boundary condition (LB) if the fuzzy implications I, J satisfy it as well.*

Proof Since I, J satisfy left boundary condition (LB), we have $I(0, y) = 1$ (1) and $J(0, y) = 1$ (2)

$$H(0, y) = \max\left\{\frac{g(y)}{g(1)} I(0, y), \frac{f(0)}{f(0)} J(0, y)\right\} \overset{(1),(2)}{=} \max\left\{\frac{g(y)}{g(1)}, 1\right\} = 1$$

Proposition 3.5.3 *The fuzzy implication of the Theorem 3.5.1 satisfies the right boundary condition (RB) if the fuzzy implications I, J satisfy it as well.*

Proof Since I, J satisfy the right boundary condition (RB), we have $I(x, 1) = 1$ (1) and $J(x, 1) = 1$ (2)

$$H(x,1)=\max\left\{\frac{g(1)}{g(1)}I(x,1),\frac{f(x)}{f(0)}J(x,1)\right\}\overset{(1),(2)}{=}\max\left\{1,\frac{f(x)}{f(0)}\right\}=1$$

Proposition 3.5.4 *The fuzzy implication of the Theorem 3.5.1 satisfies the neutrality property (NP) if the fuzzy implications I, J satisfy it as well and f or g are constant functions.*

Proof Since I, J satisfy neutrality property (NP), we have $I(1,y) = y$ (1) and $J(1,y) = y$ (2)

If g is a constant function, then $g(x) = c \geq 0 \Rightarrow g(x) = g(1) = c$. So we have the following $H(1,y)=\max\left\{\frac{g(y)}{g(1)}I(1,y),\frac{f(1)}{f(0)}J(1,y)\right\}=\max\left\{y,\frac{f(1)}{f(0)}y\right\}=y$

Similarly, if f is a constant function, we get the same results.

Proposition 3.5.5 *The fuzzy implication of the Theorem 3.5.1 satisfies the identity property (IP) if the fuzzy implications I, J satisfy it as well and f or g are constant functions.*

Proof Since I, J satisfy neutrality property (NP), we have $I(1, y) = y$ (1) and $J(1, y) = y$ (2)

If g is a constant function then $g(x) = c \geq 0 \Rightarrow g(x) = g(1) = c$. Also, f is a decreasing function, then $f(1) \leqq f(0)$. So we have the following $H(1,y) = \max\left\{\frac{g(y)}{g(1)}I(1,y),\frac{f(1)}{f(0)}J(1,y)\right\}=\max\left\{y,\frac{f(1)}{f(0)}y\right\}=y$.

Similarly, if f is a constant function and g in an increasing function we get the same results.

4 Conclusions

In this paper, five production machines of fuzzy implications are introduced. We studied certain properties of these new fuzzy implications, as neutral property (NP), exchange principle (EP), identity principle (IP), ordering property (OP), where some results are obtained if the fuzzy negations are greatest or least fuzzy negations. The advance of this method relies on the fact that we can combine fuzzy negations with fuzzy implications or we can choose increasing or decreasing functions from the rich range of literature in order to generate fuzzy implications.

References

1. M. Baczynski, B. Jayaram, *Fuzzy Implications* (Springer, Berlin, Heidelberg, 2008). https://doi.org/10.1007/978-3-540-69082-5
2. S. Massanet, J. Torrens, An overview of construction methods of fuzzy implications, in *Advances in Fuzzy Implication Functions. Studies in Fuzziness and Soft Computing*, ed. by M. Baczyński, G. Beliakov, H. Bustince Sola, A. Pradera, vol. 300, (Springer, Berlin, Heidelberg, 2013), pp. 1–30. https://doi.org/10.1007/978-3-642-35677-3

3. M. Baczynski, B. Jayaram, S. Massanet, J. Torrens, Fuzzy implications: Past, present, and future, in *Springer Handbook of Computational Intelligence*, ed. by J. Kacprzyk, W. Pedrycz, (Springer, Berlin, Heidelberg, 2015), pp. 183–202. https://doi.org/10.1007/978-3-662-43505-2_12
4. E. Sainio, E. Turunen, R. Mesiar, A characterization of fuzzy implications generated by generalized quantifiers. Fuzzy Sets Syst. **159**, 491–499. 2007 Elsevier B.V. (2008). https://doi.org/10.1016/j.fss.2007.09.018
5. M. Baczynski, B. Jayaram, On the characterization of (S, N)-implications. Fuzzy Sets Syst. **158**(15), 1713–1727 (2007). https://doi.org/10.1016/j.fss.2007.02.010
6. M. Baczynski, B. Jayaram, (S, N) - and R-implications; a state-of-the- art survey. Fuzzy Sets Syst. **159**(14), 1836–1859 (2008). https://doi.org/10.1016/j.fss.2007.11.015
7. M. Baczynski, B. Jayaram, QL-implications: Some properties and intersections. Fuzzy Sets Syst. **161**(2), 158–188 (2010). https://doi.org/10.1016/j.fss.2008.09.021
8. M. Baczynski, B. Jayaram, (U, N)-implications and their characterizations. Fuzzy Sets Syst. **160**(14), 2049–2062 (2009). https://doi.org/10.1016/j.fss.2008.11.001
9. F. Durante, E.P. Klement, R. Meriar, C. Sempi, Conjunctors and their residual implicators: Characterizations and construction methods. Mediterr. J. Math. **4**(3), 343–356 (2007). https://doi.org/10.1007/s00009-007-0122-1
10. S. Massanet, J. Torrens, Threshold generation method of construction of a new implication from two given ones. Fuzzy Sets Syst. **205**(16), 50–75 (2012). https://doi.org/10.1016/j.fss.2012.01.013
11. A.C. Bogiatzis, B.K. Papadopoulos, Producing fuzzy inclusion and entropy measures and their application on277 global image thresholding. Evol. Syst. **9**, 331–353 (2018). https://doi.org/10.1007/s12530-017-9200-1
12. A.C. Bogiatzis, B.K. Papadopoulos, Local thresholding of degraded or unevenly illuminated documents using fuzzy inclusion and entropy measures. Evol. Syst. **10**, 593–619 (2019). https://doi.org/10.1007/s12530-018-09262-5
13. A.C. Bogiatzis, B. Papadopoulos, Global image thresholding adaptive neuro- fuzzy inference system trained with fuzzy inclusion and entropy measures. Symmetry **11**, 286 (2019). https://doi.org/10.3390/sym11020286

Tree Containment and Degree Conditions

Maya Stein

Abstract We survey results and open problems relating degree conditions with tree containment in graphs, random graphs, digraphs, and hypergraphs, and their applications in Ramsey theory.

1 Introduction

A fundamental question in extremal graph theory is how to guarantee certain subgraphs by imposing a global condition on the host graph. Often, this is a condition on the degree sequence. Classical examples include Turán's theorem on containment of a complete subgraph, or Dirac's theorem on containment of a Hamilton cycle. One of the most intriguing open questions in the area is to determine degree conditions a graph G has to satisfy in order to ensure it contains a fixed tree T, or more generally, all trees of a fixed size.

Let us start with an easy observation. A greedy embedding argument yields that for $k \in \mathbb{N}$, a minimum degree $\delta(G)$ of at least k is enough to ensure that each tree T with k edges is a subgraph of G. Note that T is not necessarily an induced subgraph. Also note that although a copy of each k-edge tree is present, these copies need not be disjoint. For instance, if $|V(G)| = k + 1$, we are considering a complete graph and its spanning trees.

Although the minimum degree condition $\delta(G) \geq k$ is tight (it cannot be lowered to $k - 1$), the condition seems quite strong. It might not be necessary that *all* vertices of the host graph have large degree. For path containment, there is a famous result relying on the *average degree*: Erdős and Gallai [41] showed in 1959 that if G has average degree $d(G) > k$, then G contains a k-edge path. Erdős and Sós conjectured in 1963 (see [40]) that this bound on the average degree should in fact guarantee

M. Stein (✉)
Department of Mathematical Engineering of the University of Chile, Santiago, Chile

Center for Mathematical Modeling, UMI 2807 CNRS, Santiago, Chile

A. M. Raigorodskii, M. Th. Rassias (eds.), *Discrete Mathematics and Applications*, Springer Optimization and Its Applications 165,
https://doi.org/10.1007/978-3-030-55857-4_19

all trees with k edges to appear as subgraphs. This conjecture, its variants and generalisations, will be one of the guiding themes of this survey.

We discuss the Erdős–Sós conjecture in Section 2 and then turn to related questions. Namely, various other conditions have been suggested that might ensure the appearance of all trees of some fixed size. One well-known conjecture in this direction is the Loebl–Komós–Sós conjecture from 1995 (see [44]). This conjecture replaces the assumption on the average degree with an assumption on the median degree. We will discuss the Loebl–Komós–Sós conjecture and related results in Section 3.

More recent conjectures with the same conclusion employ a condition on a combination of the maximum and the minimum degree. The first conjecture in this direction is due to Havet et al. [72]. The idea is that a minimum degree below k may still be sufficient to find all fixed-size trees, as long as we require one vertex of large degree in the host graph. This vertex both caters for a possible large degree vertex in the tree T, and ensures we have enough space for the embedding of all of T. See Section 5 for details.

If we only wish to condition on the minimum degree of the host graph, with no assumptions on the maximum degree, and if our minimum degree condition is strictly below k, it is clearly necessary to exclude some trees, for instance, stars, from our considerations. More precisely, it will make sense to add a restriction on the maximum degree of the trees we wish to find. There is a well-known result of Komlós et al. [92] from 1995, which had been conjectured by Bollobás [18] in 1978. It states that in large graphs G, a minimum degree slightly above $\frac{|V(G)|}{2}$ is sufficient to guarantee all bounded degree spanning trees. This result will be another recurring theme of this survey. We discuss variations of Komlós, Sárközy, and Szemerédi's result in Section 4.

It has also been considered to require, apart from a minimum degree condition, additional properties in the host graph, for instance, expansion (in terms of large girth, excluded subgraphs, or neighbourhood conditions). With expansion, and for bounded degree trees, the degree bounds on the host graph can be lowered. This naturally leads to considering random graphs as well. A conjecture of Kahn [86] regarding the threshold for containment of bounded degree spanning trees was recently solved by Montgomery [101]. For several of the above-mentioned extremal results for tree containment (such as Komlós, Sárközy, and Szemerédi's result), there are resilience versions for random graphs. Also, randomly perturbed graphs have been considered as host graphs. Expansion and random graphs will be discussed in Section 6.

Some of the above conjectures have direct applications in Ramsey theory, giving upper bounds on Ramsey numbers of trees. For most trees, however, these bounds do not seem to be sharp, and it might be that the correct numbers need to take into account the relative size of the partition classes of the tree. A conjecture of Burr [23] from 1974 for Ramsey number of trees, although asymptotically confirmed for bounded degree trees in [75], has turned out to be far from correct, leaving plenty of open questions in this area. To date, not even the two-colour Ramsey number of double stars is understood. We will give an overview of the state of the art of Ramsey theory for trees in Section 7.

Tree containment is also being studied for oriented trees in oriented graphs and digraphs. It is not sufficient to simply consider the degree of the underlying graph, so even the case of the tournament as a host graph is interesting. In Section 8, we will first look at two famous conjectures on tree containment in tournaments from the 1980s, due to Sumner [115] and Burr [24], respectively. Then, we will highlight a recent conjecture from [1] which attempts to generalise the Erdős–Sós conjecture and Burr's conjecture at the same time. Finally, we turn to results generalising the theorem of Komlós, Sárközy, and Szemerédi to digraphs, and some more open questions.

Finally, tree containment problems have been translated to the hypergraph setting. We will describe this thriving area in Section 9. We cover three types of trees: Tight trees, expansions of trees, and Berge trees. Each of these notions corresponds to the respective notion for hyperpaths (and these are the most commonly studied hyperpath notions). For tight trees, Kalai's conjecture (see [48]) is widely regarded as an analogue of the Erdős–Sós conjecture for hypergraphs. There is also a generalisation of Komlós, Sárközy, and Szemerédi's theorem to tight hypertrees. For expansions of trees and Berge trees, we will present some Erdős–Sós type results, phrased in terms of their Turán numbers.

For the reader's convenience, we summarise here how the survey is organised: Section 2: Average degree; Section 3: Median degree; Section 4: Minimum degree; Section 5: Maximum and minimum degree; Section 6: Expanders and random graphs; Section 7: Ramsey theory; Section 8: Directed graphs; Section 9: Hypergraphs.

2 Average Degree

The most prominent conjecture on tree containment is a classical conjecture of Erdős and Sós from 1963 which focuses on the average degree. It appeared for the first time in [40].

Conjecture 2.1 (Erdős–Sós Conjecture, See [40]) *Every graph with average degree $d(G) > k-1$ contains every tree with k edges as a subgraph.*

A different way to state this conjecture would be in terms of the extremal number or *Turán number* of trees. Namely, we define as usual the Turán number $ex(n, H)$ of a graph H to be the largest number of edges an n-vertex graph may have without containing H as a subgraph. Then, Conjecture 2.1 states that

$$ex(n, T) \leq \frac{k-1}{2}n$$

for any k-edge tree T.

The Erdős–Sós conjecture is tight for every $k \in \mathbb{N}$: If k divides n, consider the n-vertex graph consisting of the union of $\frac{n}{k}$ disjoint copies of cliques on k vertices.

This graph has average degree $k - 1$ but it does not contain any tree with k edges since its connected components are too small. One can also consider any other $(k-1)$-regular graph, for instance, the complete bipartite graph $K_{k-1,k-1}$, which does not contain the star with k edges.

A structurally different example is given by a complete graph on n vertices, in which all edges inside a set of $\lfloor \frac{k}{2} \rfloor - 1$ vertices have been deleted. This graph does not contain any balanced tree on k edges. The graph is not extremal, however, as its average degree is slightly lower than the average degree of the examples from the previous paragraph.

Before giving an overview of the known results concerning the conjecture, let us insert here a quick observation on the minimum degree we may assume the host graph from the Erdős–Sós conjecture to have. Since every graph of average degree greater than $k - 1$ has a subgraph of minimum degree at least $\frac{k}{2}$ and average degree greater than $k - 1$ (this subgraph can be found by successively deleting vertices of too low degree), one can assume that the host graph from the Erdős–Sós conjecture has minimum degree at least $\frac{k}{2}$.

Similarly, one can argue that if we replaced the condition $d(G) > k - 1$ with the condition $d(G) > 2k - 1$, then a greedy embedding of any k-edge tree into an appropriate subgraph of G will succeed, and, therefore, such a version of Conjecture 2.1 trivially holds. The bound $d(G) > 2k - 1$ can be lowered to $d(G) > 2\frac{k}{k+1}(k - 1)$ [122].

In the early 1990s Ajtai, Komlós, Simonovits, and Szemerédi announced a proof of the Erdős–Sós conjecture for large graphs. Nevertheless, many particular cases have been settled since then, or, in some cases, earlier.

The results mainly group into four types. First, the conjecture has been verified for special types of trees. Most prominently, and as we mentioned before, a classical result of Erdős and Gallai [41] from 1959 implies that the Erdős–Sós conjecture holds for paths. The Erdős–Sós conjecture is also true for stars and double stars. Indeed, for stars this is trivial, while for double stars it suffices to establish the existence of an edge between a vertex of degree $\geq k$ and a vertex of degree $\geq \frac{k}{2}$ in the host graph. Since we can assume that the minimum degree of the host graph is at least $\frac{k}{2}$, such an edge clearly exists. Moreover, it is easy to see that the Erdős–Sós conjecture holds for all trees having a vertex adjacent to at least $\frac{k}{2}$ leaves. McLennan [100] showed the conjecture holds for all trees of diameter at most 4. Fan, Hong and Liu [45] recently proved the conjecture for all *spiders*, i.e. for all trees having at most one vertex of degree exceeding 2.

Second, the Erdős–Sós conjecture has been verified for special types of host graphs. Brandt and Dobson [21] proved in 1996 that the Erdős–Sós conjecture is true for graphs with girth at least 5. Saclé and Woźniak [120] improved on this result showing in 1997 that the Erdős–Sós conjecture holds for all graphs that do not contain C_4, the cycle on 4 vertices. The conjecture also holds if we exclude certain complete bipartite subgraphs in the host graph or its complement [7, 37, 73], and if the host graph is bipartite [122].

Third, there are results building on the relation between k and n. In particular, Conjecture 2.1 has been established for several cases when k is very close to n, the order of the host graph (note that the largest possible value of k is $k = n-1$ and then we need to find a spanning tree in an almost complete host graph). More precisely, the conjecture holds if $k+1 \leq n \leq k+4$ (for all k), and even for the case $n \leq k+c$, where c is any given constant and k is sufficiently large depending on c (see [59] and references therein). Furthermore, it is shown in [16] that if we additionally assume that $k \geq 10^6$, then the Erdős–Sós conjecture holds for all graphs G with $|V(G)| \leq (1 + 10^{-11})k$ and trees T with $\Delta(T) \leq \frac{\sqrt{k}}{1000}$.

Finally, there are some recent results building on the regularity method, thus only applying to the case when k is linear in n, and n is large. In 2019, Rozhoň [119] and independently, the authors of [16] (and [14]) gave an approximate version of the Erdős–Sós conjecture for trees with linear maximum degree and large dense host graphs.

Theorem 2.2 ([16, 119]) *For each $\delta > 0$ there are n_0, γ such that for each k and for each n-vertex graph G with $n \geq k \geq \delta n \geq \delta n_0$ the following holds.*

If G satisfies $d(G) \geq (1+\delta)k$, then G contains every k-edge tree T with $\Delta(T) \leq \gamma k$.

In [16], this is used to obtain the following sharp version of Conjecture 2.1 for large dense host graphs, which unfortunately relies on the tree having constant maximum degree.

Theorem 2.3 ([16]) *For each $\delta > 0$ and Δ there is n_0 such that for each k and for each n-vertex graph G with $n \geq k \geq \delta n \geq \delta n_0$ the following holds.*

If G satisfies $d(G) > k-1$, then G contains every k-edge tree T with $\Delta(T) \leq \Delta$.

3 Median Degree

A well-known variant of the Erdős–Sós conjecture, which replaces the assumption on the average degree with an assumption on the median degree, is the Loebl–Komlós–Sós conjecture from 1995. Two variants of this conjecture first appeared in [44].

Conjecture 3.1 (($\frac{n}{2}$-$\frac{n}{2}$-$\frac{n}{2}$)-Conjecture [44]) *Every n-vertex graph having at least $\frac{n}{2}$ vertices of degree at least $\frac{n}{2}$ contains each tree on at most $\frac{n}{2}$ vertices as a subgraph.*

The ($\frac{n}{2}$-$\frac{n}{2}$-$\frac{n}{2}$)-Conjecture has been attributed to Loebl, while according to [44], Komlós and Sós are the originators of the following variation.

Conjecture 3.2 (Komlós–Sós Conjecture [44]) *Every n-vertex graph having more than $\frac{n}{2}$ vertices of degree at least k contains each tree with k edges as a subgraph.*

The following amalgamation came to be called the Loebl–Komlós–Sós Conjecture.

Conjecture 3.3 (Loebl–Komlós–Sós Conjecture) *Every n-vertex graph having at least $\frac{n}{2}$ vertices of degree at least k contains each tree with k edges as a subgraph.*

Note that the Loebl–Komlós–Sós conjecture neither implies nor is implied by Conjecture 2.1.

Also note that the bound on the degrees in the conjecture cannot be lowered, because we might need to embed a star. Another example is the disjoint union of cliques of order k, which contains no tree with k edges.

As for the number of vertices of large degree, we do not know of any example making the bound $\frac{n}{2}$ sharp. The best example we know of is the following. If k is odd, consider the complete graph on $k+1$ vertices and delete all edges inside a set of $\frac{k+3}{2}$ vertices. It is easy to check that this graph has $\frac{k+1}{2}$ vertices of degree k, and it does not contain the k-edge path. Taking the disjoint union of several such graphs we obtain examples for other values of n (and we can add a disjoint small graph to reach any value of n). The total number of vertices of large degree is somewhat lower than $\frac{n}{2}$, and in [78], it was conjectured that the number given by this example might be the correct number.

Conjecture 3.4 ([78]) *Let G be a graph on n vertices having more than $\frac{n}{2} - \lfloor\frac{n}{k+1}\rfloor - (n \mod k+1)$ vertices of degree at least k. Then G contains each k-edge tree.*

The Loebl–Komlós–Sós conjecture (Conjecture 3.3) clearly holds for stars, and it has been proved for several other special classes of trees. One of the first results of this type is due to Bazgan, Li, and Woźniak [13], who proved the conjecture for paths in 2000. Piguet and Stein [109] proved that Conjecture 3.3 is true for trees of diameter at most 5, which improved earlier results of Barr and Johansson [12] and Sun [124] for smaller diameter.

Conjecture 3.3 has also been proved for special classes of host graphs. Soffer [121] showed that the conjecture is true if the host graph has girth at least 7. Dobson [37] proved the conjecture for host graphs whose complement does not contain the complete bipartite graph $K_{2,3}$.

The use of a different approach to the Loebl–Komlós–Sós conjecture based on the regularity method has been initiated by Ajtai et al. [3] in 1995 who solved an approximate version of Conjecture 3.1 for large graphs. Their strategy (see also [92] which appeared around the same time) relies on the regularity method, and has been replicated in similar forms in numerous articles on tree embeddings in large dense graphs. The leading idea in [3] is to cut up the tree into many tiny trees connected by a constant number of vertices (of possibly very large degree), and additionally, to find a useful matching structure in the regularised host graph. The tiny trees are then embedded into the regular pairs corresponding to the matching, while the connecting vertices are embedded in suitable clusters that see a large amount of matching edges.

Zhao [126] used a refinement of the approach from [3] plus stability arguments to prove the exact version of Conjecture 3.1 for large graphs. Also using regularity, an approximate version of Conjecture 3.3 for k linear in n was proved by Piguet and Stein [110]. Finally, adding stability arguments, Hladký and Piguet [76] and independently, Cooley [29] succeeded in proving Conjecture 3.3 for large dense graphs.

Theorem 3.5 ([29, 76]) *For every $q > 0$ there is n_0 such that for any $n > n_0$ and $k > qn$, each n-vertex graph G with at least $\frac{n}{2}$ vertices of degree at least k contains each k-edge tree.*

The regularity method described above fails in sparse host graphs. A new approach covering also this type of host graphs was explored by Hladký, Komlós, Piguet, Simonovits, Szemerédi, and the present author in [78–81] (for a 10-page overview of the proof, see [77]). These authors introduced a decomposition technique for graphs (stemming from previous work of some of the authors on the Erdős–Sós conjecture) whose output resembles the regularity lemma if applied to a dense graph but is also meaningful in the sparse setting. This enabled them to show the following approximate version of Conjecture 3.3 for large trees.

Theorem 3.6 ([78–81]) *For every $\varepsilon > 0$ there is k_0 such that for every $k \geq k_0$, every n-vertex graph having at least $(1+\varepsilon)\frac{n}{2}$ vertices of degree at least $(1+\varepsilon)k$ contains each k-edge tree as a subgraph.*

In [90], Klimošová et al. suggest an interesting generalisation of the Loebl–Komlós–Sós conjecture, inspired by a question of Simonovits. Let us say that a k-edge tree is r-skew if one of its colour classes has size at most $r(k+1)$.

Conjecture 3.7 (Skew LKS Conjecture) *Let G be a graph on n vertices having more than rn vertices of degree at least k. Then G contains each r-skew k-edge tree.*

The authors of [90] show an approximate version of this conjecture for large dense graphs. Conjecture 3.7 has also been verified for paths and trees of diameter at most five [118]. Examples similar to the ones given earlier in this section show the conjecture would be close to tight.

4 Minimum Degree

As mentioned in the introduction, any n-vertex graph G with minimum degree at least k contains every tree with k edges. Clearly, the bound on the minimum degree is tight, as we might have to embed a star. Even if we disregard for a moment stars and other trees having vertices of very large degree, it is not possible to lower the bound on the minimum degree of the host graph. In order to see this, it suffices to consider the union of several disjoint copies of K_k which does not contain any tree with k edges. However, the latter example only works if k divides n. In particular, it fails if $k > \frac{n}{2}$. So one might suspect that for $k > \frac{n}{2}$, a lower minimum degree condition could be sufficient to ensure that G contains all k-edge trees that have bounded maximum degree.

In this direction, Bollobás [18] conjectured in 1978 that any graph on n vertices and minimum degree at least $(1+o(1))\frac{n}{2}$ would contain every spanning tree whose maximum degree is bounded by a constant. This conjecture was proved by Komlós et al. [92] in 1995, giving one of the earliest applications of the Blow-up lemma.

Theorem 4.1 (Komlós et al. [92]) *For all $\delta > 0$ and $\Delta \in \mathbb{N}$, there is n_0 such that every graph G on $n \geq n_0$ vertices with $\delta(G) \geq (1+\delta)\frac{n}{2}$ contains each n-vertex tree T with $\Delta(T) \leq \Delta$.*

Subsequently, each of the two bounds in Theorem 4.1 has been improved.

Theorem 4.2 (Csaba et al. [30]) *For all $\Delta \in \mathbb{N}$, there are n_0 and c such that every graph G on $n \geq n_0$ vertices with $\delta(G) \geq \frac{n}{2} + c\log n$ contains each n-vertex tree T with $\Delta(T) \leq \Delta$.*

Theorem 4.3 (Komlós et al. [93]) *For all $\delta > 0$, there are n_0 and c such that such that every graph G on $n \geq n_0$ vertices with $\delta(G) \geq (1+\delta)\frac{n}{2}$ contains each n-vertex tree T with $\Delta(T) \leq c\frac{n}{\log n}$.*

The bound on the minimum degree in Theorem 4.2 is essentially tight (see [30]). Also the bound on the maximum degree in Theorem 4.3 is essentially best possible. This can be seen by considering the random graph with edge probability $p = 0.9$ which a.a.s. does not contain a forest of stars of order $\frac{n}{\log n}$ (and thus also does not contain any tree containing such a forest).

In contrast to the results from earlier sections, the results from [30, 92, 93] are all for the case when the tree and the host graph have the same order. In view of the examples from the beginning of the section, we know that these results cannot be generalised to non-spanning trees in host graphs of smaller minimum degree. However, if in addition we require the host graph to have a connected component of size at least $k+1$, then it does at least contain the k-edge path P_k. This is the core observation behind the Erdős–Gallai Theorem. Let us state the observation here for later reference.

Observation 4.4 (Erdős–Gallai [41], Dirac (see [41])) *If $\delta(G) \geq \frac{k}{2}$, G is connected, and $|V(G)| \geq k+1$, then $P_k \subseteq G$.*

In order to see that this observation is true, note that a variant of Dirac's theorem [35] states that every 2-connected n-vertex graph G has a cycle of length at least $\min\{n, 2\delta(G)\}$. So, if G has a 2-connected component of size at least $k+1$, then this component contains a cycle of length at least k, and thus also a k-edge path (possibly using one edge that leaves the cycle). Otherwise, we can embed either the middle vertex of the path, or a vertex adjacent to the middle edge, into any cutvertex x of G, and then greedily embed the remainder of the path into two components of $G - x$, using the minimum degree of G.

This argument, however, only seems to work for the case when the tree we are looking for is the path. Already the following tree, which has only one vertex of degree > 2, cannot be embedded into all large enough connected graphs obeying the minimum degree condition from above. Assume 3 divides k and consider the tree obtained from identifying the starting vertices of three distinct $\frac{k}{3}$-edge paths. This tree is not a subgraph of the graph obtained from adding an edge between two cliques of size $\lceil \frac{k}{2}+1 \rceil$.

Still, there is hope: It has been suggested that requiring one large degree vertex in the host graph might remedy the situation. This vertex will at the same time provide the necessary space in the host graph, and cater for a possibly existing large degree vertex of the tree. See the next section for details.

5 Maximum and Minimum Degree

As noted in Section 2, we may assume that the host graph from the Erdős–Sós conjecture has minimum degree at least $\frac{k}{2}$, and as we have seen in the previous section, this alone is not enough to force all k-edge trees as subgraphs. However, a graph H of average degree exceeding $k-1$ does not only have a subgraph H' of minimum degree $\geq \frac{k}{2}$, but this subgraph H' also maintains the average degree of H (that is, $d(H') \geq d(H)$). Therefore, H' has a vertex of degree at least k. So, in Conjecture 2.1, we may assume the host graph G to obey the following three conditions: $\delta(G) \geq \frac{k}{2}, d(G) > k-1$, and $\Delta(G) \geq k$.

Now, the conditions $\delta(G) \geq \frac{k}{2}$ and $\Delta(G) \geq k$ alone are not sufficient for guaranteeing all k-edge trees as subgraphs. This is because of a variation of the example given in the penultimate paragraph of Section 4: Adding a universal vertex to the disjoint union of two cliques of size $\lceil \frac{k}{2}+1 \rceil$, we obtain a graph G satisfying the maximum and minimum degree conditions from above. But the tree obtained from joining any three trees on roughly $\frac{k}{3}$ vertices each to a new vertex (of degree 3) is not contained in G.

However, if we elevate either the bound on $\delta(G)$ or the bound on $\Delta(G)$ sufficiently, this example ceases to work. So one may suspect there is a suitable combination of conditions on the minimum and the maximum degree of the host graph that might replace the condition on the average degree in the Erdős–Sós conjecture. In this spirit, Havet, Reed, Wood, and the present author [72] put forward the following conjecture.

Conjecture 5.1 **($\frac{2}{3}$-Conjecture [72])** *Every graph of minimum degree at least $\lfloor \frac{2k}{3} \rfloor$ and maximum degree at least k contains each k-edge tree.*

In [17], Besomi, Pavez-Signé, and the present author suggested another combination of bounds on the maximum and the minimum degree of the host graph.

Conjecture 5.2 **($2k$-$\frac{k}{2}$ Conjecture [17])** *Every graph of minimum degree at least $\frac{k}{2}$ and maximum degree at least $2k$ contains each k-edge tree.*

Conjectures 5.1 and 5.2 are asymptotically best possible, as we will discuss later in this section.

Each of the conjectures is clearly true for stars and double stars. They also hold for paths, because of Observation 4.4. In [17], both Conjectures 5.1 and 5.2 were proved in an approximate form for large dense host graphs and trees whose maximum degree is bounded by $k^{\frac{1}{49}}$ and $k^{\frac{1}{67}}$, respectively. (For trees of constant maximum degree, the maximum degree of the host graph can even be lowered slightly.)

For Conjecture 5.1, more is known. Recently, Reed and the present author [113, 114] showed that the conjecture holds if we are looking for a spanning tree in a large graph.

Theorem 5.3 ([113, 114]) *There is n_0 such that for every $n \geq n_0$, every n-vertex graph of minimum degree at least $\lfloor\frac{2(n-1)}{3}\rfloor$ and maximum degree at least $n-1$ contains each n-vertex tree.*

This theorem can also be seen as an extension of Theorem 4.3: By elevating the bound on the minimum degree of the host graph, we can dispose of the bound on the maximum degree of the tree.

Moreover, in [72], Havet, Reed, Wood, and the present author prove the following two variants of their conjecture.

Theorem 5.4 ([72]) *There are a function $f : \mathbb{N} \to \mathbb{N}$ and a constant $\gamma > 0$ such that if for a graph G, either of the following holds*

(i) $\Delta(G) > f(k)$ and $\delta(G) \geq \lfloor\frac{2m}{3}\rfloor$; or
(ii) $\Delta(G) \geq k$ and $\delta(G) \geq (1-\gamma)m$,

then G contains each k-edge tree.

Theorem 5.4 confirms that, even if the bounds suggested by Conjectures 5.1 and 5.2 should be incorrect, the idea behind the conjectures is not: It is possible to simultaneously bound the maximum and the minimum degree of a graph (with the bound on the minimum degree strictly below k) and as a result, guarantee the appearance of each tree of size k as a subgraph.

The proof of part (i) of Theorem 5.4 is relatively easy, and relies on strategically placing into a maximum degree vertex of the host graph a vertex of the tree that cuts the tree into conveniently sized components. It would be very interesting to find an extension of Theorem 5.4 (i), with the bound on the minimum degree lowered to $\frac{k}{2}$ (ideally), or to some other number strictly smaller than $\lfloor\frac{2m}{3}\rfloor$.

The following conjecture of Besomi et al. [15] tries to correlate the bounds on maximum and the minimum degree of the host graph given in Conjectures 5.1 and 5.2.

Conjecture 5.5 (Intermediate Range Conjecture [15]) *For each $\alpha \in [0, \frac{1}{3})$ every graph of minimum degree at least $(1+\alpha)\frac{k}{2}$ and maximum degree at least $2(1-\alpha)k$ contains each k-edge tree.*

As for the earlier conjectures from this section, it can be seen that Conjecture 5.5 holds for stars, double stars, and paths. Moreover, an approximate version for large dense host graphs and trees of bounded maximum degree is shown in [15].

Conjecture 5.5 is best possible for certain values of α. Namely, it is shown in [15] that for all odd $\ell \in \mathbb{N}$ with $\ell \geq 3$, and for all $\gamma > 0$ there are $k \in \mathbb{N}$, a k-edge tree T, and a graph G not containing T such that $\delta(G) \geq (1 + \frac{1}{\ell} - \gamma)\frac{k}{2}$ and $\Delta(G) \geq 2(1 - \frac{1}{\ell} - \gamma)k$. Let us give a quick description of the example from [15].

Example 5.6 Consider two complete bipartite graphs $H_i = (A_i, B_i)$ with $|A_i|$ slightly below $\frac{\ell-1}{\ell}k$ and $|B_i|$ slightly below $\frac{\ell+1}{\ell} \cdot \frac{k}{2}$. Let G^* be obtained by adding a new vertex x to $H_1 \cup H_2$, such that x is connected to all of $A_1 \cup A_2$.

Then the tree T^* formed by ℓ stars of order $\frac{k}{\ell}$ and an additional vertex v connected to the centres of the stars does not embed in G^*.

Taking $\ell = 3$, or letting ℓ converge to ∞, we obtain examples that prove the asymptotic sharpness of Conjectures 5.1 and 5.2. For Conjecture 5.1, one or both of the bipartite graphs H_i may be replaced with a conveniently sized complete graph. A structurally different example for the sharpness of Conjecture 5.1 is obtained by joining a universal vertex to a bipartite graph whose sides are both slightly below $\frac{2}{3}k$, which does not contain the tree T^* from above.

Example 5.6 only shows the asymptotic tightness of Conjecture 5.5 for values of α that are equal to $\frac{1}{\ell}$, for some odd ℓ. We are not aware of similarly good examples for other values of α, and perhaps, the degree conditions from Conjecture 5.5 could be lowered for these values, and there might be jumps. One such jump can be observed when the minimum degree bound is close to $\frac{2}{3}k$. We believe that with $\delta(G) \geq \frac{2}{3}k$, we would only need to bound $\Delta(G)$ by k (according to Conjecture 5.1). However, with the bound $\delta(G) \geq (1 - \varepsilon)\frac{2}{3}k$, for any $\varepsilon > 0$, it becomes necessary to bound $\Delta(G)$ by almost $\frac{4}{3}k$ (because of Example 5.6).

Let us remark that if we exclude host graphs that are very close to the graph G^*, a different set of maximum/minimum degree conditions might be sufficient. In [15], it is shown that any large enough graph G with $\delta(G) \geq (1+\delta)\frac{k}{2}$ and $\Delta(G) \geq (1+\delta)\frac{4}{3}k$ either looks very much like the graph G^* from Example 5.6, or contains all k-edge trees T with $\Delta(T) \leq k^{\frac{1}{67}}$.

We close the section with a new conjecture due to Klimošová, Piguet, and Rohzoň which appeared in [119]. Their conjecture combines the essence of Conjecture 3.7 with the spirit of the minimum/maximum degree conjectures from the present section.

Conjecture 5.7 ([119]) *Every n-vertex graph with $\delta(G) \geq \frac{k}{2}$ and at least $\frac{n}{2\sqrt{k}}$ vertices of degree at least k contains all k-edge trees.*

Conjecture 5.7 would be tight [119].

6 Expanders and Random Graphs

In the results and conjectures we have seen so far, the degrees in the host graph are of the same order as the tree we wish to embed. Examples showed that this is necessary, even if we bound the maximum degree of the tree. However, adding the assumption that the host graph has some expansion properties changes the situation. Different types of expansion have been considered for this problem, among these are large girth, guarantees of large neighbourhoods of small sets of vertices, and exclusion of dense bipartite subgraphs. Also random graphs fall into this category.

An early result for tree containment in expanding graphs is due to Friedman and Pippenger [49] who extended Pósa's rotation-extension technique [111] from paths to trees and showed that if each set $X \subseteq V(G)$ with $|X| \leq 2k - 2$ has at least $(\Delta + 1)|X|$ neighbours, then G contains all k-edge trees of maximum degree Δ. This result has been generalised in [8, 73].

A use of expansion in the form of large girth is the result by Brandt and Dobson [21] we cited in Section 2. They showed more generally that every graph of girth at least 5 satisfying $\delta(G) \geq \frac{k}{2}$ contains every k-edge tree with $\Delta(T) \leq \Delta(G)$. A generalisation of this was conjectured by Dobson [36], and, after preliminary results by Haxell and Łuczak [74], confirmed by Jiang [83]: For any $t \in \mathbb{N}_+$, every graph G of girth at least $2t + 1$ satisfying $\delta(G) \geq \frac{k}{t}$ contains every k-edge tree with $\Delta(T) \leq \delta(G)$. This result was greatly improved by Sudakov and Vondrák [123] who used tree-indexed random walks to show that every graph G of girth at least $2t + 1$ and $\delta(G) \geq d$ contains every tree with $\Delta(T) > (1 - \varepsilon)d$ and $|V(T)| = cd^k$ (where c is a constant depending on ε). The same authors also show the requirement of large girth may be replaced with forbidding the host graph to contain a complete bipartite graph $K_{s,t}$ for certain s, t.

Some of these results directly apply to random graphs. One natural possibility in this setting is to replace the degree conditions with probability thresholds. Then, the main problem amounts to determining the probability threshold $p = p(n)$ for the binomial random graph[1] $G(n, p)$ to contain asymptotically almost surely (a.a.s.) each tree/all trees from a given class $\mathcal{T}_n$ of trees. Clearly, as the error probabilities for missing individual trees might add up, there is a difference between containing 'each tree' and 'all trees' (the latter is often referred to as *universality*).

Most of the relevant literature for tree containment in random graphs is focused on spanning trees, or almost spanning trees, and the first case to be tackled was the path. Komlós and Szemerédi [91] and Bollobás [19] showed the threshold for spanning paths is $p = (1 + o(1))\frac{\log n}{n}$. The lower bound follows immediately from the fact that for smaller values of p, there are a.a.s. isolated vertices in $G(n, p)$.

---

[1]The graph $G(n, p)$ is defined as a probability space on the set of all graphs on n (fixed) vertices where every edge appears with probability p, independently, but we also refer to an element of this space as *the random graph* $G(n, p)$. The random graph $G(n, p)$ is said to have a property $\mathcal{P}$ asymptotically almost surely (a.a.s.) if the probability of $G(n, p)$ having $\mathcal{P}$ tends to 1 as n tends to infinity.

Kahn (see [86]) conjectured that the same threshold also applies to bounded degree trees. Namely, he conjectured that, for each Δ, there is C such that, given any sequence of n-vertex trees T_n, each with maximum degree at most Δ, the random graph $G(n, C\frac{\log n}{n})$ a.a.s. contains a copy of T. After preliminary results due to a number of authors (see, e.g., [2, 5, 47, 84, 96]), Montgomery [101] recently solved Kahn's conjecture. He showed more generally the following statement.

Theorem 6.1 ([101]) *For each $\Delta > 0$, there is a $C > 0$ such that the random graph $G(n, C\frac{\log n}{n})$ almost surely contains a copy of every n-vertex tree T with maximum degree at most Δ.*

Predating [101], some results for *almost* spanning trees appeared. Most importantly, Alon et al. [5] proved that for all ε and Δ there is C such that $G(n, \frac{C}{n})$ a.a.s. contains all trees of order $(1 - \varepsilon)n$ of maximum degree at most Δ. The value of the constant C was improved by Balogh et al. [8] by using the embedding result from [73]. Their result, as well as many of the results we cited in the last two paragraphs, also applies to other types of expanding graphs (not only random graphs).

Balogh, Csaba, and Samotij [9] showed a result in the spirit of Theorem 4.1 for *subgraphs* of random graphs. In order to appreciate their result, let us observe that Theorem 4.1 can be stated in terms of *local resilience*: If we delete some edges from the complete graph K_n, in a way that at each vertex, at least a $(\frac{1}{2} + \delta)$-fraction of its incident edges is preserved, then the resulting graph still contains all spanning trees of maximum degree Δ (where δ, n, and Δ are as in the theorem). Now, in [9] this is translated to random graphs (and *almost* spanning trees).

Theorem 6.2 ([9]) *For all Δ, ε, and δ there is C such that after deleting any set of edges from $G(n, \frac{C}{n})$ in a way that at each vertex, at least a $(\frac{1}{2} + \delta)$-fraction of the original incident edges is preserved, then the resulting graph a.a.s. contains all trees of order $(1 - \varepsilon)n$ and of maximum degree at most Δ.*

There are also some *global resilience* results for random graphs (in this type of result, a fraction of the edges is deleted without any restrictions on the number of edges deleted at each vertex). Balogh et al. [10] proved a version of the Erdős–Gallai theorem for random graphs. Namely, they determine asymptotically the number of edges a subgraph of $G(n, p)$ needs to have in order to guarantee a k-edge path, for different ranges of p and k. Araújo et al. [6] show a version of the Erdős–Sós conjecture for random graphs, and linear sized trees. More precisely, they show that for all Δ, ε, and $t \in (0, 1)$ there is C such that after deletion of at most a $(1-t-\varepsilon)$-fraction of the edges of $G(n, \frac{C}{n})$, the resulting graph still contains w.h.p. all trees of order tn and of maximum degree at most Δ. It seems not to be known whether analogues of the results from [9] and [6] for spanning trees exist.

Finally, there are some recent results for *randomly perturbed* graphs. This model relies on a graph of linear but very small minimum degree, which is 'randomly perturbed' by adding a few random edges to it. More precisely, we consider the union of a graph G_α of minimum degree at least αn and the random graph $G(n, p)$, on the same set of vertices. The study of bounded degree spanning trees in this model was initiated by Krivelevich et al. [97]. They determined the threshold $p = \frac{C}{n}$ (with C depending on α and Δ) for containment of a single tree of maximum degree Δ, and conjectured the same threshold for the corresponding universality result. This was confirmed by Böttcher et al. [20]. Joos and Kim [85] show a variant of Theorem 4.1 for randomly perturbed graphs.

7 Ramsey Numbers

Both Conjecture 2.1 (the Erdős–Sós conjecture) and Conjecture 3.3 (the Loebl–Komlós–Sós conjecture) have a direct application in Ramsey theory. Let us start with 2-colour Ramsey numbers. The *(2-colour) Ramsey number* $R(H_1, H_2)$ of a pair of graphs H_1, H_2 is the smallest integer n such that every 2-colouring of the edges of K_n contains a copy of H_1 in the first colour, or a copy of H_2 in the second colour. Generalising the notion to classes $\mathcal{H}_1, \mathcal{H}_2$ of graphs, we write $R(\mathcal{H}_1, \mathcal{H}_2)$ for smallest integer n such that every 2-colouring of the edges of K_n contains a copy of *each* $H_1 \in \mathcal{H}_1$ in the first colour, or a copy of *each* $H_2 \in \mathcal{H}_2$ in the second colour. We write short $R(H)$ ($R(\mathcal{H})$) for $R(H, H)$ ($R(\mathcal{H}, \mathcal{H})$).

Some of the earliest results on Ramsey numbers for trees were the following. In 1967, Gerencsér and Gyárfás [58] determined the Ramsey number of two paths. They showed that

$$R(P_k, P_\ell) = k + \left\lfloor \frac{\ell + 1}{2} \right\rfloor$$

for k-edge and ℓ-edge paths P_k and P_ℓ with $k \geq \ell \geq 2$. For stars, the Ramsey number is known to be larger. Harary [67] observed in 1972 that

$$R(K_{1,k}, K_{1,\ell}) = k + \ell$$

if at least one of k, ℓ is odd, and $R(K_{1,k}, K_{1,\ell}) = k + \ell - 1$ in the case that k and ℓ are both even.

Conjectures 2.1 and 3.3 can be applied as follows in the Ramsey setting. Given a 2-edge-coloured $K_{k+\ell}$, say with colours red and blue, it is easy to see that either the red graph has average degree greater than $k-1$, or the blue graph has average degree greater than $\ell - 1$. Also, either the red graph has median degree at least k, or the blue graph has median degree at least ℓ. Therefore, each of the two conjectures would imply that every 2-edge-colouring of $K_{k+\ell}$ with colours red and blue contains either *all* k-edge trees in red, or *all* ℓ-edge trees in blue, and, therefore, $R(\mathcal{T}_k, \mathcal{T}_\ell) \leq k + \ell$, where $\mathcal{T}_j$ is the class of all trees with j edges. If k and ℓ are both even, the bound we can infer from Conjecture 2.1 is even lower: In that case $R(\mathcal{T}_k, \mathcal{T}_\ell) \leq k + \ell - 1$. Accordingly, and focusing on the case $k = \ell$, Burr and Erdős [25] conjectured in 1976 that $R(\mathcal{T}_k) \leq 2k$, and $R(\mathcal{T}_k) \leq 2k - 1$ if k is even. The bound $R(\mathcal{T}_k) \leq 2k$ has been confirmed for large k, by Zhao's solution of Conjecture 3.1 for large host graphs [126].

However, the bound $R(T_k, T_\ell) \leq k + \ell$ for a k-edge tree T_k and an ℓ-edge tree T_ℓ seems to be far from best possible for non-star trees. As noted above, the Ramsey number for paths differs significantly from the Ramsey number for stars with the same number of edges. Note that paths are (almost) completely balanced trees, while stars are the most unbalanced trees. Believing this difference to be the reason for the variation in their Ramsey numbers, Burr [23] put forward the following conjecture in 1974. He suggested that if T is a tree whose bipartition classes have sizes t_1, t_2, with $t_2 \geq t_1 \geq 2$, then the Ramsey number of T is

$$R_B(T) := \max\{2t_1 + t_2, 2t_2\} - 1.$$

Standard examples show this number would be best possible, and $R_B(T)$ matches the Ramsey numbers for paths from [58].

Haxell et al. [75] confirmed Burr's conjecture asymptotically for trees with (linearly) bounded maximum degree in 2002. However, already shortly after the conjecture was posed, Grossman et al. [60] found that it was not true for certain double stars. A *double star* D_{t_1,t_2} is a union of two stars K_{1,t_1-1} and K_{1,t_2-1} whose centres are joined by an edge. The examples from [60] still allowed for the possibility that Burr's conjecture was off only by one, that is, that the Ramsey number of any tree T would be bounded by $R_B(T) + 1$. The authors of [60] conjectured this to be the truth for double stars. This has been confirmed for a range[2] of values of t_1, t_2. But recently, Norin et al. [106] disproved the conjecture from [60] in general by showing that the numbers $R(D_{t_1,t_2})$ and $R_B(D_{t_1,t_2})$ differ considerably if t_2 lies between $\frac{7}{4}t_1 + o(t_1)$ and $\frac{105}{41}t_1 + o(t_1)$. In particular, for the case $t_2 = 2t_1$ they find that

$$R(D_{t_1,t_2}) \geq 4.2t_1 - o(t_1)$$

while

$$R_B(D_{t_1,t_2}) = 4t_1 - 1.$$

The authors of [106] pose the following question.

Question 7.1 (Norin et al. [106]) *Is it true that $R(D_{t_1,t_2}) = 4.2t_1 + o(t_1)$ if $t_2 = 2t_1$?*

A question of Erdős et al. [43], who, in 1982, asked whether $R(T) = R_B(T)$ for all trees T with colours classes of sizes t_1 and $t_2 = 2t_1$, has also been answered in the negative by the above-mentioned results from [106]. The authors of [106] offer the following alternative.

[2] The current best results are $R(D_{t_1,t_2}) \leq R_B(D_{t_1,t_2}) + 1$ if $t_2 \geq 3t_1 - 2$ (obtained using ad hoc arguments [60]) and $R(D_{t_1,t_2}) \leq R_B(D_{t_1,t_2})$ if $t_2 \leq 1.699t_1 + 1$ (obtained using flag algebras [106]).

Question 7.2 (Norin et al. [106]) *Is it true that $R(T) \leq 4.2t_1 + o(t_1)$ for all trees T with colours classes of sizes t_1 and $2t_1$?*

Another natural question in this context seems to be whether there is an exact version of the asymptotic results of Haxell et al.[75] (which would be interesting even if we had to restrict the maximum degree of the tree by, say, a constant). A second question is whether their result can be extended to graphs of slightly larger maximum degree. In the main result from [75], the bound on the maximum degree of the tree is $\delta|V(T)|$, with δ depending on the approximation. On the other hand, the known counterexamples to Burr's conjecture are all double stars D of maximum degree exceeding $\frac{7}{11}|V(D)|$. So, there might be a chance that for some reasonable constant $c \leq \frac{7}{11}$, Burr's conjecture still holds for all trees T of maximum degree at most $c|V(T)|$.

Question 7.3 *Is there a constant c such that $R(T) = R_B(T)$ for all trees T with $\Delta(T) \leq c|V(T)|$?*

Let us now briefly look at results and questions for multicolour Ramsey numbers of trees. The r-colour Ramsey number $R_r(H)$ of a graph H is defined as the smallest integer n such that every r-colouring of the edges of K_n contains a monochromatic copy of H.

Multicolour Ramsey numbers for trees have not been studied much. The most studied case is the path P_k. The 3-colour Ramsey number of the k-edge path P_k has been conjectured to be $2k$ for even k and $2k + 1$ for odd k by Faudree and Schelp [46], and this is best possible. This conjecture has been confirmed for large k by Gyárfás et al. [63]. For more colours, less is known. Constructions based on affine planes show that $R_r(P_k) \geq (r - 1)k$ if $r - 1$ is a prime power. An upper bound on $R_r(P_k)$ can be obtained by applying the Erdős–Gallai theorem to the most popular colour in a given r-colouring. This yields $R_r(P_k) \leq r(k + 1)$. Recently, the latter bound has been improved to $(r - \frac{1}{4})(k + 1) + o(k)$ by Davies et al. [31].

Multicolour Ramsey numbers for k-edge stars were determined by Burr and Roberts [26] in 1973. They showed that $(k - 1)r + 1 \leq R_r(K_{1,k}) \leq (k - 1)r + 2$. The lower bound is tight if and only if both k and r are even.

General bounds for all trees have also been considered. Erdős and Graham [42] observed that an affirmative answer to the following question would follow from the Erdős–Sós conjecture. (A version for skew trees would also follow from Conjecture 3.7.)

Question 7.4 (Erdős and Graham [42]) *Is the r-colour Ramsey number for a k-edge tree T_k equal to $rk + O(1)$?*

The authors of [42] observe that $R_r(T_k)$ is bounded from above by $2rk$. This bound can be obtained using a similar argument as for the 2-colour Ramsey number, and the fact that Conjecture 2.1 holds if the average degree bound is replaced by $2k$, as we observed in Section 2.

8 Directed Graphs

In this section, we will shift our focus from trees and graphs to their oriented versions, that is, oriented trees and digraphs/oriented graphs. An *oriented tree (graph)* is a tree (graph) all whose edges have been given a direction. A *digraph* may have (at most) two edges between a pair of vertices, as long as these go in opposite directions. A *tournament* is an oriented complete graph, and a *complete digraph* is a digraph having all possible edges.

Let us start with oriented graphs and trees. Before we turn to possible generalisations of the results in the earlier sections, let us illustrate how the orientations of the edges bring new difficulties. Just considering the degree in the underlying graph is clearly not enough. Indeed, it is fairly easy to construct an orientation of a complete graph K_{k+1} that does not contain the k-edge star having all its edges directed inwards, thus preventing even the easiest observation for graphs to carry over to digraphs.

So, let us consider the class of tournaments as possible host graphs. One of the first results on oriented trees in tournaments was established by Rédei [112] in 1934. It states that every tournament on $k + 1$ vertices contains the *directed k-edge path* (i.e. the k-edge path having all its edges directed in the same direction). More results on oriented paths appeared in, e.g. [61, 117, 125], until in 2000, Havet and Thomassé [71] showed that with three exceptions, all oriented k-edge paths appear in any $(k + 1)$-vertex tournament. Similar results have been shown for some classes of oriented trees with bounded maximum degree (see [103] and references therein).

A generalisation of these results for containment of all oriented trees of some fixed size was conjectured by Sumner in the 1980s.

Conjecture 8.1 (Sumner, See [115]) *Every tournament on* $2k$ *vertices contains every oriented* k*-edge tree.*

This conjecture is best possible, which can be seen by considering a $(k - 1)$-*regular tournament* (that is, a tournament whose vertices each have in- and out-degree $k - 1$) on $2k - 1$ vertices, which does not contain the k-edge star with all edges directed inwards (or outwards).

Variants of Conjecture 8.1 replacing $2k$ with a larger number are known [39, 66, 68, 70]. The current best bound is $\frac{21}{8}k$, and was found by Dross and Havet [38]. Havet and Thomassé [70] showed that Conjecture 8.1 holds for *arborescences*, that is, oriented trees having all their edges directed away from (or towards) a specific vertex. After proving an approximate version [98], Kühn et al. [99] confirmed Sumner's conjecture for large n, using the regularity method. For oriented trees of bounded degree, the size of the host tournament can be lowered to $n + o(n)$ (see [99, 103, 105]).

In 1996, Havet and Thomassé proposed that the size of the host tournament can be smaller if we add a restriction on the number of leaves of the tree. This gives the following generalisation of Conjecture 8.1.

Conjecture 8.2 (Havet and Thomassé, See [69]) *Let T be an oriented k-edge tree with ℓ leaves. Then every tournament on $k + \ell$ vertices contains a copy of T.*

Note that for oriented stars, Conjecture 8.2 gives the same bound as Conjecture 8.1, but for other trees, the bound is lower. As we saw above, if T is a path, the tournament may be by one smaller than required by Conjecture 8.2. For progress on Conjecture 8.2, see [27, 66, 69].

Turning now to oriented graphs as possible host graphs for oriented trees, there is a natural generalisation of the results and conjectures from above, which involves the chromatic number of a digraph. An oriented graph is *n-chromatic* if the underlying graph has chromatic number n. The well-known Gallai–Hasse–Roy–Vitaver (GHRV) theorem (see, e.g., [11]) states that every $(k+1)$-chromatic oriented graph contains the directed path with k edges. As any n-vertex tournament is n-chromatic, this is a generalisation of Rédei's theorem mentioned above.

An extension of the GHRV theorem to oriented trees was suggested by Burr [24] in 1980. His conjecture would imply Sumner's conjecture.

Conjecture 8.3 (Burr [24]) *Every $2k$-chromatic oriented graph contains each oriented k-edge tree.*

A version of Conjecture 8.3 for large k, replacing oriented trees with oriented paths, and '$2k$-chromatic' with 'k-chromatic', is attributed in [11] to Bondy. There is also a generalisation of Conjecture 8.2 in the spirit of Conjecture 8.3: Every $(k + \ell)$-chromatic digraph contains each oriented k-edge tree having ℓ leaves (this was conjectured in [70]). Furthermore, Naia [105] conjectures that for every oriented k-edge tree T, the minimum n such that every tournament of order n contains T coincides with the minimum n such that every n-chromatic oriented graph contains T. This would imply that Conjecture 8.3 and Conjecture 8.1 are equivalent.

Conjecture 8.3 is only known for some specific classes of oriented paths (see [1] for references) and for all oriented stars [105]. Burr [24] showed that Conjecture 8.3 is true if we replace $2k$ with k^2, and Addario-Berry et al. [1] improved this (roughly by a factor of 2). A better bound only for oriented graphs with large chromatic number is given in [105].

The authors of [1] also propose an interesting conjecture of their own. In order to be able to state their conjecture, we need a definition. An *antidirected* tree is an oriented tree each of whose vertices either has no incoming edges or no outgoing edges.

Conjecture 8.4 (Addario-Berry et al. [1]) *Every digraph D with more than $(k-1)|V(D)|$ edges contains each antidirected k-edge tree.*

This conjecture is best possible because of the $(k-1)$-regular tournament which does not contain the k-edge out-star, or alternatively, because of the complete digraph on k vertices which does not contain *any* oriented k-edge tree. Also note that a version of Conjecture 8.4 for oriented trees that are not antidirected fails (even if we made the condition on the number of edges stronger): Consider a large complete bipartite graph $G = (A, B)$, and orient all its edges from A to B. The resulting oriented graph only has antidirected subgraphs.

The authors of [1] verify Conjecture 8.4 for antidirected trees of diameter at most 3, and they note it is not difficult to see that Conjecture 8.4 implies Conjecture 8.3 (and therefore also Conjecture 8.1) for antidirected trees. (This is because every $2k$-chromatic graph has a subgraph H of minimum degree at least $2k - 1$, and thus H has more than $(k - 1)|V(H)|$ edges.) They also note that if we restrict Conjecture 8.4 to symmetric digraphs (a digraph is *symmetric* if all its edges are bidirected), then the conjecture becomes equivalent to the Erdős–Sós conjecture (Conjecture 2.1). So one can interpret Conjecture 8.4 as a common generalisation of Conjecture 2.1 and Conjecture 8.3.

It seems, however, slightly dissatisfying that Conjecture 8.4 only applies to antidirected trees. We have seen above that this is necessary, as there are oriented graphs on $2n$ vertices with n^2 edges that do not even contain a two-edge directed path. In fact, any antidirected host graph with enough edges would serve as an example. Now, in order to avoid these examples, one might try requiring that the vertices of the host digraph, on average, had *both* large enough in-degree $d^-(v)$ *and* large enough out-degree $d^+(v)$.

That is, defining the *semidegree* of a vertex v as

$$d^0(v) := \min\{d^-(v), d^+(v)\},$$

we would require the average of the semidegrees, taken over all vertices v of the digraph D, to be larger than $k-1$ (or more generally, to be larger than some function of k). Although this would clearly exclude all antidirected host graphs, it is not sufficient to guarantee all oriented trees as subdigraphs. In order to see this, just consider an appropriate blow-up[3] of a $(k - 1)$-edge directed path. (Observe that this example also shows that a naïve extension of the Loebl–Komlós–Sós conjecture that replaces $d(v)$ with $d^0(v)$ fails.)

Another possibility is to consider the *minimum semidegree*

$$\delta^0(D) := \min\left\{d^0(v) : v \in V(D)\right\}$$

of a digraph D. Using a greedy embedding argument, it is clear that any digraph with $\delta^0(D) \geq k$ must contain each oriented k-edge tree.

This trivial bound can be lowered if $k = n$, and the tree is a path. Indeed, results from [33, 34] imply that if D is an n-vertex digraph with $\delta^0(D) \geq \frac{n}{2}$, then D contains every orientation of the path on n vertices, and this is sharp. This might extend to oriented paths of smaller size. A first question in this direction would be whether Observation 4.4 extends to oriented graphs.

[3] A *blow-up* of a digraph D is obtained by replacing each vertex with an independent set of vertices, and adding all edges from such a set X to a set Y, if X and Y originated from vertices x and y belonging to an edge $\overrightarrow{xy}$ of D.

Conjecture 8.5 *Does every oriented graph D with $\delta^0(D) > \frac{k}{2}$ contain each oriented k-edge path?*

If this conjecture is true, it would be sharp. This can be seen by considering, for even k, a blow-up of the directed triangle, replacing each vertex with an independent set of size $\frac{k}{2}$. The antidirected path with k edges is not contained in this graph. Moreover, the conjecture is true for directed paths, by a result of Jackson [82]. If we replace the bound on the minimum semidegree with $\frac{3}{4}k$, it holds for antidirected paths [89]. If the host graph is a tournament, Conjecture 8.5 follows from Conjecture 8.1.

An analogous question can be asked for digraphs. Observe that now, we need to require, in addition to the minimum semidegree condition, a lower bound on the size of the largest component (in order to prevent the digraph being the union of complete digraphs of order $\frac{k}{2} + 2$).

Question 8.6 *Does every digraph D with $\delta^0(D) > \frac{k}{2}$ having a component of size at least $k + 1$ contain each oriented k-edge path?*

If not, can we lift the bound on the minimum semidegree (to some bound strictly below k) so that the question can be answered in the affirmative?

If necessary, one might additionally require a larger component (or a large strong component).

Let us shift our attention from oriented paths to oriented bounded degree trees. Mycroft and Naia [104] used the minimum semidegree notion to give an extension of Theorem 4.1 to digraphs.

Theorem 8.7 (Mycroft and Naia [104]) *For all positive real α, Δ there exists n_0 such that for all $n \geq n_0$ every n-vertex digraph D with $\delta^0(D) \geq (\frac{1}{2} + \alpha)n$ contains every oriented n-vertex tree of maximum degree at most Δ.*

In view of their result we feel encouraged to ask whether generalisations to digraphs, using the minimum semidegree notion, of the results and conjectures from Section 5 exist. In particular, if Conjecture 8.5 (Question 8.6) is true, one might try for results in the spirit of Theorem 5.4 and Conjectures 5.1, 5.2, and 5.5.

Question 8.8 *Are there constants $c < 1$ and C such that every oriented graph (digraph) D with $\delta^0(D) \geq ck$ that has a vertex v with $d^0(v) \geq Ck$ contains each oriented k-edge tree?*

Another possibility is to substitute the semidegree with another degree notion. One natural candidate is the *total minimum degree* $\delta_{tot}(D)$, which is defined as the minimum of the sums of the in- and out-degrees of the vertices of the digraph D. Mycroft and Naia asked the following question [104, Problem 4.1].

Question 8.9 ([104]) *Does Theorem 8.7 remain true if $\delta^0(D)$ is replaced by $\frac{\delta_{tot}(D)}{2}$?*

If this is true, one could ask for similar variants of the other open questions from this section.

We close the section with a short remark on Ramsey numbers for oriented trees. There are two natural notions. The *oriented Ramsey number* $R_k^{\rightarrow}(T)$ of an oriented tree T is the smallest integer n such that every k-coloured tournament on n vertices contains a monochromatic copy of T. The *directed Ramsey number* $R_k^{\leftrightarrow}(T)$ is defined in the same way, replacing the k-coloured tournament with a k-coloured complete digraph. Early results using these notion focused on directed paths and two colours [28, 62]. An interesting insight gives a recent work of Bucič, Letzter, and Sudakov [22] who establish a difference in the order of magnitude of the two numbers, by showing that $R_k^{\rightarrow}(T) = c_k|V(T)|^k$ and $R_k^{\leftrightarrow}(T) = c_k|V(T)|^{k-1}$. As observed in [22], the former of these two equalities would also follow from Conjecture 8.3 (Burr's conjecture).

9 Hypergraphs

We will only discuss r-uniform hypergraphs, and call such hypergraphs r-graphs for short. As one might expect, there is more than one natural generalisation of trees to hypergraphs. In what follows, we will discuss *tight r-trees, linear r-trees, r-expansions* and *Berge r-trees*. We refer to [56, 87] for an overview of more Túran type results for hypergraphs.

9.1 *Tight Hypertrees*

We start our overview with tight hypergraphs. Call an r-graph a *tight r-tree* if its edges can be ordered such that except for the first edge, every edge consists of an $(r-1)$-set contained in some previous edge, and an entirely new vertex. Note that, for instance, the widely studied tight r-paths are examples of tight r-trees. (A *tight r-path* has vertices $v_1, \ldots, v_n$ and edges $\{v_i, \ldots, v_{i+k-1}\}$ for $1 \leq i \leq n-k+1$.)

For r-graphs and tight r-trees, Kalai proposed in 1984 the following natural generalisation of the Erdős–Sós conjecture (see [48]).

Conjecture 9.1 (Kalai's Conjecture, see [48]) *Let $r \geq 2$ and let H be an r-graph on n vertices with more than $\frac{k-1}{r}\binom{n}{r-1}$ edges. Then H contains every tight r-tree T having k edges.*

As already noted in [48], it follows from constructions using a result of Rödl [116] (or alternatively, one can use designs whose existence is guaranteed by Keevash's work [88]) that Conjecture 9.1 is tight as long as certain divisibility conditions are satisfied.

It is not difficult to observe (see, e.g., [51, Proposition 5.4]) that any n-vertex r-graph on n vertices with more than $(k-1)\binom{n}{r-1}$ edges contains every tight r-tree with k edges, which is a factor of r away from the conjectured bound. This bound can be proved as follows: successively delete all edges at $(r-1)$-sets of vertices that lie in few edges until arriving at a subhypergraph of large minimum ‘codegree’. Then, greedily embed the tree.

Not much is known on Kalai’s conjecture in general. Restricting the class of host r-graphs, it is known that the conjecture holds if the host r-graph H is r-partite [122].

Restrictions on the type of tight r-trees have led to the following results. In 1987, Frankl and Füredi [48] showed that Conjecture 9.1 holds for all ‘star-shaped’ tight r-trees, that is, for all tight r-trees whose first edge intersects each other edge in $r-1$ vertices. Füredi et al. show in [54] an asymptotic version of Conjecture 9.1 for a broadened concept of ‘star-shaped’ (the first c edges have to intersect all other edges in $r-1$ vertices, for a constant c), and in [55] an exact result for a class of tight 3-trees. Füredi and Jiang [51] show Conjecture 9.1 for special types of tight r-trees with many leaves.

On the opposite extreme of the spectrum of tight r-trees, there are the tight r-paths. Improving on results of Patkós [107], Füredi et al. [53] show that for tight r-paths the bound in Conjecture 9.1 can be replaced by $\frac{k-1}{2}\binom{n}{r-1}$ if r is even, and by a similar bound if r is odd. Moreover, an asymptotic version of Kalai’s conjecture for tight r-paths whose order is linear in the order n of the host r-graph has been established by Allen, Böttcher, Cooley, and Mycroft [4] for large n. The authors of [4] remark that they do not believe their result to be best possible, arguing that the constructions and designs from [88, 116] only exist when the order of the host graph is much larger than the order of the tight path.

It seems natural to seek extensions of other results for graphs to r-graphs and tight r-trees. This has been done for Theorem 4.1. As usual, for any r-graph H, let $\delta_i(H)$ ($\Delta_i(H)$) denote the minimum (maximum) number of edges any i-subset of $V(H)$ belongs to. With this notation, and using hypergraph regularity, Pavez-Signé et al. [108] show a version of Theorem 4.1 for hypergraphs. Namely, they show that for any $\gamma, \Delta > 0$, every large enough r-graph H with $\delta_{k-1}(H) \geq (\frac{1}{2}+\gamma)|V(H)|$ contains each r-tree T of the same order obeying $\Delta_1(T) \leq \Delta$.

One might also ask for generalisations of the results/conjectures from Section 5 to tight hypergraph trees.

Question 9.2 *Is there a function f such that every r-graph H with $\delta_{k-1}(H) \geq \frac{k}{2}$ and $\Delta_{k-1} \geq f(k)$ contains each k-edge tight r-tree?*

More cautiously, one could replace $\frac{k}{2}$ in Question 9.2 with $(1-\gamma)k$, for some fixed $\gamma > 0$. Perhaps it is also possible to extend Conjecture 3.3 to tight r-trees.

Question 9.3 *Let H be an r-graph such that at least $\frac{\binom{n}{r-1}}{2}$ of its $(r-1)$-tuples each belongs to at least k edges. Does H contain each k-edge tight r-tree?*

9.2 *Expansions of Trees and Linear Paths*

The r-expansion of a tree T is the r-uniform hypergraph obtained from T by adding to each edge $r-2$ new vertices. A *linear r-tree* is obtained from an edge by subsequently adding any number of new edges that each contain precisely one of the previous vertices. An r-expansion of a path is also called a *linear path*, as it satisfies the definition of a linear tree.

The *Turán number* $ex_r(n, H)$ of an r-graph H is defined (in complete analogy to the Turán number of a graph) as the maximum number of edges a hypergraph can have if it does not contain H. There is a considerable amount of literature on Turán numbers of expansions. For an overview we refer to the survey of Mubayi and Verstraëte [102]. One of the important results relevant for this survey is the determination of the Turán number of the linear r-path with k edges for fixed $r \geq 4$ and k and large n by Füredi et al. [52] using the delta-system-method. The case $r = 3$ was solved by Kostochka et al. [94] using an approach based on random sampling.

Füredi [50] asymptotically determined the Turán number for r-expansions of trees for $r \geq 4$, and conjectured the corresponding asymptotics for $r = 3$; this was confirmed by Kostochka et al. in [95]. These results relate the Turán number of an r-expansion T with the minimum size $\sigma(T)$ of a crosscut of T (where a *crosscut* is a set of vertices met by every edge of T in exactly one vertex). More precisely, for a fixed r-expansion T, and for $r \geq 3$, the Turán number $ex_r(n, T)$ is asymptotically determined as follows [50, 95]:

$$ex_r(n, T) = \big(\sigma(T) - 1 + o(n)\big)\binom{n}{r-1}.$$

That this bound is asymptotically best possible can be seen by considering the r-graph consisting of all edges containing exactly one vertex from a fixed set of size $\sigma(T) - 1$: This r-graph does not contain T. See also [51] for some related results.

9.3 Berge Hypertrees

Other recent activity has focused on Berge r-trees. A *Berge r-tree* is an r-graph H such that there is a tree T (i.e. an acyclic connected 2-graph), an injection from $V(T)$ to $V(H)$, and a bijection from $E(T)$ to $E(H)$ such that the images of the endpoints of any edge $e \in E(T)$ are contained in the image of e. This definition gives the usual definition of a Berge path if T is a path.

The Turán number for Berge r-paths $BP_k^{(r)}$ was almost completely determined by Győri et al. [64], with the last remaining case solved in [32]. The bound is $ex(n, BP_k^{(r)}) \leq \frac{n(k-1)}{r+1}$ if $r \geq k$ and $ex(n, BP_k^{(r)}) \leq \frac{n}{k}\binom{k}{r}$ if $r < k$, and extremal r-graphs are known.

Results for k-edge Berge r-trees $BT_k^{(r)}$ have been obtained by Gerbner et al. [57] and by Győri et al. [65]. If $r \geq k(k-2)$ and the tree we are looking for is not a star, then the bound for Berge r-paths from the previous paragraph applies [65]. In the case $k > r$ the best known bound is $ex(n, BT_k^{(r)}) \leq \frac{2(r-1)n}{k}\binom{k}{r}$, although this can be lowered by a factor of $2(r - 1)$, thus reaching the bound for Berge paths from the previous paragraph, if we assume the Erdős–Sós conjecture holds [57]. These bounds are sharp under certain divisibility conditions.

It would be interesting to see extensions, both to linear trees and to Berge trees, of the results we have seen in Sections 4 and 5 for graphs, that is, results that use conditions on the minimum (and maximum) degree of the host graph, instead of the average degree (i.e. number of edges).

Acknowledgments Maya Stein acknowledges support by CONICYT + PIA/Apoyo a centros científicos y tecnológicos de excelencia con financiamiento Basal, Código AFB170001, and by Fondecyt Regular Grant 1183080.

References

1. L. Addario-Berry, F. Havet, C. Linhares Sales, B. Reed, S. Thomassé, Oriented trees in digraphs. Discrete Math. **313**, 967–974 (2013)
2. M. Ajtai, J. Komlós, E. Szemerédi, The longest path in a random graph. Combinatorica **1**(1), 1–12 (1981)
3. M. Ajtai, J. Komlós, E. Szemerédi, On a conjecture of Loebl, in *Graph Theory, Combinatorics, and Algorithms, Vol. 1, 2 (Kalamazoo, MI, 1992)* (Wiley, New York, 1995), pp. 1135–1146
4. P. Allen, J. Böttcher, O. Cooley, R. Mycroft, Tight cycles and regular slices in dense hypergraphs. J. Combin. Theory Ser. A **149**, 30–100 (2017)
5. N. Alon, M. Krivelevich, B. Sudakov, Embedding nearly-spanning bounded degree trees. Combinatorica **27**(6), 629–644 (2007)
6. P. Araújo, L. Moreira, M. Pavez-Signé, Ramsey goodness of trees in random graphs (Preprint 2020). arXiv:2001.03083
7. S. Balasubramanian, E. Dobson, Constructing trees in graphs with no $K_{2,s}$. J. Graph. Theory **156**, 301–310 (2007)
8. J. Balogh, B. Csaba, M. Pei, W. Samotij, Large bounded degree trees in expanding graphs. Electron. J. Combin. **17**, 1 (2010). Research Paper 6, 9 pages
9. J. Balogh, B. Csaba, W. Samotij, Local resilience of almost spanning trees in random graphs. Random Struct. Algoritm. **38**, 121–139 (2011)
10. J. Balogh, A. Dudek, L. Li, An analogue of the Erdős-Gallai theorem for random graphs (Preprint 2019). arXiv:1909.00214
11. J. Bang-Jensen, F. Havet, *Tournaments and Semicomplete Digraphs* (Springer International Publishing, Cham, 2018), pp. 35–124
12. O. Barr, R. Johansson, Another Note on the Loebl–Komlós–Sós Conjecture. Research Reports No. 22, Umeå University, Sweden, 1997
13. C. Bazgan, H. Li, M. Woźniak, On the Loebl-Komlós-Sós conjecture. J. Graph Theory **34**(4), 269–276 (2000)
14. G. Besomi, Tree embeddings in dense graphs. Master Thesis, University of Chile (2018)
15. G. Besomi, M. Pavez-Signé, M. Stein, Minimum and maximum degree conditions for embedding trees. SIAM J. Discret. Math. (2018, accepted). arXiv:1808.09934
16. G. Besomi, M. Pavez-Signé, M. Stein, On the Erdős-Sós conjecture for bounded degree trees (Preprint 2019). arXiv:1906.10219
17. G. Besomi, M. Pavez-Signé, M. Stein, Degree conditions for embedding trees. SIAM J. Discrete Math. **33**, 1521–1555 (2019)
18. B. Bollobás, *Extremal Graph Theory*. L.M.S. Monographs (Academic, Cambridge, 1978)

19. B. Bollobás, The evolution of sparse graphs, in *Graph Theory and Combinatorics (Cambridge, 1983)* (Academic, London, 1984), pp. 35–57
20. J. Böttcher, J. Han, Y. Kohayakawa, R. Montgomery, O. Parczyk, Y. Person, Universality for bounded degree spanning trees in randomly perturbed graphs. Random Struct. Algoritm. **55**(4), 854–864 (2019)
21. S. Brandt, E. Dobson, The Erdős–Sós conjecture for graphs of girth 5. Discrete Math. **150**, 411–414 (1996)
22. M. Bucić, S. Letzter, B. Sudakov, Directed Ramsey number for trees. J. Combin. Theory Ser. B **137**, 145–177 (2019)
23. S.A. Burr, Generalized Ramsey theory for graphs—a survey, in *Graphs and Combinatorics (Proceedings of the Capital Conference on, George Washington University, Washington, DC, 1973)*. Lecture Notes in Mathematics, vol. 406 (Springer, Berlin, 1974), pp. 52–75
24. S.A. Burr, Subtrees of directed graphs and hypergraphs, in *Proceedings of the Eleventh Southeastern Conference on Combinatorics, Graph Theory and Computing (Boca Raton, Fla., 1982)*, vol. 28 (1980), pp. 227–239
25. S.A. Burr, P. Erdős, Extremal Ramsey theory for graphs. Utilitas Math. **9**, 247–258 (1976)
26. S.A. Burr, J.A. Roberts, On Ramsey numbers for stars. Utilitas Math. **4**, 217–220 (1973)
27. S. Ceroi, F. Havet, Trees with three leaves are (n+1)-unavoidable. Discrete Appl. Math. **141**(1), 19–39 (2004). Brazilian Symposium on Graphs, Algorithms and Combinatorics
28. V. Chvátal, Monochromatic paths in edge-colored graphs. J. Combin. Theory Ser. B **13**(1), 69–70 (1972)
29. O. Cooley, Proof of the Loebl–Komlós–Sós conjecture for large, dense graphs. Discrete Math. **309**(21), 6190–6228 (2009)
30. B. Csaba,, I. Levitt, J. Nagy-György, E. Szemerédi, Tight bounds for embedding bounded degree trees, in *Fête of Combinatorics and Computer Science*, ed. by G.O.H. Katona, A. Schrijver, T. Szenyi, G. Sági, vol. 20 (Springer, Berlin, 2010), pp. 95–137
31. E. Davies, M. Jenssen, B. Roberts, Multicolour Ramsey numbers of paths and even cycles. Eur. J. Combin. **63**, 124–133 (2017)
32. A. Davoodi, E. Győri, A. Methuku, C. Tompkins, An Erdős-Gallai type theorem for uniform hypergraphs (Preprint 2016). arXiv:1608.03241
33. L. DeBiasio, T. Molla, Semi-degree threshold for anti-directed Hamiltonian cycles. Electr. J. Combin. **43**, 22 (2015)
34. L. DeBiasio, D. Kühn, T. Molla, D. Osthus, A. Taylor, Arbitrary orientations of Hamilton cycles in digraphs. SIAM J. Discrete Math. **29**, 1553–1584 (2015)
35. G.A. Dirac, Some theorems on abstract graphs. Proc. London Math. Soc. **2**, 69–81 (1952)
36. E. Dobson, Ph.D. Dissertation. Louisiana State University, Baton Rouge, LA, 1995
37. E. Dobson, Constructing trees in graphs whose complement has no $K_{2,s}$. Combin. Probab. Comput. **11**(4), 343–347 (2002)
38. F. Dross, F. Havet, On unavoidability of oriented trees. Electron. Notes Theor. Comput. Sci. **346**, 425–436 (2019)
39. A. El Sahili, Trees in tournaments. J. Combin. Theory (Ser. B) **92**, 183–187 (2004)
40. P. Erdős, Extremal problems in graph theory, in *Theory of Graphs and Its Applications, Proceedings of the Symposium Smolenice* (1964), pp. 29–36
41. P. Erdős, T. Gallai, On maximal paths and circuits of graphs. Acta Mathe. Acad. Sci. Hung. **10**(3), 337–356 (1959)
42. P. Erdős, R.L. Graham, On partition theorems for finite graphs, in *Infinite and Finite Sets (Colloq., Keszthely, 1973; Dedicated to P. Erdős on his 60th birthday), Vol. I*. Colloquia Mathematica Societatis János Bolyai, vol. 10 (1975), pp. 515–527
43. P. Erdős, R.J. Faudree, C.C. Rousseau, R.H. Schelp, Ramsey numbers for brooms, in *Proceedings of the Thirteenth Southeastern Conference on Combinatorics, Graph Theory and Computing (Boca Raton, Fla., 1982)*, vol. 35 (1982), pp. 283–293
44. P. Erdős, Z. Füredi, M. Loebl, V. Sós, Discrepancy of trees. Studia Sci. Math. Hungar. **30**, 47–57 (1995)

45. G. Fan, Y. Hong, Q. Liu, The Erdős-Sós conjecture for spiders (Preprint 2018). arXiv:1804.06567
46. R.J. Faudree, R.H. Schelp, Path Ramsey numbers in multicolorings. J. Combin. Theory, Ser. B **19**, 150–160 (1975)
47. A. Ferber, R. Nenadov, U. Peter, Universality of random graphs and rainbow embedding. Random Struct. Algoritm. **48**(3), 546–564 (2016)
48. P. Frankl, Z. Füredi, Exact solution of some Turán-type problems. J. Combin. Theory Ser. A **45**(2), 226–262 (1987)
49. J. Friedman, N. Pippenger, Expanding graphs contain all small trees. Combinatorica **7**(1), 71–76 (1987)
50. Z. Füredi, Linear trees in uniform hypergraphs. Eur. J. Combin. **35**, 264–272 (2014)
51. Z. Füredi, T. Jiang, Turán numbers of hypergraph trees (Preprint 2015). arXiv:1505.03210
52. Z. Füredi, T. Jiang, R. Seiver, Exact solution of the hypergraph Turán problem for k-uniform linear paths. Combinatorica **34**, 299–322 (2014)
53. Z. Füredi, T. Jiang, A. Kostochka, D. Mubayi, J. Verstraëte, Tight paths in convex geometric hypergraphs (Preprint 2017). arXiv:1709.01173
54. Z. Füredi, T. Jiang, A. Kostochka, D. Mubayi, J. Verstraëte, Hypergraphs not containing a tight tree with a bounded trunk. SIAM J. Discrete Math. **33**(2), 862–873 (2019)
55. Z. Füredi, T. Jiang, A. Kostochka, D. Mubayi, J. Verstraëte, Hypergraphs not containing a tight tree with a bounded trunk II: 3-trees with a trunk of size 2. Discrete Appl. Math. **276**, 50–59 (2019)
56. D. Gerbner, B. Patkós, *Extremal Finite Set Theory* (Chapman and Hall/CRC, Boca Raton, 2018)
57. D. Gerbner, A. Methuku, C. Palmer, General lemmas for Berge-Turán hypergraph problems (Preprint 2018). arXiv:1808.10842
58. L. Gerencsér, A. Gyárfás, On Ramsey-type problems. Ann. Univ. Sci. Budapest. Eötvös Sect. Math. **10**, 167–170 (1967)
59. A. Goerlich, A. Zak, On Erdős-Sós conjecture for trees of large size. Electr. J. Combin. **23**(1), P1–52 (2016)
60. J.W. Grossman, F. Harary, M. Klawe, Generalized Ramsey theory for graphs. X. Double stars. Discrete Math. **28**(3), 247–254 (1979)
61. B. Grünbaum, Antidirected Hamiltonian paths in tournaments. J. Combin. Theory (Ser. B) **11**, 249–257 (1971)
62. A. Gyárfás, J. Lehel, A Ramsey type problem in in directed and bipartite graphs. Period. Math. Hung. **3**(3–4), 299–304 (1973)
63. A. Gyárfás, M. Ruszinkó, G.N. Sárközy, E. Szemerédi, Three-color Ramsey numbers for paths,. Combinatorica **27**, 35–69 (2007); Corrigendum **28**, 499–502 (2008)
64. E. Győri, G.Y. Katona, N. Lemons, Hypergraph extensions of the Erdős-Gallai theorem. Eur. J. Combin. **58**, 238–246 (2016)
65. E. Győri, A. Salia, C. Tompkins, O. Zamora, Turán numbers of Berge trees (Preprint 2019). arXiv:1904.06728
66. R. Häggkvist, A. Thomason, Trees in tournaments. Combinatorica **11**, 123–130 (1991)
67. F. Harary, Recent results on generalized Ramsey theory for graphs, in *Graph Theory and Applications (Proceedings of the Conference on Western Michigan University, Kalamazoo, Michigan, 1972; Dedicated to the Memory of J. W. T. Youngs)*. Lecture Notes in Mathematics, vol. 303 (Springer, Berlin, 1972), pp. 125–138
68. F. Havet, Trees in tournaments. Discrete Math. **243**, 121–134 (2002)
69. F. Havet, On unavoidability of trees with k leaves. Graph. Combin. **19**, 101–110 (2003)
70. F. Havet, S. Thomassé, Median orders of tournaments: a tool for the second neighbourhood problem and Sumner's conjecture. J. Graph Theory **35**, 244–256 (2000)
71. F. Havet, S. Thomassé, Oriented Hamiltonian paths in tournaments: a proof of Rosenfeld's conjecture. J. Combin. Theory (Ser. B) **78**, 243–273 (2000)
72. F. Havet, B. Reed, M. Stein, D.R. Wood, A variant of the Erdős-Sós conjecture. J. Graph Theory **94**(1), 131–158 (2020)

73. P.E. Haxell, Tree embeddings. J. Graph Theory **36**(3), 121–130 (2001)
74. P. Haxell, T. Łuczak, Embedding trees into graphs of large girth. Discrete Math. **216**(1), 273–278 (2000)
75. P.E. Haxell, T. Łuczak, P.W. Tingley, Ramsey numbers for trees of small maximum degree. Combinatorica **22**(2), 287–320 (2002). Special issue: Paul Erdős and his mathematics.
76. J. Hladký, D. Piguet, Loebl–Komlós–Sós Conjecture: dense case. J. Comb. Theory Ser. B **116**, 123–190 (2016)
77. J. Hladký, D. Piguet, M. Simonovits, M. Stein, E. Szemerédi, The approximate Loebl-Komlós-Sós conjecture and embedding trees in sparse graphs. Electr. Res. Announc. Math. Sci. **22**, 1–11 (2015)
78. J. Hladký, J. Komlós, D. Piguet, M. Simonovits, M. Stein, E. Szemerédi, The approximate Loebl–Komlós–Sós conjecture I: the sparse decomposition. SIAM J. Discrete Math. **31**(2), 945–982 (2017)
79. J. Hladký, J. Komlós, D. Piguet, M. Simonovits, M. Stein, E. Szemerédi, The approximate Loebl–Komlós–Sós conjecture II: the rough structure of LKS graphs. SIAM J. Discrete Math. **31**(2), 983–1016 (2017)
80. J. Hladký, J. Komlós, D. Piguet, M. Simonovits, M. Stein, E. Szemerédi, The approximate Loebl–Komlós–Sós conjecture III: the finer structure of LKS graphs. SIAM J. Discrete Math. **31**(2), 1017–1071 (2017)
81. J. Hladký, J. Komlós, D. Piguet, M. Simonovits, M. Stein, E. Szemerédi, The approximate Loebl–Komlós–Sós conjecture IV: embedding techniques and the proof of the main result. SIAM J. Discrete Math. **31**(2), 1072–1148 (2017)
82. B. Jackson, Long paths and cycles in oriented graphs. J. Graph Theory **5**(2), 145–157 (1981)
83. T. Jiang, On a conjecture about trees in graphs with large girth. J. Combin. Theory Ser. B **83**(2), 221–232 (2001)
84. D. Johannsen, M. Krivelevich, W. Samotij, Expanders are universal for the class of all spanning trees, in *Proceedings of the Twenty-Third Annual ACM-SIAM Symposium on Discrete Algorithms*, SODA '12, SIAM (2012), pp. 1539–1551
85. F. Joos, J. Kim, Spanning trees in randomly perturbed graphs (Preprint 2018). arXiv:1803.04958
86. J. Kahn, E. Lubetzky, N. Wormald, The threshold for combs in random graphs. Random Struct. Algoritm. **48**, 794–802 (2016)
87. P. Keevash, Turán numbers for hypergraphs. Surveys Combin. **392**, 83–140 (2011)
88. P. Keevash, The existence of designs (Preprint 2014). arXiv:1401.3665
89. T. Klimošová, M. Stein, Personal communication
90. T. Klimošova, D. Piguet, V. Rohzoň, An version of the Loebl–Komlós–Sós conjecture for skewed trees (Preprint 2018). arXiv 1802.00679
91. J. Komlós, E. Szemerédi, Limit distribution for the existence of Hamiltonian cycles in a random graph. Discrete Math. **43**(1), 55–63 (1983)
92. J. Komlós, G.N. Sárközy, E. Szemerédi, Proof of a packing conjecture of bollobás. Combin. Probab. Comput. **4**(3), 241–255 (1995)
93. J. Komlós, G.N. Sárközy, E. Szemerédi, Spanning trees in dense graphs. Combin. Probab. Comput. **10**(5), 397–416 (2001)
94. A. Kostochka, D. Mubayi, J. Verstraëte, Turán problems and shadows I: paths and cycles. J. Combin. Theory, Ser. B **129**, 57–79 (2015)
95. A. Kostochka, D. Mubayi, J. Verstraëte, Turán problems and shadows II: trees. J. Combin. Theory Ser. B **122**, 457–478 (2017)
96. M. Krivelevich, Embedding spanning trees in random graphs. SIAM J. Discrete Math. **24**(4), 1495–1500 (2010)
97. M. Krivelevich, M. Kwan, B. Sudakov, Bounded-degree spanning trees in randomly perturbed graphs. SIAM J. Discrete Math. **31**(1), 155–171 (2017)
98. D. Kühn, Mycroft, R., D. Osthus, An approximate version of Sumner's universal tournament conjecture. J. Combin. Theory Ser. B **101**(6), 415–447 (2011)

99. D. Kühn, R. Mycroft, D. Osthus, A proof of Sumner's universal tournament conjecture for large tournaments. Proc. Lond. Math. Soc. (3) **102**(4), 731–766 (2011)
100. A. McLennan, The Erdős-Sós conjecture for trees of diameter four. J. Graph Theory **49**(4), 291–301 (2005)
101. R. Montgomery, Spanning trees in random graphs (Preprint 2018). arXiv:1810.03299
102. D. Mubayi, J. Verstraëte, *A Survey of Turán Problems for Expansions*. (Springer, Berlin, 2016), pp. 117–143
103. R. Mycroft, T. Naia, Unavoidable trees in tournaments. Random Struct. Algoritm. **53**(2), 352–385 (2018)
104. R. Mycroft, T. Naia, Spanning trees of dense directed graphs. Electr. Notes Theoret. Comput. Sci. **346**, 645–654 (2019). The proceedings of Lagos 2019, the tenth Latin and American Algorithms, Graphs and Optimization Symposium (LAGOS 2019)
105. T. Naia, Large structures in dense directed graphs. PhD Thesis, University of Birmingham, 2018
106. S. Norin, Y.R. Sun, Y. Zhao, Asymptotics of Ramsey numbers of double stars (Preprint 2016). arXiv:1605.03612
107. B. Patkós, A note on traces of set families. Moscow J. Combin. Number Theory **2**, 47–55 (2012)
108. M. Pavez-Signé, N. Sanhueza-Matamala, M. Stein, Dirac-type conditions for spanning bounded-degree hypertrees (In preparation)
109. D. Piguet, M.J. Stein, The Loebl–Komlós–Sós conjecture for trees of diameter 5 and for certain caterpillars. The Electr. J. Combin. **15**, R106 (2008)
110. D. Piguet, M.J. Stein, An approximate version of the Loebl–Komlós–Sós conjecture. J. Combin. Theory Ser. B **102**(1), 102–125 (2012)
111. L. Pósa, Hamiltonian circuits in random graphs. Discrete Math. **14**(4), 359–364 (1976)
112. L. Rédei, Ein kombinatorischer Satz. Acta Litt. Sci. Szeged **7**, 39–43 (1934)
113. B. Reed, M. Stein, Spanning trees in graphs of high minimum degree with a universal vertex I: An approximate asymptotic result (Preprint 2019). arXiv 1905.09801
114. B. Reed, M. Stein, Spanning trees in graphs of high minimum degree with a universal vertex II: A tight result (Preprint 2019). arXiv 1905.09806
115. K. Reid, N. Wormald, Embedding oriented n-trees in tournaments. Studis Sci. Math. Hun. **18**, 377–387 (1983)
116. V. Rödl, On a packing and covering problem. Eur. J. Combin. **6**, 69–78 (1985)
117. M. Rosenfeld, Antidirected Hamiltonian paths in tournaments. J. Combin. Theory (Ser. B) **12**, 93–99 (1971)
118. V. Rozhoň, Sufficient conditions for embedding trees. Master Thesis, Charles University, 2018
119. V. Rozhoň, A local approach to the Erdős–Sós conjecture. SIAM J. Discrete Math. **33**(2), 643–664 (2019)
120. J.-F. Saclé, M. Woźniak, A note on the Erdős–Sós conjecture for graphs without C_4. J. Combin. Theory (Ser. B) **70**(2), 229–234 (1997)
121. S.N. Soffer, The Komlós-Sós conjecture for graphs of girth 7. Discrete Math. **214**(1–3), 279–283 (2000)
122. M. Stein, On Kalai's conjecture in r-partite graphs (Preprint 2019). arXiv 1912.11421
123. B. Sudakov, J. Vondrák, A randomized embedding algorithm for trees. Combinatorica **30**(4), 445–470 (2010)
124. L. Sun, On the Loebl-Komlós-Sós conjecture. Australas. J. Combin. **37**, 271–275 (2007)
125. A. Thomason, Antidirected Hamiltonian paths in tournaments. Trans. Amer. Math. Soc. **296**, 167–180 (1986)
126. Y. Zhao, Proof of the $(n/2 - n/2 - n/2)$ conjecture for large n. Electr. J. Comb. **18**, P27 (2011)

Extremal Singular Graphs

Irene Triantafillou

Abstract A graph is called singular if its adjacency matrix is singular. The problem of characterizing all singular graphs was posed in 1957 by L. Collatz and U. Sinogowitz. Since then, it has been studied intensely due to its great importance in mathematics and in chemistry. However, the problem has proved to be rather complicated and has not yet been solved completely. In this paper, we emphasize on some of the most important results on graphs that are extremal with regard to their nullity/rank.

1 Introduction and Preliminaries

Let $G = (V, E)$ be a simple, undirected graph with nonempty vertex set V and edge set E. The *adjacency matrix*, $A(G)$, of a graph G on n vertices is defined as the $n \times n$ symmetric matrix whose entries a_{ij} are $a_{ij} = 1$ if vertex v_i is adjacent to vertex v_j, and $a_{ij} = 0$ otherwise. The *rank* of a graph G, $rank(G)$, is defined as the rank of its adjacency matrix $A(G)$. Equivalently, the rank of a graph G is the number of the nonzero eigenvalues of $A(G)$. The *nullity*, $\eta(G)$, of a graph G is the multiplicity of the eigenvalue zero of $A(G)$. It is clear that $\eta(G) + rank(G) = n$, where n is the order of the graph. A graph G is said to be *singular* if its adjacency matrix is singular, that is, $\eta(G) \geq 1$. Otherwise, G is said to be nonsingular.

In 1957, L. Collatz and U. Sinogowitz [5] posed the problem of characterizing all singular graphs (the problem had previously appeared in [31]). The problem is of great importance in mathematics since it is related to the singularity of the graph's adjacency matrix. The characterization of singular graphs presents a great interest in chemistry, also. In [20], it was shown that the nullity of a singular bipartite graph G corresponding to an alternant hydrocarbon is connected to the instability of the molecule that is represented by the graph. In the last 20 years, there has been a

I. Triantafillou (✉)
Department of Computer Science, University of Thessaly, Lamia, Greece

A. M. Raigorodskii, M. Th. Rassias (eds.), *Discrete Mathematics and Applications*, Springer Optimization and Its Applications 165,
https://doi.org/10.1007/978-3-030-55857-4_20

considerable amount of research on the relation between the structure of a graph and the rank/nullity of the graph, and many classes and families of graphs have been taken into consideration (see, for example, [7, 8, 11, 13, 16, 18, 25–27]). However, the characterization of all singular graphs has only been partially solved and still remains an open problem.

In Section 2 of this paper, we focus on some of the most important results on graphs that are extremal with respect to their nullity and their rank. In Section 3, we study the relation between certain graph parameters such as the matching number, the chromatic index and the independence number, and the nullity/rank of the graph. These parameters are taken into consideration as the problem of characterizing singular graphs becomes more complicated.

We denote the *path* and *cycle* with n vertices by P_n and C_n, respectively. By T_n we denote the *tree* and by S_n the *star graph* of order n. The *complete r-partite graph* is denoted by $K_{n_1,n_2,\ldots,n_r}$ and the *null graph* with n vertices by N_n. Let G_n denote the set of all graphs with n vertices, and let $[0, n] = \{0, 1, 2, \ldots, n\}$. A subset N of $[0, n]$ is called the *nullity set* of G_n if for all $k \in N$, there exists at least one graph $G \in G_n$ such that $\eta(G) = k$.

2 Extremal Graphs with Regard to Their Nullity/Rank

In this section, we focus on graphs on n vertices that achieve maximum nullity and equivalently minimum rank. In particular, we study graphs of nullity $n - r$ (or rank r), for $2 \leq r \leq 5$, which have been completely characterized. The characterization of graphs with nullity $n - 6$ remains an open problem as only partial results are known for certain graphs, such as bicyclic graphs and regular bipartite graphs.

The following lemma is straightforward.

Lemma 1 *Let G be a graph on n vertices. Then $\eta(G) = n$ ($rank(G) = 0$) if and only if G is the null graph N_n.*

It is well known that, there are no graphs G of rank 1 or nullity $\eta(G) = n - 1$. It then follows that $\eta(G) \leq n - 2$, if $G \neq N_n$. In [3], graphs that achieve the upper bounds $n - 2$ and $n - 3$, respectively, were completely characterized.

Theorem 1 ([3]) *Let G be a simple graph on n vertices ($n \geq 2$), and G does not contain any isolated vertices. Then, $\eta(G) = n - 2$ if and only if G is isomorphic to the complete bipartite graph K_{n_1,n_2},, where $n_1 + n_2 = n$, $n_1, n_2 > 0$.*

Theorem 2 ([3]) *Let G be a simple graph on n vertices ($n \geq 3$) and G does not contain any isolated vertices. Then, $\eta(G) = n - 3$ if and only if G is isomorphic to the complete tripartite graph where $n_1 + n_2 + n_3 = n$, $n_1, n_2, n_3 > 0$.*

For graphs of nullity $n - 4$ (or rank 4) several results were obtained for certain classes of graphs, such as unicyclic graphs, bicyclic graphs, bipartite graphs, and graphs with pendent vertices [8, 15, 16, 33]. In [1], all graphs of rank 4 (nullity $n - 4$) were completely characterized.

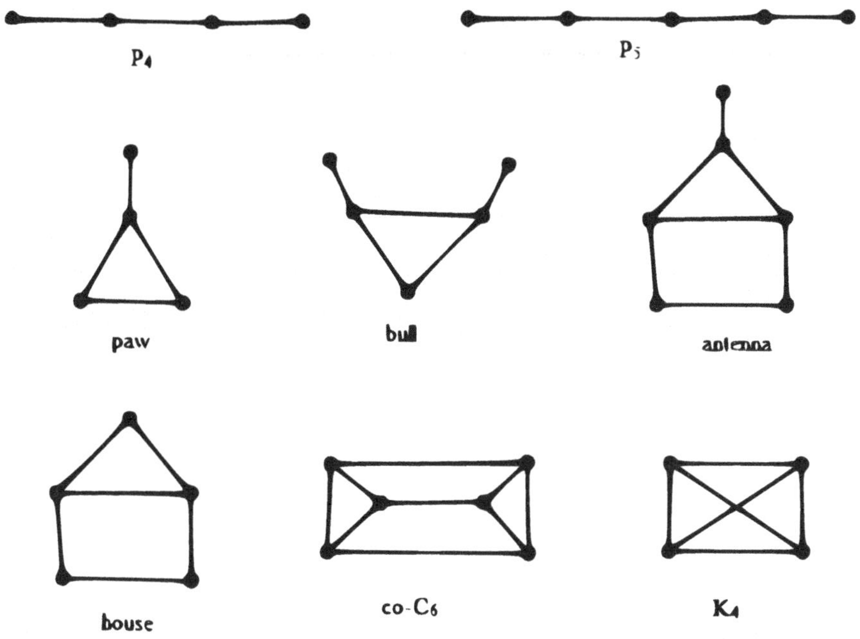

Fig. 1 Graphs of rank 4

First, we need to define a graph operation that is known as the multiplication of vertices. Let G be a graph with vertex set $V(G) = \{v_1, v_2, \ldots, v_n\}$ and let $m = (m_1, m_2, \ldots, m_n)$ be a vector of positive integers. The graph $G \circ m$ that is obtained from G by multiplication of vertices is the graph that is produced by replacing each vertex v_i of G with an independent set of m_i vertices v_i^1, $v_i^2, \ldots, v_i^{m_i}$ and by joining v_i^s and v_j^t if and only if v_i and v_j are adjacent in G. Let $M(G_1, G_2, \ldots, G_k)$ be the collection of all graphs which can be obtained from one of the graphs in $G_1, G_2, \ldots, G_k$ by multiplication of vertices.

Theorem 3 ([1]) *Let G be a connected graph. Then $rank(G) = 4$ if and only if G can be obtained from one of the graphs shown in Figure 1 by multiplication of vertices.*

In [2], all graphs of rank 5 (nullity $n - 5$) were completely characterized.

Theorem 4 ([2]) *Let G be a connected graph. Then $rank(G) = 5$ if and only if $G \in M(G_1, G_2, \ldots, G_{24})$, where $G_1, G_2, \ldots, G_{24}$ are shown in Figure 2.*

The problem of characterizing graphs of nullity $n - 6$ (rank 6) has proved to be more complicated and has only been solved for certain graphs. In the following part of this section, we emphasize on these particular families and classes of graphs which have attained the next upper bound for nullity, $n - 6$.

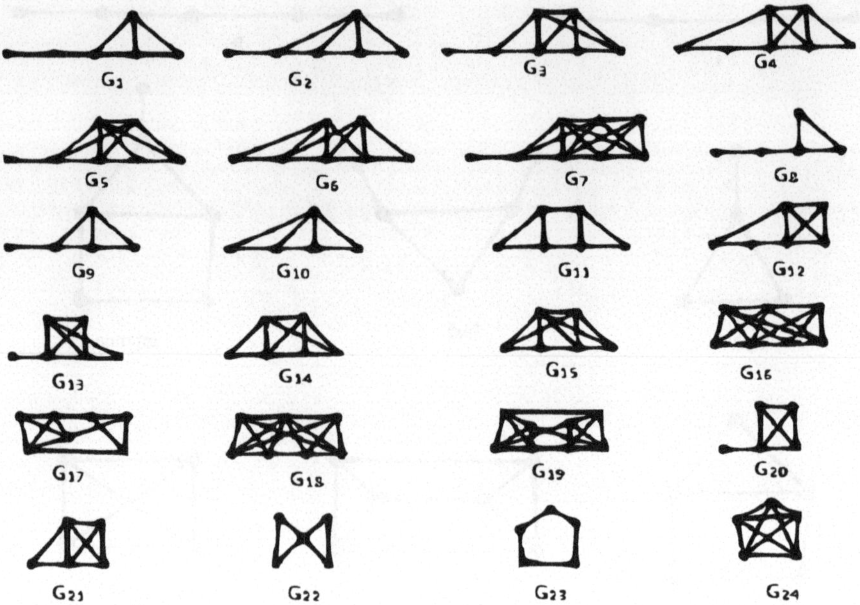

Fig. 2 Graphs of rank 5

2.1 Trees

A matching of G is a set of independent edges of G and a maximal matching is a matching that has the maximum possible number of edges. The size of a maximal matching of G, is denoted by $m = m(G)$.

The following result is well known.

Lemma 2 ([6]) *If T is a tree on n vertices and m is the size of its maximal matchings, then its nullity is equal to $\eta(T) = n - 2m$.*

From the above Lemma, it is easy to see that if n is even, then all the trees with a perfect matching have zero nullity. In the case where n is odd, then the nullity of the tree is one (minimal), if $m = (n-1)/2$.

In [16], the extremal nullity of trees and the nullity set of trees were determined.

Theorem 5 ([16]) *Let $\mathscr{T}_n$ denote the collection of all n-vertex trees.*

(i) Let $T \in \mathscr{T}_n$, then $\eta(T) \leq n-2$, and the equality holds if and only if $T \cong S_n$, where S_n denotes the star graph on n vertices.
(ii) Let $T \in \mathscr{T}_n - S_n$, then $\eta(T) \leq n-4$, and the equality holds if and only if $T \cong T_1$ or $T \cong T_2$, where T_1 and T_2 are shown in Figure 3.
(iii) Let $T \in \mathscr{T}_n - \{S_n, T_1, T_2\}$, then $\eta(T) \leq n-6$, and the equality holds if and only if $T \cong T_3$ or $T \cong T_4$ or$T \cong T_5$, where T_3, T_4, T_5 are shown in Figure 3.

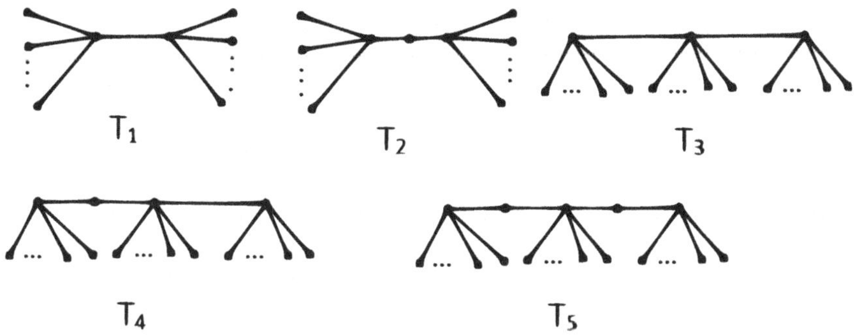

Fig. 3 Maximum nullity of trees

It is quite clear that we can use the above theorem to determine the n-vertex trees whose nullity is $n-8$, and so on.

Corollary 1 ([16]) *The nullity set of* $\mathcal{T}_n$ *is* $\{0, 2, 4, \ldots, n-4, n-2\}$ *if* n *is even, otherwise is* $\{1, 3, 5, \ldots, n-4, n-2\}$.

We will close this subsection with a known result regarding the line graphs of trees.

Theorem 6 ([13]) *If* T *is a tree, then* $L(T)$ *has either nullity zero or has nullity one.*

2.2 *Bipartite Graphs*

Let $G(V_1, V_2, E)$ be a bipartite graph in which E denotes the number of edges and V_1, V_2 are the two vertex sets having r, s number of vertices respectively. Then, the adjacency matrix of a bipartite graph can be written as:

$$A = \begin{bmatrix} 0 & B \\ B^T & 0 \end{bmatrix},$$

where B (also known as the biadjacency matrix) is an $r \times s$ matrix with $b_{i,j} = 1$ if and only if $u_i v_j \in E$ and 0 otherwise. B^T denotes the transpose matrix of B.

Theorem 7 ([32]) *For the bipartite graph* G *with* n *vertices and biadjacency matrix* B, $\eta(G) = n - 2rank(B)$.

The following theorem was shown in [14].

Theorem 8 ([14]) *If* G *is a bipartite graph on* n *vertices and* G *does not contain any cycle of length* $4s$ $(s = 1, 2, \ldots)$, *then* $\eta(G) = n - 2m$, *where* m *is the size of its maximum matching.*

In the previous section, it was shown that the complete bipartite graph achieves the maximum nullity $n - 2$. In [8], bipartite graphs with nullity $n - 4$ and regular bipartite graphs with nullity $n-6$ were completely characterized. We will start with some definitions.

Let $P_n := v_1 v_2 \cdots v_n$ $(n \geq 2)$ denote a path on n vertices. If we replace each vertex v_i by an empty graph O_{m_i} of order m_i for $i = 1, 2, \ldots, n$ and add edges between each vertex of O_i and each of O_{i+1} for $i = 1, 2, \ldots, n - 1$, we obtain a graph G of order $(m_1+m_2+\cdots+m_n)$, denoted by $\mathbb{P}_{m_1,m_2,\ldots,m_n} = O_{m_1} O_{m_2} \cdots O_{m_n}$, which is called an *expanded path* of length n. The empty graph O_{m_i} is called an *expanded vertex* of order m_i for $i = 1, 2, \ldots, n$. We can also define an *expanded cycle* of length $n(n \geq 3)$, denoted by $C_{m_1,m_2,\ldots,m_n} = O_{m_1} O_{m_2} \cdots O_{m_n} O_{m_1}$.

Denote by $\mathscr{BG}_n$ the set of all bipartite graphs of order n. Then,

Theorem 9 ([8]) *The nullity set of $\mathscr{BG}_n$ is $N = \{n - 2k : k = 0, 1, 2, \ldots, \lfloor n/2 \rfloor\}$.*

Theorem 10 ([8]) *Let G be a bipartite graph of order $n \geq 4$. Then $\eta(G) = n-4$ if and only if G is a graph H possibly adding some isolated vertices, where H is one of the following graphs: a union of two disjoint expanded paths both of length 2, an expanded path of length 4 or 5.*

Theorem 11 ([8]) *Let G be a regular bipartite graph of order $n \geq 6$. Then, $\eta(G) = n - 6$ if and only if G is one of the following graphs: a union of three disjoint expanded paths of length 2, an expanded cycle of length 6 or 8, and in each graph all expanded vertices have the same orders.*

2.3 Unicyclic, Bicyclic, and Tricyclic Graphs

A unicyclic graph is a simple connected graph that has equal number of vertices and edges (contains a unique cycle). Let $\mathscr{U}_n$ denote the set of all unicyclic graphs of order n. In [33], the nullity set as well as a characterization of unicyclic graphs with extremal nullity was given.

Theorem 12 ([33]) *The nullity set of $\mathscr{U}_n$ $(n \geq 5)$ is $[0, n - 4]$.*

Theorem 13 ([33]) *Let $U \in \mathscr{U}_n (n \geq 5)$. Then $\eta(U) = n-4$ if and only if $U \cong U_1^*$ or $U \cong U_2^*$ or $U \cong U_3^*$, where U_1^*, U_2^*, U_3^* are depicted in Figure 4 and S_{n_i} is the star graph K_{1,n_i-1}.*

The complete characterization of unicyclic graphs with nullity $n - 5$ was given in [11].

Theorem 14 ([11]) *Let G be a unicyclic graph with n vertices $(n \geq 5)$ and with nullity $\eta(G) = n - 5$. Then G must have the form of U_4^* shown in Figure 5 or $G = C_5$, where $r > 0$.*

In [11], the following theorem was also proven.

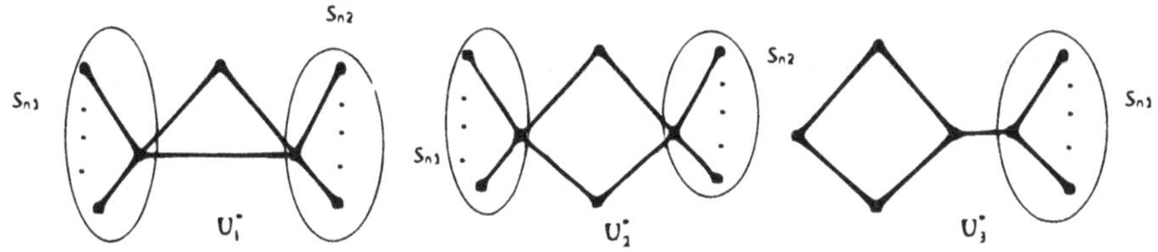

Fig. 4 The graphs U_1^*, U_2^*, U_3^*

Fig. 5 The graph U_4^*

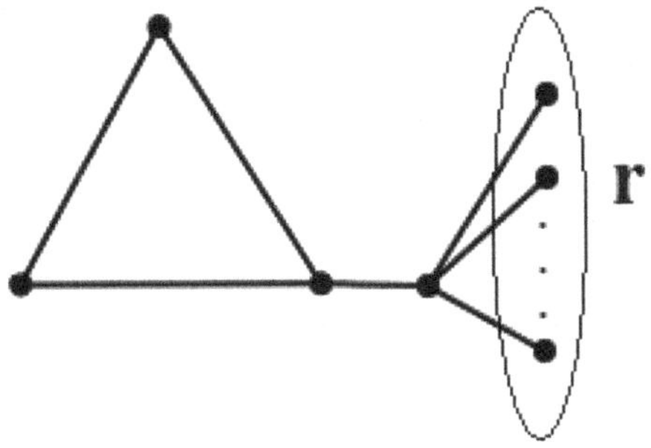

Theorem 15 ([11]) *Let G be a unicyclic graph with n vertices and m be the size of a maximum matching of G. Then the nullity of G is* $n - 2m - 1$, $n - 2m$, *or* $n - 2m + 2$.

A bicyclic graph is a simple connected graph in which the number of edges equals the number of vertices plus one. We denote the set of all bicyclic graphs of order n by $\mathscr{B}_n$.

Theorem 16 ([15]) *Let B be a bicyclic graph of order n. Then*

(i) $rank(B) = 2$ *(nullity* $\eta(G) = n - 2$*) if and only if* $B = K_{2,3}$.
(ii) $rank(B) = 3$ *(nullity* $\eta(G) = n - 3$*) if and only if* $B = K_4 - e$, *where* $e \in E(K_4)$.
(iii) $rank(B) = 4$ *(nullity* $\eta(G) = n - 4$*) if and only if* $B \cong B_i (1 \leq i \leq 7)$, *where* B_i *are shown in Figure 6.*

The nullity set of the bicyclic graph was also given in [15].

Theorem 17 ([15]) *The nullity set of* $\mathscr{B}_n$ $(n \geq 5)$ *is* $[0, n - 2]$.

In [18], two types of bicyclic graphs of nullity $\eta(G) = n - 6$ were identified. First, we need some definitions.

Let u_1, u_q be two vertices in C_k and C_l respectively where C_k, C_l are two vertex-disjoint cycles. The graph that is obtained by joining u_1, uq by a path $u_1u_2 \cdots u_q$ of length $q - 1$, where $q \geq 1$ is called an ∞-*graph*. In the case where $q = 1$ we identify vertices u_1, u_q. Let B_n^+ denote the set of all graphs that are ∞-graphs with trees attached when $q > 1$ and let B_n^{++} denote the set of all graphs that are ∞-graph with trees attached when $q = 1$.

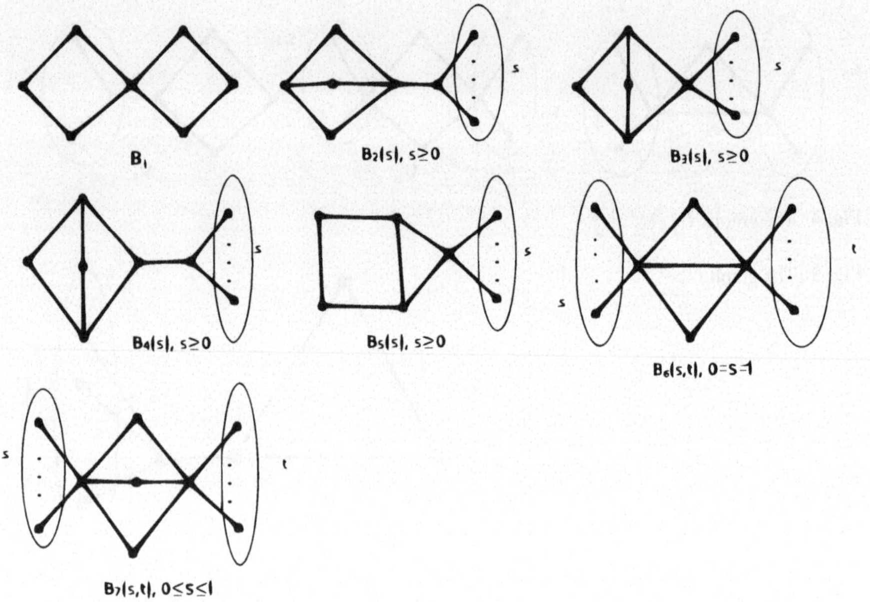

Fig. 6 The graphs B_i

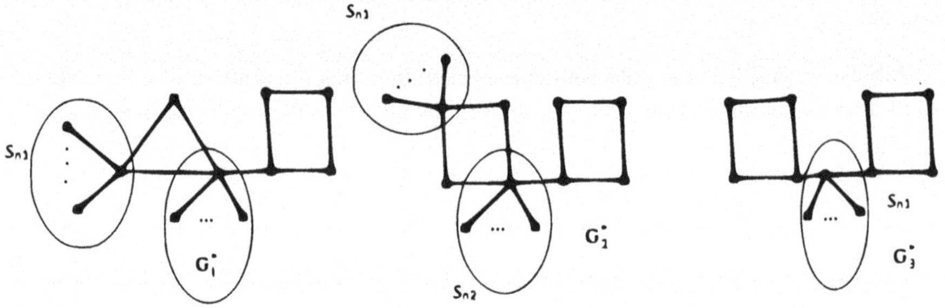

Fig. 7 The graphs G_1^*, G_2^*, G_3^*

Theorem 18 ([18]) *The nullity set of* B_n^+ $(n \geq 7)$ *is* $[0, n-6]$.

Theorem 19 ([18]) *Let* $G \in B_n^+$ $(n \geq 10)$. *Then* $\eta(G) = n - 6$ *if and only if* $G \cong G_1^*$ *or* $G \cong G_2^*$ *or* $G \cong G_3^*$, *where* G_1^*, G_2^*, G_3^* *are shown in Figure 7.*

Theorem 20 ([18]) *The nullity set of* B_n^{++} $(n \geq 8)$ *is* $[0, n-6]$.

Theorem 21 ([18]) *Let* $G \in B_n^{++}$ $(n \geq 8)$. *Then* $\eta(G) = n - 6$ *if and only if* $G \cong G_i^*$, $i \leq i \leq 8$, *where* G_i^* *are shown in Figure 8.*

A tricyclic graph is a simple connected graph in which the number of edges equals the number of vertices plus two. In [4], the tricyclic graphs with maximum

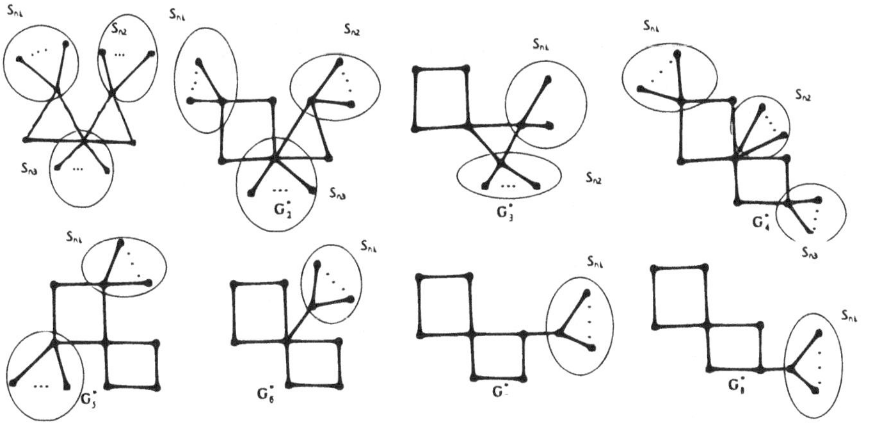

Fig. 8 The graphs G_i^*

nullity $n-4$ were completely characterized. We will start with some definitions. Let G be an r-partite graph and $X_1, X_2, \ldots, X_r$ the r classes of the graph. A chain-like r-partite graph is a simple r-partite graph with partition $X_1, X_2, \ldots, X_r$ such that each vertex of class X_i is joined to each vertex of class X_{i+1} for $i = 1, \ldots, r-1$. We denote a chain-like r-partite graph by $P(X_1, X_2, \ldots, X_r)$ or $P_{n_1,n_2,\ldots,n_r}$ if $\mid X_i \mid = n_i$. The graph obtained from $P(X_1, X_2, \ldots, X_r)$, (where $r \geq 4$), by joining each vertex of X_2 to each vertex of X_4, will be denoted by $P^{(1)}(X_1, X_2, \ldots, X_r)$. If $\mid X_i \mid = n_i$, the graph will be denoted by $P^{(1)}_{n_1,n_2,\ldots,n_r}$. For $r \geq 5$, $P^{(2)}(X_1, X_2, \ldots, X_r)$ will denote the graph obtained from $P(X_1, X_2, \ldots, X_r)$ by joining each vertex of X_1 to each vertex of X_4 and X_5. If $\mid X_i \mid = n_i$, the graph will be denoted by $P^{(2)}_{n_1,n_2,\ldots,n_r}$.

Theorem 22 ([4]) *For $n \geq 8$, let Tr_n be the set of tricyclic graphs on n vertices. If $G \in Tr_n$, then $\eta(G) \leq n-4$. The equality holds if and only if G is isomorphic to one of the graphs: $P^{(1)}_{m_1,1,3,1,m_2}$, $P^{(1)}_{n-5,1,3,1}$, $P_{m_1',1,4,1,m_2'}$, $P_{1,4,1,n-6}$, $P^{(1)}_{n-5,1,1,2,1}$, $P^{(2)}_{1,2,1,1,1,n-6}$, $P_{1,2,2,1,n-6}$, $P_{3,2,1,n-6}$, $P_{1,4,1,1,n-7}$, $P_{3,2,1,1,n-7}$, where $m_1 + m_2 = n-5$ and $m_1' + m_2' = n-6$.*

3 Characterization of Singular Graphs with Other Given Parameters

In the previous section, graphs with maximum nullity and minimum rank were characterized in regard to the order of the graph. However, as the nullity decreases (the rank increases) the problem of characterizing singular graphs becomes more complex. In the research on singular graphs, other graph parameters were taken into consideration (several upper bounds on nullity can be found in [3] and [12]). In this section, we focus on the characterization of singular graphs if a parameter is given, such as the matching number, the chromatic index, the independence number, the diameter, and the maximum degree of the graph.

The chromatic number of a graph G in relation to its rank has been studied by various authors [10, 22]. In [30], the matching number, the chromatic index, and the independence number were considered in relation to the rank of the graph.

Let $c(G)$ denote $e(G)-n(G)+\theta(G)$, where $e(G)$ is the number of edges, $n(G)$ the number of vertices, and $\theta(G)$ is the number of connected components of G.

Theorem 23 ([30]) *Suppose G is a simple graph with rank $rank(G)$, matching number $m(G)$, and chromatic index $\chi'(G)$. Then*

(i) $\lceil \frac{rank(G)-c(G)}{2} \rceil \leq m(G) \leq \lfloor \frac{rank(G)+2c(G)}{2} \rfloor$.
(ii) $\lceil \frac{2e(G)}{rank(G)+2c(G)} \rceil \leq \chi'(G)$.

Theorem 24 ([30]) *Let G be a simple graph and let $\alpha(G)$ be the independence number of G. Then*

$$n(G) - \lfloor \frac{rank(G)}{2} \rfloor - c(G) \leq \alpha(G) \leq n(G) - \lceil \frac{rank(G)}{2} \rceil$$

The relation of Theorem 25(i) can be written as an expression of the nullity of G:

Theorem 25 ([24]) $n(G) - 2m(G) - c(G) \leq \eta(G) \leq n(G) - 2m(G) + 2c(G)$.

Graphs that achieve the right equality in the above theorem were determined in [28].

Theorem 26 ([28]) *For any graph G, $\eta(G) = n(G) - 2m(G) + 2c(G)$ if and only if the following three conditions are all satisfied:*

(i) All cycles of G are pairwise vertex-disjoint.
(ii) The length of each cycle of G is a multiple of 4.
(iii) $m(T_G) = m(T_G - W_G)$, where T_G is obtained from G by contracting each cycle into a vertex, and W_G consists of new vertices obtained by contracting each cycle of G.

In [24], all graphs G with nullity $\eta(G) = n(G) - 2m(G) - c(G)$ were completely characterized.

Theorem 27 ([24]) *For any graph G, $\eta(G) = n(G) - 2m(G) - c(G)$ if and only if the following three conditions are all satisfied:*

(i) All cycles of G (if any) are pairwise vertex-disjoint.
(ii) Each cycle of G (if any) is an odd cycle.
(iii) $m(T_G) = m(T_G - W_G)$.

In [21], the nullity of the graph was written in relation to the dimension of the cycle space and the number of pendant vertices of the graph. Graphs attaining the upper bound were also determined.

Theorem 28 ([21]) *Let G be a graph and every connected component of G contains at least two vertices. Then $\eta(G) \leq 2c(G) + p(G)$, where $c(G)$ is the dimension of the cycle space and $p(G)$ is the number of pendant vertices of G. The equality is attained if and only if every component of G is a cycle of size a multiple of 4.*

An upper bound of a tree on n vertices in which no vertex has degree greater than a fixed value D was given in [9].

Theorem 29 ([9]) *Let D be a fixed value and T be a tree with all vertices having degree less than or equal to D. Then, for all $n \geq 1$ and $D \geq 3$, $\eta(T) \leq n - 2\lceil (n-1)/D \rceil$. For all $n \geq 1$ and $D \geq 3$, there exist trees T with vertex degree less than or equal to D, such that $\eta(T) = n - 2\lceil (n-1)/D \rceil$.*

The complete characterization of trees of maximum nullity among n-vertex trees in which no vertex has degree greater than a fixed value D was given in [9] and [17].

In [23], the result of the above theorem was generalized for bipartite graphs.

Let u, v be two vertices of G which have the same neighbourhood, that is, $N(u) = N(v)$. The induced subgraph $G - u$ of G that is obtained by deleting vertex u is called a *reduction* of G. If no reductions can be applied to G, then the graph is called *reduced*.

The result $\eta(G) \leq n - 2\lceil (n-1)/D \rceil$ was improved for reduced bipartite graphs in [29].

Theorem 30 ([29]) *Let G be a reduced bipartite graph with n vertices and let D be the maximum degree of G. Then $\eta(G) \leq n - 2 - 2ln_2 D$.*

The graphs attaining equality were also determined in the same paper.

In [19], the diameter of the graph was taken into consideration and the bipartite graphs achieving minimum rank among bipartite graphs with a given diameter were determined.

We close this section with the result of Theorem 29 improved for any arbitrary graph. More specifically, it was proven in [34] that: $\eta(G) \leq \frac{D-1}{D} n$ for any graph G of order n and maximum degree D. It was also shown that the equality holds if and only if G is the disjoint union of some copies of $K_{D,D}$.

References

1. G.J. Chang, L.H. Huang, H.G. Yeh, A characterization of graphs with rank 4. Linear Algebra Appl. **434**(8), 1793–1798 (2011)
2. G.J. Chang, L.H. Huang, H.G. Yeh, A characterization of graphs with rank 5. Linear Algebra Appl. **436**(11), 4241–4250 (2012)

3. B. Cheng, B. Liu, On the nullity of graphs. Electron. J. Linear Algebra **16**(1), 5 (2007)
4. B. Cheng, B. Liu, On the nullity of tricyclic graphs. Linear Algebra Appl. **434**(8), 1799–1810 (2011)
5. L. Collatz, U. Sinogowitz, Spektren endlicher grafen, in *Abhandlungen aus dem Mathematischen Seminar der Universität Hamburg*, vol. 21, No. 1 (Springer, Berlin, 1957), pp. 63–77
6. D.M. Cvetković, I.M. Gutman, The algebraic multiplicity of the number zero in the spectrum of a bipartite graph. Matematički vesnik **9**(56), 141–150 (1972)
7. D.M. Cvetkovic, M. Doob, H. Sachs, *Spectra of Graphs*, vol. 10 (Academic, New York, 1980)
8. Y.Z. Fan, K.S. Qian, On the nullity of bipartite graphs. Linear Algebra Appl. **430**(11–12), 2943–2949 (2009)
9. S. Fiorini, I. Gutman, I. Sciriha, Trees with maximum nullity. Linear Algebra Appl. **397**, 245–251 (2005)
10. C.D. Godsil, G.F. Royle, Chromatic number and the 2-rank of a graph. J. Combin. Theory Ser. B **81**(1), 142–149 (2001)
11. J.M. Guo, W. Yan, Y.N. Yeh, On the nullity and the matching number of unicyclic graphs. Linear Algebra Appl. **431**, 1293–1301 (2009)
12. I. Gutman, B. Borovicanin, Nullity of graphs: an updated survey, in *Selected Topics on Applications of Graph Spectra* (Mathematical Institute, Belgrade, 2011), pp. 137–154
13. I. Gutman, I. Sciriha, On the nullity of line graphs of trees. Discrete Math. **232**(1–3), 35–45 (2001)
14. I. Gutman, N. Trinajstić, Graph theory and molecular orbitals. Total -electron energy of alternant hydrocarbons. Chem. Phys. Lett. **17**(4), 535–538 (1972)
15. S. Hu, T. Xuezhong, B. Liu, On the nullity of bicyclic graphs. Linear Algebra Appl. **429**(7), 1387–1391 (2008)
16. S. Li, On the nullity of graphs with pendent vertices. Linear Algebra Appl. **429**(7), 1619–1628 (2008)
17. W. Li, A. Chang, On the trees with maximum nullity. MATCH Commun. Math. Comput. Chem. **56**(3), 501–508 (2006)
18. J. Li, A. Chang, W.C. Shiu, On the nullity of bicyclic graphs. MATCH Commun. Math. Comput. Chem. **60**(1), 21–36 (2008)
19. H. Li, L. Su, H.X. Sun, On bipartite graphs which attain minimum rank among bipartite graphs with given diameter. Electron. J. Linear Algebra **23**(1), 9 (2012)
20. H.C. Longuet-Higgins, Resonance structures and molecular orbitals in unsaturated hydrocarbons. J. Chem. Phys. **18**, 265–274 (1950)
21. X. Ma, D. Wong, F. Tian, Nullity of a graph in terms of the dimension of cycle space and the number of pendant vertices. Discrete Appl. Math. **215**, 171–176 (2016)
22. C. Nuffelen, A bound for the chromatic number of a graph. Am. Math. Mon. **83**(4), 265–266 (1976)
23. G.R. Omidi, On the nullity of bipartite graphs. Graphs Combin. **25**(1), 111–114 (2009)
24. S. Rula, A. Chang, Y. Zheng, The extremal graphs with respect to their nullity. J. Inequal. Appl. **2016**(1), 71 (2016)
25. I. Sciriha, On the construction of graphs of nullity one. Discrete Math. **181**, 193–211 (1998)
26. I. Sciriha, On the rank of graphs, in *Combinatorics, Graph Theory, and Algorithms*, vol. II, ed. by Y. Alavi, D.R. Lick, A. Schwenk (New Issue Press, Western Michigan University, Kalamazoo, 1999), pp. 769–778
27. I. Sciriha, A characterization of singular graphs. Electron. J. Linear Algebra **16**, 451–462 (2007)
28. Y. Song, X. Song, B. Tam, A characterization of graphs G with nullity $\mid V(G) \mid -2m(G) - c(G)$. Linear Algebra Appl. **465**, 363–375 (2015)
29. Y.Z. Song, X.Q. Song, M. Zhang, An upper bound for the nullity of a bipartite graph in terms of its maximum degree. Linear Multilinear Algebra **64**(6), 1107–1112 (2016)
30. L. Wang, D. Wong, Bounds for the matching number, the edge chromatic number and the independence number of a graph in terms of rank. Discrete Appl. Math. **166**, 276–281 (2014)
31. H. Whitney, On the classification of graphs. Am. J. Math. **55**, 236–244 (1933)

32. R. Wilson, Singular graphs, in *Recent Studies in Graph Theory*, ed. by V.R. Kulli (Vishwa International Publications, Gulbarga, 1989), pp. 228–236
33. T. Xuezhong, B. Liu, On the nullity of unicyclic graphs. Linear Algebra Appl. **408**, 212–220 (2005)
34. Q. Zhou, D. Wong, D. Sun, An upper bound of the nullity of a graph in terms of order and maximum degree. Linear Algebra Appl. **555**, 314–320 (2018)

The manufacturer's authorised representative in the EU is Springer Nature Customer Service Centre GmbH, Europaplatz 3, 69115 Heidelberg, Germany. If you have any concerns regarding our products, please contact ProductSafety@springernature.com

Printed and bound by CPI Group (UK) Ltd, Croydon, CR0 4YY
30/09/2026
02225462-0017